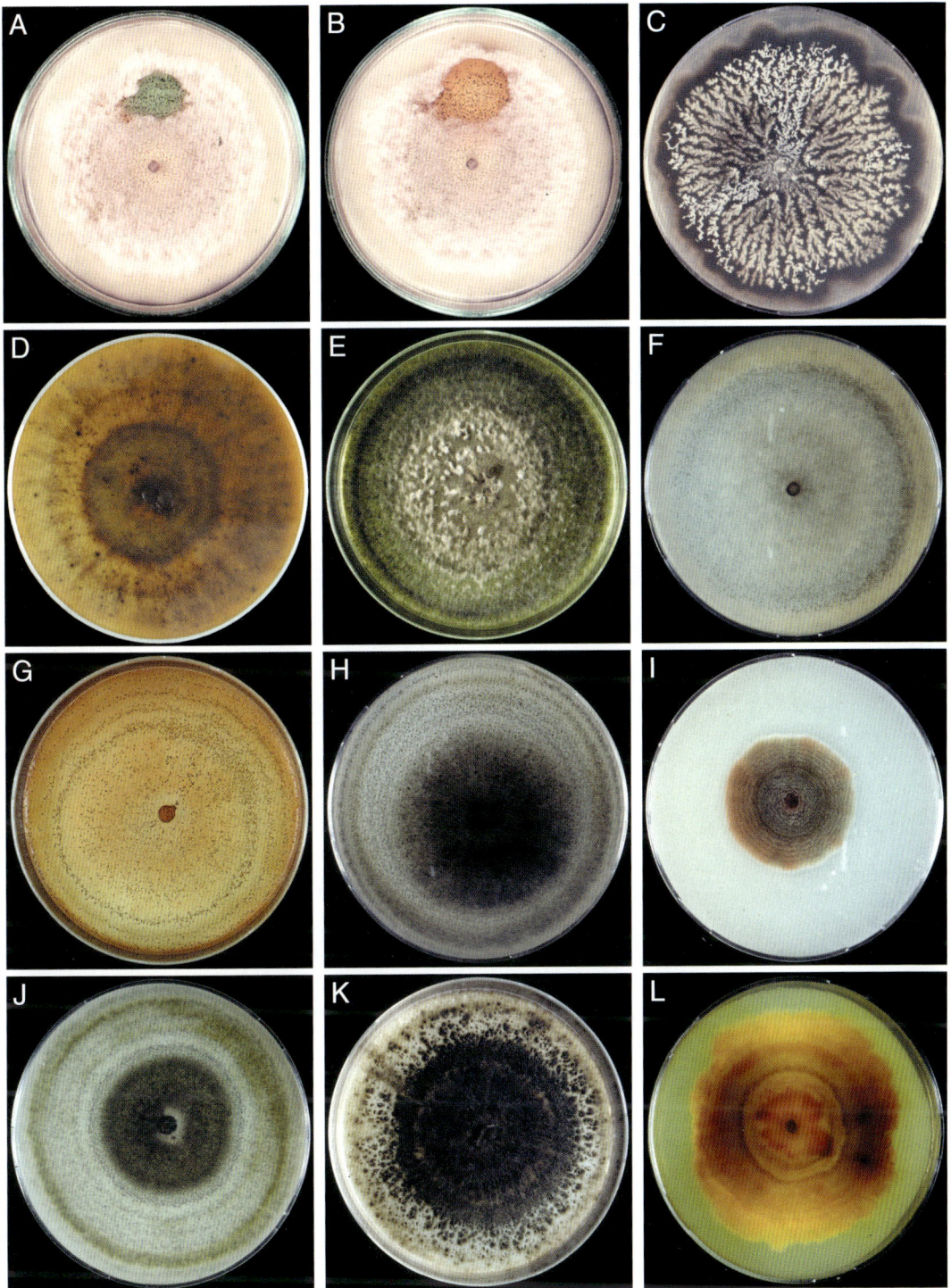

Plate 1. A, B The NaOH spot test on metabolite 'E', see text (*Ph. exigua* var. *exigua*). **C, D** Crystal formation on MA: dendritic 'pinodellalides' crystals (*Ph. dorenboschii*), yellow/green anthraquinone pigments as 'spickle' or needle-shaped crystals (*Ph. foveata*). **E, F** Colony characters, aerial mycelium on OA: coarsely floccose (*Ph. polemonii*), sparse, felty (*Ph. crystalliniformis*). **G–L** Colony colours on OA: apricot/fulvous to ochraceous by diffusible pigments (*Ph. adianticola*), olivaceous/dull green to olivaceous black (*Ph. eupyrena*), apricot/olivaceous grey (*Ph. multipora*), colourless to citrine/greenish olivaceous at centre (*Ph. nebulosa*), colourless with olivaceous to olivaceous black sectors (*Ph. hedericola*), straw/amber/luteous to fulvous/umber (*Phoma*-anam. of *Leptosphaeria biglobosa*).

PHOMA IDENTIFICATION MANUAL

Differentiation of Specific and Infra-specific Taxa in Culture

[Result of many years' experience at the diagnostic mycological section of the Dutch Plant Protection Service (PD)]

G.H. Boerema, J. de Gruyter, M.E. Noordeloos and Maria E.C. Hamers

PHOMA IDENTIFICATION MANUAL

Differentiation of Specific and Infra-specific Taxa in Culture

G.H. Boerema

Zandvoort
The Netherlands

J. de Gruyter

Plant Protection Service
Wageningen
The Netherlands

M.E. Noordeloos

National Herbarium of The Netherlands
Leiden
The Netherlands

and

Maria E.C. Hamers

Plant Protection Service
Wageningen
The Netherlands

CABI Publishing

CABI Publishing is a division of CAB International

CABI Publishing	CABI Publishing
CAB International	875 Massachusetts Avenue
Wallingford	7th Floor
Oxfordshire OX10 8DE	Cambridge, MA 02139
UK	USA
Tel: +44 (0)1491 832111	Tel: +1 617 395 4056
Fax: +44 (0)1491 833508	Fax: +1 617 354 6875
E-mail: cabi@cabi.org	E-mail: cabi-nao@cabi.org
Website: www.cabi-publishing.org	

A catalogue record for this book is available from the British Library, London, UK.

Library of Congress Cataloging-in-Publication Data
Phoma identification manual : differentiation of specific and infra-specific taxa in culture / G.H. Boerema ... [et al.].
     p. cm.
Includes bibliographical references and index.
  ISBN 0-85199-743-0 (alk. paper)
  1. Phoma--Identification. 2. Phoma--Pictorial works. I. Boerema, G. H. (Gerhard H.) II. Title.
  QK625.S5P56 2004
  579.5'5--dc21
              2003012556

ISBN 0 85199 743 0

B 2141400

Typeset in 10pt Souvenir by Columns Design Ltd, Reading.
Printed and bound in the UK by Biddles Ltd, King's Lynn.

Contents

Preface

This monographic study involves the cultural descriptions of 223 specific and infraspecific taxa of *Phoma* Sacc. emend. Boerema & G.J. Bollen, with 1146 synonyms in various Coelomycetes genera. They are divided into nine sections, which assists in their identification. Three of these sections include species with a *Didymella* teleomorph and one section includes two species with a *Mycosphaerella* teleomorph; another section shows a specific metagenetic relation with the genus *Leptosphaeria* and another with the genus *Pleospora*.

Most of the species are described from isolates of European origin and refer to parasites as well as ubiquitous saprophytes. A smaller number of species is described from isolates made in other continents and refers to both parasitic and saprophytic species.

Descriptions *in vivo* are only added when important for identification; species with a teleomorph are always described *in vivo*. References on diagnostic literature and short notes on ecology and distribution are given for each species. With one exception the keys refer only to the characteristics *in vitro*. Various tables on host relations or connections are given as well as indices to all fungal taxa and hosts mentioned.

All well-known plant pathogenic and economically important *Phoma* species come up. Some examples:

Ph. andigena (black potato blight) and *Ph. foveata* (potato gangrene), both included in the lists of quarantine organisms of the European Plant Protection Organization (EPPO), *Ph. betae* (black leg of beet and spinach), a common seed-borne pathogen, *Ph. apiicola* (phoma root and crown rot of celeriac and celery), *Ph. terrestris* (pink root of onion) and *Ph. lingam* (dry rot and canker or black leg of brassicas). Pathogens of pulse and forage crops such as *Ph. subboltshauseri* (blotch or leaf spot of beans and cowpea), *Ph. medicaginis* (black stem of lucerne) and *Ph. pinodella* (foot rot and black stem of pea and clover).

On tomato plants *Ph. destructiva* (fruit rot, leaf and stem blight) and *Ph. lycopersici* (canker, stem and fruit rot). On florists' chrysanthemums *Ph. chrysanthemicola* (root rot and basal stem rot) and *Ph. ligulicola* (ray blight, leaf and
stem spot). On Amaryllidaceae *Ph. narcissi* (red spot diseases, leaf scorch).

Important plurivorous opportunistic plant pathogens are also treated, e.g.
Ph. exigua, *Ph. macrostoma*, *Ph. multirostrata* and *Ph. sorghina* (common in
the tropics and subtropics).

1 Introduction

This study had an unplanned start at the diagnostic mycological section of the Dutch Plant Protection Service at Wageningen (PD) in the early 1960s with the examination of a sample of *Brassica* seed suspected of being infected with *Phoma lingam* (Tode: Fr.) Desm., the anamorph of the causal fungus of dry rot and canker or black leg. However, it appeared that the seed coat was contaminated with two other species of *Phoma* (Boerema, 1962). This was established with a comparative study of isolates on agar media.

The dominating species proved to be *Phoma herbarum* Westend. nominated as type species of *Phoma* sensu Saccardo (*nomen genericum conservandum*) at the 8th International Botanical Congress at Paris in 1954. According to Saccardo's taxonomic system, the genus *Phoma* refers to 'pycnidia with one-celled hyaline conidia occurring on herbaceous stems' (Saccardo, 1884). However, our study showed *Ph. herbarum* to be a ubiquitous saprophyte, occurring on very diverse substrata, not only on all parts of dead and dying herbaceous plants, but also on woody plants, in soil, water, milk, butter, paint, paper, etc. (Boerema, 1964). Since the experimental study by Hughes (1953) on Hyphomycetes, the conidiogenesis has been recognized as the most important taxonomic criterion in all conidial fungi. In mature pycnidia of *Ph. herbarum* and other similar *Phoma* species the hyaline unicellular conidia arise from less differentiated cells lining the pycnidial cavity in a kind of repetitive monopolar budding. Apart from some wall-thickening at the top of the conidiogenous cells, the light microscopy gave few details. In cooperation with the Agricultural Highschool at Wageningen (LH, now Wageningen University and Research, WUR) it was possible to study the conidiogenesis in *Phoma* spp. with transmission electron microscopy (Brewer & Boerema, 1965; Boerema & Bollen, 1975). The conidiogenesis appeared to begin with the formation of a papilla and breakage of the conidiogenous cell to leave a 'collarette' (periclinal thickening). The double layered wall of the first conidium arises as new, its maturation synchronous with the ontogeny; the wall building of successively produced conidia starts with the basal layer of the septum remaining after secession of the previous conidium, maturation still being synchronous with ontogeny (for details see Figs 1–3). The conidia are produced at about the same level by percurrent proliferation. This conidiogenesis, associated with abundant production of mucilage, is analogous with that in Hyphomycetes now recognized as phialidic (Kendrick, 1971) with blastic conidial ontogeny. This is more precisely termed false- or no-chain phialidic (Minter *et al.*, 1982) as the conidia are produced in mucilaginous masses or chains (Boerema, 1964) that are not held together by wall material as in true phialides.

On agar media *Ph. herbarum* shows specific cultural characteristics which very much simplify its identification. This appeared to be true also for other species of *Phoma*. In addition to characteristics of pycnidia and conidia, cultures give information about mycelial characters such as development of chlamydospores, and biochemical properties such as production of pigments and crystals. The importance of cultural characteristics for identifying and differentiating *Phoma* species was already noted by Wollenweber & Hochapfel (1936) in their study on fruit-rotting, and by Dennis (1946) in a study of *Phoma* isolates from various sources in comparison with those found in association with rots of potato tubers. Their *in vitro* studies provided valuable information about various plurivorous species of *Phoma*, described under many different names; see Boerema & Dorenbosch (1973) and Boerema (1976).

Pycnidia of *Ph. herbarum* are usually glabrous, but may have some hyphal outgrowths. Some other species, however, produce hairy or setose pycnidia. This explains why some *Phoma* species according to the classic Saccardoan criteria have been classified in *Pyrenochaeta* De Not., characterized by setose pycnidia. However, conidia conforming to the type species of *Pyrenochaeta* are produced on elongated, branched conidiophores, see Fig. 5C and Schneider (1976). Many species of *Phoma* have been formerly classified in *Phyllosticta* Pers., a leaf spot analogue of *Phoma* according to the Saccardoan system. However, the type species of *Phyllosticta* is quite different as it produces large conidia with an appendage, see van der Aa (1973) and van der Aa & Vanev (2002). Some conidia of *Ph. herbarum* produced a septum preceding germination, but such secondary septation appeared to occur commonly in various other species of *Phoma in vivo*. However, it proved to be characteristic and diagnostic of *Phoma* spp. that *in vitro* on agar media under normal laboratory conditions most conidia always remain aseptate. On account of the Saccardoan taxonomic criteria *Phoma* species producing a variable number of 1-septate conidia *in vivo* were formerly often classified in *Ascochyta* Lib. ('pycnidia with 2-celled hyaline conidia occurring on leaves'). For example the fungus causing black stem of lucerne was known as either *Phoma medicaginis* Malbr. & Roum. or *Ascochyta imperfecta* (!) Peck (Boerema *et al.*, 1965b). Extreme variability of conidial septation *in vivo* (0–95%) is shown by the wound and weak parasite *Phoma exigua* Desm. This most commonly found *Phoma* species in Europe has been described under more than 300 different names. Žerbele (1962) in Russia proved experimentally that the host influences the rate of conidial septation. However, fluctuating humidity and desiccation also play a role. It appeared (Žerbele, 1962) that in true species of *Ascochyta*

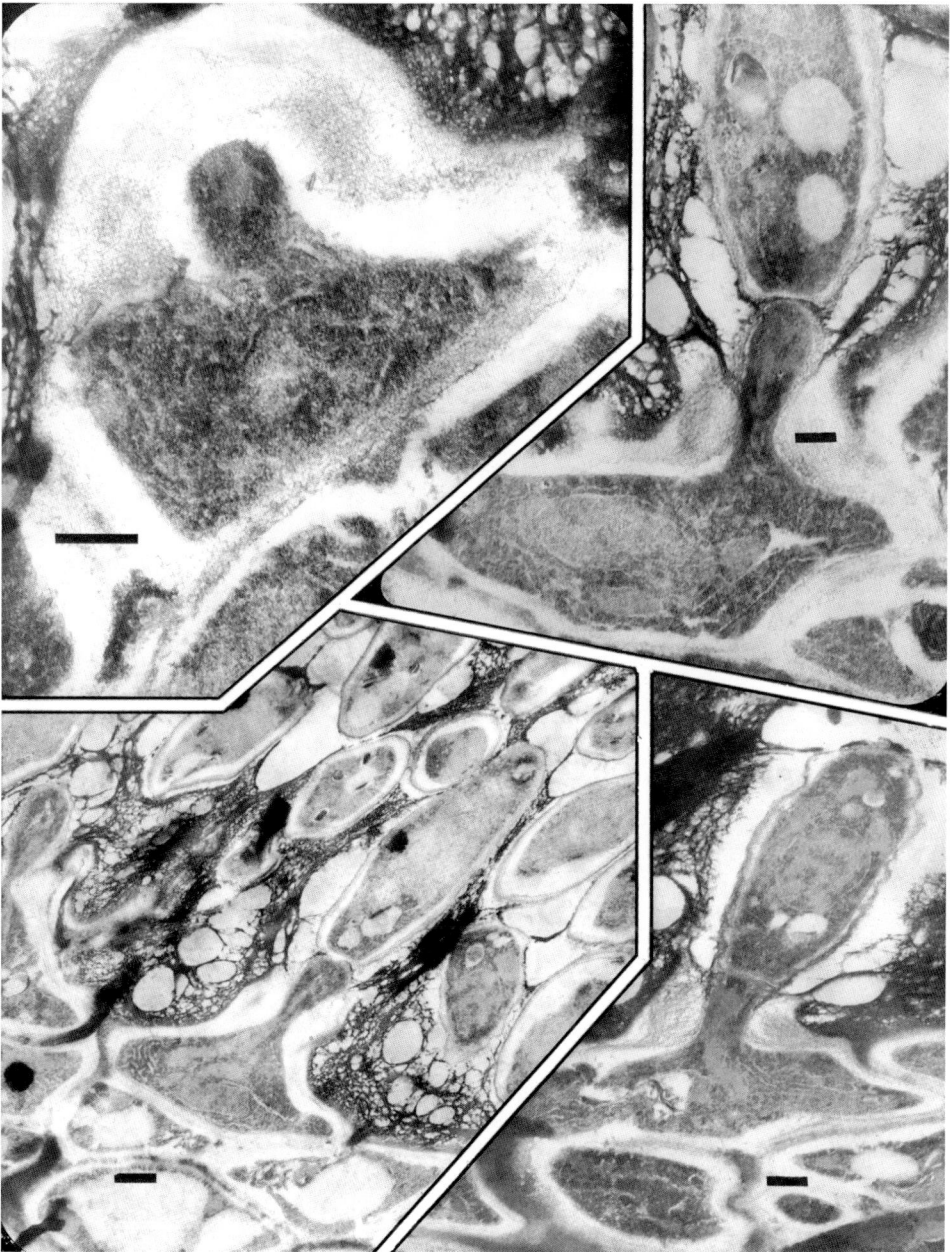

Fig. 1. *Phoma herbarum* Westend., transmission electron micrographs of the conidiogenesis (Technical and Physical Engineering Research Service, Wageningen, 1972); compare the diagrams in Figs 2 and 3. Left above the characteristic thick-walled papillate extension preceding the bud-like initiation of the first conidium. The other three micrographs show that in secession three layers are involved: the basal conidial wall, the separation-plate and the wall of the next conidial initial. In the collarette the different layers of the original papilla can be distinguished. For the other fine-structural correlates see Boerema & Bollen (1975). Bars 0.5 μm. Reproduction from Boerema (1997).

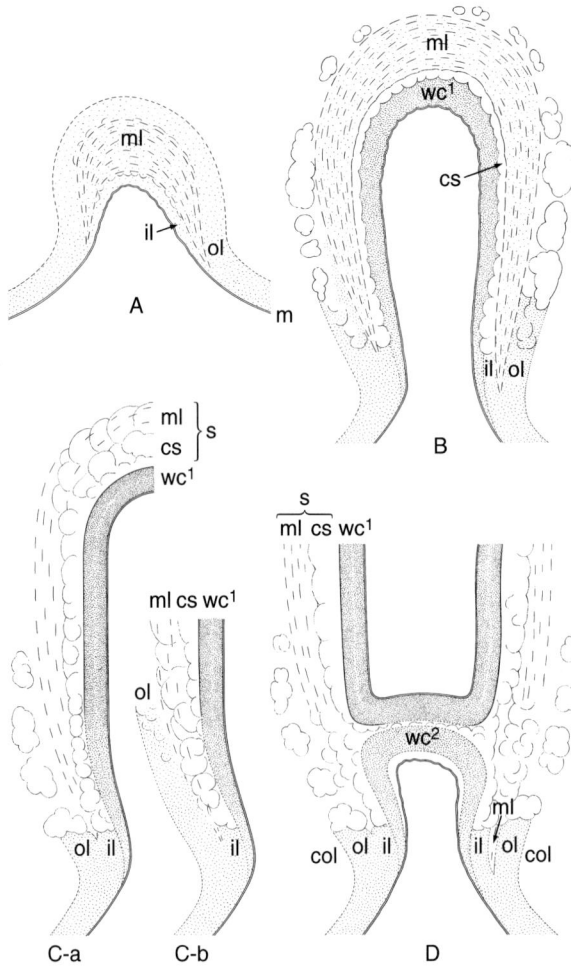

Fig. 2. *Phoma* spp., diagrammatic representation of electron microscope observations on the formation of the first conidium (**A→D**). It shows first (**A**) the characteristic three-layered papillate wall-building on the conidiogenous cell before the initiation of the first conidium; m = plasma membrane. When the apical new wall-building of the first conidium starts (wc¹ in **B**) the outer and inner layers of the papilla (ol and il) dissolve. The conspicuous swollen middle layer (ml) seems to function as an 'opener' which later also dissolves into mucilage.* The differentiation of the conidial wall into an inner and outer layer by diffuse wall-building (**→C** a and b; conidial maturation synchronous with conidium ontogeny) is associated with the production of a cloudy substance (cs; not equally abundantly formed in all the *Phoma* spp.). Together with the mucilage of the middle layer it forms the mucilaginous sheath(s) around the first conidium. After delimitation and secession by a double wall (**D**, see also Fig. 3) the basal part of the original papilla remains as a collarette (col); sometimes the different layers of the papilla (ol, ml, il) are still recognizable in the collarette (wc² = wall of second conidial initial, see Fig. 3). Drawings after Boerema & Bollen (1975). *Note that the middle layer of the papilla is not observed in the less conspicuous collarette of the type species of section *Pilosa*.

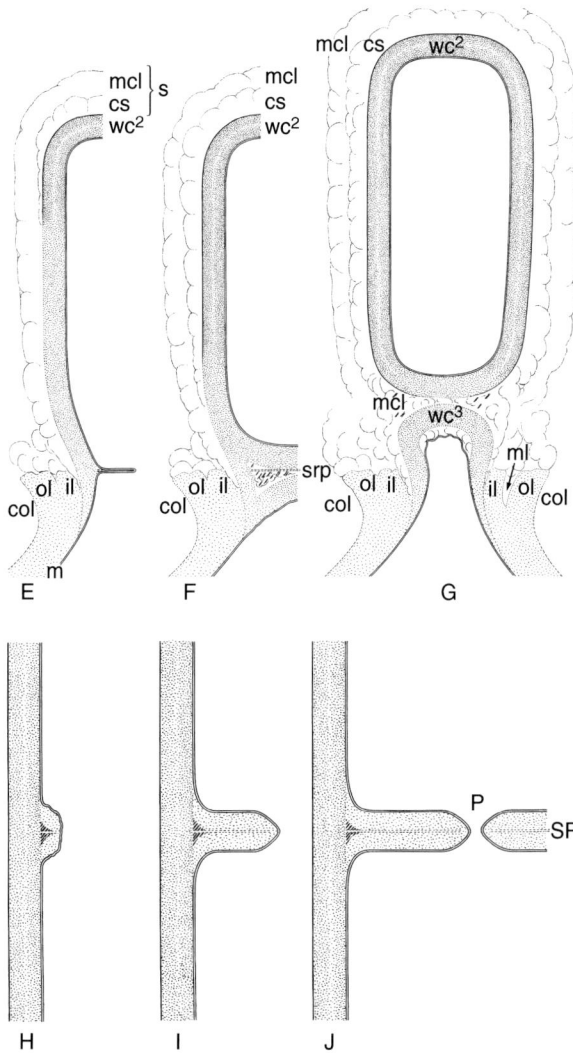

Fig. 3. *Phoma* spp., diagrammatic representation of electron microscope observations on second and successive conidial ontogeny (**E→G**) and the frequently occurring secondary septation of the conidia (**H→J**). The production of successive conidia (wc², wc³, etc.) depends on ontogeny by replacement wall-building. The wall-building locus is more or less static by enteroblastic-percurrent proliferation and surrounded by the collarette (col) formed during the wall-building activity before the initiation of the first conidium (ol, ml, il; see Fig. 2). Differentiation of the conidial wall with an inner and outer layer by diffuse wall-building is associated with the production of a mucilaginous cloudy substance (cs). Delimitation and secession starts with the development of a separation plate (srp) immediately followed by the formation of wall layers at both sides. The separation plate and the periclinal wall parts disintegrate into mucilage (mcl), which then covers the next primordium and will form the outer layer of the mucilaginous sheath (s) of the next conidium (m = plasma membrane). Secondary septation of the conidia (**H→J**) occurs as an annular ingrowth from the lateral wall leaving a pore (P) in the centre. The septum consist of a middle lamella, the septal-plate (SP), covered on both surfaces with wall layers which are decurrent with the lateral conidium wall. Drawings after Boerema & Bollen (1975).

the conidia become mainly septate under all circumstances, both *in vivo* and *in vitro*. Transmission electron microscope studies of *Ascochyta* spp. (Brewer & Boerema, 1965; Boerema & Bollen, 1975) showed that in true species of this genus the conidial septation is an essential part of the completion (maturation) of conidia: wall-thickening septation; see Fig. 5E and Boerema (1984a).

Occasionally *Phoma* species with some conidial septation *in vivo* have been placed in the genus *Diplodina* Westend., a stem-analogue of *Ascochyta* according to Saccardo's system. However, the type species of *Diplodina* is quite different because its pycnidia open by splitting, thus becoming cupulate (Sutton, 1980).

The daily mycological diagnostic work at the PD has made it possible to compare the cultural characteristics of a steadily growing number of *Phoma* species, iso-lated from agricultural and horticultural plants, i.e. saprophytes, opportunistic parasites and well-known pathogens. Publication of the results (e.g. Boerema, 1964, 1967b, 1972, 1976; Boerema *et al.*, 1965a, 1968, 1971, 1973, 1977, 1981b; Maas, 1965; Boerema & Höweler, 1967; Boerema & Dorenbosch, 1968, 1970, 1973; Dorenbosch, 1970; Boerema & Loerakker, 1981, 1985) has led to several *Phoma* isolates being forwarded from other parts of the world. Finally all *Phoma*-isolates sent for identification to the 'Centraalbureau voor Schimmelcultures' (CBS) at Baarn (now Utrecht) were also incorporated in the comparative cultural studies. To provide a workable method of identification of the various *Phoma* species obtained in culture they were grouped as time went on into sections according to pycnidial, conidial and chlamydospore characters. Some of these sections (see the 'Nomenclator' on p. 9) are completely artificial, but others represent natural units, sometimes connected with different teleomor-phic genera (van der Aa *et al.*, 1990; Boerema, 1997).

In the early 1990s we started to make comparative cultural descriptions section-by-section of all *Phoma* species so far studied and preserved as freeze-dried cultures in the PD and CBS collections. These descriptions were pub-lished with keys and host indices in a series of 'Contributions towards a monograph of *Phoma*' (de Gruyter & Noordeloos, 1992; de Gruyter *et al.*, 1993, 1998, 2002; Boerema, 1993, 1997, 2003; Boerema *et al.*, 1994, 1996, 1997, 1999; Boerema & de Gruyter, 1999; van der Aa *et al.*, 2000; de Gruyter & Boerema, 2002; de Gruyter, 2002). These 'Contributions' were the prepara-tory studies for this book, which is also only a 'contribution' to the knowledge of the anamorph genus *Phoma*. It must be realized that nearly 2800 taxa have been described in *Phoma* in literature from all over the world (cf. 'Preliminary list of names in *Phoma*, 24 July 1998', CMI).

This book deals with 223 specific and infraspecific taxa of *Phoma*, most of which are widespread in western Europe, but are often also recorded elsewhere in Eurasia, in Australasia and North America. A few are cosmopolitan and some are known especially from subtropical and tropical regions. Others are known only so far from Australasia and the Americas.

In the course of this study the terminology and classification of fungi have been changed considerably (ref. Nag Raj, 1993). The misleading terms 'imper-fect fungi' and 'perfect fungi' have been replaced by anamorph and teleomorph, respectively. 'Form genus' and 'form species' have been replaced by anamorph genus and anamorph species. Different anamorphs of a single fungus are now called synanamorphs and the term holomorph has been introduced to denote a fungus in all its morphs. Since only a small number of *Phoma* species has been correlated with a teleomorph, most *Phoma* species may be termed anamorphic holomorphs. As it was shown that Coelomycetes include a continuum from stromatic via acervulus to pycnidium, the collective term conidioma(ta) has been introduced for the conidia-bearing structures. With this terminology *Phoma* species have pycnidial conidiomata. However, for convenience we have continued to use 'pycnidium' and 'pycnidia' in the descriptions. Most important has been the recognition of the significance of conidiogenesis in classification of anamorphic fungi (see above). It helped us greatly with the genus concept of *Phoma*.

Seven Coelomycetes genera are now considered to be congeneric with *Phoma* as they have the same kind of conidiogenesis and often also show secondary conidial septation. Some of these genera, however, refer to species with typical differentiating or additional morphological characteristics; these phenomena are used to subdivide the genus into sections.

2 Nomenclator of the Genus and its Sections

Phoma Sacc. emend. Boerema & G.J. Bollen

Phoma Sacc. *in* Michelia **2**(1): 4. 1880 [as 'Fr. em.'; *nomen genericum conservandum*, 8th Intern. Bot. Congr., Paris 1954], emend. Boerema & G.J. Bollen *in* Persoonia **8**(2): 134–135. 1975.

Lectotype species (8th Intern. Bot. Congr., Paris 1954): *Phoma herbarum* Westend. *in* Bull. Acad. r. Belg. Cl. Sci. **19**(3): 118. 1852; lectotype [Boerema *in* Persoonia **3**(1): 10, pl. 1 figs 1, 2. 1964] in herbarium Westendorp & Wallays (BR): exs. Herb. crypt. Belg., Ed. Beyaerts-Feys, Fasc. 20, No. 965. 1854, on stems of *Onobrychis viciifolia*.

Teleomorphic connections
>   **Didymella Sacc. ex Sacc.** *in* Michelia **2**(1): 57. 1880.
>   **Leptosphaeria Ces. & De Not.** *in* Comment. Soc. crittogam. ital. **1**(4) [= Schema Sfer.]: 234. 1863.
>   **Mycosphaerella Johanson** *in* Öfvers. K. VetenskAkad. Förh. **41**(9): 163. 1884.
>   **Pleospora Rabenh. ex Ces. & De Not.** *in* Comment. Soc. crittogam. ital. **1**(4) [= Schema Sfer.]: 217. 1863.

Connected synanamorphs
>   [uni- and multicellular chlamydospores]
>   species of **Epicoccum Link** *in* Mag. Ges. naturf. Freunde, Berl. **7**: 32. 1816.
>   species of **Phialophora Medlar** *in* Mycologia **7**: 202. 1915.
>   species of **Stagonosporopsis Died.** *in* Annls mycol. **10**: 142. 1912.
>   species of **Sclerotium Tode**, Flora Meckl. **1**: 2. 1790 : Fr., Syst. Mycol. **2**: 246. 1823.

Subdivision in Sections (arrangement in this book)

The nine differentiated sections are listed with their generic synonyms in alphabetical order of the connected teleomorphic genera.

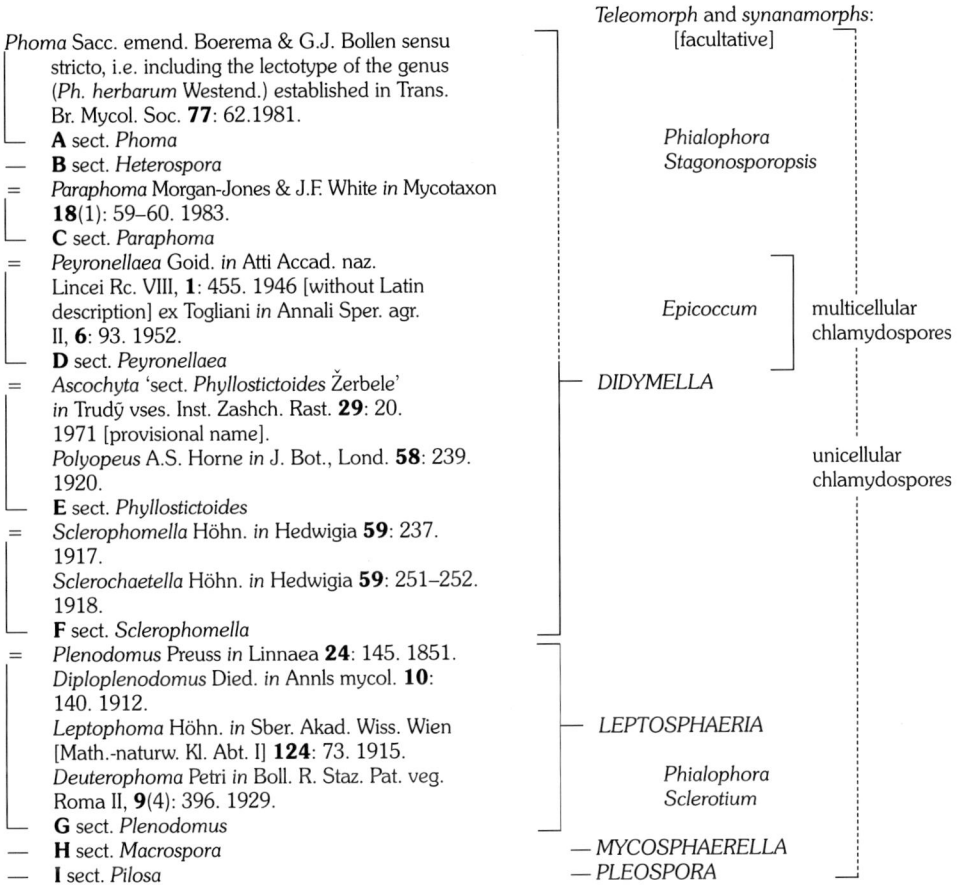

Teleomorph and synanamorphs: [facultative]

Phoma Sacc. emend. Boerema & G.J. Bollen sensu stricto, i.e. including the lectotype of the genus (*Ph. herbarum* Westend.) established in Trans. Br. Mycol. Soc. **77**: 62.1981.
— **A** sect. *Phoma*
— **B** sect. *Heterospora*
= *Paraphoma* Morgan-Jones & J.F. White *in* Mycotaxon **18**(1): 59–60. 1983.
— **C** sect. *Paraphoma*
= *Peyronellaea* Goid. *in* Atti Accad. naz. Lincei Rc. VIII, **1**: 455. 1946 [without Latin description] ex Togliani *in* Annali Sper. agr. II, **6**: 93. 1952.
— **D** sect. *Peyronellaea*
= *Ascochyta* 'sect. *Phyllostictoides* Žerbele' *in* Trudy vses. Inst. Zashch. Rast. **29**: 20. 1971 [provisional name]. *Polyopeus* A.S. Horne *in* J. Bot., Lond. **58**: 239. 1920.
— **E** sect. *Phyllostictoides*
= *Sclerophomella* Höhn. *in* Hedwigia **59**: 237. 1917. *Sclerochaetella* Höhn. *in* Hedwigia **59**: 251–252. 1918.
— **F** sect. *Sclerophomella*
= *Plenodomus* Preuss *in* Linnaea **24**: 145. 1851. *Diploplenodomus* Died. *in* Annls mycol. **10**: 140. 1912. *Leptophoma* Höhn. *in* Sber. Akad. Wiss. Wien [Math.-naturw. Kl. Abt. I] **124**: 73. 1915. *Deuterophoma* Petri *in* Boll. R. Staz. Pat. veg. Roma II, **9**(4): 396. 1929.
— **G** sect. *Plenodomus*
— **H** sect. *Macrospora*
— **I** sect. *Pilosa*

Phialophora
Stagonosporopsis

Epiccocum — multicellular chlamydospores

— DIDYMELLA

unicellular chlamydospores

— LEPTOSPHAERIA

Phialophora
Sclerotium

— MYCOSPHAERELLA
— PLEOSPORA

This subdivision has been introduced to facilitate identification on species level. The diagram shows that various sections are based on generic synonyms. Further it indicates that most sections include some species with a *Didymella* teleomorph. Three sections, however, are characterized by their connection with other teleomorphic genera. The diagram also indicates that unicellular chlamydospores may occur in all sections, whereas multicellular chlamydospores are distinctive for section *Peyronellaea* (**D**). The synanamorphic facultative genera listed are mainly important for identification at species level, but synanamorphs in the genus *Stagonosporopsis* Died. are characteristic for the section *Heterospora* (**B**).

The distribution of species within the various sections of *Phoma*, the relationships between the sections and their similarity with some other coelomycetous genera are shown diagrammatically in Fig. 4.

Introduction Data of the Sections

Phoma Sacc. sect. *Phoma*   → **A** (Pho)
Phoma sect. *Heterospora* Boerema *et al. in* Persoonia **16**(3):
336. 1997   → **B** (Het)
Phoma sect. *Paraphoma* (Morgan-Jones & J.F. White) Boerema *in*
Stud. Mycol. **32**: 7. 1990   → **C** (Par)
Phoma sect. *Peyronellaea* (Goid. ex Togliani) Boerema *in* Stud. Mycol.
32: 6. 1990   → **D** (Pey)
Phoma sect. *Phyllostictoides* Žerbele ex Boerema *in* Mycotaxon **64**:
331. 1997   → **E** (Phy)
Phoma sect. *Sclerophomella* (Höhn.) Boerema *et al. in* Mycotaxon
64: 331. 1997   → **F** (Scl)
Phoma sect. *Plenodomus* (Preuss) Boerema *et al. in* Trans. Br.
mycol. Soc. **77**: 61. 1981   → **G** (Ple)
Phoma sect. *Macrospora* Boerema *et al. in* Mycotaxon **64**:
332. 1997   → **H** (Mac)
Phoma sect. *Pilosa* Boerema *et al. in* Mycotaxon **64**: 332. 1997   → **I** (Pil)

For documentation of the *generic synonymy* see von Arx (1970), Boerema
& Bollen (1975), Sutton (1977), Boerema (1993, 1997), Boerema *et al.*
(1994, 1997).

3   The Generic Characters

Conidiomata

Pycnidial, considerably variable in shape and size (influenced by substratum and medium), immersed or superficial (when *in vitro* and also in aerial mycelium they are sometimes very small, micropycnidia), mostly subglobose-papillate, also obpyriform-flask shaped to typical rostrate, occasionally in another definite shape, solitary, aggregated or confluent (Figs 6A, 24A, 33A, 36–46) or coalesced to large irregular pycnidioid structures (Fig. 37B, C); brown-olivaceous to olivaceous black, initially sometimes nearly colourless or yellowish, occasionally ferrugineous or bluish (colours influenced by light and temperature); glabrous or with some hyphal outgrowths, sometimes hairy, pilose, or with straight or flexuous setae, setose, often mainly at the apex (Figs 21, 25B) (sometimes the occurrence of pilose and setose pycnidia is linked with the growing conditions and then different *in vivo* and *in vitro*).

Pycnidial primordia mostly arise by the meristogenous method, i.e. repeated division and growth of one or more cells of a single hypha or even by division and growth of a single conidium. The symphogenous method, i.e. the intermingling of several hyphal branches, may occur along with the meristogenous method. It seems to be the common method in some species producing pycnidia which remain closed for a long time (*in vitro* it is often observed in the merger zone between two colonies, 'conjunctive' pycnidia).

Pycnidial opening(s) usually an ostiolum, i.e. structural provisions for opening are already present in the pycnidial primordia, but sometimes they remain closed, e.g. as a pycnosclerotium (Figs 33C, 36III), or open after a long period, occasionally by rupture, but usually by a process of disintegration and growth resulting in a porus; the circular or oval ostioles and pores are surrounded by dark cells and internally lined with papillate or elongated hyaline cells (Figs 29B, 33A, 36I–II) (in mature pycnidia differentiation between ostiolate and poroid openings is usually impossible).

Pycnidial wall variable in thickness, with different types of pseudo-parenchyma, outer layers dark and relatively thick-walled, inner layers pale brown to hyaline and thin-walled (Figs 6D, 29A, 33A), secondary thickening of the cell-walls in the inner layers may occur, then becoming scleroplectenchymatous (Fig. 36); cavity usually unilocular, sometimes irregular by cellular protrusions from the wall, and occasionally plurilocular by coalescing of pycnidia (Figs 36, 37B, C).

Conidiogenous cells generally discrete, lining the cavity, in young pycnidia usually more or less globose, later becoming bottle-shaped, i.e. variations between ampulliform and doliiform; sometimes in irregular groups due to proliferation and regeneration (displacement of defunct cells, e.g. after desiccation) (Figs 6G, 29C, 33A).

Conidiogenesis

Phialidic with blastic conidia produced in mucilaginous masses (false- or no-chain phialidic); development of conidia preceded by formation of a thick-walled papilla, the top of which dissolves into mucilage at the initiation of the first conidium. Its basal part remains as a collarette (periclinal thickening), sometimes very conspicuous (Figs 1, 2A→B, 6G, 29C).

Ontogeny by apical wall-building in the first conidium and by replacement wall-building in the subsequent conidia, which are produced at about the same level by enteroblastic–percurrent proliferation (Figs 2B→D, 6G, 3E→G). Walls of the developing conidia immediately obtain their final thickness with an inner and outer layer, maturation being synchronous with ontogeny due to diffuse wall-building. Delimitation by a double wall, secession schizolytic.

Conidia

Always unicellular when released (e.g. Figs 7–13), but a proportion sometimes becoming two- or even more-celled by secondary septation (e.g. Figs 30–32), highly variable in shape and size, mostly ellipsoidal to cylindrical, also ovoid or obovoid, sometimes subglobose or irregular, colourless-hyaline, but sometimes becoming brownish, smooth walled and often guttulate as they age; mostly once or twice as long as wide, generally measuring (2–)3–11(–13) × (0.5–)1–4(–5) μm, but sometimes also considerably larger, (7–)11–25(–30 or even longer) × (2–)2.5–7(–8 or even wider) μm.

Guttules variable in size and number, sometimes random, but usually bipolar in position (see e.g. Figs 7–13), yellowish-greenish in contrast with the colourless hyaline plasma (which may become brownish with age).

Growing conditions *in vivo* may greatly influence the size of conidia in some species; large, long conidia then often become two- or occasionally more-celled by secondary septation (up to 95%); this may also occur in smaller conidia just before germination.

Secondary septation of mature conidia occurring as an annular ingrowth of the inner wall layer, immediately reaching final thickness, consisting of two similar layers with a thin middle lamella, and leaving a pore in the centre (Fig. 3H→J). Wall of septate conidia not substantially thicker than that of aseptate conidia. Type of septation resembles 'euseptation' in dark, large-spored Hyphomycetes (cf. Reisinger, 1972).

4   The Methods Used for Differentiation and Identification

Observations *In Vivo*: Wall Structures, Synanamorphs and Teleomorphs

Examination of the characteristics of a *Phoma* species *in vivo* remains of course the first step in identification of the species. It usually gives an immediate indication of the section(s) in which a species may be grouped, see p. 23.

The wall structure and pycnidial shape *in vivo* are essential for species differentiation in section *Plenodomus* (**G**) (Boerema *et al.*, 1994). Pycnidia *in vivo* were also used for the study of conidiogenesis and conidial septation in *Phoma* spp. (Boerema & Bollen, 1975). The pseudoparenchymatous or scleroplectenchymatous wall structure of pycnidia was studied in thin and thick sections made with a freezing microtome. The thick sections were used for staining with Lugol's iodine (JKJ), the walls of scleroplectenchyma then became red by adsorption of the iodine (blotting paper effect; first noticed by von Höhnel, 1918: 250). Species with a pseudoparenchymatous wall structure do not show this wall staining, but sometimes the contents of the cells become red (often the case in species of section *Sclerophomella* (**F**), Boerema & de Gruyter, 1998).

The descriptions of the *Stagonosporopsis* synanamorphs of species in section *Heterospora* (**B**) and the *Didymella*, *Leptosphaeria*, *Mycosphaerella* and *Pleospora* teleomorphs of species in other sections could be made only from observations *in vivo*. Their relations to the *Phoma* anamorphs developing in culture were always based on a comparative study of isolates.

Isolation and Preservation

Isolations were usually made on cherry-decoction agar (CA), a special Dutch (CBS) medium: 0.1 l cherry juice, 20 g oxoid agar no. 3 and 0.9 l tapwater. The use of this medium with low pH suppresses bacterial growth during isolation, but plum agar is as effective. Most *Phoma* species grow very well at a pH around 5.5. Subsequently the isolates were transferred to oatmeal agar (OA) to stimulate sporulation, see below under 'Agar Plates' and 'Inoculation and Incubation'. In the past a serious problem has been maintaining the *Phoma*-isolates in optimal sporulating condition for a long time (this study covers a period of more than 40 years!). Periodic transfer in test tubes with different media or substrata could not prevent some isolates from degenerating and becoming sterile. However, in the early 1970s the PD's *Phoma* collection was lyophilized (freeze-dried). The technique is the same as that used for the CBS-collection (Gams *et al.*, 1998: 102). It appears that most species of *Phoma* then maintain their sporulation capacity, pathogenicity and biochemical characteristics over several decades.

Agar Plates

The fungi are compared in cultures on oatmeal agar (OA) and malt-extract agar (MA) of *c.* 3 mm thick (20 ml) in standard 9 cm plastic Petri dishes (16 mm high with ridges in the lid). On OA the production of pycnidia is usually abundant, whereas MA stimulates pigment production and crystal formation. The OA used contained 20 g oat-flakes, boiled in 0.5 l tapwater, filtered through cheese cloth and made up to 1 l with tapwater. The MA contained 40 g malt extract-oxoid L39, 15 g oxoid agar no. 1 and 1 l tapwater.

Inoculation and Incubation

The agar plates were inoculated in the centre with a small disc (0.5 cm diam.) cut from the margin of a fresh, active growing plate culture. This method has the advantage that transfers are more likely to show an 'average' type of growth than are single-spore cultures (Dennis, 1946). Nevertheless it is possible that extreme mycelial and pycnidial types segregate (Hansen, 1938). The inoculated plates were incubated in complete darkness at 20–22°C for one week. During a second week the Petri dishes were placed in an incubator with a day–night regime of 13 h NUV light and 11 h darkness to stimulate the pigmentation of the colonies and the formation of pycnidia.

Measurements and Descriptions of Colonies

After the first week (7 days) the diameter of the colonies was measured and the gross morphology of the colonies was described. The outline of the colony was determined as either 'regular' or 'irregular'. The colour nomenclature of Rayner (1970) was used in order to give a precise colour description of the aerial mycelium, the colonies and the reverse, as well as the conidial matrix. Variability in cultural appearance is inherent in most species of *Phoma*, and some deviation from the cultural characters noted by us is normal. This holds especially for the colours of the colonies and conidial matrices. These colours are generally duller and darker in old cultures. Two weeks after inoculation the colonies were measured and described again, and the morphology of pycnidia, conidia and other structures were studied from OA cultures, see below.

NaOH Spot Test

Next the cultures were always tested with a drop of NaOH near the growing margins. Various species of *Phoma* produce a colourless diffusible antibiotic metabolite 'E' (after *Phoma exigua*). On addition of a drop of concentrated NaOH, this metabolite oxidizes successively into the pigments 'α' and 'β'. Pigment α is red-purple at pH < 10.5 and blue-green at pH > 12.5. Pigment β is yellow at pH < 3.5 and red at pH > 5.5. This means that application of a drop of NaOH on metabolite E-producing cultures results within about 10 min in a greenish spot or ring (pigment α), which changes to red (pigment β) after about 1 h, see Colour Plate IA, B (Boerema & Höweler, 1967). Logan & O'Neill (1970) showed the metabolite to have bactericidal and fungicidal properties; its production is stimulated by light.

Pigment and Crystal Formation

The above NaOH spot test may (also) cause a conspicuous colour change of naturally produced pigments in *Phoma* cultures (see e.g. Dorenbosch, 1970). Pigments are sometimes restricted to the cytoplasm or guttules in the hyphae, but usually metabolites diffusing into the agar are involved. In many cases these pigments are composed of anthraquinone components such as pachybasin, chrysophanol, emodin and phomarin (Bick & Rhee, 1966). The colours of the pigments may be pH-dependent and therefore change by application of NaOH: yellow then often becomes red and red may become purple or blue. In old cultures the pigments may crystallize out, e.g. as yellow green 'spickle'-crystals or as red needle-shaped crystals. Some species of *Phoma* produce very characteristic dendritic crystals in the agar media. At least four different chemical components are responsible for this phenomenon: brefeldin A, pinodellalide A and B, and radicin (Noordeloos *et al.*, 1993). Finally it should be noted that some metabolites produced by *Phoma* spp. are fluorescent (Boerema & Loerakker, 1985).

Pycnidia

Gross characters of the pycnidia – surface: glabrous, hairy (pilose) or setose; occurrence of an ostiole or pycnidium initially closed and later opened by a pore; and wall thickness and structure – are used for differentiation into sections. At species level pycnidia usually vary considerably in shape, size and colour. In the keys on growth characteristics *in vitro*, those characteristics are therefore only used in extreme cases. One species may produce both large and small pycnidia; micropycnidia often occur in aerial mycelium (common in sect. *Peyronellaea*, **D**). The pycnidia may be regularly globose or very irregular and compound in shape, with or without necks and with one or more ostioles or pores. The colour of pycnidia may change with time *in vitro*, e.g. from nearly colourless or yellow-tinged to brown or black. The pycnidia may be submerged or superficial on the medium or in aerial mycelium. Strains may differ in this aspect within a species (Boerema, 1970).

It should be remembered that characteristics of pycnidia *in vivo* are important for species differentiation in sect. *Plenodomus* (**G**).

Sections of pycnidia were made with a freezing microtome and drawings were prepared from micrographs, or with the help of a drawing tube.

Conidia

Shape, dimensions and septation of the conidia are important criteria for species identification. However, conidium dimensions may differ considerably between different strains of one species. This means that the most common sizes given cannot be representative for all strains of the species. Our data refer to 30 measurements with oil-immersion ×1250. Drawings were made with the help of a drawing tube. Measurements should always be taken of conidia in fresh mature pycnidia on OA, because conidia from immature pycnidia are often smaller, and very old pycnidia often contain abnormally swollen, dark and septate conidia.

The conidia of most *Phoma* species are guttulate, especially near both ends (bipolary). The number and size of the guttules are often specific (see the figures), but the exception proves the rule.

Chlamydospores

The production of thick-walled, more or less dark chlamydospores on OA and MA is characteristic of several species of *Phoma* in different sections. Their shape, size and position in the mycelium, terminal or intercalary, are often very specific. A dark sector in a culture may indicate the presence of chlamydospores.

Multicellular chlamydospores, characteristic for sect. *Peyronellaea* (**D**) (Boerema, 1993), usually develop in aerial mycelium. The use of cellophane tape (Butler & Mann, 1959) is recommended for mounting aerial chlamydospores.

Drawings of chlamydospores were made with the help of a drawing tube or drawn from micrographs.

Swollen Cells

Sometimes conspicuous thin-walled swollen cells occur in the aerial mycelium. They may be septate and darker in colour than the hyphae. These 'pseudochlamydospores' occur particularly in strains with curly aerial mycelia.

Other Morphological Phenomena

Cultures are always carefully examined for the presence of other phenomena, such as fragmentation of hyphae (production of arthrospores), production of hyphal conidia (*Phialophora* synanamorphs), small or large sclerotia

(*Sclerotium* synanamorph), sclerotium-like structures (pseudosclerotioid aggregates of chlamydospores and pycnosclerotia, i.e. sclerotized pycnidial initials) and holdfasts (clusters of swollen multibranched hyphal tips produced on the bottom of the Petri dishes).

Genetic Studies

The separate status of some morphologically very similar species and varieties is supported by amplified fragment length polymorphism (AFLP) studies (Abeln *et al.*, 2002).

References to Additional Data

Scleroplectenchyma (pycnidial dimorphism): Boerema & van Kesteren (1964). *Techniques of isolation and preservation*: Gams *et al.* (1998). *NaOH spot test*, colour plates of oxidation reaction (E + O$_2$ → α → β) in purified extract and in agar plate cultures. *Pigments*, colours in agar plate cultures and colour change with addition of a drop of NaOH: Boerema & Höweler (1967), Dorenbosch (1970). *Crystals*, colour plate of anthraquinone crystals in agar plate cultures: Tichelaar (1974); production of dendritic crystals in agar plate cultures: Noordeloos *et al.* (1993).

5 Entries to the Differentiation of the Sections

For a direct first impression of the most significant characteristics of the nine sections see Figs 6–51.

The treatment starts (**A**) with section *Phoma*, followed by the other sections: **B** *Heterospora*, **C** *Paraphoma*, **D** *Peyronellaea*, **E** *Phyllostictoides*, **F** *Sclerophomella*, **G** *Plenodomus*, **H** *Macrospora*, **I** *Pilosa* (compare the 'Nomenclator' on p. 9).

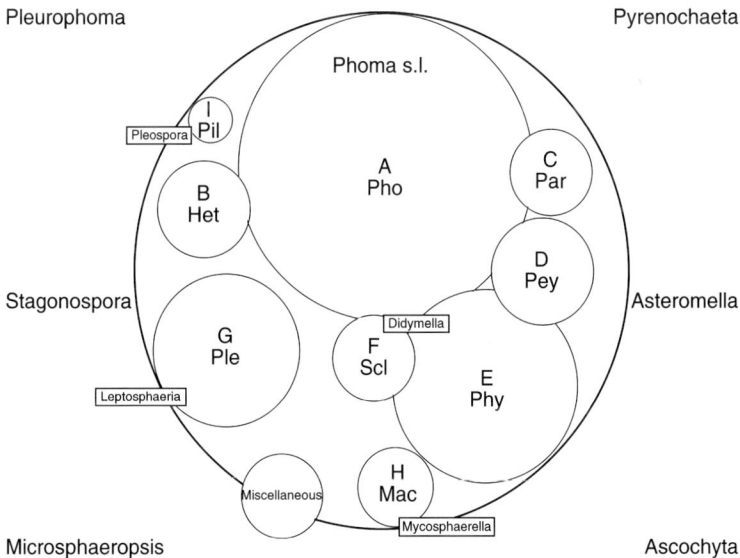

Fig. 4. The diameter of the circles in this diagram reflect the relative species numbers. Arrangement, touching and overlap of the circles indicate resemblance. Apart from the connected teleomorphic genera some 'adjacent' or 'convergent' anamorphic genera (p. 24) are also noted.

© G.H. Boerema, J. de Gruyter, M.E. Noordeloos and M.E.C. Hamers 2004.
Phoma *Identification Manual* (G.H. Boerema *et al.*)

Key to the Sections Based on Characteristics *In Vitro*

1. a. Fresh *Phoma* isolates readily develop pycnidia *in vitro* **2**
    b. Isolates do not readily develop pycnidia *in vitro*; conidia may be produced directly from the mycelium; sometimes only multicellular chlamydospores develop **11**

2. a. Colonies produce typical multicellular chlamydospore type structures, unicellular chlamydospores may also be present **11a**
    b. Colonies lacking multicellular chlamydospore type structures, but unicellular chlamydospores may be present **3**

3. a. Pycnidia scleroplectenchymatous **15b**
    b. Pycnidia pseudoparenchymatous **4**

4. a. Pycnidia thin-walled **5**
    b. Pycnidia rather thick-walled **14**

5. a. Pycnidia with predetermined ostiole (initiated in primordium) **6**
    b. Pycnidia at first closed, then opened by a secondary pore; often thin-walled cellular protrusions in pycnidial cavity **15a**

6. a. Pycnidia glabrous, but often with some hyphal outgrowths (semi-pilose) **7**
    b. Pycnidia distinctly pilose (hairy by many protruding hyphae) or setose (with straight or slightly flexuous setae) **10**

7. a. Conidia: all relatively small, (2–)3–11(–13) × (0.5–)1–4(–5) µm, aseptate or secondarily septate **8**
    b. Conidia: some or all distinctly large, (7–)11–25(–30→) × (2–)2.5–7(–8→) µm, aseptate or septate **9**

8. a. Conidia all aseptate [also *in vivo*, but germination may be associated with secondary septation], mostly 3–11 × 1.5–4 µm. In a single species conidia are produced directly from the mycelium (*Phialophora* synanamorph) ....**A** sect. *Phoma*
        NB: If septate conidia occur *in vivo*, a species of sect. *Heterospora* (this key no. 9a), or sect. *Phyllostictoides* (this key no. 13a) may be involved.
    b. Some conidia 1-septate (occasionally 2-septate) [septation often more prominent and sometimes even dominant *in vivo*] **13**

9. a. Conidia usually dimorphic, but often only a few extremely large conidia occur; mostly becoming 1-septate, but in some species remaining aseptate. Sometimes only small and aseptate conidia [*in vivo* and also occasionally in fresh cultures a high percentage of the conidia may become large and 1-septate or even 2–3-septate (*Stagonosporopsis* synanamorphs)] ....**B** sect. *Heterospora*
        NB: Distinctly large septate conidia also occur in some species of sect. *Sclerophomella* (this key no. 14a), and also occasionally in species of sect. *Phyllostictoides* (this key no. 13a).
    b. Always many large conidia, aseptate or 1-septate **16**

10. a. Pycnidia setose with distinct ostiole; the well differentiated thick-walled setae may be scattered over entire surface or concentrated around the ostiole. ....**C** sect. *Paraphoma*

NB: Setose pycnidia also occur in some species of sect. *Plenodomus* (this key no. 15b).
b. Pycnidia pilose (hairy) with inconspicuous ostiole **17**

11. a. Production of multicellular chlamydospores, mostly alternarioid to irregular-botryoid, sometimes epicoccoid (*Epicoccum* synanamorph) or resembling pseudosclerotia; often unicellular chlamydospores also occur. Conidia mostly aseptate, but occasionally longer ones become 1-septate ......**D** sect. *Peyronellaea*
NB: Species with multicellular chlamydospores also occur in sect. *Heterospora* (this key no. 9a) and sect. *Macrospora* (no. 16a).
b. Only sterile mycelium or production of conidia directly from the mycelium (*Phialophora* synanamorph) **12**

12. a. Pycnidial-type *in vivo* (source of isolate) scleroplectenchymatous ('*Plenodomus*-like') **15b**
b. Pycnidial-type *in vivo* (source of isolate) pseudoparenchymatous ('*Phoma*-like') **8a**

13. a. The 1-septate conidia only a little larger than the aseptate conidia. Majority of the conidia always aseptate [*in vivo* 5 to 95% of the conidia may become 1(–2)-septate] ....**E** sect. *Phyllostictoides*
NB: Species with septate conidia also occur in sect. *Peyronellaea* (this key no. 11a) and sect. *Plenodomus* (this key no. 15b).
b. The 1-septate conidia sometimes extremely large **9a**

14. a. Pycnidia initially closed, thick-walled with pseudoparenchymatous wall structure. Conidia mostly aseptate, but usually some are 1(–2)-septate. These may be normal in size and also extremely large [*in vivo* a high percentage of both sizes may be septate, ascochytoid]. .... **F** sect. *Sclerophomella*
b. Pycnidia and epicoccoid multichlamydospores occur (*Epicoccoid* synanamorph) **11a**

15. a. Pycnidia poroid, thin-walled and pseudoparenchymatous, often with a series of cells protruding into the pycnidial cavity [*in vivo* these pseudoparenchymatous pycnidia (Type I) may become scleroplectenchymatous (Type II)]................**15b**
b. Pycnidia poroid, thick-walled and scleroplectenchymatous [Type II], cavity often irregular by invaginations from the walls. Sometimes only pycnosclerotia occur [III]. All species make scleroplectenchyma but some produce only thin-walled pseudoparenchymatous pycnidia *in vitro* [Type I, this key no. 15a]. One species usually only produces conidia directly from the mycelium *in vitro* (*Phialophora* synanamorph). A single species produces true sclerotia under certain conditions (*Sclerotium* synanamorph). Conidia mostly aseptate, but some of the longer ones may become 1-septate [*in vivo* such species may produce a high percentage of 1-septate conidia]. ....**G** sect. *Plenodomus*
NB: Often associated *in vivo* with teleomorph, *Leptosphaeria*.

16. a. Conidia for the most part large, aseptate or secondarily septate [*in vivo* conidia are also large, aseptate or septate]. ....**H** sect. *Macrospora*
[b. Many large conidia also occur occasionally in fresh cultures of species belonging to sect. *Heterospora*; this key 9a.]

17. a. Pycnidia distinctly pilose *in vitro* (hairy by a great number of protruding hyphae) with inconspicuous ostiole. Conidia always aseptate [also always aseptate *in vivo*] ....**I** sect. *Pilosa*
      NB: Pycnidia *in vivo* glabrous, often associated with teleomorph, *Pleospora*.
   [b. Semi-pilose pycnidia, i.e. with some hyphal outgrowths, especially near the opening, may occur in most sections.]

Table 1. Collective and differential features of the sections *in vitro*.
+ = characteristic for all or most species. ++ = principal determining character.
± = only found in some species

Features of:	A Pho	B Het	C Par	D Pey	E Phy	F Scl	G Ple	H Mac	I Pil
Pycnidia									
Glabrous	+	+		+	+	+	+	+	
Pilose	±		±	±	±		±		++
Setose			++				±		
Thin-walled	+	+	+	+	+		+	+	+
Thick-walled				±		++	++		
Pseudoparenchymatous	+	+	+	+	+	+	±	+	+
Scleroplectenchymatous							++		
Conspicuous ostiole	+	+	+	+	+			+	
Inconspicuous ostiole	±	±			±				+
With pore or ruptured (secondary opening)				±		++	++		+
Hyphae									
With phialoconidia	±						±		
With dark unicellular chlamydospores	±	±	±	±	±	±	±	±	±
With dark multicellular chlamydospores		±		++				±	
Conidia									
Of 'common size' [(2–)3–11(–13) x (0.5–)1–4(–5) μm]	++	++	++	++	+	+	++		++
Some or all distinctly large [(7–)8–25(–28→) x (2.5–)3–6(–9→) μm]		++			±	±		++	
Aseptate	++	+	++	+	+	+	+	+	++
Also secondarily septate		±		±	++	±	±	±	
Teleomorph									
Didymella	±	±			±	±			
Leptosphaeria							±		
Mycosphaerella								±	
Pleospora									++

Synopsis of the Collective and Differential Characteristics of the Sections *In Vivo*

Pycnidia

With discrete, mostly ampulliform–doliiform conidiogenous cells, phialidic usually with a conspicuous collarette, the basal part of a papilla that precedes the conidiogenesis.
— Mostly glabrous, or with some hyphal outgrowths [a few species produce distinctly hairy pycnidia *in vitro*: sect. *Pilosa* **I**]
  - sometimes setose, especially around the ostiole: sect. *Paraphoma* **C**
  - occasionally with dictyochlamydospore(s) on pycnidial wall: sect. *Peyronellaea* **D**.
— Mostly thin-walled with pseudoparenchymatous wall structure
  - but sometimes obviously thick-walled and pseudoparenchymatous: sect. *Sclerophomella* **F**
  - or scleroplectenchymatous: sect. *Plenodomus* **G**.

Conidia

Hyaline or subhyaline
  - In old pycnidia sometimes becoming light brown: common in sect. *Peyronellaea* **D**.
— Mostly relatively small (2–13 × 0.5–5 μm)
  - all aseptate: determining for sect. *Phoma* **A**; also characteristic for sect. *Paraphoma* **C** and sect. *Pilosa* **I**
  - or partly 1–2(–3)-septate by secondary septation: determining for sect. *Phyllostictoides* **E**, but also found in sect. *Peyronellaea* **D**, sect. *Sclerophomella* **F** and sect. *Plenodomus* **G**.
— Sometimes (also) distinctly large (8–28→ × 2.5–8→ μm)
  - aseptate or septate: determining for sect. *Macrospora* **H**; characteristic of the conidial dimorph in sect. *Heterospora* **B** (when septate conidia dominant, classified in the synanamorphic genus *Stagonosporopsis*), occasionally also in sect. *Phyllostictoides* **E** and sect. *Sclerophomella* **F**.

6 Notes on 'Adjacent' or 'Convergent' Genera

As indicated diagrammatically in Fig. 4, various coelomycetous genera show some similarity with *Phoma*, but they are not necessarily related due to the possibility of evolutionary convergence.

The sections *Phoma* (**A**), *Pilosa* (**I**) and *Paraphoma* (**C**; setose pycnidia), which have species with only aseptate conidia, superficially resemble other similar pycnidial genera with aseptate hyaline conidia commonly borne on conidiogenous cells integrated in septate conidiophores:

Pleurophoma Höhn. (Fig. 5A)

Globose to collabent ostiolate pycnidia, thick-walled with characteristic cell structure (Sutton, 1980; Boerema & Dorenbosch, 1973: second example below), conspicuous filiform septate conidiophores, frequently occurring in pairs. The conidia are produced at the apex of the conidiophore and just below its transverse septa. However, the pycnidia often also contain a number of *Phoma*-like discrete conidiogenous cells.

Examples: **_Pleurophoma pleurospora_ (Sacc.) Höhn.**
≡ *Phoma pleurospora* Sacc.
Type species of the genus (Sutton, 1980), isolated from branches and bare wood of all kinds of trees and shrubs in Europe (Boerema *et al.*, 1996: 176)
Pleurophoma cava (Schulzer) Boerema et al.
≡ *Phoma cava* Schulzer
= *Phoma collabens* Schulzer & Sacc. (later homonym of *Ph. collabens* Durieu & Mont.)
= *Phoma aposphaerioides* Briard & Har.
Also common on woody plants, in Europe and North America. For synonymy see Boerema & Dorenbosch (1973: 23–24, with isolate sources) and van der Aa & Vanev (2002:

Phoma *Identification Manual* (G.H. Boerema *et al.*)

9–10). For illustrations of conidiophores and conidiogenous cells see Boerema *et al.* (1996: 172).

Asteromella Pass. & Thüm. (Fig. 5B)

Globose thick-walled pycnidia with more or less papillate ostioles and characterized by very small cylindrical or ellipsoidal conidia arising at the tapered apex and just below the transverse septa of the conidiophores (Sutton, 1980). *Asteromella* species are generally regarded as spermogonial states of *Mycosphaerella* species. So far there are more than 220 taxa described in the genus *Asteromella*, see Vanev & van der Aa (1998) and van der Aa & Vanev (2002).

Examples: **Asteromella brassicae (Chevall.) Boerema & Kesteren**
= e.g. *Phoma siliquastrum* Desm.
Phyllosticta brassicicola McAlpine
Spermogonial state of *Mycosphaerella brassicicola* (Duby) Lind, a worldwide common leaf spot pathogen of *Brassica* spp. For synonymy of both morphs see Boerema & van Kesteren (1964). For descriptions see Dring (1961).
Asteromella ulmi Boerema
≡ e.g. *Phyllosticta bellunensis* Martelli
Spermogonial state of *Mycosphaerella ulmi* Kleb., a common leaf spot pathogen of elms, *Ulmus* spp., in Europe and North America. Conidial anamorph: *Phloeospora ulmi* (Fr. : Fr.) Wallr. For synonymy of the spermogonial anamorph see Boerema (2003). For descriptions see Klebahn (1905).

Pyrenochaeta De Not. (Fig. 5C)

Globose ostiolate pycnidia, thick- or thin-walled with abundant setae around the ostiole, but fewer over the rest of the pycnidium. Conidia always arising from long filiform conidiophores, branched at the base and multiseptate. Teleomorphs, where known, usually belong to *Herpotricha* Fuckel; see Schneider (1976) and Sutton (1980). Most of the more than 120 taxa described in the genus *Pyrenochaeta* have only simple, discrete conidiogenous cells and belong to *Phoma* sect. *Paraphoma* (**C**); see Schneider (1979).

Example: **Pyrenochaeta fallax Bres.**
A specific necrophyte of nettle, *Urtica* spp., with more or less scleroplectenchymatous pycnidia. For description see Sutton & Pirozynski (1963). The scleroplectenchymatous wall structure of this anamorph fully agrees with that of *Phoma acuta* (Hoffm. : Fr.) Fuckel subsp. *acuta* (sect. *Plenodomus*, **G**). Both anamorphs may occur on the same nettle stem, which has caused much confusion; see Boerema *et al.* (1996: 181).

Some species classified in *Pyrenochaeta* produce a number of discrete conidiogenous cells, in addition to those integrated in conidiophores and thus may also be interpreted as setose members of *Pleurophoma*; see de Gruyter & Boerema (2002: note on p. 542).

Example: **Pyrenochaeta corni (Bat. & A.F. Vital) Boerema et al.**
≡ *Plenodomus corni* Bat. & A.F. Vital
A species often found in Europe in association with bacterial knot (canker) of ash, *Fraxinus* spp.; see Janse (1981) sub *Phoma riggenbachii* Boerema & J.D. Janse. For description and illustrations see Boerema *et al.* (1996: 158–160).

Old pycnidia of species in section *Peyronellaea* (**D**; distinguished by multicellular chlamydospores) often contain relatively dark conidia (occasionally also occurring in species of other sections). Such specimens are sometimes interpreted as a species of the following phialidic dark-spored analogue of *Phoma*:

Microsphaeropsis Höhn. (= *Coniothyrium* Auct. p.p.) (Fig. 5D)

The conidium ontogeny in this genus agrees with that in *Phoma* (cf. transmission electron micrographs; Jones, 1976). However, the dark pigmentation and differentiation of the conidium wall into an outer and inner layer (diffuse wall-building) takes place after conidial secession, maturation being asynchronous with conidium ontogeny. This means that young pycnidia of *Microsphaeropsis* species with immature, still colourless conidia, may be easily mistaken for a *Phoma* species.

Examples: **Microsphaeropsis fuckelii (Sacc.) Boerema**
≡ *Coniothyrium fuckelii* Sacc.
Teleomorph: *Leptosphaeria coniothyrium* (Fuckel) Sacc.
Isolates of this worldwide occurring pathogen of trees and shrubs were frequently received by the PD as presumed species of *Phoma*. For description and synonyms of the anamorph see Wollenweber & Hochapfel (1937). See also the documentation in Boerema (2003) and Boerema & Verhoeven (1972).
Microsphaeropsis glumarum (Ellis & Tracy) Boerema
≡ *Phoma glumarum* Ellis & Tracy
≡ *Phyllosticta glumarum* (Ellis & Tracy) I. Miyake
This anamorph, reported as a pathogen on kernels of rice, *Oryza sativa*, in the USA and China, has been confused with the plurivorous *Phoma sorghina* (Sacc.) Boerema *et al.* (sect. *Peyronellaea*, **D**); see Boerema *et al.* (1973: 136) and Boerema (2003).

In the introduction it was noted that species with secondary conidial septation [universal in section *Phyllostictoides* (**E**), but also occurring in other sections, especially in the large-sized conidial dimorph of species in section *Heterospora* (**B**) (synanamorph *Stagonosporopsis*)], were often formerly placed in a pycnidial genus that always produces septate conidia, *in vivo* and *in vitro*:

Ascochyta Lib. (Fig. 5E)

Conidiogenesis in species of this genus is not preceded by a thick-walled papilla as in *Phoma*. The first and successive conidia arise as a thin-walled protrusion, whose wall after secession (occasionally also just before secession) thickens by a new inner wall layer, concurrently dividing the conidium into two or more cells: wall thickening septation (analogue with distoseptation in dark, large spored Hyphomycetes, cf. Reisinger, 1972). Ontogeny (holo)blastic by apical wall building in the first conidium and by replacement wall building in subsequent conidia. Enteroblastic percurrent proliferation may result in conidia produced either at a higher level leaving an accumulation of annellations (annellidic appearance) or approximately at the same level with a gradually thickening collar of periclinal annellations. Such an annulate collar is different from, but under the light microscope looks very much like, the collarette in *Phoma* spp. (i.e. the basal part of the papilla preceding conidiogenesis). Wall thickening septation ('distoseptation'), i.e. a conidium maturation stage by diffuse wall-building asynchronous with conidium ontogeny, explains why mature conidia in true *Ascochyta* spp. are nearly always septate, both *in vivo* and *in vitro*: the septation in *Ascochyta* spp. is an essential part of the conidium maturation.[1] The conidiogenesis and septation described above have been established by transmission electron micrographs in three species of *Ascochyta* (Brewer & Boerema, 1965; Boerema & Bollen, 1975), namely:

> **Ascochyta pisi Lib.**
> Type species of the genus. Leaf and pod spot of field and garden peas, *Pisum sativum*.
> **Ascochyta fabae Speg.**
> One of the fungi causing leaf spot of field (broad) beans, *Vicia faba*.
> **Ascochyta pinodes L.K. Jones**
> Teleomorph: *Mycosphaerella pinodes* (Berk. & A. Bloxam) Vestergr.
> One of the fungi causing foot rot of field and garden peas, *Pisum sativum*.

It is obvious that the wall thickening septation ('distoseptation') of the conidia can only be established in micrographs of perpendicular sections. The micrographs used by Punithalingam (1979a) to reject the 'distoseptation' in *Ascochyta pisi* are not at all perpendicular. For descriptions and phytopathological literature of these pathogens see Boerema *et al.* (1993a).

---

[1] For identification and differentiation of *Ascochyta* spp. study *in vitro* is also most important. This is not only to distinguish them from *Phoma* spp. with secondary conidial septation, but also because some species of *Ascochyta* produce quite different septate conidia depending on whether they are in a parasitic or saprophytic phase (see van der Aa & van Kesteren, 1979; Boerema, 1984b).

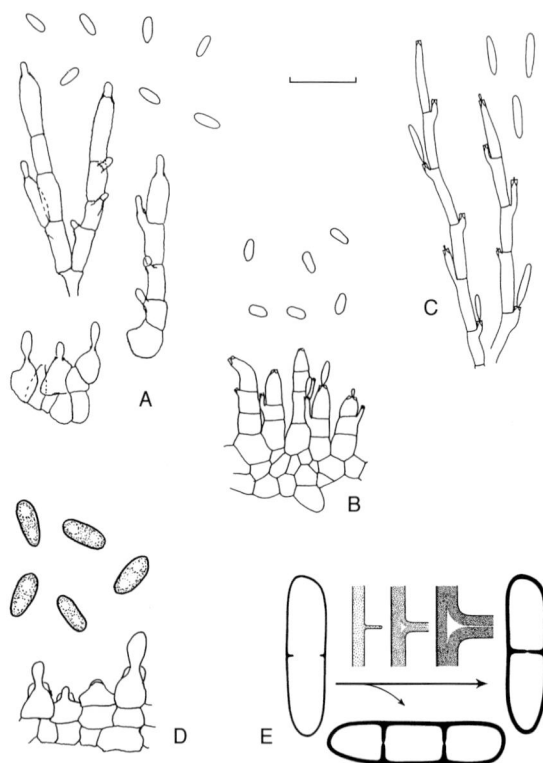

Fig. 5. Some characteristics of 'adjacent' or 'convergent' pycnidial genera. **A** *Pleurophoma*, conidiophores and some discrete conidiogenous cells from a fresh culture of the type species *Pl. pleurospora* (Sacc.) Höhn. **B** *Asteromella*, conidiophores of the type species *A. ovata* Thüm. (cf. Sutton, 1980). **C** *Pyrenochaeta*, conidiophores of *Pyr. ilicis* Wilson, closely related to the type species *Pyr. nobilis* De Not. (cf. Schneider, 1976; Sutton, 1980). **D** *Microsphaeropsis*, brown conidia and hyaline conidiogenous cells of *M. olivaceous* (Bonord.) Höhn. (cf. Jones, 1976; Sutton, 1980). **E** *Ascochyta*, diagram of the characteristic wall thickening septation ('distoseptation') of the conidia (cf. electron micrographs, Boerema, 1984a). Bar 10 μm.

Conidial dimorphism, distinctive for section *Heterospora* (**B**), is also found in pycnidial anamorphs of the genera *Leptosphaeria* Ces. & De Not. and *Phaeosphaeria* I. Miyake. However, in that case the large multiseptate conidial phenotype dominates not only *in vivo*, but also *in vitro*. The septation process in those large conidia needs further study. They are commonly classified in:

Stagonospora (Sacc.) Sacc. emend. Leuchtmann (1984)

> The connected '*Phoma*-like microconidial forms' are mostly unnamed, see 'Miscellaneous', p. 411.

Finally we must mention two coelomycetous genera which are now well known to be different from *Phoma*:

Phomopsis (Sacc.) Sacc.

Characteristics of this genus, i.e. pycnidial structure (eustromatic), conidiogenesis (elongated phialidic cells) and conidial morphology (two basic types: fusiform biguttulate conidia and filiform conidia), are easy to distinguish from those of *Phoma* spp. (Sutton, 1980; van der Aa *et al.*, 1990). Many species of *Phomopsis* have a teleomorph in the genus *Diaporthe* Nitschke.

On stems or branches of trees and shrubs the stromatic pycnidia are usually aggregated and confluent, but on leaf spots, living stems or roots of herbaceous plants they are mostly separate and of reduced size. In the past, many species with these single pycnostromata were described as *Phyllosticta* (on leaves, see van der Aa & Vanev, 2002) and as *Phoma* or *Plenodomus* (on stems and roots).

Examples: **Phomopsis asparagi (Sacc.) Bubák**
≡ *Phoma asparagi* Sacc.
Saccardo's description of this fungus refers to various European collections on putrifying stems of *Asparagus officinalis*. The name *Phoma asparagi* was until recently used for the causal organism of a serious stem blight of asparagus in Taiwan and Brazil (Hsu & Sun, 1969; Reifschneider & Lopes, 1982).
Phomopsis destruens (Harter) Boerema et al.
≡ *Plenodomus destruens* Harter
In North America (USA) and South America (Argentina) this fungus is frequently recorded as the causal agent of foot rot of sweet potatoes, *Ipomoea batatas* (Harter, 1913). The name *Plenodomus destruens* was used until recently in phytopathological literature (*see* Boerema *et al.*, 1996).

Phyllosticta Pers.

Saccardo's interpretation of *Phyllosticta* as a leaf spot analogue of *Phoma* (see introduction) explains why nearly all sections include *Phoma* species with *Phyllosticta* synonyms.

True species of *Phyllosticta* are quite different from *Phoma* and are readily distinguished by their relatively large conidia enclosed in a mucilaginous sheath and a mucoid apical appendage (van der Aa, 1973; van der Aa & Vanev, 2002). Teleomorphs in *Guignardia* Viala & Ravaz (syn. *Discochora* Höhn.).

Additional data on the pycnidial genera with conidiogenous cells integrated in conidiophores are available in: Sutton (1980): p. 395 (*Pyrenochaeta*), p. 397 (*Pleurophoma*), p. 405 (*Asteromella*). Characteristics of *Microsphaeropsis/Coniothyrium* in: Sutton (1980): pp. 123, 423, and Boerema (2003). Characteristics of *Phomopsis* in: Sutton (1980): p. 569 and van der Aa & Vanev (2002): pp. 11–12. Electron micrographs of the conidiogenesis and conidial septation in *Ascochyta* spp. in: Boerema & Bollen (1975): fig. 3, pls 24–29. Conidiogenesis and appendage morphogenesis in *Phyllosticta* spp. in: Nag Raj (1993): pp. 37–41.

7 Entries to the Species per Section

Abbreviations and Symbols Used in the Nomenclatural Parts
(cf. Boerema *et al.*, 1993a)

Authors' names

    The names of authors, their abbreviations and the use of initials are in accordance with Brummitt & Powell (1992).

Serials, exsiccata works, books

    Citations of serial publications are abbreviated as in the World List of scientific periodicals published in 1900–1960 (Brown & Stratton, 1963–1965) and in succeeding annual volumes.

in   The abbreviated citations are always preceded by the word '*in*' (italicized). If the name of a periodical has changed, the abbreviations used conform with the title of the journal at the time the article was published.

    Volumes are indicated by an arabic numeral, in bold. When volumes are not numbered the years on the title pages are used as volume numbers. If a volume consists of parts paged separately the number of the part has been inserted in parentheses; this is also done when the parts have been issued in different years. If a periodical has appeared in more than one series in which the numbers of the volumes are repeated, the later series are designated by a roman capital figure, 'II', 'III', etc. Pages are given in arabic numerals, except those otherwise designated in the original.

    Distributed exsiccata series accompanied by printed matter are considered as serial publications and coded in conformance with abbreviations in the World List. They can always be recognized by the designation:

No.   'No. ...' instead of the page number.

Fasc.,   Fascicle 'Fasc.', and Centurion 'Cent.' indications are always given in arabic
Cent.   numerals.

    The titles of books are abbreviated as for serials and further adapted to the title abbreviations of publications by Hawksworth (1974) and to the list of abbreviations published by the Commonwealth Mycological Institute (Anonymous, 1969). Volumes are indicated by an arabic numeral in bold.

ed., Ed. If a book or *exsiccata* work has appeared in more than one edition, after the first one these are designated 'ed. 2', 'ed. 3', etc. The abbreviation 'Ed.' between square brackets means editor(s).

Various abbreviations and signs

[ ]   Square brackets are always used to enclose informative notes on the (original) citation of binomials, authors, papers, books, and so forth.

l.c.   *Loco citato*; refers to literature cited previously for the same fungus.

apud   When a binomial of one author is published in the work of another the word 'apud' is used if the work appeared in a serial publication or exsiccata series

in   and the word 'in' if the work appeared in a book. When a name has been proposed but not validly published by one author and thereafter validly published

ex   and ascribed to him by another author the word 'ex' is used to connect the names of the two authors.

emend. *Emendavit* (emended, changed by). Followed by the author(s) who emended or changed the circumscription of a taxon without excluding its type.

=   Equal. This sign is used for taxonomic synonyms based on different types.

≡   Triple equal. Used for nomenclatural synonyms based on the same type.

: Fr.   The colon in this citation indicates that Fries has sanctioned the name published by the original author(s) (Botanical Code, Art. 15). It means that the name attains a special priority status (at its own rank) over any earlier homonym or synonym. In addition, such a name has a special typification status.

V   Means that a correction in the spelling has been made (according to Code, Art. 60).

H   Illegitimate later homonym (Code, Art. 53).

Θ   *Nomen nudum*; a naked name, i.e. a name published with no description or diagnosis, nor any reference to one.

‡   Non rite published, i.e. invalid according to Code Art. 33 (no basionym reference), or Art. 36 (no Latin diagnosis), or Art. 37 (no full type indication), or another annotated Article of the Code.

8   A *Phoma* sect. *Phoma*

Phoma Sacc. sect. *Phoma*

> *Phoma* Sacc. sect. *Phoma*; automatically established with the differentia-
> tion of *Phoma* sect. *Plenodomus*, see Boerema, van Kesteren & Loerakker
> *in* Trans. Br. Mycol. Soc. **77**: 62. 1981.
> Type: *Phoma herbarum* Westend.

Within the heterogenous genus *Phoma* the section *Phoma* embraces a number
of species that have much in common with the lectotype species *Ph. herbarum*
(Figs 1 and 6). Pycnidia simple or complex, generally glabrous but sometimes
bearing scattered hyphal outgrowths (semi-pilose), thin-walled, pseudoparenchymatous and distinctly ostiolate, producing only aseptate conidia *in
vivo* and *in vitro*. However, with initiation of germination conidial swelling and
elongation may be occasionally associated with septation.

The conidial dimensions vary within this section; there are species with
very small conidia, length on average not exceeding 3 μm, and species with
relatively large conidia, up to 13 μm in length (comp. Fig. 12B and C). The
morphology of pycnidia and conidia does not differ essentially *in vivo* and *in
vitro*. However, the range of variability in conidial shape and size is often wider
in vivo. A few species produce significantly larger conidia *in vivo* than *in vitro*.
Conidia production freely on hyphae (*Phialophora*-like) and fragmentation of
hyphae (arthrospores) are occasionally observed *in vitro*.

Included in the section are further species producing simple chlamydospores, solitary or in series, and species which do not produce chlamydospores. The type species, *Ph. herbarum*, belongs to the latter category.
Swollen cells are common in many species.

It should be noted that species which are also characterized by the production of multicellular chlamydospores have been classified separately in
sect. *Peyronellaea* (**D**).

The majority of the numerous species described in the genus probably
belong to sect. *Phoma*. As noted before, the species studied in culture so far

mainly concern saprophytes or parasites commonly occurring in western Europe and furthermore any selection of species from other parts of the world.

Most members of the section do not apparently develop a teleomorph, but two species are known which in addition to pycnidia produce pseudothecia *in vitro* that belong to the genus *Didymella* Sacc. ex Sacc.

Diagn. lit.: Boerema (1964, 1970) (type species, history, characteristics, synonymy and misapplications), Morgan-Jones (1988a) (type species, history and description), van der Aa *et al.* (1990) and Boerema (1997) (explanation of the concept of the present section), de Gruyter & Noordeloos (1992), de Gruyter *et al.* (1993, 1998) (precursory contributions to this chapter).

Synopsis of the Section Characteristics

- Pycnidia simple or complex, thin-walled, pseudoparenchymatous and distinctly ostiolate, glabrous but sometimes with hyphal outgrowths (semi-pilose).
- Conidia always aseptate *in vitro* and *in vivo*; production of conidia on hyphae (*Phialophora*-like) and fragmentation of hyphae (arthrospores) may occur.
- Chlamydospores, if present, unicellular; swollen cells common in many species.
- Teleomorph usually absent; if present belonging to *Didymella*.

Key to the Species and Varieties Treated of Section *Phoma*

Differentiation based on the characteristics *in vitro*

One insufficiently known species is not incorporated in this key, but can be identified by its specific host relation and characteristics *in vivo*, see 'Possibilities of identification *in vivo*', p. 45.

1. a. Species with relatively small conidia, length not exceeding 5.5 μm **3**
    b. Species with larger conidia **2**

2. a. Species producing conidia up to 7 μm in length **31**
    b. Species able to produce conidia longer than 7 μm **46**

3. a. Conidia between 2 and 4 μm in length...**4**
    b. Conidia between 3 and 5.5 μm in length **12**

4. a. Conidia subglobose to broadly ellipsoidal **5**
    b. Conidia ellipsoidal to subcylindrical **6**

5. a. Conidia 2–3 × 1.5–2 μm, with one large guttule. Colonies on MA with distinct pinkish/reddish or apricot colours; chlamydospores present; growth rate slow, on OA *c.* 2 cm after 7 days, on MA *c.* 2.5 cm ....*Ph. minutispora* [Probably common saprophytic soil fungus in south-west Asia. Also reported as an opportunistic human pathogen.]
    b. Conidia 3–4 × 2–2.5 μm, with mostly 2 or 3 conspicuous greenish guttules. Colonies producing a yellow pigment that stains the agar (reverse) honey to umber. ....*Ph. putaminum*

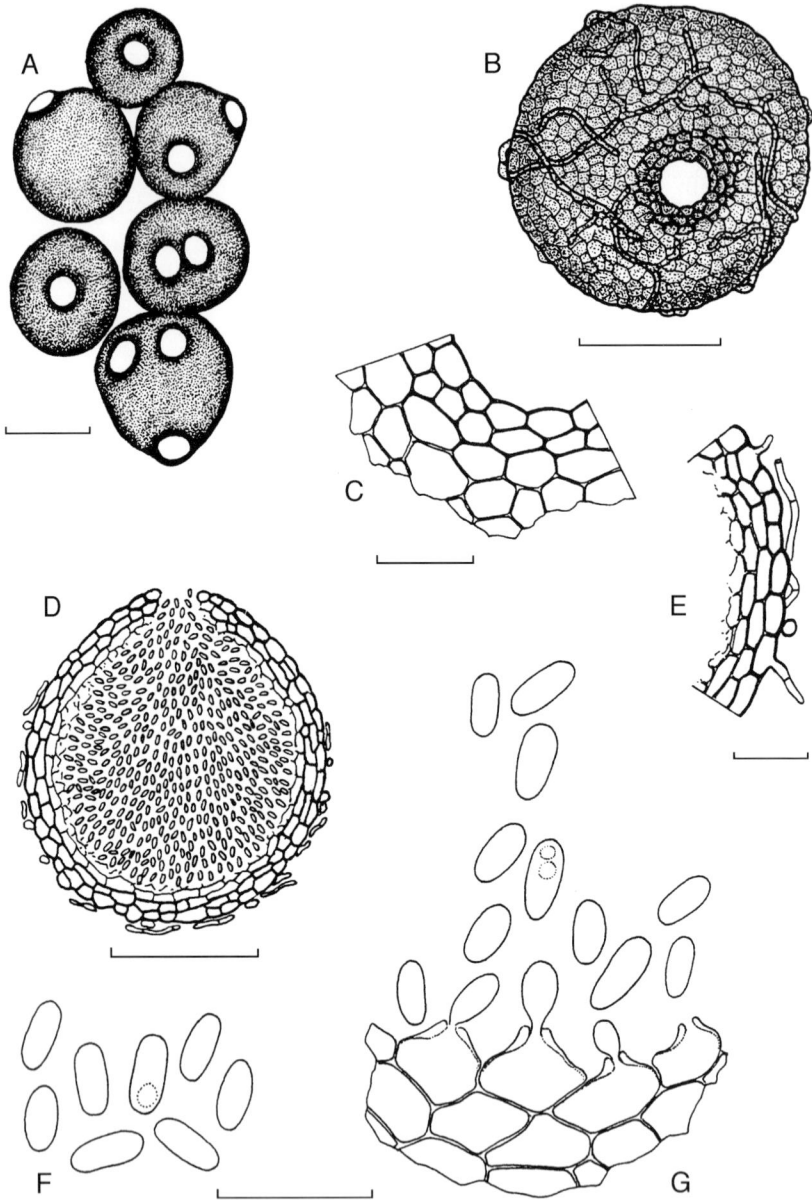

Fig. 6. *Phoma herbarum*, type species of the genus and section *Phoma*. **A, B** Pycnidia from 14-day-old colonies. **C** Surface view of pycnidial wall in vicinity of ostiole. **D** Vertical section of pycnidium. **E** Portion of pycnidial wall in section. **F** Conidia. **G** Conidiogenous cells. Drawings after Morgan-Jones (1988a; with permission). Bars pycnidia 100 μm, bars pycnidial wall, conidia and conidiogenous cells 10 μm.

[Common saprophytic soil fungus in Europe and North America, occasionally also recorded from the Southern hemisphere. May act as an opportunistic parasite on roots, etc.]

6. a. Colonies slow growing, 2–3.5 cm after 7 days...**7**
   b. Colonies faster growing, 3–6.5 cm after 7 days...**8**

7. a. Colonies exuding yellow pigment(s), discolouring the agar; not changing with NaOH. Conidia more ellipsoidal than cylindrical, 2–4 × 1–2 µm, mostly eguttulate, sometimes with 1–2 very small guttules........................*Ph. flavigena*
   [Once isolated from fresh water, eastern Europe (Romania).]
   b. Colonies exuding variable reddish or greenish pigments, discolouring the agar; on application of a drop of NaOH the reddish pigments turn blue. Conidia more cylindrical than ellipsoidal, 2–3(–3.5) × (0.5–)1–1.5 µm, eguttulate. Aerial mycelium may produce hyphal conidia..................................
   ....*Ph. vasinfecta*, synanam. *Phialophora chrysanthemi*
   [Vascular pathogen of *Dendranthema*–Grandiflorum hybrids. Slow wilt, phoma decline disease. Recorded in Europe, North America and Australia.]

8. a. Colonies on OA fast growing, *c.* 6.5 cm after 7 days, rather slow on MA, *c.* 3.5 cm; distinctly dark olivaceous grey to olivaceous black. Conidia mostly 3–4 × 2 µm, with two, distinct, polar guttules.....................................*Ph. valerianae*
   [Common on dead stems of *Valeriana* spp. in Europe; occasionally also isolated from other unrelated herbaceous plants.]
   b. Growth rate on OA 3–4.5 cm after 7 days, faster on MA, 4–5.5 cm. Conidia eguttulate or with some small guttules...**9**

9. a. NaOH reaction on OA greenish, then red: E$^+$ reaction. Colonies with distinct bright yellow/green tinges (citrine). Conidia mostly 3–4 × 1.5–2 µm; sometimes with small polar guttules ....*Ph. anigozanthi*
   [Pathogen of kangaroo-paw, *Anigozanthus* spp.: leaf blotch. Probably indigenous to Australia, but so far only recorded in Europe.]
   b. NaOH spot test: usually resulting only in a slight sienna discoloration............**10**

10. a. Colonies growing relatively fast, growth rate on OA *c.* 3 cm after 7 days, on MA *c.* 4 cm, with olivaceous grey and citrine tinges. The sienna discoloration with NaOH slowly turns bluish green. Conidia 2.5–3.5 × 1–1.5 µm, usually eguttulate ....*Ph. opuntiae*
   [Pathogen of *Opuntia* spp.: necrotic spot. Indigenous to South America, but probably also elsewhere on the hosts.]
   b. Growth rate *c.* 4.5 cm on OA after 7days, and 5–5.5 cm on MA....................**11**

11. a. Colonies greenish olivaceous. Pycnidia solitary, globose to flask-shaped, usually with 1 ostiole. Conidia mostly 3–4 × 1.5–2 µm (*in vivo* larger), sometimes with small guttules ....*Ph. costarricensis*
   [Pathogen of coffee, *Coffea arabica*; coffee blight. Until recently only known from South and Central America, now also recorded in other parts of the world.]
   b. Colonies olivaceous grey, olivaceous, or olivaceous black with citrine tinges. Pycnidia solitary or in clusters, globose with 1–5 ostioles, papillate or with

distinct neck. Conidia 3–4 × 1–2 μm, usually eguttulate ....*Ph. eucalyptica*
[Seed-borne pathogen of *Eucalyptus* and *Eugenia* spp. in Australasia: shoot
wilt, leaf necrosis, blister disease. Also recorded in southern Asia and Europe.]

12. a. Conidia broadly or shortly ellipsoidal **13**
    b. Conidia ellipsoidal to oblong **16**

13. a. Growth rate very slow, 2–2.5 cm on OA after 7 days....................................**14**
    b. Growth rate faster..**15**

14. a. Colonies with distinct apricot or scarlet tinges on OA, with NaOH spot test
        quickly changing to purplish/blue. Conidia mostly 3.5–5 × 1.5–2.5 μm,
        sometimes with a few guttules ....*Ph. multipora*
        [Probably a common 'halophilic' soil fungus in south-west Asia and northern
        Africa.]
    b. Colonies grey olivaceous, and follow greenish olivaceous or medium yellow on
        OA; not changing with NaOH spot test. Conidia mostly 3–5 × 2–3 μm, with 2
        or more polar guttules ....*Ph. fimeti*
        [Probably a worldwide occurring saprophytic soil fungus.]

15. a. Growth rate 3.5 cm on OA after 7 days, on MA *c.* 4 cm. Colonies peach
        to salmon, reverse similar. Conidia mostly 3–4.5 × 2–3 μm, with 1–2
        guttules...*Ph. capitulum*
        [Probably a common 'halophilic' soil inhabiting fungus in south-west Asia.]
    b. Growth rate *c.* 5 cm on OA after 7 days, on MA *c.* 7 cm. Conidia usually 2.5–3
        × 2–5 μm, occasionally up to 5.5 × 3 μm with one to three acentric guttules;
        colonies on MA with grey olivaceous tinges....................................*Ph. anserina*
        [Apparently a common saprophytic soil fungus in temperate Eurasia and North
        America; also recorded from northern Africa: 'Omnivorous', e.g. isolated from
        meat, gelatine and frequently from cysts of *Globodera rostochiensis.*]

16. a. Growth rate very slow, 2–2.5 cm on OA after 7 days..................................**17**
    b. Growth rate faster, 5–7.5 cm on OA after 7 days **18**

17. a. Growth rate about 2 cm. Pycnidia usually 175–240 μm diam. Conidia usually
        with 2 small guttules, mostly 3.5–4.5 × 1.5–2(–2.5) μm, but occasionally larger,
        6.5–9.5 × 2–4 μm. In the mycelium swollen cells may occur ....*Ph. apiicola*
        [Widespread on celeriac and celery, *Apium graveolens*, in temperate regions of
        Europe and North America: root rot, scab, crown rot. Soil-borne, sometimes
        also found on seeds (seedling canker).]
    b. Growth rate about 2.5 cm (14 days *c.* 3.5 cm). Pycnidia 60–130 μm diam.
        Conidia 4–4.5 × 1.5(–2) μm; without guttules or with 1–2 inconspicuous
        guttules ....*Ph. haematocycla*
        [Common pathogen of New Zealand flax, *Phormium tenax*, causing red-
        encircled lesions: leaf spot.]

18. a. Chlamydospores present **19**
    b. Chlamydospores absent **20**

19. a. Growth rate 5–5.5 cm on OA after 7 days. Colonies dull green or herbage
        green to olivaceous black. Pycnidial wall consisting of about 3 layers of cells.

Conidia 4–5.5 × 2–2.5 μm, with 2 polar guttules. Chlamydospores single or in chains, 4–15 μm diam., with greenish guttules ....*Ph. eupyrena*
[Cosmopolitan soil-inhabiting fungus, which may cause damping-off of seedlings of herbaceous and woody plants.]

    b.   Growth rate *c.* 7 cm on OA after 7 days. Colonies grey olivaceous to olivaceous grey. Pycnidial wall consistent of 5–8 layers of cells. Conidia 3.5–5 × 1.5–2 μm with 2 polar guttules. Chlamydospores single or in chains, 6–9 μm diam., with greenish guttules..*Ph. pereupyrena*
[Apparently a plurivorous opportunistic parasite in Eurasia, isolated from quite different herbaceous and woody plants.]

20. a.   Crystals present **21**
    b.   Crystals absent **23**

21. a.   Dendritic whitish crystal figures are readily formed within two weeks on MA. Colonies on OA pale buff to buff, dull green or olivaceous grey, finally olivaceous grey. Conidia 3–5.5 × 1.5–2(–2.5) μm, usually with 2 small polar guttules ....*Ph. dorenboschii*
[Recorded in The Netherlands on *Physostegia virginiana* and *Callistephus* sp. in association with leaf spots and stem lesions.]
    b.   Crystals not dendritically arranged..**22**

22. a.   White to buff needle-like crystals present on all media. Colonies on OA colourless without aerial mycelium. Conidia mostly 3–5 × 1.5 μm, with 2 small polar guttules .... *Ph. piperis*
[Causal organism of leaf spots on Indian long pepper, *Piper longus*, and other species of *Piper*. The South American genus *Peperomia* may also be attacked: leaf spot.]
    b.   Citrine-green crystals present, associated with yellow anthrachinon pigment(s) **29a**

23. a.   NaOH reaction negative (no reaction). Growth rate 5–5.5 cm after 7 days. Colonies on OA colourless to (olivaceous) buff; with fine velvety aerial mycelium. Pycnidia 125–330 μm diam., with one usually non-papillate ostiole. Conidia mostly 3–4.5 × 1.5 μm (sometimes curved), with 2–3 guttules ....*Ph. selaginellicola*
[Recorded in association with wild and cultivated plants of *Selaginella helvetica*: leaf necroses and wilting, Europe.]
    b.   NaOH reaction positive, at least on MA...**24**

24. a.   NaOH addition causes the E-oxidation reaction: initially yellow/green discoloration, gradually changing to red **25**
    b.   NaOH spot test results in a distinct orange/purple discoloration, or in a slight reddish or sienna discolouring **28**

25. a.   Growth rate 5 cm on OA after 7 days, on MA *c.* 6 cm. Colonies colourless, but general impression salmon by abundant conidial exudates. Conidia 4–5.5 × 1.5–2.5 μm, with 2 polar guttules..*Ph. eupatorii*
[Apparently a specific necrophyte of *Eupatorium cannabinum*, Europe.]
    b.   Growth rate faster..**26**

26. a. Growth rate 7–7.5 cm on OA after 7 days. Conidia uniform ellipsoidal, 3.5–5.5 × 1.5–2.5 μm, with two distinct guttules. Colonies colourless with grey olivaceous or olivaceous grey shade ....*Ph. ajacis*
[Probably a widely occurring seed-borne pathogen of annual species of *Delphinium* in southern Eurasia: leaf spot, stem rot.]
    b. Growth rate 6–7 cm on OA after 7 days...**27**

27. a. Colonies almost colourless with whitish aerial mycelium. Pycnidia after some transfers often non-ostiolate. Conidia 3.5–5 × 1.5–2 μm, usually with 2 inconspicuous polar guttules....................................*Ph. poolensis* var. *poolensis*
[A worldwide recorded pathogen of cultivated varieties of snapdragon, *Antirrhinum majus*: leaf spot, (basal) stem rot. Frequently encountered on seed of this host (damping-off in seedlings). Also isolated from other Scrophulariaceae.]
    b. A variety with stable production of distinctly ostiolate pycnidia and somewhat wider size range of conidia, 3.5–5.5 × 1.5–2.5 μm, has been differentiated as ....*Ph. poolensis* var. *verbascicola*
[Pathogen of *Verbascum* spp., causing necrotic lesions on leaves and stems and damping-off of young plants. Probably widely distributed in Eurasia and North America.]

28. a. Growth rate 4.5–5 cm on OA after 7 days **29**
    b. Growth rate faster...**30**

29. a. NaOH causes slight reddish discolouring of citrine agar zones (occurrence of citrine green crystals in old and acid media points to anthrachinon pigments). Colonies colourless to buff with grey olivaceous centre, partly developed white to greenish grey aerial mycelium. Pycnidia solitary or in rows, usually with a papillate ostiole. Conidia mostly 3–5 × 2–2.5 μm, without or with two indistinct polar guttules..*Ph. arachidis-hypogaeae*
[Plurivorous pathogen, frequently found on living leaves of groundnut, *Arachis hypogaea* in India: leaf spot. Probably indigenous to that subcontinent. Leguminosae seem to be very susceptible.]
    b. NaOH causes usually a slight sienna discolouring. Colonies with olivaceous grey to olivaceous black colour. Pycnidia with 1–3 papillate ostioles, which in a later stage develop into necks. Conidia 3.5–5.5 × 1.5–2 μm, with several small, usually polar guttules ....*Ph. viburnicola*
[A widespread occurring opportunistic parasite of woody plants in Europe, e.g. common on *Viburnum* spp.]

30. a. Growth rate 6.5 cm on OA after 7 days. NaOH discolours yellow pigment on MA to orange, later purplish. Conidia mostly 3.5–5.5 × 1.5–2 μm, occasionally with 1 or 2 polar guttules ....*Ph. lini*
[Apparently a common necrophyte on *Linum* spp. in Europe; also isolated from water used for flax retting. Probably also on other herbaceous plants.]
    b. Growth rate 7–7.5 cm on OA after 7 days. NaOH causes a slight reddish discoloration, not associated with a specific pigment. Colonies pale buff to grey olivaceous (OA). In mycelium swollen cells, intercalary or terminal, solitary or in short chains, 5–12 μm diam. Conidia mostly 3.5–5 × 1.5–2.5 μm, with 2 or more polar guttules...*Ph. bismarckii*

[Worldwide recorded on dead branches of *Malus domestica*: also found in association with lenticel rot of apple.]

31. a. Growth rate very slow on OA, only 1.5–3 cm after 7 days **32**
    b. Growth rate faster **33**

32. a. Colonies with floccose or finely floccose aerial mycelium, dull green to dark herbage green on OA. Conidia mostly 3–5(–6.5) × 1.5–2 μm, sometimes with 1(–2) inconspicuous guttules ....*Ph. lupini*
    [Noxious pathogen of *Lupinus mutabilis* and other *Lupinus* spp. in North and South America.]
    b. Colonies without aerial mycelium, colourless becoming luteous in a later stage due to the production of a yellow pigment (not changing on addition of a drop of NaOH). Conidia 4–7 × 2–3.5 μm with 2 large polar guttules ....*Ph. flavescens*
    [So far only known from soil in The Netherlands.]

33. a. NaOH reaction positive, at least on MA..**34**
    b. NaOH reaction negative[2]...**39**

34. a. NaOH discolouring diffusible pigments on MA to blue/purplish, red, yellow/green or orange/purplish...**35**
    b. NaOH causing initially a yellow green discoloration, gradually changing to red (E$^+$ reaction) **37**

35. a. Growth rate 6–6.5 cm on OA after 7 days. MA staining scarlet, NaOH reaction yellow/green. Conidia mostly 4–6.5 × 1.5–2.5 μm, usually with 2 polar guttules ....*Ph. adianticola*, teleom. *Did. adianticola*
    [Pathogen of *Polypodiaceae*, probably widespread distributed in (sub-)tropical America: leaf necroses.]
    b. Growth rate slower, up to 5 cm on OA after 7 days **36**

36. a. MA usually staining reddish, NaOH reaction blue or purplish. Colonies on OA without aerial mycelium, flesh coloured with greenish tinge by abundant production of pycnidia. Pycnidia finally olivaceous black. Conidia (3.5–)4–5.5(–8) × 1.5–2(–3) μm, mostly eguttulate, but sometimes with inconspicuous polar guttules..*Ph. herbarum*
    [A saprophytic soil fungus isolated from dead material of all kinds of herbaceous and woody plants, as well as from animal and inorganic material. Especially common on surface of seeds. Cosmopolitan.]
    b. MA staining yellow, NaOH reaction red. Colonies on OA colourless with sparse aerial mycelium bearing abundant pycnidia. Pycnidia honey with olivaceous tinge around ostiole. Conidia mostly 4–6.5 × 1.5–2.5 μm, eguttulate or biguttulate...*Ph. senecionis*

[2] Fading of reddish or yellow pigments by addition of NaOH is characteristic for a species which frequently produces multichlamydosporal structures and therefore has been treated under sect. *Peyronellaea* (**D**). The pigments are associated with orange-red crystals. Conidia 3.5–6.5 × 1.5–2.5 μm, with 2 guttules..*Ph. chrysanthemicola*
[Common saprophytic soil fungus in most temperate regions. A specific form on florists' chrysanthemums is known from western Europe and North America, root rot and basal stem rot.]

[Apparently a specific necrophyte of *Senecio* spp. Europe, New Zealand, probably also elsewhere.]

37. a.   Growth rate 5.5–6 cm on OA after 7 days. Colonies colourless with olivaceous sectors or centre and sometimes sparse whitish aerial mycelium. Conidia mostly 4–6 × 2–3 µm, eguttulate or with 2 or more polar guttules...........*Ph. hedericola*
[A pathogen of *Hedera* spp. causing necroses on leaves and stems: leaf spot. Probably cosmopolitan.]

   b.   Growth rate 6.5–8 cm on OA after 7 days **38**

38. a.   Colonies colourless on OA, but general impression salmon, caused by abundant conidial exudates. Conidia mostly 4–6.5 × 2–2.5 µm, usually biguttulate ....*Ph. bellidis*
[Pathogen of daisy, *Bellis perennis*, seed-borne: damping-off. Also recorded in association with leaf necroses. Eurasia.]

   b.   Colonies colourless with dull green tinges on OA. Conidia 4–6.5 × 2–3 µm, with 2–5, usually polar guttules ....*Ph. crinicola*
[Pathogen of *Crinum* and *Nerine* spp.: leaf spot, bulb rot. So far only found in Europe.]

39. a.   Growth rate *c.* 8 cm on OA and MA after 7 days. Colonies staining flesh to rust coloured due to the release of a pigment (not changing with NaOH). Conidia 4–5(–6.5) × 1.5–2 µm, eguttulate ....*Ph. subherbarum*
[A plurivorous saprophyte, so far only known from North and South America. Frequently recorded on the surface of seeds.]

   b.   Colonies not so fast growing and without such pigment production **40**

40. a.   Chlamydospores present **41**
   b.   Chlamydospores absent **42**

41. a.   Colonies olivaceous with colourless concentric ring near margin. Pycnidia non-papillate or distinct papillate. Conidia oblong to ellipsoidal, 3–5.5(–7) × 1.5–2.5 µm, with 2(3), usually polar guttules ....*Ph. insulana*
[Associated with discolouring leaves and ripening fruits of olive, *Olea europaea*. So far only known from southern Europe.]

   b.   Colonies colourless to weakly olivaceous on OA, on MA often more distinctly pigmented; pycnidia with 1 or several ostioles, often with some variously shaped necks. Conidia ellipsoidal, variable in dimension (2.5–)3–6(–7.5) × 1.5–2.5(–3) µm ....vars of *Ph. multirostrata*
[Apparently worldwide distributed in subtropical regions; most records from India. The vars *macrospora* and *microspora* may be characterized as plurivorous opportunistic plant pathogens.]

42. a.   Growth rate 6–6.5 cm after 7 days. Colonies with distinct grey olivaceous to olivaceous colour. Addition of NaOH causes a reddish brown discoloration. Pycnidia usually with one papillate ostiole. Conidia 4–6.5 × 2–3 µm, oblong ellipsoidal with two or more polar guttules...*Ph. labilis*
[Soil-inhabiting fungus esp. known from southern Eurasia. Apparently a warmth preferring necrophyte isolated from various herbaceous and woody plants. In The Netherlands recorded in glasshouses.]

   b.   Growth rate 4.5–5.5 cm on OA after 7 days **43**

43. a.   Crystals present esp. on OA and MA; colonies on OA colourless to buff, with
         radially arranged rows of greyish pycnidia. Conidia 4–6 × 1.5–2 μm,
         eguttulate or with some inconspicuous polar guttules.................*Ph. crystallifera*
         [Apparently a common necrophyte of Leguminosae in Central Europe.]
    b.   Crystals absent **44**

44. a.   Colonies colourless to greenish olivaceous to dull green, outer margin light
         green. Usually abundant production of pycnidia with 1–5 non-papillate ostioles.
         Conidia 3–6 × 1–2 μm, ellipsoidal with 2 distinct polar guttules.........*Ph. tropica*
         [Necrophyte from tropical origin (South America?), commonly found in warm
         greenhouses in Europe.]
    b.   Colonies with whitish aerial mycelium **45**

45. a.   Colonies with ochre, primrose or yellow tinges (golden coloured), esp. on
         MA. Pycnidia with one ostiole, non-papillate or slightly papillate. Conidia
         4.5–7 × 2–3 μm, oblong to ellipsoidal, often acuminate with two or many
         small guttules ....*Ph. aurea*
         [Isolated from a dead stem of *Medicago polymorpha* in New Zealand. Possibly
         a common saprophyte in Australasia.]
    b.   Colonies grey olivaceous to greenish olivaceous, with citrine tinges by
         pigmentation of the agar. Pycnidia with 1–2 papillate ostioles, which in a later
         stage may develop in distinct necks. Conidia 3.5–6.5 × 1.5–2 μm, oblong to
         ellipsoidal with two or many, usually polar-oriented guttules ....*Ph. nebulosa*
         [Worldwide recorded plurivorous necrophyte; in Europe particularly common
         on dead stems of *Urtica* and *Scrophularia* spp.]

46. a.   Conidia broadly ellipsoidal to ovoid or cylindrical, width up to 5 μm **47**
    b.   Conidia (sub)cylindrical to allantoid, width (1–)2–3(–4) μm **60**

47. a.   Growth rate very slow on OA, up to 2.5 cm after 7 days **48**
    b.   Growth rate moderate to fast on OA, at least 3.5 cm after 7 days **49**

48. a.   NaOH causing a red-brown discoloration esp. on MA. Colonies on OA dull
         green to greenish glaucous. Conidia 5–7.5(–10.5) × 2–4 μm, sometimes with
         several indistinct guttules...*Ph. fallens*
         [Probably widespread in olive, *Olea europaea*, -growing regions of the world,
         particularly southern Europe; associated with spots on fruit and leaves.]
    b.   NaOH reaction negative. Colonies smoke grey to grey olivaceous/dull green on
         OA. Conidia oblong to ellipsoidal, 5–7(–8) × 2.5–4.5 μm, hyaline with a
         greenish/yellow tinge by several small guttules ....*Ph. glaucispora*
         [Common and widespread on oleander, *Nerium oleander*, in southern Europe:
         leaf spot. Also reported from glasshouse ornamental cultures.]

49. a.   Colonies producing a diffusible pigment on MA, staining the agar yellowish
         or ochre..**50**
    b.   Colonies not producing a diffusible pigment on MA.....................................**53**

50. a.   On MA diffusible pigment staining the agar yellowish to greenish **51**
    b.   On MA diffusible pigment staining the agar ochre.......................................**52**

51. a. Yellow-green crystals are formed on MA, NaOH reaction on OA and MA rosy
        vinaceous to coral (not an E$^+$ reaction). Colonies on OA pale luteous without
        aerial mycelium; growth rate moderate, 4–4.5 cm after 7 days. Conidia 7–12 ×
        2.5–4 μm, eguttulate ....*Ph. humicola*
        [Soil fungus in North and South America isolated from a variety of organic
        substrates. May act as an opportunistic parasite on roots, etc.]
   b.   Crystals absent, NaOH reaction weak greenish/yellow on MA (not an E$^+$
        reaction). Colonies on OA colourless to rosy buff, with or without whitish aerial
        mycelium; growth rate fast, 6–7 cm after 7 days. Conidia 3.5–10.5 × 2–4 μm
        with a few, small guttules ....*Ph. obtusa*
        [Common on dead stems of *Daucus carota*, but also as necrophyte on other
        herbaceous plants, Europe.]

52. a. NaOH reaction on OA greenish, then red (E$^+$ reaction), on MA an orange
        discoloration of the original ochraceous pigment also occurs. Colonies on OA
        colourless to rosy buff/honey. Conidia 4–8.5 × 2–4 μm, with numerous large
        guttules ....*Ph. draconis*
        [Pathogen of *Dracaena* and *Cordyline* spp. recorded from Africa, Europe, India
        and North America.]
   b.   NaOH reaction on OA and MA red, then blue (not an E$^+$ reaction). Colonies
        on OA olivaceous buff to primrose. Conidia 4–12 × 2.5–4.5 μm, with or
        without polar guttules ....*Ph. huancayensis*
        [A common saprophyte in the Andes region of South America, also recorded
        on various dicotyledonous and monocotyledonous plants in New Zealand.]

53. a. NaOH reaction on OA and MA greenish, then red (E$^+$ reaction). Growth rate
        very fast, completely filling a Petri dish in 7 days, with coarsely floccose to
        woolly, white to olivaceous grey aerial mycelium on OA and MA. Pycnidia
        scattered, sometimes on the agar, but mostly partly submerged in the agar.
        Conidia 5.5–8.5(–11) × 2.5–4 μm, mostly eguttulate, but sometimes with 1–3
        small polar guttules ....*Ph. paspali*
        [Soil-borne saprophyte or weak parasite recorded on various Gramineae
        (*Paspalum*, *Dactylis* and *Lolium* spp.) in New Zealand. Probably widespread in
        Australasia.]
   b.   NaOH reaction negative **54**

54. a. Growth rate fast on OA, 7–8 cm after 7 days. Chlamydospores abundant
        on OA^3, mainly in the aerial mycelium, with a typical distinct 'envelope'.
        Conidia 3.5–7.5(–12) × 2–3.5(–4.5) μm, sometimes with small polar
        guttules ....*Ph. heteroderae*
        [Soil-inhabiting saprophyte, recorded in Eurasia and North America from
        various organic and inorganic substrata. Probably weakly parasitic on
        Heterodera glycines eggs.]
   b.   Growth rate moderate on OA, 4–6 cm after 7 days **55**

^{3} Similar growth rate but absence of chlamydospores is found in strains of a species which produces normally dictyochlamydospores and therefore has been treated under sect. *Peyronellaea*
(**D**). Conidia (3–)4–7(–8) × (2–)2.5–3.5(–4) μm, eguttulate ....*Ph. zantedeschiae*
[Specific pathogen of the arum or calla lily, recorded from South Africa, western Europe, North
and South America, leaf blotch.]

55. a. Conidia long cylindrical to allantoid (distorted shapes occur), 7.5–12.5 ×
        1.5–3.5 μm[4], with numerous small guttules. Colonies on OA colourless to
        olivaceous grey or greenish grey..*Ph. astragali*
        [Probably an opportunistic parasite of *Astragalus* in North America: stem rot.]
    b. Conidial shape different..**56**

56. a. Chlamydospores present. Growth rate on OA between 4 and 4.5 cm. Colonies on
        OA colourless without aerial mycelium, only a short white-felted zone near
        margin. Conidia 5–7.5 × 2–3 μm, with 2(3) large polar guttules ......*Ph. henningsii*
        [In East Africa known as a harmful wound parasite of *Acacia* spp.; North
        American records may be questioned.]
    b. Chlamydospores absent **57**

57. a. Growth rate on OA and MA about similar **58**
    b. Growth rate on OA and MA different...**59**

58. a. Growth rate on OA and MA 4–5 cm after 7 days. Colonies on OA colourless
        to grey olivaceous, without or with sparse grey olivaceous aerial mycelium;
        on MA olivaceous buff/greenish olivaceous to olivaceous; reverse similar.
        Pycnidia olivaceous black. Conidia 3.5–8 × 1.5–3 μm, with several distinct
        guttules ....*Ph. plurivora*
        [Plurivorous saprophyte or opportunistic parasite in New Zealand and
        Australia.]
    b. Growth rate on OA 4–5.5 cm after 7 days, on MA 4–6 cm. Colonies colourless
        to grey olivaceous or olivaceous grey, often with ochre tinges, without or with
        whitish aerial mycelium. Pycnidia initially honey/citrine. Conidia 4–7(–10) ×
        1.5–2.5(–4) μm, with 2 or many small guttules ....*Ph. chenopodiicola*
        [Specific necrophyte of *Chenopodium* spp., especially *C. album* and *C.
        quinoa*. Europe, South America.]

59. a. Growth rate 5–5.5 cm on OA after 7 days, on MA growth rate faster,
        6–7 cm. Colonies on OA smoke grey, on MA greenish, with rosy/buff/honey
        sectors at the margin. Conidia 4–12 × 2.5–4.5 μm, sometimes with some
        small guttules ....*Ph. aliena*
        [An opportunistic parasite on various woody plants, e.g. *Berberis*, *Mahonia*
        and *Cotoneaster* spp., Europe.]
    b. Growth rate on OA 3.5–4.5 cm after 7 days, but less on MA, 2.5–3.5 cm.
        Colonies on OA greenish olivaceous with colourless margin at first, later
        greenish by pycnidia. Conidia 4.5–10.5 × 2–4 μm, with several distinct
        guttules ....*Ph. negriana*
        [An opportunistic parasite of *Vitis vinifera* in southern Europe, associated with
        necroses on leaves, fruits and stems.]

---

[4] Subcylindrical conidia, 5–7(–10.5) × 1.5–4 μm, may concern strains of a species treated under
sect. *Phyllostictoides* (**E**), but which often do not produce septate conidia. Distinctive for that fungus
are white dendritic crystals produced on MA after 7 days.......................*Ph. medicaginis*
[Pathogen of lucerne, known from Eurasia and North America, black stem disease.]

60. a. Growth rate very slow, 2–2.5 cm on OA after 7 days. Colonies dark herbage green on OA. Conidia cylindrical, 3.5–8.5 × 1–2 μm with distinct polar guttules...*Ph. pratorum*
    [Saprophytic on leaves of Gramineae in New Zealand, recently also recorded in Europe. Probably also common in Australia.]
   b. Growth rate moderate to fast, 3–8 cm on OA after 7 days............................**61**

61. a. NaOH causing the green-red discoloration (E$^+$ reaction). Growth rate fast, 6–8 cm after 7 days, with finely floccose, white aerial mycelium on OA. Pycnidia abundant both on and in the agar; after 3 weeks the colony on OA discolours to saffron as pycnidia ripen. Conidia 4–7.5(–9) × 1.5–2(–2.5) μm, sometimes with a few, small guttules ....*Ph. loticola*
    [Pathogen of cultivated *Lotus* spp., in Europe and New Zealand: stem and leaf spot.]
   b. NaOH reaction negative...**62**

62. a. Growth rate fast, 7–8 cm after 7 days. Chlamydospores absent. Conidia (3.5–)4.5–8(–10) × (1.5–)2–3(–4) μm, sometimes with a few small guttules. Colonies on OA grey olivaceous, rosy buff at margin ....*Ph. nigricans*
    [Saprophyte, New Zealand, probably also Australia.]
   b. Growth rate moderate, 3–6 cm after 7 days **63**

63. a. Conidia cylindrical to subcylindrical in shape, 5–7(–8.5) × 2–3 μm, always eguttulate. Colonies on OA 4.5–5 cm diam. after 7 days, colourless with darker concentric zones due to pycnidia ....*Ph. herbicola*
    [A saprophytic soil-fungus in North America, recorded from dead stems of different herbaceous plants. Also isolated from polluted lake water.]
   b. Conidia different in shape, with or without guttules **64**

64. a. Growth rate on OA 5.5–6 cm after 7 days. Colonies on OA colourless, sometimes with dark herbage green sectors, often fluffy whitish aerial mycelium which discolours in the light to saffron/salmon. Conidia (3–)4–6.5(–8.5) × (1.5–)2–3(–3.5) μm, sometimes with small guttules. Pseudothecia may be present ....*Ph. urticicola*, teleom. *Did. urticicola*
    [On dead stems of species of *Urtica*, esp. *U. dioica*, common in Europe.]
   b. Growth rate on OA slower, 3–5 cm after 7 days...**65**

65. a. Conidia variable, oblong to ellipsoidal, 2.5–8 × 1.5–3 μm, mostly 3.5–6 × 2–2.5 μm, eguttulate or with several small, mostly polar guttules. Colonies on OA 3–5 cm diam. after 7 days, dull green at centre with olivaceous grey aerial mycelium ....*Ph. destructiva* var. *destructiva*
    [Common American pathogen of *Lycopersicon esculentum* and *Capsicum annuum*, also recorded in other continents: fruit rot, leaf and stem blight.]
   b. Conidia cylindrical to allantoid, 5–10 × 1.5–3 μm, mostly 6–7.5 × 1.5–3 μm, with some small polar guttules. Colonies on OA 4–4.5 cm diam. after 7 days, grey olivaceous to dull green, with white to olivaceaous grey aerial mycelium ....*Ph. aubrietiae*
    [Seed-borne pathogen on *Aubrietia* hybrids in Europe: damping-off, decay.]

Possibilities of identification *in vivo*

The preceding pages show that *Phoma* sect. *Phoma* includes many species with a restricted host range as well as a number of plurivorous species.

The latter – mostly acting as necrophytes but sometimes also as opportunistic parasites – can usually only be exactly identified by study *in vitro*. It is exceptional for pycnidial and/or conidial characteristics of a plurivorous species to be so specific that direct identification *in vivo* is possible. Species which appear morphologically very similar *in vivo* may, *in vitro*, appear quite different in growth habit, production of metabolites, chlamydospores, etc.

The host-specific necrophytic or parasitic species of sect. *Phoma* are often easy to identify on the basis of their conidial characteristics *in vivo*, provided that they can be readily differentiated from plurivorous species. In the case of true parasitic species the disease symptoms may be an important help with identification *in vivo*. However, a problem occurs when no mature pycnidia develop in association with specific disease symptoms, an example of this being scab symptoms on celeriac and celery caused by *Ph. apiicola*.

In addition to the species listed, various other *Phoma* taxa described in the literature may also belong to sect. *Phoma*. However, most of those taxa are so far known only from a single collection and therefore cannot be checked *in vitro*.

It is a great pity that a well-known pathogenic representative of section *Phoma* in India could not be obtained *in vitro*. Therefore it has been added with a description *in vivo*:

Pycnidia subglobose with a slightly papillate ostiole, mostly 50–250 μm diam. Conidia ovoidal to ellipsoidal, 3–6.5(–8) × 2–2.5(–3) μm ....*Ph. cajanicola* [Specific pathogen of pigeon pea, *Cajanus cajan* (Leguminosae): leaf blight, stem canker. Common in India, probably also elsewhere where the host is grown.]

Distribution and Facultative Host and Substratum Relations in Section *Phoma*

Plurivorous
(But often with special
host or substratum
relations)

America, esp. temperate and/or montainous regions:
  Ph. herbicola
[Saprophytic.]
  Ph. heteroderae
[Saprophytic; but probably parasitic on eggs of *Heterodera glycines*.]
  Ph. huancayensis
[Saprophytic.]
  Ph. humicola
[Saprophytic, but may act as opportunistic parasite on roots, etc.]
  Ph. subherbarum
[Saprophytic.]

<u>Australasia, esp. temperate regions:</u>
Ph. huancayensis
[Saprophytic.]
Ph. nigricans
[Saprophytic.]
Ph. plurivora
[Saprophytic, but may act as opportunistic parasite.]

<u>Eurasia, temperate region:</u>
Ph. aliena
[Opportunistic parasite of woody plants.]
Ph. heteroderae
[Saprophytic.]
Ph. obtusa
[Saprophytic.]
Ph. pereupyrena
[Opportunistic parasite on herbaceous and woody plants.]
Ph. valerianae
[Saprophytic.]
Ph. viburnicola
[Opportunistic parasite of woody plants.]

<u>Eurasia, subtropical region:</u>
Ph. minutispora
[Saprophytic; also reported as opportunistic human pathogen.]
Ph. pereupyrena
[Opportunistic parasite of herbaceous and woody plants.]

<u>Worldwide, but esp. in temperate regions:</u>
Ph. anserina
[Saprophytic.]
Ph. eupyrena
[Saprophytic; may cause damping-off of seedlings herbaceous and woody plants.]
Ph. fimeti
[Saprophytic.]
Ph. herbarum
[Saprophytic.]
Ph. nebulosa
[Saprophytic.]
Ph. putaminum
[Saprophyte, but may act as opportunistic parasite on roots, etc.]

<u>Worldwide, but esp. in (sub-)tropical regions, occasionally in glasshouses:</u>
Ph. multirostrata
    var. *macrospora*
    var. *microspora*

[May be characterized as opportunistic parasites.]
Ph. tropica
[Saprophytic.]

With special host and/or substratum relation:

Agavaceae
Cordyline and *Dracaena* spp.   *Ph. draconis*
[Worldwide recorded on cultivated plants in association
with leaf spots and dieback; indigenous to Africa.]

Amaryllidaceae
Anigozanthus spp.   *Ph. anigozanthi*
[In Europe recorded on cultivated plants in association
with leaf spots and dieback; probably indigenous to
Australia.]
Crinum and *Nerine* spp.   *Ph. crinicola*
[A pathogen so far only known from Europe: leaf spot
and bulb rot.]

Apocynaceae
Nerium oleander   *Ph. glaucispora*
[In southern Europe frequently recorded pathogen: leaf
spot. Also reported from glasshouse ornamental cultures.]

Araceae
Zantedeschia aethiopica   *Ph. zantedeschiae* (on account of frequently occurring multicellular chlamydospores treated under
sect. *Peyronellaea*, **D**)
[Worldwide recorded on cultivated plants in association
with brown blotches on leaves and spathes: leaf blight.
Indigenous to southern Africa.]

Araliaceae
Hedera spp.   *Ph. hedericola*
[In Europe frequently recorded on leaves and stems: leaf
spot. Probably cosmopolitan.]

Berberidaceae
Berberis and *Mahonia* spp.   *Ph. aliena*
[In Europe frequently recorded opportunistic parasite of
quite different woody plants; associated with leaf necroses
and wood discoloration.]

Cactaceae
Opuntia spp.   *Ph. opuntiae*
[In South America found in association with dead spots
on the cladodes: necrotic spot. Probably also elsewhere
on the hosts.]

Cannabaceae
Humulus lupulus   *Ph. aliena*
[In Europe frequently recorded opportunistic parasite of
quite different woody plants; associated with leaf necroses
and wood discoloration.]

Caprifoliaceae
Lonicera periclymenum   *Ph. aliena*
[In Europe frequently recorded opportunistic parasite of
quite different woody plants; associated with leaf necroses
and wood discoloration.]

Viburnum spp.   *Ph. viburnicola*
[In Europe recorded opportunistic parasite which frequently also has been isolated from other trees and shrubs
in association with leaf spots and stem lesions.]

Celastraceae
Euonymus europeus   *Ph. aliena*
[In Europe frequently recorded opportunistic parasite of
quite different woody plants; associated with leaf necroses
and wood discoloration.]

Chenopodiaceae
Chenopodium spp.   *Ph. chenopodiicola*
[In Europe and South America recorded as specific necrophyte of *Ch. album* and *Ch. quinoa*; probably indigenous
to Europe.]

Compositae
Bellis perennis   *Ph. bellidis*
[In Europe common on seeds, causing reduction in seed
germination and damping-off. Also recorded in association with leaf necroses.]

Callistephus sp.   *Ph. dorenboschii*
[So far only recorded in The Netherlands, but probably of
foreign origin. Associated with leaf spots and stem lesions.
Apparently plurivorous.]

Dendranthema–Grandiflorum   *Ph. vasinfecta*
hybrids (formerly known   (synanam. *Phialophora chrysanthemi*)
as, e.g. *Chrysanthemum*
morifolium and *C. indicum*) [Occasionally recorded in Europe, North America and
Australia; a vascular pathogen: slow wilt, phoma decline
disease.]
                                                 Ph. chrysanthemicola
                                                         f. sp. *chrysanthemicola* (on account of fre-
                                                         quently occurring aggregates of chlamy-
                                                         dospores treated under sect. *Peyronellaea*, **D**)
                                                 [In western Europe and North America a common
                                                 pathogen of florists' chrysanthemum: root rot and (basal)
                                                 stem rot.]

Eupatorium cannabinum	*Ph. eupatorii* [In Europe a common specific necrophyte of the plant.]
Senecio jacobea and *Senecio* spp.	*Ph. senecionis* [In Europe and New Zealand a common and widespread specific necrophyte of these plants. Probably occurring wherever the hosts are growing.]

Cruciferae
Aubrietia spp.   *Ph. aubrietiae*
                                          [In Europe common on seeds: damping-off and decay.]

Gramineae
Plurivorous, e.g. on *Dactylis,*   *Ph. paspali*
Lolium and *Paspalum* spp.   *Ph. pratorum*
                                          [In New Zealand recorded as necrophyte or wound para-
                                          site of wild grasses; probably widespread in Australasia.
                                          Ph. pratorum is recently also reported in Europe.]

Juglandaceae
Carya pecan   *Ph. bismarckii*
                                          [Occasionally recorded in New Zealand, better known
                                          from apple branches and fruits.]

Labiatae
Physostegia virginiana   *Ph. dorenboschii*
                                          [So far only recorded in The Netherlands, but probably of
                                          foreign origin. Associated with leaf spots and stem lesions.
                                          Apparently plurivorous.]

Leguminosae
Plurivorous   *Ph. crystallifera*
                                          [In Central Europe apparently a common necrophyte of
                                          leguminous plants.]
Acacia spp.   *Ph. henningsii*
                                          [In East Africa recorded as an harmful wound parasite;
                                          North American records may be questioned.]
Arachis hypogaea   *Ph. arachidis-hypogaeae*
                                          [In India frequently recorded in association with leaf
                                          necroses: leaf spot. A plurivorous pathogen probably
                                          indigenous to that subcontinent. Leguminosae seem to be
                                          very susceptible.]
                                          Ph. multirostrata (esp. var. *microspora*)
                                          [In India a common opportunistic parasite, e.g. frequently
                                          isolated from leaf spots of groundnut.]
Astragalus spp.   *Ph. astragali*
                                          [In North America reported from various wild species:
                                          stem rot; probably an opportunistic parasite.]
Cajanus cajan   *Ph. cajanicola*
                                          [In India a common pathogen of pigeon pea: leaf blight,
                                          stem canker.]

Chamaespartium sp.   *Ph. crystallifera*
[Host of the type specimen; apparently a common necrophyte of leguminous plants in Central Europe.]

Lotus spp.   *Ph. loticola*
[In Europe and New Zealand recorded pathogen of cultivated plants: stem and leaf spot.]

Lupinus mutabilis and   *Ph. lupini*
other *Lupinus* spp.   [In North and South America known as pathogen of
indigenous plants: leaf, stem and pod spot.]

Medicago polymorpha   *Ph. aurea*
[In New Zealand isolated from dead stems; possibly a
common saprophyte in Australasia.]

Medicago sativa   *Ph. medicaginis* (on account of the occasional
occurrence of 1-septate conidia treated under sect.
Phyllostictoides, **E**)
[In Eurasia and North America involved with the black
stem disease.]

Liliaceae
Phormium tenax   *Ph. haematocycla*
[In New Zealand commonly occurring pathogen causing
red-encircled lesions: leaf spot.]

Linaceae
Linum usitatissimum and   *Ph. lini*
other *Linum* spp.   [In Europe frequently recorded on dead stems of the
plants; also isolated from water used for flax retting.
Probably also occurring on other herbaceous plants.]

Myrtaceae
Eucalyptus and *Eugenia* spp. *Ph. eucalyptica*
[In Australasia a common seed-borne pathogen: shoot
wilt, leaf necroses and blister disease. Also recorded in
southern Asia and Europe.]

Oleaceae
Olea europaea   *Ph. fallens*
[In southern Europe frequently recorded in association
with fruit and leaf spots; probably also elsewhere in olive-
growing regions.]
           Ph. insulana
[In southern Europe common on discolouring leaves and
ripening fruits.]

Piperaceae
Piper and *Peperomia* spp. *Ph. piperis*
[Worldwide recorded in association with irregular lesions
on the leaves: leaf spot.]

Polypodiaceae,
e.g. *Adiantum, Polystichum*   *Ph. adianticola*
and *Pteris* spp.   (teleom. *Did. adianticola*)
                    [In (sub-)tropical America probably widespread distributed: leaf necroses.]

Ranunculaceae
Delphinium spp.   *Ph. ajacis*
                    [In southern Eurasia a common seed-borne pathogen of annual species: leaf spot and stem rot.]

Rosaceae
Cotoneaster spp.   *Ph. aliena*
                    [In Europe frequently recorded opportunistic parasite of quite different woody plants; associated with leaf necroses and wood discoloration.]
Malus pumila   *Ph. bismarckii*
                    [Worldwide recorded on dead branches; also found in association with lenticel rot. Probably plurivorous.]

Rubiaceae
Coffea arabica   *Ph. costarricensis*
                    [Recently recorded in various coffee-growing regions of the world: coffee blight; formerly only known from South and Central America.]

Scrophulariaceae
Antirrhinum majus   *Ph. poolensis* var. *poolensis*
                    [Worldwide on cultivated plants, leaf spot, (basal) stem rot and damping-off (seed-borne). Occasionally also isolated from other Scrophulariaceae.]
Scrophularia spp.   *Ph. nebulosa*
                    [In Eurasia common on dead stems; a worldwide recorded necrophyte of herbaceous plants.]
Verbascum spp.   *Ph. poolensis* var. *verbascicola*
                    [In Eurasia and North America probably widespread distributed; it may cause necrotic lesions on leaves or stems and damping-off of young plants.]

Selaginellaceae
Selaginella helvetica   *Ph. selaginellicola*
                    [In Europe recorded on wild and cultivated plants in association with wilting and leaf necroses.]

Solanaceae
Capsicum annuum and   *Ph. destructiva* var. *destructiva*
Lycopersicon esculentum [In North and South America a common pathogen; frequently also recorded in other continents: fruit rot, leaf- and stem blight.]

Umbelliferae
Apium graveolens   *Ph. apiicola*
                                        [In temperate Europe and North America widespread
                                        recorded on celeriac and celery: root rot, scab, crown rot
                                        (black neck), seedling canker.]
Daucus carota   *Ph. obtusa*
                                        [In Europe common on dead stems, occasionally also as
                                        necrophyte on other herbaceous plants.]

Urticaceae
Urtica dioica and sometimes   *Ph. urticicola*
other *Urtica* spp.   (teleom. *Did. urticicola*)
                                        Ph. nebulosa
                                        [In Europe both common on dead stems.]

Valerianaceae
Valeriana spp.   *Ph. valerianae*
                                        [In Europe common on dead stems; also isolated from
                                        dead stems of other herbaceous plants.]

Vitaceae
Vitis vinifera   *Ph. negriana*
                                        [In southern Europe widespread on vine. Opportunistic
                                        parasite associated with necroses on leaves, fruits and
                                        stems.]

Seeds
                                        Ph. herbarum
                                        Ph. subherbarum
                                        [Saprophytic on surface of seeds; *Ph. herbarum* esp. in
                                        Europe, *Ph. subherbarum* thus far only known from
                                        North and South America.]
(for possible seed-borne pathogens see the host list above)

Soil
                                        Ph. anserina
                                        [May act as an opportunistic parasite on roots, etc.]
                                                [Ph. chrysanthemicola]
                                        Ph. eupyrena
                                        [May cause damping-off of seedlings.]
                                                Ph. fimeti
                                                Ph. flavescens
                                                Ph. herbarum
                                                Ph. herbicola – America
                                                Ph. humicola – America
                                        [May act as an opportunistic parasite on roots, etc.]
                                                Ph. paspali – Australasia
                                                Ph. putaminum
                                        [May act as an opportunistic parasite on roots, etc.]

(all frequently isolated from soil in temperate regions)

> *Ph. capitulum*
> *Ph. heteroderae*
> *Ph. minutispora*
> *Ph. multipora*
> *Ph. multirostrata*
> [The vars *macrospora* and *microspora* can be characterized as opportunistic plant parasites.]
> *Ph. labilis*

(isolates obtained from soil in tropical and subtropical regions)

Substrata of animal (incl. human) and nutritional / inorganic origin

> *Ph. anserina*
> [Isolated from meat, gelatine and frequently from cysts of *Globodera rostochiensis*, Europe, temperate region.]
> *Ph. eucalyptica*
> [Isolated from sea water Mediterranean zone, planting area of *Eucalyptus globulus*.]
> *Ph. flavigena*
> [Isolated from fresh water source, eastern Europe, temperate region.]
> *Ph. herbarum*
> [Isolated from human peripheral long tissue (asthma patients), from air-bladder in fishes (lethal bladder-disease of salmon and trout), from air, asbestos, butter, carpets, cement, cream, paint, plaster, rubber, water, etc., Eurasia, North and South America, temperate region.]
> *Ph. herbicola*
> [Isolated from polluted lake water, North America.]
> *Ph. heteroderae*
> [Isolated from packaged food, wall plaster and eggs of *Heterodera glycines*, Europe, North America.]
> *Ph. lini*
> [Isolated from stagnant water, used for retting of flax.]
> *Ph. minutispora*
> [Recorded as an opportunistic human pathogen.]
> *Ph. multirostrata* var. *macrospora*
> [Repeatedly found in association with cyst of *Heterodera glycines*, North America.]

Descriptions of the Taxa

Characteristics based on study in culture.

Phoma adianticola (E. Young) Boerema, Fig. 7A, Colour Plate I-G
Teleomorph: *Didymella adianticola* Aa & Boerema

> *Phoma adianticola* (E. Young) Boerema, *in* Versl. Meded. plziektenk. Dienst Wageningen **159** (Jaarb. 1982): 25. 1983.
>> ≡ *Phyllosticta adianticola* E. Young *in* Mycologia **7**: 144. 1915.
>> = *Phyllosticta polypodiorum* Bat. *in* Bolm Secr. Agric. Ind. Com. Est. Pernambuco **19**: 47. 1952 [cf. van der Aa & Vanev (2002)].

Diagn. data: Boerema (1983a) (fig. 8: culture on OA, drawing of conidia), Boerema *et al.* (1984) (history, occurrence teleomorph *in vitro*), de Gruyter *et al.* (1993) (fig. 7, conidial shape, reproduced in present Fig. 7A; cultural descriptions, partly adopted below).

Description in vitro

Pycnidia globose to subglobose or irregular shaped with 1–5 ostioles, sometimes with a short neck, 80–250 μm diam., glabrous or with hyphal outgrowths, often confluent. Conidial matrix buff, pale luteus or peach.

Conidia ellipsoidal, usually with two, relatively large, polar guttules, mostly 4–6.5 × 1.5–2.5 μm.

Pseudothecia (sometimes intermixed with the pycnidia) subglobose or pyriform, 120–180 × 150–200 μm, usually solitary with a distinct ostiolum, sometimes with hyphal outgrowths. Asci 40–45 × 7–8 μm, 8-spored. Ascospores 12–15.5 × 3.5–5 μm, 2-celled, the lower cell sybcylindrical with a rounded or slightly truncate base, the upper cell widest near the septum, tapering gradually to a round apex.

Colonies on OA 6–6.5 cm diam. after 7 days, regular, occasionally with thin felted white aerial mycelium, colourless, but the agar staining flesh, saffron, apricot/fulvous or ochraceous by the release of diffusible pigments; on addition of a drop of NaOH the colour becomes yellow-green. In colonies on MA the agar staining sienna to scarlet at the centre and yellow in a marginal zone; reverse similar or darker scarlet to bay or blood-colour. Pycnidia are usually produced in concentric zones, sometimes together with pseudothecia. Representative culture CBS 187.83.

Ecology and distribution

A pathogen of Polypodiaceae, probably widely distributed in (sub-)tropical America. First described from *Adiantum tenerum* on the island of Puerto Rico. In The Netherlands isolated from leaves of *Polystichum adiantiforme* imported from Florida, USA, and Puerto Rico. Later also from diseased prothallia of

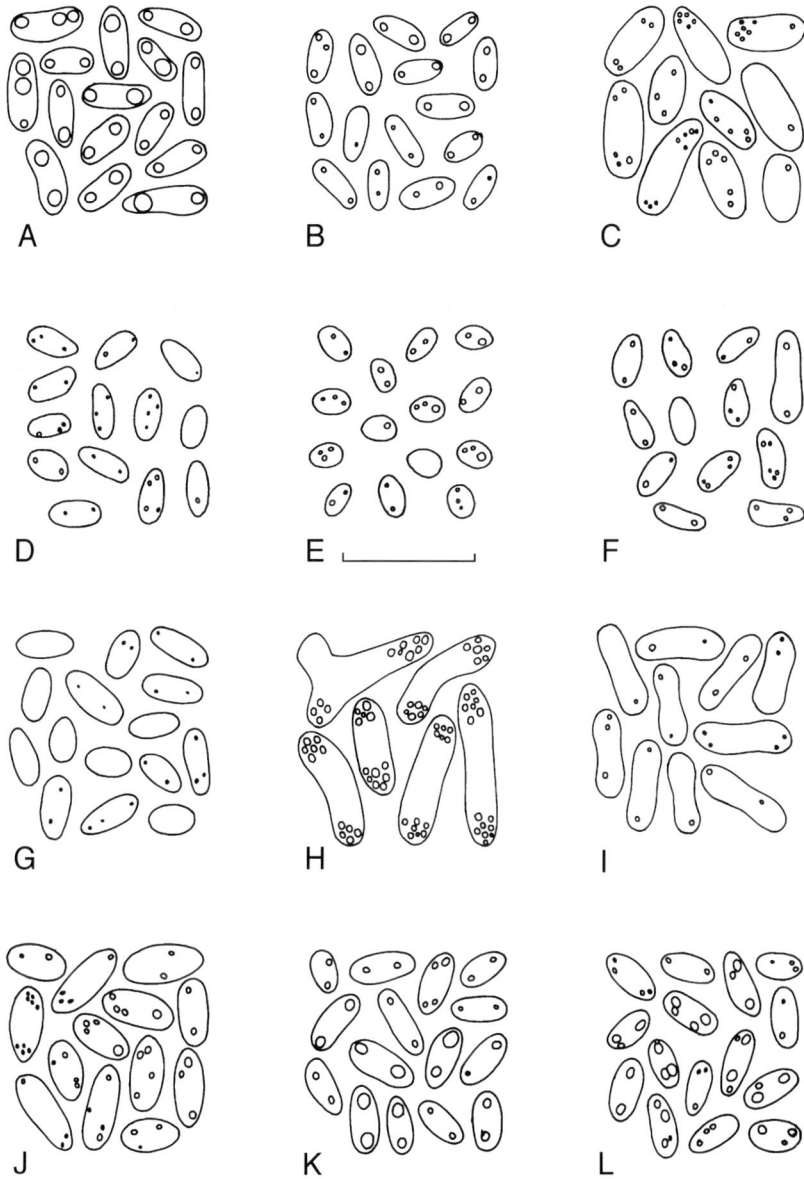

Fig. 7. Variation of conidia in size and shape. **A** *Phoma adianticola.* **B** *Phoma ajacis.* **C** *Phoma aliena.* **D** *Phoma anigozanthi.* **E** *Phoma anserina.* **F** *Phoma apiicola.* **G** *Phoma arachidis-hypogaeae.* **H** *Phoma astragali.* **I** *Phoma aubrietiae.* **J** *Phoma aurea.* **K** *Phoma bellidis.* **L** *Phoma bismarckii.* Bar 10 μm.

Pteris ensiformis grown in a Dutch nursery: leaf necroses. The teleomorph is so far only known from observations in vitro.

Phoma ajacis (Thüm.) Aa & Boerema, Fig. 7B

>Phoma ajacis (Thüm.) Aa & Boerema apud de Gruyter, Noordeloos & Boerema in Persoonia **15**(3): 383–384. 1993.
>>≡ Phyllosticta ajacis Thüm. apud Bolle & von Thümen in Boll. Soc. adriat. Sci. nat. **6**: 329. 1880.

Diagn. data: de Gruyter et al. (1993) (fig. 12: conidial shape, reproduced in present Fig. 7B; designation of neotype; cultural descriptions, partly adopted below).

Description in vitro

Pycnidia globose, with 1–2 ostioles, occasionally papillate, 130–300 μm diam., glabrous, solitary or confluent. Conidial matrix whitish.

Conidia oblong to ellipsoidal with usually 2 distinct polar guttules, 3.5–5.5 × 1.5–2.5 μm.

Colonies on OA 7–7.5 cm diam. after 7 days, regular, colourless with floccose, white to pale olivaceous grey aerial mycelium, reverse colourless with grey or olivaceous shade. Pycnidia are produced abundantly, mainly on, sometimes partly in the agar, hardly in aerial mycelium. The strains tested showed a positive reaction on the sodium hydroxide test: on application of a drop of NaOH a green spot turning red (E$^+$ reaction). Representative culture CBS 177.93 [isolate of neotype].

Ecology and distribution

In southern Eurasia probably a widely occurring seed-borne pathogen of annual species of Delphinium (Ranunculaceae): leaf spot. The fungus has been confused with Phoma delphinii (Rabenh.) Cooke, synanam. Stagonosporopsis delphinii Lebedeva, which may cause similar leaf spots on Delphinium spp., see sect. Heterospora (**B**).

Phoma aliena (Fr. : Fr.) Aa & Boerema, Fig. 7C

>Phoma aliena (Fr. : Fr.) Aa & Boerema in Persoonia **16**(4): 486. 1998.
>>≡ Sphaeria aliena Fr. : Fr., Syst. mycol. **2** [Sect. 2]: 502. 1823.
>>≡ Perisporium alienum (Fr. : Fr.) Fr., Syst. mycol. **3** [Sect. 1]: 252. 1829.
>>≡ Phyllosticta aliena (Fr. : Fr.) Sacc. in Michelia **2**(2): 342. 1881.
>>= Phyllosticta asiatica Cooke in Grevillea **13**: 91. 1885.
>>>≡ Phoma asiatica (Cooke) Aa, Revision Phyllosticta, CBS: 87. 2002.

= *Phoma berberidicola* Vestergr. *in* Öfvers. K. Svensk Vet.-
Akad. Förh. **1897**: 38. 1897 [Jan.]; not *Phoma berberidi-
cola* Brunaud *in* Act. Soc. Linn. Bordeaux **1898**: 12. 1898
[= *Phoma enteroleuca* Sacc. var. *enteroleuca*, sect.
Plenodomus, **G**].
= *Phyllosticta cotoneastri* Allesch. apud P. Sydow *in* Beibl.
Hedwigia **36**: 158. 1897.

Diagn. data: Neergaard (1956) (first record as seed-borne pathogen), de
Gruyter *et al.* (1998) (fig. 15: conidial shape, reproduced in present Fig. 7C;
nomenclature; cultural descriptions, partly adopted below), van der Aa &
Vanev (2002) (description *in vivo* and *in vitro* sub *Phyllosticta/Phoma asiatica*).

Description in vitro

Pycnidia globose to irregular, without or with 1(–2) non-papillate ostioles,
80–260 μm diam., glabrous, usually solitary. Conidial matrix rosy buff to salmon.
    Conidia ellipsoidal to slightly ovoid, usually with some small guttules,
(4–)5.5–10.5(–12) × 2.5–4.5 μm.
    Colonies on OA 5–5.5 cm diam. after 7 days, regular, smoke grey, with or
without scanty floccose, pale olivaceous grey aerial mycelium; reverse smoke
grey with olivaceous grey tinge. Pycnidia on and in the agar, also in aerial
mycelium. On MA faster growing, 6–7 cm diam., greenish with rosy buff/honey
sectors at the margin. Representative culture CBS 628.68.

Ecology and distribution

Isolated from quite different woody plants, esp. shrubs (including also ever-
greens and conifers) in Europe. The fungus is most frequently encountered
on *Berberis vulgaris*, *Mahonia aquifolium* (Berberidaceae), *Euonymus
europaeus* (Celastraceae), *Lonicera periclymenum* (Caprifoliaceae),
Humulus lupulus (Cannabaceae) [commonly growing together as wild
plants in the Dutch dunes] and *Cotoneaster* spp. (Rosaceae).
Characteristically it is an opportunistic parasite often occurring in association with leaf necroses and wood discoloration (sometimes flesh coloured).
It is commonly seed-borne on *Berberis*-species.

Phoma anigozanthi Tassi, Fig. 7D

Phoma anigozanthi Tassi *in* Boll. R. Orto bot. Siena **3**(2)['1899']: 148.
1900.
≡ *Phyllosticta anigozanthi* (Tassi) Allesch. *in* Rabenh. Krypt.-
Flora [ed. 2], Pilze **7** [Lief. 86]: 754. 1903.

Diagn. data: de Gruyter & Noordeloos (1992) (fig. 13: conidial shape, reproduced in present Fig. 7D; cultural descriptions, partly adopted below).

Description in vitro

Pycnidia irregular globose, with 1–3 ostioles, usually on a short neck (that may develop into a longer neck at a later stage), mostly 120–240 μm diam., glabrous, solitary or confluent. Conidial matrix pale vinaceous to salmon/saffron.

Conidia ellipsoidal, sometimes with very small polar guttules, 3–4(–5) × 1.5–2 μm.

Colonies on OA *c.* 4.5 cm diam. after 7 days, regular, flat, without aerial mycelium, distinctly concentric zonate, citrine to greenish olivaceous with paler margin; reverse similar. On MA somewhat faster growing *c.* 5 cm diam. with abundant hairy to floccose woolly white/grey aerial mycelium. Pycnidia developing in distinct concentric rings on the agar and in aerial mycelium. NaOH spot test positive on OA and MA: greenish, then red (E$^+$ reaction). Representative culture CBS 381.91.

Ecology and distribution

In Europe (Italy, The Netherlands) found on cultivated *Anigozanthus* spp. (Amaryllidaceae): leaf blotch. The pathogen is probably indigenous to Australia, where 'kangaroo paws' are popular wild flowers. In the original description has been stated that it represents the spermogonial state of *Sphaerella millepunctata* Tassi (= *Didymella* sp.?). However, our study *in vitro* showed a common conidial function of the anamorph, but a metagenetic relation with the teleomorph mentioned is quite well possible but could not be checked.

Phoma anserina Marchal, Fig. 7E

> *Phoma anserina* Marchal, Champ. copr. 11. 1891.
>   = *Phoma marchali* Sacc. in Sylloge Fung. **10**: 188. 1892.
>   = *Phoma bulbicola* Tassi in Boll. R. Orto bot. Siena '**1900**': 124. 1900.
>   = *Aposphaeria humicola* Oudem. in Ned. kruidk. Archf III, **2**(3): 721. 1902; not *Phoma humicola* J.C. Gilman & E.V. Abbott [this section].
>   = *Phoma radicalis* Sacc. & Traverso in Sylloge Fung. 20: 358. 1911.
>   H ≡ *Phoma radicicola* Maubl. in Bull. Soc. mycol. Fr. **21**: 90. 1905; not *Phoma radicicola* McAlpine, Fung. Dis. Stone-fruit trees Melb. 126. 1902 [= *Phoma putaminum* Speg., this section].
>   H = *Phoma bulbicola* Hollós in Annls hist.-nat. Mus. natn. hung. **5**: 457. 1907 [not *Phoma bulbicola* Tassi, see above].
>   = *Phoma radicis-callunae* R.W. Rayner in Bot. Gaz. **73**: 231. 1922, with reference to description in Ann. Bot. **29**: 128–130. 1915.

= *Phoma suecica* J.F.H. Beyma *in* Antonie van Leeuwenhoek
  8: 110. 1942.
= *Pyrenochaeta origani* S. Ahmad *in* Biologia, Lahore **10**: 5.
  1964; not *Phoma origani* Mark.-Let. *in* Bolez. Rast. **16**: 194.
  1927 [= *Phoma doliolum* P. Karst., sect. *Plenodomus*, **G**].

Diagn. data: Maublanc (1905) (pl. 6 fig. 8: pycnidial and conidial shape *in vivo*
sub *Phoma radicicola*), Rayner (1915) (pl. 6 figs 8–9, 19–20 sub 'endophyte':
characteristics of mycelium, pycnidia and conidia in culture on rice and tapwater), Kranz (1963) (figs 3, 5 sub isol. 6 '*Phoma tuberosa*': pycnidial and conidial shape; morphological and cultural characteristics *in vitro* on potato stems
and on agar media), Boerema (1967a) (synonymy), Boerema (1985) (fig. 10A:
pycnidial and conidial shape; variability *in vitro*, synonymy, ecology and history), de Gruyter & Noordeloos (1992) (fig. 2: conidial shape, reproduced in
present Fig. 7E; cultural descriptions, partly adopted below; variability).

Description in vitro

Pycnidia subglobose, occasionally papillate, with 1–3 ostioles, 100–175 μm
diam., sometimes with hyphal outgrowths (semi-pilose), solitary. Conidial
matrix whitish to rosy buff.
    Conidia broadly ellipsoidal with 1 or 2 (3) acentric guttules, variable in
dimensions, mostly 2.5–3 × 2–2.5 μm, but sometimes larger, up to 5.5 ×
3 μm (*in vivo* often only macro-conidial forms occur).
    Colonies on OA 5–5.5 cm diam. after 7 days, regular, usually without aer-
ial mycelium, pale to dark grey/olivaceous black; reverse similar but with lead
grey tinge. On MA faster growing, 6–7 cm after 7 days, with floccose pale olivaceous grey aerial mycelium. Pycnidia scattered, mostly on, but sometimes in
the agar (partly or entirely). Some strains produce swollen cells. The mycelium
is often anastomosing. Representative cultures CBS 360.84, CBS 363.91.

Ecology and distribution

A ubiquitous soil fungus in temperate Eurasia which also has been found in
northern Africa and North America. Frequently occurring on roots of herbaceous and woody plants, but also recorded on a variety of other organic substrates (e.g. meat, gelatine). The fungus may act as an opportunistic root
parasite and may cause secondary root decay. It appeared to be parasitic to
cysts of *Globodera* (*Heterodera*) *rostochiensis* (golden nematode of potatoes).
Most records from warm regions refer to the 'macro-conidial' phenotype.

Phoma apiicola Kleb., Fig. 7F

Phoma apiicola Kleb. *in* Z. PflKrankh. **20**: 22. 1910.
    ‡ = *Phoma apiicola* '*macroforma*' and '*microforma*', cultural types dis-
       tinguished by Goossens *in* Tijdschr. PlZiekt. **34**: 285–288. 1928
       [now referred to as the dual phenomenon of Hansen (1938)].

Diagn. data: Goossens (1928) (with 3 plates and 2 graphs; disease symptoms on rootstocks and petiole bases of celeriac on sandy soil and clay soil; study of the pathogen *in vitro* revealed two cultural types, now known as the dual phenomenon of Hansen (1938): the type with copious mycelial growth was referred to as *Ph. apiicola* 'macro-forma', the type developing numerous, relatively small pycnidia as 'micro-forma', when grown together both types produced at the line of contact of their mycelia numerous 'conjunct pycnidia' resembling those of the 'macro-forma'; inoculation experiments (fully described) showed both types to be capable of attacking the roots and bases of the petioles of celeriac), Neergaard (1977) (occurrence on seed), de Gruyter & Noordeloos (1992) (fig. 11: conidial shape, reproduced in present Fig. 7F; cultural descriptions, partly adopted below).

Description in vitro

Pycnidia globose with a distinct short ostiolated neck, 160–240 μm diam., glabrous, solitary or confluent. Conidial matrix yellowish white.

Conidia ellipsoidal, sometimes slightly constricted in the middle, usually with 2(–3) small guttules, mostly 3.5–4.5 × 1.5–2(–2.5) μm. (Occasionally some macroconidia occurred, see de Gruyter & Noordeloos (1992).)

Colonies on OA 1.5–2 cm diam. after 7 days, regular but also with irregular margin, pale olivaceous grey with darker concentrical zones; aerial mycelium scarse, compact, greyish, reverse smoke grey to grey olivaceous, sometimes with greenish or buff olivaceous tinges [on MA colonies densely woolly hairy at centre, marginal zone almost without aerial mycelium]. Pycnidia in concentric zones, darkening the colonies. In the mycelium swollen cells may occur. On MA sometimes crystals are formed. Representative culture CBS 285.72.

Ecology and distribution

Widespread on celeriac and celery, *Apium graveolens* (Umbelliferae), in temperate Europe, Australasia and North America: root rot, scab, *Phoma* rot, crown rot (black neck), seedling canker. Serious in areas where the host is constantly grown. Occasionally recorded from other Umbelliferae. Mainly soil-borne, but also found on seed [pycnidia found on celery seed in germination usually refer to non-parasitic species, e.g. *Phoma herbarum* Westend., this section].

Phoma arachidis-hypogaeae (V.G. Rao) Aa & Boerema, Fig. 7G

> *Phoma arachidis-hypogaeae* (V.G. Rao) Aa & Boerema apud de Gruyter, Noordeloos & Boerema *in* Persoonia **15**(3): 388. 1993.
> > ≡ *Phyllosticta arachidis-hypogaeae* V.G. Rao *in* Sydowia **16** ['1962']: 275. 1963.

Diagn. data: Patil (1987) (tabs 1–3; disease symptoms, pathogenicity–host range, effect of pH, temp., C/N ratio on growth and sporulation, and cultural behaviour on various media), de Gruyter *et al.* (1993) (fig. 16: conidial shape, reproduced in present Fig. 7G; cultural descriptions, partly adopted below).

Description in vitro

Pycnidia globose-papillate or flask-shaped, ostiolate, 80–200 μm diam., glabrous, solitary or in rows along hyphal strands, not confluent. Conidial matrix whitish.

Conidia oblong to ellipsoidal, eguttulate or with 2(–3) indistinct polar guttules, mostly 3–5 × 2–2.5 μm.

Colonies on OA 4.5–5 cm diam. after 7 days, regular with undulating margin, colourless to buff with grey olivaceous centre, aerial mycelium poorly developed, flat, finely floccose, white to greenish grey, reverse olivaceous at centre, towards margin buff [in cherry decoct agar the fungus produced after three weeks citrine crystals]. Pycnidia developing abundantly on and in the agar, but also in aerial mycelium. Representative culture CBS 125.93.

Ecology and distribution

In India frequently found on living leaves of groundnut, *Arachis hypogaea* (Leguminosae): leaf spot. The fungus is probably indigenous to that subcontinent (the groundnut itself is native in South America). Inoculation experiments have shown the 'non-specific nature' of the fungus; it could also infect various other plants, esp. Leguminosae. It is quite well possible that some of the records of the plurivorous *Phoma medicaginis* Malbr. & Roum. (sect. *Phyllostictoides*, **E**) on Indian leguminous plants refer in fact to this fungus. Further it should be noted that in India also *Phoma multirostrata* (P.N. Mathur *et al.*) Dorenb. & Boerema (this section) repeatedly has been isolated from groundnut plants.

Phoma astragali Cooke & Harkn., Fig. 7H

Phoma astragali Cooke & Harkn. *in* Grevillea **13**: 111. 1885; not '*Phoma astragali* Ellis & Kellerm.', nomen nudum [herbarium name, Mycol. Coll., NY].

Diagn. data: Anonymous (1960) (occurrence in USA), de Gruyter *et al.* (1998) (fig. 12: conidial shape, reproduced in present Fig. 7H; cultural descriptions, partly adopted below).

Description in vitro

Pycnidia mainly globose, sometimes irregular, with 1 rarely up to 3 non-papillate or papillate ostioles, 80–180 μm diam., glabrous, solitary or confluent. Conidial matrix white to buff.

Conidia cylindrical to allantoid, distorted shapes occur, with numerous polar guttules, 7.5–12.5 × 2–3.5 μm.

Colonies on OA 5–5.5 cm after 7 days, regular, colourless to olivaceous grey or greenish grey, with or without scarce, finely floccose, white aerial mycelium; reverse also colourless to olivaceous greenish grey. Pycnidia scattered all over the colony, on and (partly) in the agar. Representative culture CBS 178.25.

Ecology and distribution

This fungus is often recorded in North America (USA, Canada) on stems of various species of *Astragalus* (Leguminosae): stem rot. Probably an opportunistic parasite.

Phoma aubrietiae (Moesz) Boerema, Fig. 7I

> *Phoma aubrietiae* (Moesz) Boerema apud Boerema & Valckx *in* Gewasbescherming **1**: 66. 1970.
>
> ≡ *Sclerophomella aubrietiae* Moesz *in* Balkan-Kutat. Tud. Eredm. **3**: 144. 1926.

Diagn. data: Boerema & Valckx (1970) (disease symptoms), Neergaard (1977) (occurrence on seed), de Gruyter *et al.* (1998) (fig. 18: conidial shape, reproduced in present Fig. 7I; cultural descriptions, partly adopted below).

Description in vitro

Pycnidia globose to subglobose, with 1 non-papillate or papillate ostiole, 80–160 μm diam., glabrous, solitary. Conidial matrix white.
    Conidia cylindrical to allantoid, with some small polar guttules, (5–)6–7.5(–10) × 1.5–3 μm.
    Colonies on OA 4–4.5 cm diam. after 7 days, regular, grey olivaceous to dull green with appressed-felted to finely floccose, white to olivaceous grey aerial mycelium; reverse also grey olivaceous to dull green. Pycnidia abundant, mainly on the agar. Representative cultures CBS 383.67, CBS 627.97.

Ecology and distribution

A common seed-borne pathogen of *Aubrietia* hybrids (Cruciferae) in Europe. The fungus causes damping-off of seedlings and decay of stems and leaves of older plants.

Phoma aurea Gruyter *et al.*, Fig. 7J

> *Phoma aurea* Gruyter, Noordel. & Boerema *in* Persoonia **15**(3): 394. 1993.

Diagn. data: de Gruyter *et al.* 1993 (fig. 23: conidial shape, reproduced in present Fig. 7J; cultural descriptions, partly adopted below).

Description in vitro

Pycnidia globose with 1 non-papillate or slightly papillate ostiole, 50–150 μm diam., glabrous, solitary. Conidial matrix whitish grey.

Conidia oblong to ellipsoidal, often acuminate, with 2–many small polar guttules, 4.5–7 × 2–3 μm.

Colonies on OA 4.5–5 cm diam. after 7 days, regular, weakly olivaceous buff to greenish olivaceous, aerial mycelium pure white, poorly developed, flat, in scattered floccules over whole dish, reverse gradually becoming distinctly olivaceous tinged. On MA colonies faster growing, up to 6 cm diam. after 7 days and more or less golden coloured (aureate). Pycnidia abundant, mainly on and in the agar, but also rarely in aerial mycelium. Representative culture CBS 269.93.

Ecology and distribution

This fungus is so far known only of an isolate from dead stems of *Medicago polymorpha* (Leguminosae) in New Zealand. It may be a common saprophyte in Australasia.

Phoma bellidis Neerg., Fig. 7K

Phoma bellidis Neerg. *in* Friesia **4**: 74. 1950 [as '*Phoma (Phyllosticta)*'].
‡ ≡ *Phoma bellidis* Neerg. *in* Aarsberetn. J.E. Ohlsens plpatol. Lab. **6**: 9. 1941 [Danish diagnosis; as '*Phoma (Phyllosticta)*'].
= *Phyllosticta bellidis* Bond.-Mont. *in* Bolez. Rast. **12**: 70. 1923; not *Phyllosticta bellidis* Hollós *in* Mat. Természettud. Közl. [Magy. Tudom. Akad.] **35**(1): 45. 1926 [= *Phoma exigua* Desm. var. *exigua*, sect. *Phyllostictoides*, **E**], nor *Phyllosticta bellidis* Sawada, listed below.
H = *Phyllosticta bellidis* Sawada *in* Bull. Govt Forest Exp. Sta. Meguro **105**: 38. 1958 [see above].
= *Phyllosticta bellidicola* Nelen *in* Nov. Sist. Niz. Rast. **1966**: 224. 1966 [cf. holotype; conidial data in description not correct].

Diagn. data: Neergaard (1950) (tab. 1: infection percentages of seed lots of *Bellis perennis* grown in various countries during the period 1939–48; Latin description, type indication and notes on pathogenicity), (1977) ('common on seeds of *Bellis perennis*'), de Gruyter *et al.* (1993) (fig. 9: conidial shape, reproduced in present Fig. 7K; cultural descriptions, partly adopted below), van der Aa & Vanev (2002) (*Phyllosticta* synonyms).

Description *in* vitro

Pycnidia globose to irregularly composed with 1–5 non-papillate or slightly papillate ostioles, 50–260 μm diam., glabrous, solitary or confluent. Conidial matrix salmon to saffron.

Conidia ellipsoidal, with usually 2 polar guttules, mostly 4–6.5 × 2–2.5 μm.

Colonies on OA 6.5–7 cm diam. after 7 days, regular, with poorly developed, felted, white aerial mycelium, colourless but in centre salmon to saffron
coloured by the conidial masses from pycnidia. Pycnidia scattered but most
numerous in the centre, on or partly in the agar, rarely submerged. The sodium
hydroxide test showed a positive reaction on MA: on application of a drop of
NaOH green → red (E⁺). Representative culture CBS 714.85.

Ecology and distribution

This fungus has been recorded on seed of daisy, *Bellis perennis* (Compositae),
from various European countries (Denmark, England, Italy, The Netherlands,
Switzerland). It may cause the death of infected seeds and damping-off of
seedlings. It is also repeatedly found in association with leaf necroses.

Phoma bismarckii Kidd & Beaumont, Fig. 7L

Phoma bismarckii Kidd & Beaumont *in* Trans. Br. mycol. Soc. **10**:
104–105. 1924.

Diagn. data: de Gruyter *et al.* (1993) (fig. 26: conidial shape, reproduced in
present Fig. 7L; cultural descriptions, partly adopted below).

Description in vitro

Pycnidia globose or irregular with 1 distinct, sometimes papillate ostiole, mostly
50–250 μm diam., glabrous, sometimes with hyphal outgrowths, solitary or
confluent and many-ostiolated. Conidial matrix whitish, pale vinaceous or buff.
   Conidia oblong to ellipsoidal, with 2–4 polar guttules, mostly 3.5–5 ×
1.5–2.5 μm.
   Colonies on OA *c.* 7–7.5 cm diam. after 7 days, regular, very pale buff to
grey olivaceous, usually with grey tone by the presence of abundant pycnidia;
aerial mycelium, if present, scanty pruinose; reverse pale buff to grey olivaceous or slightly more greyish olivaceous to olivaceous. Pycnidia may also
occur in the aerial mycelium. Globose swollen cells, about 5–12 μm diam. are
usually present, intercalary or terminal, solitary or in short chains. NaOH
causes a slight reddish discoloration, not associated with a specific pigment.
Representative culture CBS 119.93.

Ecology and distribution

This species has occasionally been recorded worldwide on dead branches of
apple trees, *Malus pumila* (Rosaceae). Its occurrence in association with lenticel
rot of apples 'Bismarck' has led to the assumption that it was conspecific with
Phoma pomorum Thüm. ('dictyochlamydosporic' section *Peyronellaea*, **D**), but
Ph. bismarckii does not produce any chlamydospore in culture. In New
Zealand the fungus also has been isolated from leaf spots on *Carya pecan*

(Juglandaceae), which points to a plurivorous behaviour. *In vivo* (esp. on apple) pycnidia not confluent and conidia often longer and compressed, i.e. when seen edgewise, allantoid 5–6 × 1–1.5 μm.

Phoma cajanicola (Trotter) Aa & Boerema

> *Phoma cajanicola* (Trotter) Aa & Boerema apud van der Aa & Vanev, Revision Phyllosticta, CBS: 118. 2002.
>> ≡ *Phyllosticta cajanicola* Trotter *in* Sylloge Fung. **25**: 47. 1931 [as '(Syd.) Trotter'; name change].
>> H ≡ *Phyllosticta cajani* P. Syd. apud Sydow, Sydow & Butler *in* Annls mycol. **14**: 178. 1916; not *Phyllosticta cajani* Rangel *in* Bolm Agric., S. Paulo **16**: 154. 1915 [= *Phomopsis* sp., but erroneously also interpreted as identical with *Ph. cajanicola*: *Phoma cajani* (Rangel) Khune & J.N. Kapoor *in* Indian Phytopath. **34**: 259. 1981].

Diagn. data: Khune & Kapoor (1981) sub the misapplied combination *Phoma cajani* (figs 1, 2: disease symptoms on leaves and stems), van der Aa & Vanev (2002) (description of type specimen; nomenclature and misapplications).

Description in vivo *(Cajanus cajan; cultures were not at our disposal)*

Pycnidia (aggregated in the centres of brownish leaf spots with dark purplish margin, and in cankerous grey to brown stem lesions) subglobose with 1 slightly papillate ostiole, (100–)150–250 μm diam., brownish, sometimes darker around the ostiole, glabrous. Wall made up of about 3 layers of cells, outer layer pigmented.

   Conidia ovoidal to ellipsoidal with usually 2 polar guttules 3–6.5 × 2–3 μm.

Note: *Phoma cajanicola* may be identical with a *Phoma*-isolate from leaf spots on *Cajanus cajan* described by Rai (1998) sub *Ph. herbarum* isol. 2 (Govt Sci. Coll. garden, Jabalpur, India). Should that be true, the fungus produces red anthraquinone pigments which on application of a drop of NaOH change in colour to violet-blue, growth rate on OA 4–4.5 cm after 7 days, pycnidia *in vitro* c. 70–150 μm diam., conidial dimensions 3.5–7.5 × 1.5–3 μm.

Ecology and distribution

This fungus is apparently a common pathogen of *Cajanus cajan* (= *C. indicus*; Leguminosae), in India: leaf blight, stem canker. Probably also elsewhere on the host. It concerns a leguminous shrub known as pigeon pea. As indicated in the synonymy, the fungus has been confused with a *Phomopsis* species in Brazil recorded on leaves and stems of *Cajanus cajan*.

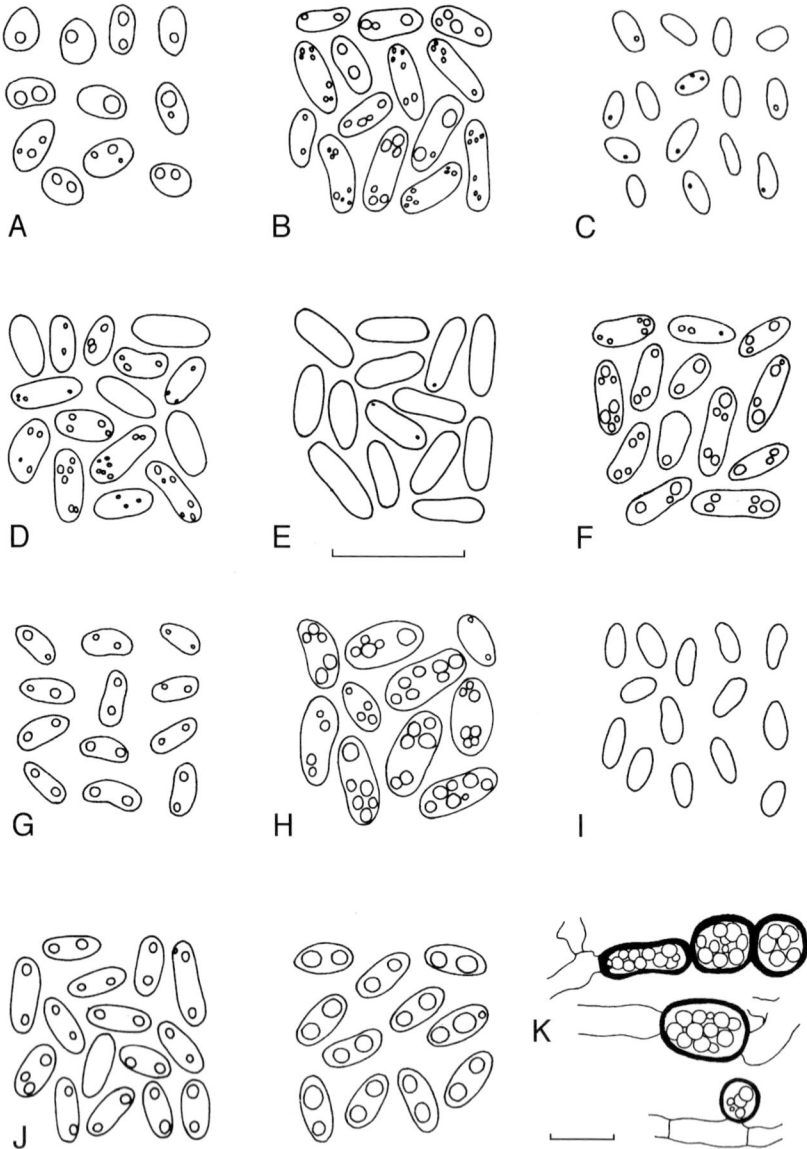

Fig. 8. Shape of conidia and some hyphal chlamydospores. **A** *Phoma capitulum.* **B** *Phoma chenopodiicola.* **C** *Phoma costarricensis.* **D** *Phoma crinicola.* **E** *Phoma crystallifera.* **F** *Phoma destructiva* var. *destructiva.* **G** *Phoma dorenboschii.* **H** *Phoma draconis.* **I** *Phoma eucalyptica.* **J** *Phoma eupatorii.* **K** *Phoma eupyrena.* Bars 10 μm.

Phoma capitulum V.H. Pawar *et al.*, Fig. 8A

> *Phoma capitulum* V.H. Pawar, P.N. Mathur & Thirum. *in* Trans. Br. mycol.
> Soc. **50**: 261. 1967.
>> V ⊖ ≡ *Phoma capitulum* V.H. Pawar & Thirum. *in* Nova Hedwigia
>> **12**: 502. 1966 [as '*capitula*'; without description].
>> = *Phoma ostiolata* V.H. Pawar, P.N. Mathur & Thirum. *in*
>> Trans. Br. mycol. Soc. **50**: 262. 1967, var. *ostiolata*.
>> ⊖ ≡ *Phoma ostiolata* V.H. Pawar & Thirum. *in* Nova
>> Hedwigia **12**: 502. 1966 [without description].
>> = *Phoma ostiolata* var. *brunnea* V.H. Pawar, P.N. Mathur &
>> Thirum. *in* Trans. Br. mycol. Soc. **50**: 263. 1967.
>> ⊖ ≡ *Phoma ostiolata* var. *brunnea* V.H. Pawar & Thirum. *in*
>> Nova Hedwigia **12**: 502. 1966 [without description].

Diagn. data: Boerema & Dorenbosch (1968) (variability; synonymy), Pawar *et al.* (1967) (figs 1E–H, 2A–G: drawings of pycnidia, conidiogenous cells and conidia sub *Ph. capitulum* and *Ph. ostiolata*; cultural characteristics), de Gruyter & Noordeloos (1992) (fig. 6: conidial shape, reproduced in present Fig. 8A; cultural descriptions, partly adopted below), Rai & Rajak (1993) (cultural characteristics of type isolate; NB: the descriptions sub *Ph. capitulum* in Rai (1998) include isolates of *Ph. multipora* V.H. Pawar *et al.*, also this section, and of *Ph. pimprina* P.N. Mathur *et al.*, sect. *Peyronellaea*, **D**).

Description *in vitro*

Pycnidia globose with 1–3 ostioles on a short neck, variable in dimensions, mostly 50–100 μm diam., glabrous, usually occurring in large clusters. Conidial matrix grey to saffron.

Conidia broadly or shortly ellipsoidal, with 1–2(–3) large or small guttules, 3–4.5(–5) × 2–3 μm.

Colonies on OA *c.* 3.5 mm diam. after 7 days [on MA *c.* 4 cm], regular, salmon to flesh coloured with fine velvety floccose white aerial mycelium; reverse also salmon to flesh [on MA peach to salmon]. Pycnidia in concentric zones or in sectors, mostly on the agar but also in the agar (partly or entirely). Representative culture CBS 337.65.

Ecology and distribution

Isolated from saline soil, marine environment and oak forest soil in India. Probably a common 'halophilic' soil inhabiting fungus in south-west Asia.

Phoma chenopodiicola Gruyter *et al.*, Fig. 8B

> *Phoma chenopodiicola* Gruyter Noordel. & Boerema *in* Persoonia **15**(3):
> 395. 1993.

≡ *Gloeosporium chenopodii* P. Karst. & Hariot *in* J. Bot.,
Paris **3**: 209. 1889; not *Phoma chenopodii* S. Ahmad *in*
Sydowia **2**: 79. 1948; not *Phoma chenopodii* Pavgi & U.P.
Singh *in* Mycopath. Mycol. appl. **30**: 265. 1966 [both sect.
Macrospora, **H**].

≡ *Plenodomus chenopodii* (P. Karst. & Har.) Arx *in* Verh. K.
ned. Akad. Wet. [Afd. Natuurk.] reeks 2, **51**(3) [= Revis.
Gloeosporium, ed. 1]: 73. 1957.

Diagn. data: von Arx (1957) (char. of type specimen, PC), de Gruyter *et al.*
(1993) (fig. 25: conidial shape, reproduced in present Fig. 8B; cultural descriptions, partly adopted below), Boerema *et al.* (1996: 156) (history).

Description in vitro

Pycnidia more or less globose, with 1 or 2 distinct ostioles, sometimes papillated or with elongated neck, 100–250 µm diam., glabrous, often confluent.
Conidial matrix white to saffron.

Conidia oblong to ellipsoidal, with 2 to many small or large guttules,
4–7(–10) × 1.5–2.5(–4) µm.

Colonies on OA 4–5.5 cm diam. after 7 days [on MA 4–6 cm], regular,
colourless to grey olivaceous or greenish olivaceous, or olivaceous grey; without any aerial mycelium or with scanty compact white aerial mycelium; reverse
olivaceous black with ochre margin, or grey to greenish olivaceous with vinaceous buff margin. Pycnidia are usually produced in abundance on the agar,
but also (partly) submerged in the agar. Representative culture CBS 128.93.

Ecology and distribution

In Europe a common necrophyte on *Chenopodium album* (Chenopodiaceae).
It probably has a worldwide distribution on this host by seed transmission. In
South America it is repeatedly found on a cultivated variety of the related
Chenopodium quinoa. *In vivo* the pycnidia are often deeply immersed, which
may explain the original classification in *Gloeosporium*. The fungus has been
confused with the plurivorous *Phoma exigua* Desm. var. *exigua* (sect.
Phyllostictoides, **E**) which in Europe also commonly occurs on dead stems of
Ch. album.

Phoma costarricensis Echandi, Fig. 8C

Phoma costarricensis Echandi *in* Rev. Biol. Trop. **5**: 83. 1957.
[= '*Phyllosticta coffeicola*' sensu F. Stevens *in* Ill. biol.
Monographs **11**(2): 52–53. 1927; not *Phyllosticta coffe-
icola* Speg. *in* Rev. Fac. Agron. Veter. La Plata **1896**:
345. 1896, as '*coffaeicola*'; possibly a species of
Phomopsis, see van der Aa & Vanev (2002).]

Diagn. data: Echandi (1957) (figs 1–7: photographs of disease symptoms on coffee, pycnidia and cultures on potato dextrose agar; original description; variability *in vitro*; relation to temperature; pathogenicity), de Gruyter & Noordeloos (1992) (fig. 7: conidial shape, reproduced in present Fig. 8C; cultural descriptions, partly adopted below).

Description in vitro

Pycnidia globose to flask-shaped, usually with one ostiole, 50–150 μm diam., glabrous, solitary. Conidial matrix whitish.

Conidia ellipsoidal to subcylindrical, eguttulate or sometimes with 1(–2) small guttules, (1.5–)3–4 × (1–)1.5–2 μm [*in vivo* the conidia should be larger, mostly 5–6 × 2–3 μm].

Colonies on OA 4–4.5 cm diam. after 7 days, regular, greenish olivaceous with tufts of whitish aerial mycelium in the margin zone, reverse grey olivaceous [on MA growth rate 5–5.5 cm, regular with floccose olivaceous aerial mycelium]. Pycnidia scattered, on and partly in the agar. With addition of a drop of NaOH a slightly sienna discoloration of the agar may occur. Representative culture CBS 506.91.

Ecology and distribution

This fungus causes lesions on the young leaves, stems and fruits of coffee, *Coffea arabica* (Rubiaceae). The disease, known as coffee blight ('La Quema') was until recently only known from South and Central America (Brazil, Costa Rica, Nicaragua), but is now also recorded in other parts of the world (e.g. India, Papua New Guinea: coffee seedling dieback). Wounds are required for infection. Insect damage seems to be an important factor in the epidemiology of the disease.

Phoma crinicola (Siemaszko) Boerema, Fig. 8D

Phoma crinicola (Siemaszko) Boerema apud Boerema & Dorenbosch *in* Versl. Meded. plziektenk. Dienst Wageningen **153** (Jaarb. 1978): 18. 1979.
≡ *Phyllosticta crinicola* Siemaszko *in* Acta Soc. Bot. Pol. **1**: 22. 1923.

Diagn. data: Siemaszko (1923) (original description from dead leaves of a wild *Crinum* sp., occurrence in association with pseudothecia), Boerema & Dorenbosch (1979) (fig. 6, leaf spots and bulbrot on *Crinum powellii*; disease first recorded in The Netherlands in 1970; no indication of a teleomorph), de Gruyter *et al.* (1993) (fig. 11: conidial shape, reproduced in present Fig. 8D; cultural descriptions, partly adopted below).

Description in vitro

Pycnidia globose to irregular, usually with 1 distinct non-papillate ostiole, but sometimes without visible ostiole, 80–185 × 70–180 μm, glabrous or covered by hyphae, solitary or confluent. Conidial matrix whitish.

Conidia ellipsoidal to cylindrical with 2–6 small, usually polar guttules, sometimes eguttulate, 4–6.5 × 2–3 μm.

Colonies on OA fast growing, *c.* 7 cm diam. after 7 days, regular, with coarsely floccose dull green to olivaceous aerial mycelium; reverse colourless to dull green or olivaceous. Pycnidia abundant on and in the agar, often also in aerial mycelium. In the mycelium sometimes clusters of short swollen cells. All strains tested showed a positive reaction with the sodium hydroxide test: on application of a drop of NaOH green → red (E$^+$ reaction). Representative culture CBS 109.79.

Ecology and distribution

A pathogen of Amaryllidaceae so far only known from Europe. Originally described from a wild species of *Crinum* in Poland. In The Netherlands frequently isolated from the hybrid *Crinum powellii*; also from *Nerine bowdenii*. Leaf spot, bulb rot.

Phoma crystallifera Gruyter et al., Fig. 8E

V   *Phoma crystallifera* Gruyter, Noordel. & Boerema *in* Persoonia **15**(3): 393. 1993 [as 'crystallifer'].

Diagn. data: de Gruyter *et al.* (1993) (fig. 22: conidial shape, reproduced in present Fig. 8E; nomenclator; cultural descriptions, partly adopted below).

Description in vitro

Pycnidia globose to subglobose with 1–2 papillate ostioles, 80–240 μm diam., glabrous or with short hyphal outgrowths, solitary or confluent. In the mycelium also micropycnidia may occur, 40–60 μm diam. Conidial matrix whitish.

Conidia oblong to ellipsoidal, eguttulate or with some inconspicuous polar guttules, 4–6 × 1.5–2 μm.

Colonies on OA 5–5.5 cm diam. after 7 days, regular, colourless with radially arranged rows of greyish pycnidia; aerial mycelium, if present, scanty, pale olivaceous grey or smoke grey; reverse buff. In the mycelium swollen cells may occur, sometimes clustered. In the agar [OA and MA] always numerous hyaline crystals were present (not dendritic as in various other *Phoma* species). Pycnidial development on and in the agar. Representative culture CBS 193.82.

Ecology and distribution

In Central Europe this species is apparently a common necrophyte of Leguminosae. The representative culture was isolated from *Chamaespartium* sp. in Austria. On the dead stems of the hosts it has often been referred to as 'Phoma melaena (Fr. : Fr.) Mont. & Durieu', an old collective name for species of *Phoma* associated with 'very black patches' on the host [the identity of the basionym *Sphaeria melaena* Fr. : Fr. is not yet fixed. Sutton (1980) listed it as *Podoplaconema melaenum* (Fr. : Fr.) Petr., referring to stromatic non-ostiolate small-spored pycnidia found on *Silene nutans* (Caryophyllaceae)].

Phoma destructiva Plowr. var. *destructiva*, Fig. 8F

> *Phoma destructiva* Plowr. *in* Gdnrs' Chron. II [New Series], **16**: 621. 1881 [sometimes erroneously listed as '(Plowr.) Jamieson'], var. *destructiva* [autonym versus var. *diversispora* de Gruyter & Boerema treated under sect. *Phyllostictoides*, **E**].
> > ≡ *Diplodina destructiva* (Plowr.) Petr. *in* Annls mycol. **10**: 19. 1921 [misapplied].
> > = *Phyllosticta lycopersici* Peck *in* Bull. N.Y. St. Mus. nat. Hist. **40**: 55. 1887 [sometimes erroneously listed as *Ph. l.* 'House'].

Diagn. data: Boerema & Dorenbosch (1973) (pl. 2a, b: cultures on OA; tab. 1: synoptic table on characters *in vitro* with drawing of conidia; diagn. literature data), Neergaard (1977) (occurrence on seed), Morgan-Jones & Burch (1988c) (fig. 1A–E: drawings of pycnidium, pycnidial primordium, conidia, wall cells and toruloid hyphae; pls 1A–F, 2A–G: photographs of agar plate cultures, pycnidial initial, wall cells, hyphal strands and toruloid hyphae; history, cultural and morphological description; distinguishing characters in culture against *Phoma lycopersici* and *Ph. exigua* var. *exigua* [both sect. *Phyllostictoides*, **E**]), Rai & Rajak (1993) (cultural characteristics; toruloid hyphae interpreted as chains of chlamydospores), Rai (1998) (effect of different physical and nutritional conditions on the morphology and cultural characters), de Gruyter *et al.* (2002) (differentiation of var. *diversispora*, producing also 1-septate conidia).

Description in vitro

Pycnidia globose to irregular, solitary or confluent, glabrous with 1–3 papillate ostioles, honey/citrine to olivaceous, later olivaceous black. Conidial matrix buff to saffron coloured.

Conidia variable, oblong to ellipsoidal, 2.5–8 × 1.5–3 μm, mostly 3.5–6 × 2–2.5 μm, eguttulate or with several small, mainly polar, guttules.

Colonies on OA 3–5 cm diam. after 7 days, regular, dull green at centre, with floccose olivaceous grey aerial mycelium. Individual cells in dark thick-walled hyphae frequently become inflated, 'toruloid hyphae'. Pycnidia abundant, immersed or superficial, and in aerial mycelium, scattered or in concentric zones. Representative culture CBS 378.73.

Ecology and distribution

In tropical North and South America a common pathogen of tomato, *Lycopersicon esculentum*, which also often occurs on paprika, *Capsicum annuum* (both Solanaceae): fruit rot, leaf and stem blight. The fungus is

frequently also recorded in greenhouses and tropical areas of the other continents. Seed transmission is important in paprika, but only occasionally found in
tomato. A variety producing apart from aseptate conidia a number of somewhat larger 1-septate conidia is recently distinguished as var. *diversispora* and
treated under sect. *Phyllostictoides* (**E**).

Phoma dorenboschii Noordel. & Gruyter, Fig. 8G, Colour Plate I-C

> *Phoma dorenboschii* Noordel. & Gruyter apud de Gruyter & Noordeloos
> *in* Persoonia **15**(1): 83. 1992.

Diagn. data: de Gruyter & Noordeloos (1992) (fig. 12: conidial shape, reproduced in present Fig. 8G; cultural descriptions, partly adopted below),
Noordeloos *et al.* (1993) (fig. 3: conidia; cultural characteristics; dendritic crystals consisting of 'pinodellalides' A and B).

Description in vitro

Pycnidia subglobose or elongate, with 1 or up to 5 ostiolated necks,
80–360 μm diam., glabrous, solitary or confluent. Conidial matrix buff.
    Conidia oblong to subcylindrical, usually with 2–4 small polar guttules,
3–5.5 × 1.5–2(–2.5) μm.
    Colonies on OA 5–5.5 cm diam., regular with sinuate outline, sparse felty
to floccose aerial mycelium; initially pale buff to buff or dull green, finally olivaceous grey; reverse similar. Pycnidia scattered in aerial mycelium and on the
agar [on MA the fungus is characterized by the production of dendritic whitish
crystals in the agar below the centre of the colony: pinodellalides A and B].
Representative cultures CBS 320.90, CBS 426.90.

Ecology and distribution

This fungus is in The Netherlands isolated from leaf spots and stem lesions on
plants belonging to quite different plant families (*Physostegia virginiana*,
Labiatae, and *Callistephus* sp., Compositae). It seems to be a plurivorous
opportunistic parasite, probably not endemic in western Europe.

Phoma draconis (Berk. ex Cooke) Boerema, Fig. 8H

> *Phoma draconis* (Berk. ex Cooke) Boerema *in* Versl. Meded. plziektenk.
> Dienst Wageningen **159** (Jaarb. 1982): 24. 1983.
>> ≡ *Phyllosticta draconis* Berk. ex Cooke *in* Grevillea **19**: 8.
>> 1891.
>> Θ ≡ *Phyllosticta draconis* Berk., Welw. Crypt. Lusit. 51. 1853
>> [without description].

≡ *Macrophoma draconis* (Berk. ex Cooke) Allesch. *in* Rabenh. Krypt.-Flora [ed. 2], Pilze **7** [Lief. 88]: 836. 1903 [based on misidentification].

= *Phyllosticta maculicola* Halst. *in* Rep. New. Jers. St. agric. Exp. Stn **14**: 412. 1894.

H = *Phyllosticta dracaena* Griffon & Maubl. *in* Bull. Soc. mycol. Fr. **25**: 238. 1909; not *Phyllosticta dracaena* Henn. *in* Hedwigia **48**: 111. 1908 [= *Asteromella dracaena* (Henn.) Aa, see van der Aa & Vanev (2002)].

Diagn. data: Boerema (1983a) (fig. 7: culture on OA, drawing of conidia), de Gruyter *et al.* (1998) (fig. 6: conidial shape, reproduced in present Fig. 8H; cultural descriptions, partly adopted below).

Description in vitro

Pycnidia globose with 1(–2) non-papillate ostioles, 90–220 μm diam., glabrous, solitary or confluent. Conidial matrix white to buff.

Conidia ellipsoidal to ovoid with numerous, large polar guttules, 4–8.5 × 2–4 μm.

Colonies on OA 5–6 cm diam. after 7 days, regular, colourless to rosy buff/honey and weak greenish tinge near margin, with tufted, finely floccose, white to grey olivaceous aerial mycelium; reverse also colourless to rosy buff honey with greenish tinge near margin. On MA the agar staining ochraceous due to the release of a diffusible pigment, which on addition of NaOH discolours to orange. On both media also E$^+$ reaction with the NaOH spot test: greenish, then red. Pycnidia abundant, concentrically zoned, mostly partly in the agar. Representative culture CBS 186.83.

Ecology and distribution

Recorded from wild *Dracaena* spp. (Agavaceae) in Africa (Rwanda) and from cultivated species of *Dracaena* in Europe, India and North America. Also reported from *Cordyline* spp. (also Agavaceae): leaf spot, dieback.

Phoma eucalyptica Sacc., Fig. 8I

Phoma eucalyptica Sacc. *in* Sylloge Fung. **3**: 78. 1884 [as '(Thüm.)', but nom. nov.].

≡ *Coniothyrium eucalypti* Thüm. *in* Instituto, Coimbra **27** sub Contr. Fl. myc. Lusit. II n. 341. 1880 ['1879 *e* 1880']; quoted *in* Hedwigia **19**: 151. 1880; not *Phoma eucalypti* Cooke & J.J. Kickx *in* Sylloge Fung. **3**: 78. 1884.

Diagn. data: Neergaard (1977) (occurrence on seed), de Gruyter & Noordeloos (1992) (fig. 15: conidial shape, reproduced in present Fig. 8I; cultural descriptions, partly adopted below).

Description in vitro

Pycnidia globose with 1–5 ostioles, usually on distinct necks, 80–200 μm diam., glabrous, solitary or confluent in clusters of 2–5 pycnidia. Conidial matrix variable in colour, whitish to vinaceous buff or saffron.

Conidia ellipsoidal, eguttulate, (2–)2.5–4 × 1–2 μm.

Colonies on OA *c*. 4.5 cm diam. after 7 days, regular, olivaceous grey with citrine outer margin, sometimes with olivaceous grey woolly aerial mycelium and/or distinct radiating hyphal strands [on MA somewhat faster growing, 5–5.5 cm diam; old cultures on MA may show globose or ellipsoidal swollen cells in the mycelium]. Pycnidia scattered or in concentric zones on and in the agar. With additions of a drop of NaOH a slightly sienna discoloration of the agar may occur. Representative cultures CBS 377.91, CBS 378.91.

Ecology and distribution

An opportunistic parasite of Myrtaceae found in Australasia, South-east and south-west Asia, and in southern Europe. It is recorded in association with shoot wilt of *Eucalyptus* spp., esp. *E. globulus* (Australian gumtree), and with leaf necroses and blister symptoms on *Eugenia* spp., esp. *E. aromatica* (clove-tree). It has also been isolated from sea water in the mediterranean zone (planting area of *E. globulus*). The fungus appeared to be seed-borne (in USA intercepted on seed of a considerable range of *Eucalyptus* spp. imported from Australia and New Zealand). In the past it has been confused with an *Asteromella*-spermatial state, known as *Phyllosticta eucalypti* Thüm., see van der Aa & Vanev (2002).

Phoma eupatorii Died., Fig. 8J

Phoma eupatorii Died. in Annls mycol. **10**: 447. 1912.

Diagn. data: de Gruyter, Noordeloos & Boerema (1993) (fig. 10: conidial shape, reproduced in present Fig. 8J; cultural descriptions, partly adopted below).

Description in vitro

Pycnidia subglobose or irregular with 1–5 non-papillate or slightly papillate ostioles, 50–200 μm diam., glabrous, solitary or confluent. Conidial matrix salmon.

Conidia ellipsoidal usually with two conspicuous polar guttules, 4–5.5 × 1.5–2.5 μm.

Colonies on OA 5 cm diam. after 7 days [on MA 6 cm], regular, without aerial mycelium, colourless but general impression salmon because of abundant conidial mass exuding from pycnidia; reverse similar. Pycnidia covering the whole plate or occurring in concentric rings on or partly in the agar, rarely submerged, often associated with dense, olivaceous hyphal strands.

Application of a drop of NaOH resulted in a green spot turning red (E⁺ reaction). Representative culture CBS 123.93.

Ecology and distribution

This fungus seems to be a common necrophyte of *Eupatorium cannabinum* (Compositae), found in different parts of Europe. It frequently occurs on dead flower heads and may be seed-borne. On dead stems the pycnidia are connected by a subepidermal mycelial network comparable with the hyphal strands found in cultures on OA.

Phoma eupyrena Sacc., Fig. 8K, Colour Plate I-H

> *Phoma eupyrena* Sacc. *in* Michelia **1**(5): 525. 1879 [formerly often listed with the superfluous qualification 'sensu Wollenweber'].

Diagn. data: Dorenbosch (1970) (pl. 4 figs 1–2, pl. 6 fig. 3: cultures on OA and MA and chlamydospores; tab. 1: synoptic table with drawings of conidia and chlamydospores; nomenclatural note), Boerema (1976) (tab. 4, sources of isolates), Domsch *et al.* (1980) (fig. 280: shape of conidia and chlamydospores; occurrence in soil, literature references), Sutton (1980) (fig. 227B: conidia and chlamydospores; specimens in IMI collection), Morgan-Jones & Burch (1988a) (fig. 1A–F: drawings of pycnidium, conidiogenous cells, conidia, wall cells and chlamydospores; pls 1A–F, 2A–G: photographs of agar plate cultures, pycnidium, wall cells, chlamydospores, hyphal strands and developing pycnidia; history, cultural and morphological description), de Gruyter & Noordeloos (1992) (figs 7, 8, 22, shape of conidia and chlamydospores, reproduced in present Fig. 8K; cultural descriptions, partly adopted below), Rai & Rajak (1993) (cultural characteristics), Rai (1998) (effect of different physical and nutritional conditions on the morphology and cultural characters).

Description in vitro

Pycnidia subglobose-conical or papillate, indistinctly uniostiolate, variable in size, 100–260 μm diam., glabrous, solitary or sometimes confluent. Conidial matrix whitish to pale grey.
　　Conidia ellipsoidal, with 2 large polar guttules, (3.5–)4–5.5(–6) × (1.5–)2–2.5(–3) μm.
　　Chlamydospores barrel shaped or subglobose-ellipsoidal, 4–15 μm diam., brown ochraceous with numerous green guttules, single or in chains, intercalary or terminal. They develop in culture after about one week, darkening the colonies gradually.
　　Colonies on OA, 5–5.5 cm diam. after 7 days, regular, variable in colour from dull green or herbage green to olivaceous black with some olivaceous aerial mycelium at centre; reverse green to olivaceous black [on MA abundant woolly whitish grey to olivaceous grey aerial mycelium]. Pycnidia scattered or

arranged in concentric rings on and in the agar (partly). Representative cultures CBS 527.66, CBS 374.91.

Ecology and distribution

A cosmopolitan soil-inhabiting fungus, which may cause damping-off symptoms on seedlings of herbaceous and woody plants. Formerly considered as a specific fungus of potato (*Solanum tuberosum*; Solanaceae), but at present preserved collections from at least 30 different plant genera are known.

Phoma fallens Sacc., Fig. 9A

Phoma fallens Sacc. in Sylloge Fung. **10**: 146. 1892.

Diagn. data: de Gruyter *et al.* (1998) (fig. 1: conidial shape, reproduced in present Fig. 9A; cultural descriptions, partly adopted below).

Description in vitro

Pycnidia globose to irregularly shaped, with 1 papillate ostiole, 100–210 µm diam., glabrous, solitary or confluent. Conidial matrix white.
    Conidia oblong to ellipsoidal with or without several indistinct polar guttules, 5–7.5(–10.5) × 2–4 µm.
    Colonies on OA slow growing *c.* 2 cm after 7 days [3.5–4 cm after 14 days], somewhat irregular, dull green to greenish glaucous, with velvety, pale olivaceous grey aerial mycelium; reverse also green to greenish glaucous. On application of a drop of NaOH a weak red discoloration [on MA distinct redbrown]. Pycnidia scattered, mainly in centre of colony, on or partly in the agar. Representative culture CBS 161.78.

Ecology and distribution

Recorded in association with spots on fruits and leaves of olive, *Olea europaea* (Oleaceae), but so far no pathogenicity tests have been done. Probably widespread in olive-growing regions of the world, particularly southern Europe.

Phoma fimeti Brunaud,[5] Fig. 9B

Phoma fimeti Brunaud in Bull. Soc. bot. Fr. **36** [= II, **11**]: 338. 1889.

Diagn. data: Dorenbosch (1970) (pl. 1 fig. 3: colour plate of culture on OA showing yellow pigmentation; pls 2 figs 1, 2, 5 figs 5, 6: photographs of

---

[5] So far we have never isolated this soil fungus from leaves of living herbaceous plants, therefore we have not adopted a synonymy with *Phyllosticta epimedii* Sacc., as suggested by van der Aa & Vanev (2002).

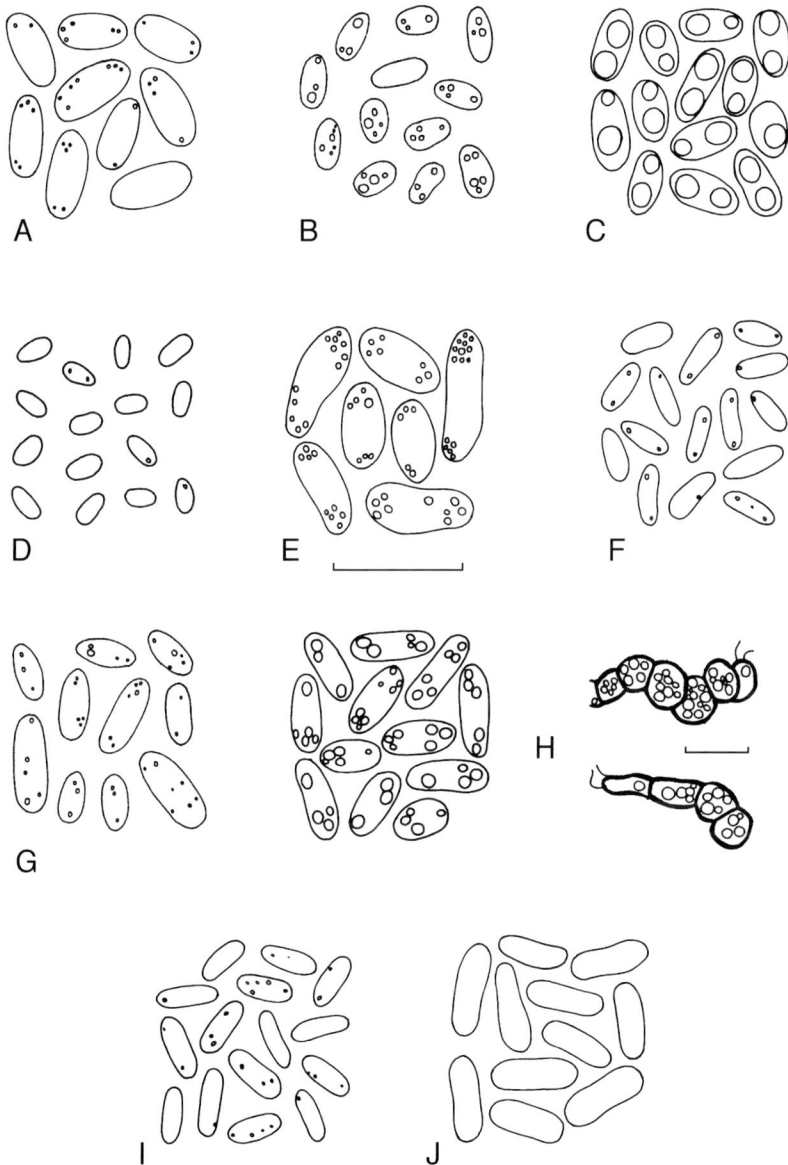

Fig. 9. Shape of conidia and some hyphal chlamydospores. **A** *Phoma fallens*. **B** *Phoma fimeti*.
C *Phoma flavescens*. **D** *Phoma flavigena*. **E** *Phoma glaucispora*. **F** *Phoma haematocycla*.
G *Phoma hedericola*. **H** *Phoma henningsii*. **I** *Phoma herbarum*. **J** *Phoma herbicola*. Bars 10 μm.

cultures on OA and MA and pycnidia *in vitro*; tab. 1: synoptic table with drawing of conidia), Boerema & Dorenbosch (1973) (pl. 3c, d: cultures on OA; tab.
1: synoptic table on characters *in vitro* with drawing of conidia), Sutton (1980)
(fig. 227D: conidia; specimens in CMI collection), Johnston (1981) (tab. 1: cultural and microscopic characteristics; fig. 3: cultures on OA and MA), Tuset &

Portilla (1985) (fig. 4: pycnidial and conidial shape; tab. 3: dimensions), de Gruyter & Noordeloos (1992) (fig. 10: conidial shape, reproduced in present Fig. 9B; cultural descriptions, partly adopted below), Rai & Rajak (1993) (cultural characteristics), Rai (1998) ('always slow growing under different physical and nutritional conditions').

Description in vitro

Pycnidia subglobose with 1 conspicuous somewhat papillated ostiole, 60–200 μm diam., glabrous, solitary or confluent. Conidial matrix whitish to ochraceous.
    Conidia broadly ellipsoidal with 2 or more polar guttules, mostly (2–)2.5–4(–5) × (1.5–)2–2.5(–3) μm.
    Colonies on OA slow growing, c. 2 cm diam. after 7 days [4–4.5 cm after 14 days], regular with velvety pale olivaceous grey aerial mycelium, colony colour grey olivaceous to greenish olivaceous, reverse shows that the agar below the colony becomes gradually discoloured by a yellow pigment. The yellow colour does not change on addition of a drop of NaOH. Pycnidia develop in concentric zones on the agar, sometimes entirely in the agar or in the aerial mycelium. Representative cultures CBS 170.70, CBS 368.91.

Ecology and distribution

A saprophytic soil fungus which has been isolated from dead tissue of various herbaceous and woody plants. Most isolates of this fungus are made in Europe (Germany, the UK, The Netherlands, Spain), but there are also records from Africa (Zambia), Australasia (Australia, New Zealand), Asia (India, Myanmar) and North America (Alaska, Canada, USA).

Phoma flavescens Gruyter *et al.*, Fig. 9C

> *Phoma flavescens* Gruyter, Noordel. & Boerema *in* Persoonia **15**(3): 375. 1993.

Diagn. data: de Gruyter *et al.* (1993) (fig. 3: conidial shape, reproduced in present Fig. 9C; cultural descriptions, partly adopted below).

Description in vitro

Pycnidia globose with a rather indistinct, non-papillate ostiole, mostly 20–140 μm diam., glabrous or covered by hyphae, solitary or confluent. Conidial exudate not observed.
    Conidia ellipsoidal with 2 very large polar guttules, 4–7 × 2–3.5 μm.
    Colonies slow growing, c. 1.5 cm after 7 days [c. 2.5 cm after 14 days], regular, without any aerial mycelium, colourless but the agar becomes gradually discoloured by a yellow pigment. The yellow colour does not change on addition of a drop of NaOH. The pycnidia develop abundantly on and in the agar. Representative culture CBS 178.93.

Ecology and distribution

This species is thus far known only from an isolate of a soil sample drawn from a potato field in the N.E. Polder, The Netherlands.

Phoma flavigena Constant. & Aa, Fig. 9D

Phoma flavigena Constant. & Aa *in* Trans. Br. mycol. Soc. **79**: 343. 1982.

Diagn. data: Constantinescu & van der Aa (1982) (fig. 1a–f: drawings of pycnidia, wall structure, ostiole, conidiogenous layer and conidia; original detailed description), de Gruyter & Noordeloos (1992) (fig. 4: conidial shape, reproduced in present Fig. 9D; cultural description adopted from original).

Description in vitro

Pycnidia globose, often with conspicuous neck and distinct ostiole, 50–210 μm diam., glabrous, solitary or confluent. Conidial matrix buff or rosy buff.

Conidia cylindrical-ellipsoidal, mostly eguttulate, sometimes with 1–2 small guttules, 2–4 × 1–2 μm.

Colonies on OA 2–2.5 cm diam. after 7 days, regular, colourless with sparse whitish aerial mycelium; the agar medium shows a conspicuous discoloration by a yellow pigment. On MA very slow growing, 1–1.5 cm, with abundant grey/yellowish aerial mycelium; agar medium also strongly discoloured by a yellow pigment. Pycnidia in concentric zones, immersed in the agar or in aerial mycelium. The yellow colour does not change on addition of a drop of NaOH. Representative culture CBS 314.80.

Ecology and distribution

Once isolated from a freshwater source in Romania. The name refers to the characteristic production of a yellow pigment.

Phoma glaucispora (Delacr.) Noordel. & Boerema, Fig. 9E

Phoma glaucispora (Delacr.) Noordel. & Boerema *in* Versl. Meded. plziektenk. Dienst Wageningen **166** (Jaarb. 1987): 108. 1989 ['1988'].

≡ *Phyllosticta glaucispora* Delacr. *in* Bull. Soc. mycol. Fr. **9**: 266. 1893.

= *Phyllosticta oleandri* Gutner *in* Trudỹ bot. Inst. Akad. Nauk. SSSR Leningrad Ser. II [= Plantae Cryptogamae], **1**: 306. 1933.

Diagn. data: Gutner (1933) (disease symptoms, detailed description *in vivo*), Noordeloos & Boerema (1989a) (nomenclature, disease symptoms), de Gruyter *et al.* (1998) (fig. 3: conidial shape, reproduced in present Fig. 9E; cultural descriptions, partly adopted below).

Description in vitro

Pycnidia globose to subglobose, often non-ostiolate, 80–240 µm diam., glabrous, solitary or confluent. Conidial matrix rosy.

Conidia oblong to ellipsoidal, with several small polar guttules (yellowish: '*glaucispora*'), 5–7(–8) × 2.5–4.5 µm.

Colonies on OA *c.* 2 cm after 7 days (4–4.5 cm after 14 days), regular, smoke grey to grey olivaceous or dull green, with finely woolly to finely floccose, smoke to dull green aerial mycelium; reverse with olivaceous black sectors. Pycnidia scattered, more obviously concentrically arranged at margin, partly submerged in the agar. Representative culture CBS 284.70.

Ecology and distribution

Common and widespread on oleander, *Nerium oleander* (Apocynaceae), in southern Europe (e.g. Italy, Spain): leaf spot. Also reported from glasshouse ornamental cultures (Russia, The Netherlands).

Phoma haematocycla (Berk.) Aa & Boerema, Fig. 9F

> *Phoma haematocycla* (Berk.) Aa & Boerema apud de Gruyter, Noordeloos & Boerema *in* Persoonia **15**(3): 377. 1993.
>> ≡ *Phyllosticta haematocycla* Berk., Enum. Fungi. coll. Portugal by Welw. 5. 1853 [sometimes erroneously quoted with the author-citation 'Berk. & Welw.'].

Diagn. data: de Gruyter *et al.* (1993) (fig. 4: shape of conidia, reproduced in present Fig. 9F; designation of neotype; cultural descriptions, partly adopted below).

Description in vitro

Pycnidia globose to subglobose, with 1(–3) papillate ostioles, mostly 70–120 µm diam., glabrous, solitary or confluent. Conidial matrix white to rosy buff.

Conidia oblong to ellipsoidal, sometimes provided with 1–2 inconspicuous polar guttules, 4–4.5 × 1.5(–2) µm.

Colonies on OA slow growing, *c.* 2.5 cm diam. after 7 days [*c.* 3.5 cm after 14 days], regular, colourless to rosy buff, with velvety/felted white aerial mycelium; reverse colourless to rosy buff. Pycnidia scattered, mainly on the agar. In the mycelium usual swollen cells, about 8–9 µm diam., intercalary or terminal. Representative culture CBS 175.93 (culture of neotype).

Ecology and distribution

A common specific pathogen of New Zealand flax, *Phormium tenax* (Liliaceae), causing red-encircled ('*haematocycla*') spots on the leaves: leaf spot.

Phoma hedericola (Durieu & Mont.) Boerema, Fig. 9G, Colour Plate I-K

> *Phoma hedericola* (Durieu & Mont.) Boerema *in* Trans. Br. mycol. Soc. **67**: 295. 1976.
>> V ≡ *Phyllosticta hedericola* Durieu & Mont. *in* Flore d'Algérie crypt. **1**: 611. 1849 [as 'hederaecola'].
>> H ≡ *Phyllosticta destructiva* var. *hederae* (Durieu & Mont.) Oudem. *in* Ned. kruidk. Archf II, **1**(3): 257. 1873 [name change]; not *Phyllosticta destructiva* var. *hederae* Desm. *in* Pl. cryptog. Fr. II [ed. 3] Fasc. 14, No. 680. 1859 [not seen, probably also identical with *Ph. hedericola*].

Diagn. data: Dennis (1946) (cultural characters sub *Phyllosticta hedericola*), Boerema (1976) (fig. 3: cultures on OA; tab. 9: synoptic table on cultural characters with drawing of conidia), Sutton (1980) (fig. 227E: drawing of conidia; specimens in CMI collection), de Gruyter & Noordeloos (1992) (fig. 17: conidial shape, reproduced in present Fig. 9G; cultural descriptions, partly adopted below), van der Aa & Vanev (2002) (*Phyllosticta* synonyms).

Description *in vitro*

Pycnidia globose or irregular shaped, with 1–2, sometimes 3, more or less papillated ostioles, 80–140 μm diam., glabrous or with hyphal strands, solitary or confluent. Conidial matrix whitish.

Conidia broadly ellipsoidal, usually with 2 or more very small polar guttules, (3.5–)4–6 × (1.5–)2–3 μm.

Colonies on OA 5.5–6 cm diam. after 7 days, regular, colourless with olivaceous sectors or centre, aerial mycelium, if present, sparse and whitish, reverse also colourless with olivaceous sectors or centre [on MA slower growing, 4–5 cm diam., reverse pale luteous]. Pycnidia in clusters scattered over whole colony mostly on the agar, sometimes in the agar or in the aerial mycelium. Most strains showed a positive reaction with the sodium hydroxide test: on application of a drop of NaOH green → red (E^+ reaction). Representative cultures CBS 618.75, CBS 367.91.

Ecology and distribution

In Europe commonly found in association with necroses on leaves and stems of ivy, *Hedera helix* (Araliaceae): leaf spot. The fungus is also recorded on the himalaya-ivy (*Hedera himalaica*) and may occur everywhere ivy is grown.

Phoma henningsii Sacc., Fig. 9H

> *Phoma henningsii* Sacc. *in* Sylloge Fung. **10**: 139. 1892.
>> H ≡ *Phoma acaciae* Henn. *in* Bot. Jb. **14**: 368. 1892; not *Phoma acaciae* Penz. & Sacc. *in* Atti Ist. ven. Sci. VI, **2**: 650 [= Fung. Mortol. n. 23]. 1884 [= *Phomopsis* sp.].

Diagn. data: Olembo (1972) (as '*Ph. herbarum*'; tabs I–III, figs 1, 2: effect of temperature and conidial density on germination of conidia, pathogenicity test on *Acacia mearnsii*: conidia 'are capable of infecting the host through wounds only'), Boerema & Dorenbosch (1980) (identification), de Gruyter *et al.* (1993) (figs 17, 29: shape of conidia and chlamydospores, reproduced in present Fig. 9H; cultural descriptions, partly adopted below).

Description in vitro

Pycnidia globose, with usually 1 non-papillate ostiole, 70–220 µm diam., glabrous, solitary or confluent. Conidial matrix salmon to pale vinaceous.

Conidia ellipsoidal with (1–)2–3(–4) large polar guttules, 5–7.5 × 2–3 µm.

Colonies on OA *c.* 4.5 cm diam. after 7 days, regular, colourless; aerial mycelium only as a white-felted zone near margin; reverse colourless to pale olivaceous. Pycnidia develop abundantly in concentric rings (partly) in the agar. Representative culture CBS 104.80.

Ecology and distribution

This fungus is in East Africa known as a harmful wound parasite of *Acacia* spp. (Leguminosae). On account of its saprophytic vigour and the conidial dimensions *in vivo* it has been confused with the ubiquitous *Phoma herbarum* Westend. (this section). North American records of *Ph. henningsii* may be questioned.

Phoma herbarum Westend., Figs 6A–G, 9I

Phoma herbarum Westend. in Bull. Acad. r. Belg. Cl. Sci. **19**(3): 118.1852.

= *Phoma exigua* var. ['b'] *minor* Desm. in Annls Sci. nat. (Bot.) III, **11**: 283. 1849.

≡ *Phoma exigua* var. *ranunculorum* Desm. ex Sacc. in Sylloge Fung. **3**: 134. 1884 [name change].

= *Phoma leguminum* Westend., in Herb. crypt. [Ed. Beyaert & Feys] Fasc. **23**: No. 1135. 1857; in Bull. Acad. r. Belg. Cl. Sci. II, **11**: 645. 1861 [as collective name often misapplied].

= *Phyllosticta crastophila* Sacc. in Michelia **1**(2): 153. 1878.

≡ *Phoma crastophila* (Sacc.) Punith. in Nova Hedwigia **34**: 145. 1981.

= *Phoma charticola* Speg. in An. Soc. cient. argent. **10**: 153[–154]. 1880.

= *Phoma urticae* Schulzer & Sacc. in Hedwigia **23**: 91. 1884.

≡ *Leptophoma urticae* (Schulzer & Sacc.) Höhn. in Hedwigia **59**: 262. 1918 [misapplied].

= *Phoma oleracea* Sacc. in Michelia **2**(1): 91. 1880.

= *Phoma oleracea* var. *dipsaci* Sacc. *in* Michelia **2**(2): 337. 1881.

= *Phoma oleracea* var. *helianthi-tuberosi* Sacc. *in* Michelia **2**(2): 337. 1881.

= *Phoma oleracea* var. *scrophulariae* Sacc. *in* Michelia **2**(2): 337. 1881.

= *Phoma oleracea* var. *urticae* Sacc. *in* Michelia **2**(2): 337. 1881.

V Θ = *Phoma herbarum* var. *erysimi* Roum. *in* Revue mycol. **3**(9): 30. 1881 [as 'crysimi'].

Θ = *Phoma herbarum* var. *sambuci* Roum. *in* Revue mycol. **3**(9): 30. 1881.

Θ = *Phoma herbarum* f. *chenopodii-albi* Roum. *in* Revue mycol. **5**: 28. 1883.

= *Phoma herbarum* var. *tetragoniae* Sacc. & Berlese *in* Revue mycol. **8**: 35. 1886 [as 'Phoma herbarum* Ph. Tetragoniae'].

= *Phyllosticta betonicae* Brunaud *in* Act. Soc. linn. Bordeaux **44**: 242. 1890.

H = *Phyllosticta betonicae* Allesch. *in* Hedwigia **33**: 70. 1894; not *Phyllosticta betonicae* Brunaud, see above.

= *Aposphaeria violacea* Bertel *in* Öst. bot. Z. **54**: 205 [233, 288]. 1904.

≡ *Phoma violacea* (Bertel) Eveleigh *in* Trans. Br. mycol. Soc. **44**: 577. 1961.

= *Phoma oleracea* f. *bryoniae* Sacc. *in* Annls mycol. **7**: 435. 1909.

= *Phoma pigmentivora* Massee *in* Bull. misc. Inf. R. bot. Gdns Kew **8**: 326. 1911.

= *Phyllosticta collinsoniae* Sacc. & Dearness apud Saccardo *in* Annls mycol. **12**: 299. 1914.

= *Phyllosticta ruscigena* Sacc. *in* Nuovo G. bot. ital. II, **22**: 45. 1915.

= *Phyllosticta euchlaenae* Sacc. *in* Nuovo G. bot. ital. **23**: 207. 1916.

H = *Phoma herbarum* f. *humuli* Gonz. Frag. *in* Trab. Mus. nac. Cienc. nat., Madr., Ser. bot. **12**: 30. 1917; not *Phoma herbarum* f. *humuli* Sacc. *in* Michelia **2**(1): 92. 1880 [= *Phoma exigua* var. *exigua*, sect. *Phyllostictoides*, **E**].

= *Phoma herbarum* f. *minor* Unamuno *in* An. Jard. bot. Madr. **2** ['1941']: 56. 1942.

= *Phoma herbarum* var. *lactaria* B. Sutton *in* Trans. Br. mycol. Soc. **47**: 501. 1964.

= *Phoma hibernica* Grimes, M. O'Connor & H.A. Cummins *in* Trans. Br. mycol. Soc. **17**: 99[–101]. 1932.

= *Phoma lignicola* Rennerf. *in* Svenska SkogsvFör. Tidskr. **35**: 60. 1936.

= *Phyllosticta panici-miliacei* Săvul. & Sandu *in* Hedwigia **57**: 183. 1936.

= *Pyrenochaeta mali* M.A. Sm. *in* Phytopathology **53**: 591. 1963.

Diagn. data: Boerema (1964) (pl. 1 figs 1–6, pl. 2 figs 1–6; typification; description and illustrations; synonymy and variability; habitat) (1970) (tab. I: variable position of the pycnidia on various agar media, superficial, erumpent and immersed; 'budding-like conidiogenesis'; full synonymy; misapplications), (1976) (tab. 7: sources of isolates; additional synonym), Dorenbosch (1970) (pl. 1 figs 4–5: colourplate of mycelial and pycnidial cultural types showing red discoloration of the medium, which turns blue with addition of a drop of NaOH; tab. I: synoptic table on characters *in vitro*, with drawing of conidia), Boerema & Bollen (1975) (pls 19 fig. B, 21 fig. A: electron microscopy of conidium development; misinterpretation as *Pyrenochaeta*), Domsch *et al.* (1980) (occurrence in soil, literature references), Sutton (1980) (fig. 228B: drawing of conidia; 'represented in herb. CMI by more than 100 collections'), Morgan-Jones (1988a) (fig. 1A–G: drawings of pycnidia, conidiogenous cells, conidia, wall cells; pls 1A–F, 2A–G: photographs of agar plate cultures, pycnidia, ostiole, wall cells, seta-like hyphal outgrowths; history, cultural and morphological description), Johnston (1981) (tab. 1: cultural and microscopic characteristics; fig. 5: photographs of cultures on OA and MA; occurrence in New Zealand), Rai & Rajak (1993) (cultural characteristics; NB!: the studies sub *Ph. herbarum* in Rai (1998) must be questioned as they include various pathogenic isolates), de Gruyter *et al.* (1993) (figs 1, 5, drawing after Morgan-Jones (1988a) and shape of conidia, reproduced in present Figs 6 and 9I; cultural descriptions, partly adopted below), Boerema *et al.* (1996) (misinterpretation as *Plenodomus/Leptophoma*), van der Aa & Vanev (2002) (*Phyllosticta* synonyms).

Description in vitro

Pycnidia subglobose to elongated, usually with one ostiole, but sometimes with up to four ostioles, non-papillate or very slightly papillate, up to 350 μm diam., glabrous but often covered by a loose weft of hyphae (in old pycnidia sometimes seta-like outgrowths from outer wall cells occur) solitary, frequently densely gregarious. Conidial matrix white to buff or rosa vinaceous.

Conidia oblong to ellipsoidal, mostly eguttulate but sometimes with small inconspicuous polar guttules, (3.5–)4–5.5(–8) × 1.5–2(–3) μm (rarely reaching a length of 9 μm).

Colonies on OA *c.* 4–5 cm after 7 days, regular, mostly without any aerial mycelium and flesh coloured with greenish tinge by the conidial masses on abundant pycnidia; reverse greenish olivaceous to olivaceous; mycelial strains or sectors usually producing a yellowish brown to red diffusible pigment, which on application of a drop of NaOH changes in colour to blue [the reddish pigment is usually well developed in colonies on MA; in old cultures the red pigment changes to a bluish hue naturally]. Only occasionally isolates of the fungus fail to produce any pigment. Representative culture CBS 615.75.

Ecology and distribution

This worldwide recorded fungus is unique in having a very wide substratum range. It has been isolated from dead material of all kinds of herbaceous and woody plants (esp. in spring), as well as from animal (incl. human), nutritional and inorganic material. Its common occurrence on dead seed coats explains why it has often been confused with specific seed-borne pathogens (as *Phoma lingam* Tode : Fr., sect. *Plenodomus*, **G**). The fungus is reported to be the causal agent of a lethal disease of the air-bladder in salmon and trout[6]. In humans, the fungus is known from peripheral lung tissue with asthma patients. Other sources of repeated isolates are air, asbestos, butter, carpets, cement, cream, oil-paint, paper, plaster, rubber, soil, woodpulp and water.

Phoma herbicola Wehm., Fig. 9J

> *Phoma herbicola* Wehm. *in* Mycologia **38**: 319–320. 1946.
>> H = Phoma *pedicularis* Wehm. *in* Mycologia **38**: 319. 1946; not *Phoma pedicularis* Fuckel apud von Heuglin *in* Reisen Nordpolarmeer III Beitr. Fauna Fl. Geol. 318. 1874 [as 'pedicularidis'; belongs to sect. *Plenodomus*, **G**].

Diagn. data: Wehmeyer (1946) (figs 11, 14: shape of conidia sub *Ph. pedicularis* and *Ph. herbicola*; original descriptions *in vivo* on *Syntheris* and *Pedicularis* spp.), de Gruyter *et al.* (1998) (fig. 13: conidial shape, reproduced in present Fig. 9J; cultural descriptions, partly adopted below).

Description in vitro

Pycnidia globose to irregular with 1–2 ostioles on an elongated neck, 120–340 μm diam., glabrous, solitary or confluent. Conidial matrix buff to saffron.
    Conidia cylindrical to subcylindrical, eguttulate, 5–7(–8.5) × 2–3 μm.
    Colonies on OA 4.5–5 cm diam. after 7 days, regular, colourless with darker concentric zones due to pycnidia, aerial mycelium scanty, particular in marginal zone, fluffy, white; reverse colourless. Pycnidia abundant on and partly in the agar. Representative culture CBS 629.97.

Ecology and distribution

A saprophytic soil fungus indigenous to North America. The records so far are from northwestern USA (Montana and Wyoming) and refer to old dead stems of quite different herbaceous plants and to polluted lake water. The conidia *in vivo* are usually longer than those *in vitro* (up to 10.5 μm).

---

[6] Fish-mycosis may also be caused by an undescribed species of *Phoma*, differing from *Ph. herbarum* by the production of unicellular chlamydospores and faster growth on OA and MA, see Hatai *et al.* (1986).

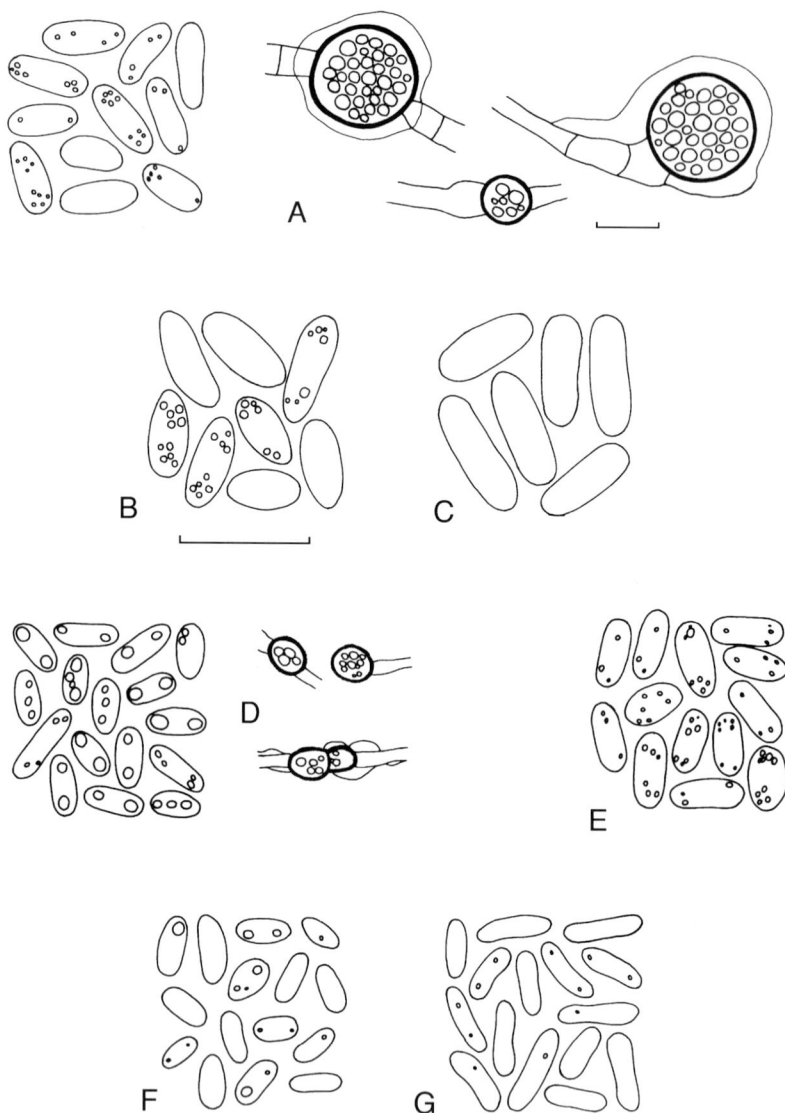

Fig. 10. Variation of conidia and some hyphal chlamydospores. **A** *Phoma heteroderae.*
B *Phoma huancayensis.* **C** *Phoma humicola.* **D** *Phoma insulana.* **E** *Phoma labilis.* **F** *Phoma lini.*
G *Phoma loticola.* Bars 10 μm.

Phoma heteroderae S.Y. Chen _et al.,_ Fig. 10A

> *Phoma heteroderae* S.Y. Chen, D.W. Dickson & Kimbrough *in* Mycologia
> **88**: 885. 1996.

Diagn. data: Chen *et al.* (1996) (study on isolate from the soybean cyst nematode with many photographs; figs 1–4: cultures on potato dextrose agar, corn-

meal agar, MA and OA; figs 5–6: mycelia; figs 7–10: chlamydospores; figs 13–25: morphology of pycnidia, conidiogenous cells and conidia; detailed description; comparison with other species of *Phoma* isolated from cysts), de Gruyter *et al.* (1998) (figs 10a, b: shape of conidia and chlamydospores, reproduced in present Fig. 10A; cultural descriptions, partly adopted below).

Description in vitro

Pycnidia globose to irregular with 1–4 non-papillate or slightly papillate ostioles, 70–250 µm diam., glabrous or with setae-like hyphal outgrowths (semipilose), solitary or confluent. Conidial matrix buff to vinaceous buff.

Conidia ellipsoidal to ovoid to cylindrical, eguttulate or with 1–5 small, polar guttules, 3.5–7.5(–12) × 2–3.5(–4.5) µm.

Chlamydospores [abundant on OA] unicellular, intercalary, solitary or in chains, globose, 7.5–25.5 µm with a distinct 'envelope'.

Colonies on OA 7–8 cm diam. after 7 days, colourless to grey olivaceous or citrine, with or without velvety grey olivaceous aerial mycelium; reverse also colourless to grey olivaceous or citrine. Pycnidia abundant, mainly on, partly also in the agar. Representative culture CBS 96/2022.

Ecology and distribution

A saprophyte, isolated from various organic and inorganic substrata in Eurasia and North America, *e.g.* packaged food and wall plaster. Apparently it is a soil-inhabiting fungus (fitting well with the abundant production of chlamydospores). Recorded as a probable weak parasite on eggs of *Heterodera glycines*.

Phoma huancayensis Turkenst., Fig. 10B

V   *Phoma huancayensis* Turkenst. *in* Fitopatología **13**: 68. 1978 [as 'huancayense'].

Diagn. data: Turkensteen (1978) (original description *in vitro*; occurrence in Peru), Boerema & Dorenbosch (1980) (identification history), Johnston (1981) (tab. 1: cultural and microscopic characteristics; fig. 6: cultures on OA and MA; occurrence in New Zealand), de Gruyter *et al.* (1998) (fig. 7: conidial shape, reproduced in present Fig. 10B; cultural descriptions, partly adopted below).

Description in vitro

Pycnidia globose to irregular with 1, sometimes papillate ostiole, occasionally with an elongated neck, 95–240 µm diam., glabrous, solitary or confluent. Conidial matrix white to salmon.

Conidia ellipsoidal to ovoid to subcylindrical, eguttulate or with 1–5 polar guttules, (4–)5–8(–12) × 2.5–4.5 µm.

Colonies on OA 4–5.5 cm diam. after 7 days, regular, olivaceous buff to primrose or umber to bay, due to the production of a diffusible pigment, which on application of NaOH changes to red and then blue; with finely to coarsely floccose, smoke grey to olivaceous aerial mycelium; reverse olivaceous buff to primrose or umber to bay, or with peach at the margin. On MA pigmentation ochraceous to fulvus with buff to primrose patches; reverse primrose to ochraceous and fulvus/bay to sepia, with NaOH also discolouring from reddish to blue. Pycnidia scarce to abundant both on and in the agar. Representative cultures CBS 105.80, CBS 390.93.

Ecology and distribution

A saprophytic fungus, probably indigenous to South America. It is originally described from dead leaves of *Chenopodium quinoa* (Chenopodiaceae) and wild species of potato, *Solanum* spp. series *Tuberosa* (Solanaceae), in the Andes region of Peru (prov. Huancayo). The fungus is further recorded in New Zealand where it has been isolated from necrotic tissue of various dicotyledonous and monocotyledonous plants, e.g. often from *Paspalum dilatatum* (Gramineae), a grass originating from South America.

Phoma humicola J.C. Gilman & E.V. Abbott, Fig. 10C

> *Phoma humicola* J.C. Gilman & E.V. Abbott *in* Iowa State Coll. J. Sci. **1**(3): 266. 1927.

Diagn. data: Gilman & Abbott (1927) (fig. 12a–d: drawings of pycnidia and conidia; original description *in vitro*), Boerema (1985) (fig. 10C: drawing of pycnidia and conidia; production anthraquinone pigments; isolate sources), de Gruyter *et al.* (1998) (fig. 4: conidial shape, reproduced in present Fig. 10C; cultural descriptions, partly adopted below).

Description in vitro

Pycnidia globose with 1–2 non-papillate ostioles, 80–150 μm diam., glabrous, solitary, rarely confluent. Conidial matrix white to buff.
  Conidia oblong to cylindrical, eguttulate, (7–)8–10.5(–12) × 2.5–4 μm.
  Colonies on OA 4–4.5 cm diam. after 7 days, regular, without aerial mycelium, pale luteous due to the release of a diffusible pigment which on application of a drop of NaOH changes to rosy vinaceous or coral; reverse also luteous to grey olivaceous. On MA colony citrine with yellowish marginal zone, and characterized by the production of citrine green crystals: anthraquinones. Pycnidia abundant, mainly in the agar. Representative culture CBS 220.85.

Ecology and distribution

Soil fungus in North America (Utah, Nevada, Wyoming) and South America (Andes, Peru). Isolated from a variety of cultivated land and other substrates

(Boerema, 1985). A saprophytic fungus which may be an opportunistic pathogen on roots, etc. The fungus produces the same anthraquinones as *Phoma foveata* Foister (section *Phyllostictoides*, **E**), namely pachybasin, chrysophanol, emodin and phomarin.

Phoma insulana (Mont.) Boerema & Malathr., Fig. 10D

> *Phoma insulana* (Mont.) Boerema & Malathr. apud Boerema *in* Versl. Meded. plziektenk. Dienst Wageningen **158** (Jaarb. 1981): 28. 1982.
> > ≡ *Phyllosticta insulana* Mont. *in* Annls Sci. nat. (Bot.) IV, **5**: 343. 1856.

Diagn. data: Boerema (1982a) (identification history), de Gruyter *et al.* (1993) (figs 18, 30: shape of conidia and chlamydospores, reproduced in present Fig. 10D; cultural descriptions, partly adopted below).

Description in vitro

Pycnidia globose with 1 to several (–5) ostioles, non-papillate or with a distinctly papillate neck, mostly 140–270 μm diam., but sometimes reaching up to 470 μm diam., glabrous, solitary, but in centre of colony often confluent. Conidial matrix whitish.

Conidia ellipsoidal with 2–3(–4), small or large, usually polar guttules, mostly 3–5.5(–7) × 1.5–2.5 μm.

Chlamydospores unicellular, intercalary or terminal, solitary or in chains, globose, olivaceous with green guttules, mostly 5.5–10.5 μm diam.

Colonies on OA *c.* 7 cm diam. after 7 days, regular, olivaceous with colourless concentric ring near margin, nearly without any aerial mycelium; reverse olivaceous with greenish olivaceous outer margin. Pycnidia scattered on or partly in the agar. Representative culture CBS 252.92.

Ecology and distribution

In southern Europe this fungus in autumn commonly occurs on discolouring leaves and ripening fruits of olive, *Olea europaea* (Oleaceae). In Crete it appeared to be one of the fungi involved in the natural processing of olive fruits (such fruits are greatly estimated by local people as table fruits). In the past the fungus sometimes has been confused with the quite different olive pathogen *Phoma oleae* (DC.) Sacc. ≡ *Coleophoma oleae* (DC.) Petr. & Syd.

Phoma labilis Sacc., Fig. 10E

> *Phoma labilis* Sacc. *in* Michelia **2**(2): 341. 1881, with reference to description *in* Michelia **1**(2): 258. 1878 [sub '*Phoma malvacearum* Westend.'

(misapplied; ≡ *Phomopsis malvacearum* (Westend.) Grove)]; *in* Sylloge Fung. **3**: 122. 1884.

> = *Plenodomus brachysporus* Petr. *in* Annls mycol. **21**: 197[–199]. 1923.

Diagn. data: de Gruyter *et al.* (1993) (fig. 21: conidial shape, reproduced in present Fig. 10E; cultural descriptions, partly adopted below), Boerema *et al.* (1996: 152–153) (fig. 2: conidia and conidiogenous layer; nomenclature and synonymy).

Description in vitro

Pycnidia globose, usually with 1 distinctly papillate ostiole, 70–250 μm diam., glabrous, solitary or confluent. Conidial matrix whitish, pale luteous or ochraceous.

Conidia oblong to ellipsoidal with 2 (or more) small polar guttules, mostly 4–6.5 × 2–3 μm.

Colonies on OA *c.* 6–6.5 cm diam. after 7 days, grey olivaceous to olivaceous, with scanty whitish to grey olivaceous aerial mycelium; reverse also more or less grey olivaceous. The pycnidia develop abundantly on and in the agar, but sterile sectors often occur. In the agar of the sterile sectors a yellow pigment has been observed. Addition of a drop of NaOH results in a reddish/brown discolouring. Representative culture CBS 124.93.

Ecology and distribution

Isolates of this fungus have been obtained from soil and various different herbaceous and woody plants in southern Eurasia (Italy, Turkey, Israel, Kuwait, Indonesia). It seems to be a warmth-preferring fungus, which in The Netherlands has been recorded in glasshouses. It is apparently a necrophyte which tends to loose the ability to produce pycnidia quite quickly. Most isolates have not survived the freeze drying treatment.

Phoma lini Pass., Fig. 10F

> *Phoma lini* Pass., Diagn. Funghi nuov. **4**: no. 81. 1890.

Diagn. data: de Gruyter *et al.* (1993) (fig. 8: conidial shape, reproduced in present Fig. 10F; cultural descriptions, partly adopted below).

Description in vitro

Pycnidia globose to irregularly shaped with 1 distinct non-papillate ostiole, 60–210 μm diam., glabrous, solitary or confluent. Conidial matrix sienna.

Conidia ellipsoidal, occasionally with one or two polar guttules, mostly 3.5–5.5 × 1.5–2 μm.

Colonies on OA *c.* 6.5 cm diam. after 7 days, regular, colourless or very pale greenish olivaceous without any aerial mycelium or with very scanty colourless aerial mycelium; reverse also colourless. On MA the agar staining yellow due to the release of a verdigris pigment, which on addition of a drop of NaOH changes to orange, later purplish. Representative culture CBS 253.92.

Ecology and distribution

A saprophyte in Europe frequently recorded on dead stems of *Linum* spp. (Linaceae), but probably also occurring on other herbaceous plants. Occasional isolates of the fungus from stagnant water (The Netherlands, Yugoslavia) were associated with experiments on, or the practice of retting of flax, *Linum usitatissimum*. The fungus has been repeatedly confused with a foot rot pathogen of flax, *Phoma exigua* var. *linicola* (Naumov & Vassiljevsky) P.W.T. Maas (sect. *Phyllostictoides*, **E**: producing always some septate conidia), see Maas (1965).

Phoma loticola Died., Fig. 10G

> *Phoma loticola* Died. in Krypt.-Flora Mark Brandenb. 9, Pilze **7** [Heft 1]:152. 1912 [vol. dated '1915'].
> > = *Phoma herbarum* var. *loti-cretici* Nann. in Atti Accad. Fisiocr. Siena X, **2**: 11. 1927.
> > = *Phoma lotivora* P.R. Johnst. in N.Z. Jl Bot. **19**: 178. 1981.

Diagn. data: Boerema (1970) (occurrence in Europe), Johnston (1981) (tab. 1: cultural and microscopic characteristics sub *Ph. lotivora*; fig. 7: cultures on OA and MA; description *in vitro*; occurrence in New Zealand), de Gruyter *et al.* (1998) (fig. 9: conidial shape, reproduced in present Fig. 10G; cultural descriptions, partly adopted below).

Description in vitro

Pycnidia globose to irregular with 1(–2) non-papillate or slightly papillate ostioles, 110–250 μm diam., glabrous, solitary. Conidial matrix white to pale buff.

Conidia allantoid to subcylindrical, eguttulate or with 1–2, small polar guttules, 4–7.5(–9) × 1.5–2(–2.5) μm.

Colonies on OA 6–8 cm diam. after 7 days, regular, colourless or with olivaceous grey spots, with finely floccose, white aerial mycelium; reverse also colourless or with olivaceous grey spots. Pycnidia abundant, both on and in the agar; after 3 weeks the colony develops a saffron colour due to ripe pycnidia. The strains tested showed on application of a drop of NaOH the green → red discoloration (E$^+$ reaction). Representative cultures CBS 562.81, CBS 628.97.

Ecology and distribution

On account of a deviating old Dutch *Phoma*-isolate from *Lotus* sp. (Leguminosae), the 'European' *Ph. loticola* has been considered to be different from *Ph. lotivora* found in New Zealand on *Lotus* spp. However, additional comparative study did not yield essential differences between both species. The fungus appears as a specific pathogen: stem and leaf spots.

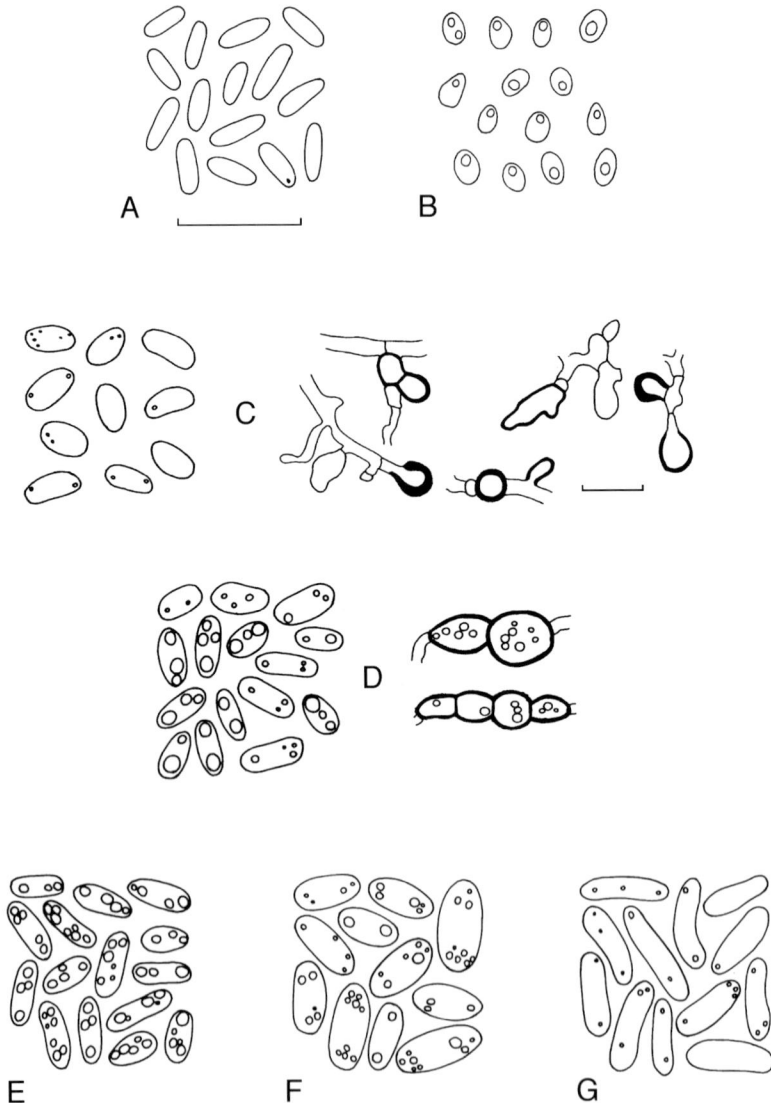

Fig. 11. Conidial variation, shape of some hyphal cells and chlamydospores. **A** *Phoma lupini.* **B** *Phoma minutispora.* **C** *Phoma multipora.* **D** *Phoma multirostrata* vars. **E** *Phoma nebulosa.* **F** *Phoma negriana.* **G** *Phoma nigricans.* Bars 10 μm.

Phoma lupini Ellis & Everh., Fig. 11A

> *Phoma lupini* Ellis & Everh. *in* Bull. Washburn [Coll.] Lab. nat. Hist. **1**: 6.
> 1884; not *Phoma lupini* N.F. Buchw. *in* Møller, Fungi Faeröes **2**: 153. 1958
> [possibly identical with *Phoma pedicularis* Fuckel, sect. *Plenodomus*, **G**].
> > ≡ *Sphaeropsis lupini* (Ellis & Everh.) Kuntze, Revisio Gen. Pl.
> > **3**(2): 526. 1898.
> > *H* ≡ *Stictochorella lupini* (Ellis & Everh.) Syd. apud Sydow &
> > Petrak *in* Annls mycol. **22**: 397. 1924; later homonym, see
> > below.
> > ≡ *Asteromella lupini* (Ellis & Everh.) Petr. *in* Sydowia **9**: 495.
> > 1955.
> > = *Phyllosticta ferax* Ellis & Everh. *in* Proc. Acad. Phil. **1894**:
> > 355. 1894.
> > = *Phaeosphaeria lupini* Syd. & P. Syd. *in* Annls mycol. **2**:
> > 172. 1904 [cf. Sydow sub *Stictochorella lupini*, see below].
> > = *Stictochorella lupini* Syd. apud Sydow & Petrak *in* Annls
> > mycol. **20**: 202. 1922.
> > = *Phyllosticta lupini* Bonar *in* Mycologia **20**: 297. 1928.

Diagn. data: Boerema & Dorenbosch (1980) (identification and history), Frey
& Yabar (1983) (figs 14, 15: colour plate disease symptoms on *L. mutabilis*,
Peru), Salas (1987) (pathogenicity to *L. mutabilis*), Paulitz & Cote (1991) (disease symptoms on cultivated *L. albus*, Canada), de Gruyter *et al.* (1993) (fig.
2: shape of conidia, reproduced in present Fig. 11A; cultural descriptions,
partly adopted below), van der Aa & Vanev (2002) (*Phyllosticta* synonyms).

Description in vitro

Pycnidia globose to irregular with 1(–3) non-papillate or slightly papillate ostioles, 40–125 μm diam., glabrous, solitary or confluent. Conidial matrix rosy
vinaceous to pale vinaceous.

    Conidia ellipsoidal, mostly eguttulate, but sometimes with 1(2) inconspicu-
ous guttules, often 4–4.5 × 1.5–2 μm.

    Colonies on OA *c.* 2.5–3 cm diam. after 7 days (5–6 cm after 14 days),
regular, dull green to dark herbage green, with floccose white to grey or some
dull green aerial mycelium; reverse dull green with olivaceous or dark herbage
green centre [on MA reverse concentrically zonated sepia and greyish blue,
outer margin hazel to buff]. Pycnidia on and in the agar, often also in aerial
mycelium. Representative culture CBS 248.92.

Ecology and distribution

A seed-borne pathogen known from various *Lupinus* spp. (Leguminosae)
indigenous to North and South America. On account of its small conidia this
pathogen has erroneously been interpreted as only a spermatial state. The fungus may infect all above ground parts of the lupins: leaf, stem and pod spot. In

the Andean highlands of South America, above 3800 m, this appeared to be the most common disease of *Lupinus mutabilis*, an important albuminous food crop of the Indians. In North and South America similar disease symptoms on lupins also may be associated with *Stagonosporopsis lupini* (Boerema & R. Schneid.) Boerema *et al.*, synanamorph of *Phoma schneiderae* Boerema *et al.*, sect. *Heterospora* (**B**).

Phoma minutispora P.N. Mathur, Fig. 11B

> *Phoma minutispora* P.N. Mathur apud de Gruyter & Noordeloos *in* Persoonia **15**(1): 75.1992 [as 'collection name' originally also referred to Thirumalachar; = depositor].
>> *H* ≡ *Phoma oryzae* Cooke & Massee *in* Grevillea **16**: 15. 1887; not *Phoma oryzae* Catt. *in* Arch. Bot. crittog. Pavia **2–3**: 118. 1877.
>> ≡ *Phyllosticta oryzae* (Cooke & Massee) I. Miyake *in* J. Coll. Agric. imp. Univ. Tokyo **2**(4): 252. 1910.

Diagn. data: Padwick (1950) (occurrence on rice), Shukla *et al.* (1984) (association with human skin lesions), de Gruyter & Noordeloos (1992) (fig. 1: shape of conidia, reproduced in present Fig. 11B; cultural descriptions, partly adopted below).

Description in vitro

Pycnidia subglobose, with 1–4 indistinct ostioles, 80–200 μm diam., glabrous, solitary. Conidial matrix salmony.

Conidia subglobose to broadly ellipsoidal, usually with one large guttule, 2–2.5(–3) × 1.5–2 μm.

Chlamydospores subglobose or irregular, 6–15 μm diam., are occasionally found in old cultures, solitary and usually formed at tips of hyphae.

Colonies on OA *c.* 2 cm diam. after 7 days [3–3.5 cm after 14 days], more or less regular, predominantly grey olivaceous with abundant greyish floccose aerial mycelium; reverse olivaceous black or grey to vinaceous buff. On MA somewhat faster growing, *c.* 2.5 cm, pinkish red (flesh to peach) with thin velvety aerial mycelium; reverse apricot with darker sectors. Pycnidia in centre of colony and in concentric zones, both on and in the agar and in aerial mycelium. Representative cultures CBS 711.76, CBS 509.91.

Ecology and distribution

Probably a common saprophytic soil fungus in south-west Asia, characterized by its extremely small conidia. First found in rice straw, *Oryza sativa* (Gramineae). Isolates made in India refer to, for example, dead leaf-sheaths of rice, bark of *Ficus* sp. (Moraceae) and soil. The fungus is also reported as an opportunistic human pathogen.

Phoma multipora V.H. Pawar *et al.,* Fig. 11C, Colour Plate I-I

> *Phoma multipora* V.H. Pawar, P.N. Mathur & Thirum. *in* Trans. Br. mycol.
> Soc. **50**: 260. 1967 [first introduced as nomen nudum by Pawar &
> Thirumalachar *in* Nova Hedwigia **12**: 501. 1966].

Diagn. data: Pawar *et al.* 1967 (fig. 1A–D: pycnidial and conidial morphology;
cultural characters), Boerema (1985) (fig. 9: characterization as soil-*Phoma*
with 'setose' pycnidia), de Gruyter & Noordeloos (1992) (figs 9, 21: conidial
shape and swollen mycelial cells, reproduced in present Fig. 11C; cultural
descriptions, partly adopted below).

Description in vitro

Pycnidia globose, usually with one or two wide ostioles (only incidentally
'multi'-ostiolate pycnidia occur), 120–300 μm in diam., glabrous, but the ostioles sometimes surrounded by short hyphae swollen at the tips ('setose'
appearance), solitary or confluent in groups of 2–5 pycnidia. Conidial matrix
rosy buff or vinaceous.
    Conidia broadly ellipsoidal, eguttulate or with a few small polar guttules,
mostly 3.5–5 × 1.5–2.5 μm.
    Colonies on OA *c.* 2–2.5 cm diam. after 7 days, regular with weakly undu-
lating outline, with scarce, more or less felted white to olivaceous grey aerial
mycelium, colony characterized by apricot to scarlet pigment; reverse similar.
With the NaOH test the colour quickly changes to purplish blue. In the agar
small orange/yellow crystals may be present. Pycnidia scattered in and on the
agar. In the mycelium somewhat pigmented swollen cells ('pseudochlamydospores') may occur. Representative cultures CBS 353.65 (type), CBS 501.91.

Ecology and distribution

This fungus is recorded in south-west Asia (India) and northern Africa (Egypt).
Probably a halophilic soil-borne saprophyte in those subtropical regions. The
reports from India refer to soil in mangrove vegetations near the Bombay coast
and dead herbaceous stems near Lahore. Rai (1998) erroneously treated this
species as identical with *Phoma capitulum* V.H. Pawar *et al.* (this section).

Phoma multirostrata (P.N. Mathur *et al.*) Dorenb. & Boerema var. *multirostrata,* Fig. 11D

> *Phoma multirostrata* (P.N. Mathur *et al.*) Dorenb. & Boerema *in* Mycopath.
> Mycol. appl. **50**: 255. 1973, var. *multirostrata.*
>   *V* ≡ *Sphaeronaema multirostratum* P.N. Mathur, S.K. Menon &
>   Thirum. apud Mathur & Thirumalachar *in* Sydowia **13**:
>   146. 1959 [as '*Sphaeronema multirostrata*'].
>   = *Phoma ushtrina* T.R.N. Rai & J.K. Misra *in* Curr. Sci. **50**:
>   377. 1981.

Diagn. data: Mathur & Thirumalachar (1959) (original description *in vitro* from soil in India), Dorenbosch & Boerema (1973) (similarity with isolates from leaves, stems and roots of different plants grown in warm climates or under greenhouse conditions), Sutton (1980) (fig. 228I: drawing of conidia; the specimens listed probably refer to var. *macrospora*, see below), Boerema (1985) (tab. 2: synonymy of soil isolates *Ph. multirostrata* s.l.), Boerema (1986b) (fig. 7: variability in conidial dimensions; varietal differentiation, synonymy), Morgan-Jones (1988b) (nomenclatural history; figures and description of isolates conform to var. *macrospora*, see below), de Gruyter *et al.* (1993) (figs 20, 32: shape of conidia and chlamydospores, reproduced in present Fig. 11D; degeneration *in vitro*; collective cultural characters, partly adopted below), Rai (1998) (effect of different physical and nutritional conditions on the morphology and cultural characters).

Description in vitro

Pycnidia globose to subglobose or irregular, in this type variety (soil isolates) frequently with some variously shaped necks (up to 240 μm) and several ostioles, where not located at the tips of necks, papillate or non-papillate, always relatively large, usually more than 550 μm diam., glabrous, solitary and confluent. Conidial matrix whitish to cream or buff coloured.

Conidia oblong to ellipsoidal, sometimes eguttulate but usually with 2–3 small or large polar guttules, variable in dimensions, mostly 4.5–6.5 × 2–2.5 μm.

Chlamydospores (common in older cultures) oblong to ellipsoidal, in chains or clustered, often intercalary, but also terminal, olivaceous with green guttules, 5–15 μm diam.

Colonies on OA 6.5–7 cm diam. after 7 days, regular, colourless to weak olivaceous, with poorly developed, felted, white to grey olivaceous aerial mycelium or without any; reverse olivaceous. Colonies on MA olivaceous to olivaceous buff with felty to floccose or woolly aerial mycelium, attaining a diameter of *c.* 8 cm; reverse leaden grey to olivaceous black. Original type culture CBS 274.60.

Ecology and distribution

The above synonyms of this worldwide recorded thermotolerant fungus (still growth at 35°C) refer to soil isolates made in India. Numerous other isolates of the fungus in India made from all kinds of plants and occasionally also from soil differ by smaller pycnidia (mostly 150–300 μm diam.) and display a much wider range in conidial dimensions (compare Dorenbosch & Höweler (1968) sub '*Ph. liliana*' and Boerema (1986). This has led to the differentiation of the vars *macrospora* and *microspora*, see below. However, intermediate variants also commonly occur; the pH of the substrata (media) appeared to have much influence on the size of the conidia in this species (Rai, 1998).

Phoma multirostrata var. *macrospora* Boerema

> *Phoma multirostrata* var. *macrospora* Boerema *in* Versl. Meded. plziektenk. Dienst Wageningen **164** (Jaarb. 1985): 29. 1986.
>
> > *V = Sphaeronaema indicum* P.N. Mathur, S.K. Menon & Thirum. *in* Sydowia **13**: 146a. 1959 [as '*Sphaeronema indica*'].
> >
> > = *Phoma lucknowensis* R.K. Saksena, Nand & A.K. Sarbhoy *in* Mycopath. Mycol. appl. **34**: 93. 1968; not *Phoma lucknowensis* S.C. Agarwal & J.K. Misra *in* Curr. Sci. **50**: 66. 1981 [= *Phoma multirostrata* var. *microspora* (Allesch.) Boerema].
> >
> > *H =* *Phoma terrestris* R.K. Saksena, Nand & A.K. Sarbhoy *in* Mycopath. Mycol. appl. **29**: 86. 1966 [as '*terrestre*']; not *Phoma terrestris* H.N. Hansen *in* Phytopathology **19**: 699. 1929 [sect. *Paraphoma*, **C**].
> >
> > = *Phoma polyanthis* Died. apud Sydow, Sydow & Butler *in* Annls mycol. **14**: 186. 1916.
> >
> > = *Phoma pardanthi* Died. apud Sydow, Sydow & Butler *in* Annls mycol. **14**: 186. 1916.
> >
> > = *Phoma garflorida* S. Chandra & Tandon *in* Curr. Sci. **34**: 258. 1965.
> >
> > = *Phoma ehretiae* S. Chandra & Tandon *in* Mycopath. Mycol. appl. **29**: 275–276. 1966.
> >
> > = *Phyllosticta gerbericola* Bat. *in* Bolm Secr. Agric. Ind. Com. Est. Pernambuco **19**: 30. 1952.
> >
> > = *Phyllosticta gladioloides* Bat. *in* Bolm Secr. Agric. Ind. Com. Est. Pernambuco **19**: 32. 1952.
> >
> > = *Ascochyta nyctanthis* V.P. Sahni *in* Mycopath. Mycol. appl. **36**: 278. 1968 [globules suggested 'septum': artefact].
> >
> > Θ = '*Phoma multirostrata* var. *macrospora* P.N. Mathur & Thirum.' [in 1965 deposited in CBS-collection, but never published: 'collection name'].

Diagn. data: Dorenbosch & Höweler (1968) (isolates from different plants conform to both, var. *macrospora* and var. *microspora*; fig. 2a–d: morphology of pycnidia, chlamydospores and conidia; total range of conidial dimensions), Patil & Rao (1974) (pathogenicity and cultural characteristics sub syn. *Phyllosticta gerberidicola*), Boerema (1986b) (fig. 7, tab. 2: variability in conidial dimensions and host range; varietal differentiation and synonymy), Morgan-Jones (1988b) (figs 1A–G, 2A–D: drawings of pycnidia, ostiole, conidiogenous cells, conidia, wall cells, hyphae and chlamydospores; pls 1A–G, 2A–D: photographs of agar plate cultures, pycnidia, hyphal strands, wall cells and chlamydospores; history, cultural and morphological description; occurrence on cysts of *Heterodera glycines*), Rai & Rajak (1993) (cultural characteristics), de Gruyter *et al.* (1993) (figs 20, 32: collective characters of pycnidia, conidia and chlamydospores; degeneration *in vitro*), Rai (1998) (descriptions of several isolates from India mainly conform to var. *macrospora*; effect of different physical and nutritional conditions on morphology and cultural characters).

Description in vitro

Pycnidia quite variable in size and shape, frequently rostrate and with a plural-
ity of ostioles, resembling the pycnidia of the type var. *multirostrata*, but explicitly smaller, mostly 150–300 μm, glabrous, solitary or confluent. Conidial
matrix usually rosy buff.

Conidia in comparison with those of the type var. relatively large, mostly
5–7.5(–8.5) × 2–2.5(–3) μm.

Chlamydospores like those of the type var., usually readily produced in old
colonies on OA.

Colonies: the cultural characteristics of all strains of *Phoma multirostrata*
are very similar and therefore not usable for varietal differentiation. Original
type-culture CBS 368.65.

Ecology and distribution

Most records of this relatively large spored variety are from India and refer to
leaf spots. Inoculation experiments with different isolates have shown that it can
be characterized as a plurivorous soil-borne opportunistic plant pathogen,
which may also cause blight and dieback symptoms. The variety is repeatedly
found in association with cysts of phytonematodes and probably worldwide
distributed in subtropical regions. *In vitro* the fungus quite quickly tends to
degenerate. Intermediate forms, between var. *macrospora* and the variety
microspora listed below frequently occur.

Phoma multirostrata var. *microspora* (Allesch.) Boerema

> *Phoma multirostrata* var. *microspora* (Allesch.) Boerema *in* Versl. Meded.
> plziektenk. Dienst Wageningen **164** (Jaarb. 1985): 30. 1986.
>> ≡ *Phoma decorticans* var. *microspora* Allesch. *in* Rabenh.
>> Krypt.-Flora [ed. 2], Pilze **6** [Lief. 63]: 284. 1898 [vol.
>> dated '1901'].
>> = *Phyllosticta ambrosioides* Thüm. *in* Instituto, Coimbra **28**
>> sub Contr. Fl. myc. Lusit. III n. 45. 1881 ['1880–81'].
>> = *Phyllosticta geranii* Ellis & Everh. *in* J. Mycol. **3**: 130. 1887.
>> = *Phyllosticta plumierae* Tassi *in* Bull. Lab. Ort. Siena **4**: 8.
>> 1901.
>> = *Phoma liliana* S. Chandra & Tandon *in* Curr. Sci. **31**: 566.
>> 1965.
>> H = *Phoma microspora* P. Balas. & Narayanas. *in* Indian
>> Phytopath. **33**: 136. 1980; not *Phoma microspora* Sacc.
>> apud Roum. *in* Revue Mycol. **1885**: 158. 1885; not
>> *Phoma microspora* Pat. *in* Hariot, Champ. Cap Horn 196.
>> 1889.
>> = *Phoma ambrosioides* Kamal & P. Kumar *in* Envir. India
>> **3**(1&2): 1. 1980.
>> = *Phoma jaunpurensis* R.C. Srivast. *in* Zentbl. Bakt. II, **136**:
>> 268. 1981.

> H = *Phoma lucknowensis* S.C. Agarwal & J.K. Misra *in* Curr. Sci.
> **50**: 66. 1981 [pycnidial primordia misinterpreted as 'dicty-
> ochlamydospores']; not *Phoma lucknowensis* R.K. Saksena,
> Nand & A.K. Sarbhoy *in* Mycopath. Mycol. appl. **34**: 93.
> 1968 [= *Phoma multirostrata* var. *macrospora* Boerema].
> Θ = '*Sphaeronaema roseum* P.N. Mathur & Thirum.' [in 1965
> deposited in CBS-collection, but never published, 'collec-
> tion-name'].

Diagn. data: Dorenbosch & Höweler (1968) (sub syn. *Ph. liliana* but including
also var. *macrospora*; fig. 2a–d: morphology of pycnidia; chlamydospores and
conidia), Balasubramanian & Narayanasany (1980) (sub syn. *Ph. microspora*:
morphology of pycnidia, chlamydospores and conidia of isolate from leaf spots
of groundnut, *Arachis hypogaea*), Boerema (1986b) (fig. 7, tab. 2: variability in
conidial dimensions and host range; varietal differentiation and synonymy;
pathogenicity; confusion with *Phoma tracheiphila*), de Gruyter *et al.* (1993) (figs
20, 32: collective characters of pycnidia, conidia and chlamydospores; degeneration *in vitro*), Boerema *et al.* (1996: 149) (confusion with *Phoma lingam*).

Description in vitro

Pycnidia in shape and size similar to those of var. *macrospora*, i.e. with respect
to the type var. small, mostly 150–300 μm, glabrous, solitary or confluent.
Conidial matrix usually rosy buff.
 Conidia mostly (3.5–)4–5(–5.5) × 1.5–2.5 μm, thus relatively small, espe-
cially in comparison with those of var. *macrospora*.
 Chlamydospores similar to those of var. *macrospora* and the type var.
 Colonies on agar media resemble very much those of var. *macrospora* and
the type var. Representative culture CBS 110.79.

Ecology and distribution

This variety with relatively small conidia is just like the var. *macrospora* most
frequently recorded in India. It also can be characterized as a plurivorous soil-
borne opportunistic plant pathogen, which often occurs in association with leaf
spots, but also may cause blight and dieback symptoms. The fungus is probably worldwide distributed in subtropical regions, but also recorded in greenhouses, all over the world. Intermediate forms between the vars *microspora*
and *macrospora* frequently occur.

Phoma nebulosa (Pers. : Fr.) Berk., Fig. 11E, Colour Plate I-J

> *Phoma nebulosa* (Pers. : Fr.) Berk., Outl. Br. Fung. 314. 1860 [as *Ph.nebu-
> losa* 'Mont.'].
>> ≡ *Sphaeria nebulosa* Pers., Obs. mycol. **2**: 69. 1799; Syn.
>> meth. Fung. 31. 1801.
>> : Fr., Syst. mycol. **2** [Sect. 2]: 430. 1823.

V ≡ *Exormatostoma nebulosa* (Pers. : Fr.) Gray, Nat. Arr. Br. Pl.
1: 522. 1821 [as '*nebulosum*'].
≡ *Sphaeropsis nebulosa* (Pers. : Fr.) Fr., Summa Veg. Scand.
2 [Sectio posterior]: 419. 1849.
≡ *Sphaerella nebulosa* (Pers. : Fr.) Sacc. *in* Michelia **2**(1): 56.
1880 [misapplied].
≡ *Mycosphaerella nebulosa* (Pers. : Fr.) Johanson ex Oudem.,
Rev. champ. Pays-bas **2**: 213. 1897 [misapplied].
= *Leptophoma paeoniae* Höhn. *in* Sber. Akad. Wiss. Wien
[Math.-naturw. Kl., Abt. I] **124**: 75. 1915.
V Θ= *Sphaeronaema paeoniae* Höhn. apud Strasser *in* Verh.
zoöl.-bot. Ges. Wien **60**: 312. 1912 [as '*Sphaeronema*';
without description].

Diagn. data: Boerema (1976) (fig. 12: cultures on OA; tabs 8, 9: isolate sources and synoptic table on cultural characters with drawing of conidia; nomenclator and history), Sutton (1980) (fig. 229A: drawing of conidia; specimens in IMI collection), de Gruyter *et al.* (1993) (fig. 24: conidial shape, reproduced in present Fig. 11E; cultural descriptions, partly adopted below), Boerema *et al.* (1996: 174) (additional synonyms).

Description in vitro

Pycnidia subglobose with 1–2 ostioles, often with distinct necks, especially when submerged, 100–250 μm diam., glabrous, solitary or confluent. Micropycnidia also occur, usually submerged, 45–80 μm diam. Conidial matrix white to buff.

Conidia oblong to ellipsoidal with 2–many large, polar-oriented guttules, mostly 3.5–6.5 × 1.5–2 μm.

Colonies on OA *c.* 4.5–5.5 cm diam. after 7 days, regular, grey olivaceous at centre, towards margin colourless to citrine or greenish, without any aerial mycelium, or, mainly in central part, with rather coarsely floccose white aerial mycelium; reverse also grey olivaceous at centre and towards margin colourless to citrine or greenish. Pycnidia abundant, mostly on the agar in greyish concentric zones [colonies on MA colourless to grey olivaceous with citrine pigment production in the agar; at centre as agglutinated gelatinous clots; NaOH spot test negative]. Representative culture CBS 503.75.

Ecology and distribution

In Europe a common soil-borne saprophyte, occurring on dead stems of various herbaceous plants, particularly *Urtica* (Urticaceae) and *Scrophularia* spp. (Scrophulariaceae) (cf. isolates from Austria, Belgium, Germany, the UK, and The Netherlands). The fungus has also been recorded in New Zealand, North America (USA) and India. The history of this species is a concatenation of different interpretations. The present *Phoma*-concept is fixed by a lectotype (L 910.269–51).

Phoma negriana Thüm., Fig. 11F

V   *Phoma negriana* Thüm., Pilze Weinst. 185. 1878 [as 'negrianum'].
    ≡ *Phyllosticta negriana* (Thüm.) Allesch. *in* Rabenh. Krypt.-
       Flora [ed. 2], Pilze **6** [Lief. 60]: 98. 1898 [vol. dated
       '1901'].
    = *Phyllosticta vitis* Sacc. *in* Michelia **1**(2): 135. 1878; not
       Phoma vitis Bonord. *in* Abh. naturforsch. Ges. Halle **8**: 14.
       1864.

Diagn. data: Boerema & Dorenbosch (1979) (identification, history), de
Gruyter *et al.* (1998) (fig. 16: conidial shape, reproduced in present Fig. 11F;
cultural descriptions, partly adopted below).

Description in vitro

Pycnidia globose or irregular with 1–2(–4) papillate ostioles, 70–220 μm diam.,
glabrous, solitary or confluent. Conidial matrix saffron to pale vinaceous.
    Conidia broadly ellipsoidal to oblong with several, small and large, distinct
polar guttules, 4.5–8.5(–10.5) × 2–4 μm.
    Colonies on OA 3.5–4.5 cm diam. after 7 days [on MA 2.5–3.5 cm], regu-
lar to irregular, greenish olivaceous with colourless margin at first, later greenish
due to the development of pycnidia, with scanty woolly, pale olivaceous grey
aerial mycelium; reverse greenish olivaceous. Pycnidia abundant, in concentric
rings mainly on the agar, sometimes (partly) in the agar. Additions of a drop of
NaOH on MA results in a weak reddish/brown discoloration. Representative
culture CBS 358.71.

Ecology and distribution

A common opportunistic pathogen of vine, *Vitis vinifera* (Vitaceae), in southern
Europe. It may be associated with necroses on leaves, fruits or stems. On stems
it has often been misidentified as *Phoma viticola* Sacc., a name referring to a
quite different pathogen of vine: *Phomopsis viticola* (Sacc.) Sacc., believed by
some workers to be the cause of dead arm disease.

Phoma nigricans P.R. Johnst. & Boerema, Fig. 11G

Phoma nigricans P.R. Johnst. & Boerema *in* N.Z. Jl Bot. **19**: 394. 1981.

Diagn. data: Johnston & Boerema (1981) (figs 1, 3: cultures on OA and MA,
drawing of conidia; original description *in vitro*, New Zealand), de Gruyter *et
al.* (1998) (fig. 11: conidial shape, reproduced in present Fig. 11G; cultural
descriptions, partly adopted below).

Description in vitro

Pycnidia globose to irregular with 1 papillate ostiole, 80–310 μm diam.,
glabrous, solitary to confluent. Conidial matrix white to saffron.

Conidia allantoid to subcylindrical, eguttulate or with 1–3 small polar guttules, (3.5–)4.5–8(–10) × (1.5–)2–3(–4) μm.

Colonies on OA 7–8 cm diam. after 7 days, regular, grey olivaceous, rosy buff
at margin, with finely floccose, white to pale olivaceous grey aerial mycelium;
reverse also grey olivaceous and rosy buff at margin. Pycnidia abundant, mostly
on, but also immersed in the agar. Representative culture CBS 444.81.

Ecology and distribution

Isolated from various woody and herbaceous plants in New Zealand apparently
as an ubiquitous saprophyte; probably also common in Australia. The epithet
refers to the conspicuous mature black pycnidia.

Phoma obtusa Fuckel, Fig. 12A

> *Phoma obtusa* Fuckel *in* Jb. nassau. Ver. Naturk. **23–24** [= Symb.
> mycol.]: 378. 1870 ['1869 und 1870'].
> > = *Phoma carotae* Died. *in* Krypt.-Flora Mark Brandenb. 9,
> > Pilze **7** [Heft 1]: 136. 1912 [vol. dated '1915'].

Diagn. data: de Gruyter *et al.* (1998) (fig. 5: conidial shape, reproduced in present Fig. 12A; cultural descriptions, partly adopted below).

Description in vitro

Pycnidia globose with 1 non-papillate ostiole, 50–180 μm diam., covered with
mycelial outgrowths, solitary. Conidial matrix white to dirty white.

Conidia ellipsoidal, ovoid, usually with 1–5 small polar guttules,
(3.5–)5–8.5(–10.5) × 2–4 μm.

Colonies on OA 6–7 cm after 7 days, regular, colourless to rosy buff, with
or without scarce, finely floccose, white aerial mycelium; reverse also colourless
to rosy buff. On MA colony olivaceous buff by a diffusible pigment, discolouring greenish-yellow with NaOH. Pycnidia abundant, on and in the agar.
Representative cultures CBS 391.93 and CBS 377.93.

Ecology and distribution

In Europe this fungus is repeatedly recorded on dead stems of carrots, *Daucus
carota* (Umbelliferae). Occasionally it has also been isolated from necrotic tissue
of other herbaceous plants.

Phoma opuntiae Boerema *et al.*, Fig. 12B

> *Phoma opuntiae* Boerema, Gruyter & Noordel. *in* Persoonia **16**(1):
> 131.1995 [not *Phoma opuntiae* Ellis apud Baker *in* Bull. S. Calif. Acad. **4**:
> 57.1905, nomen nudum].

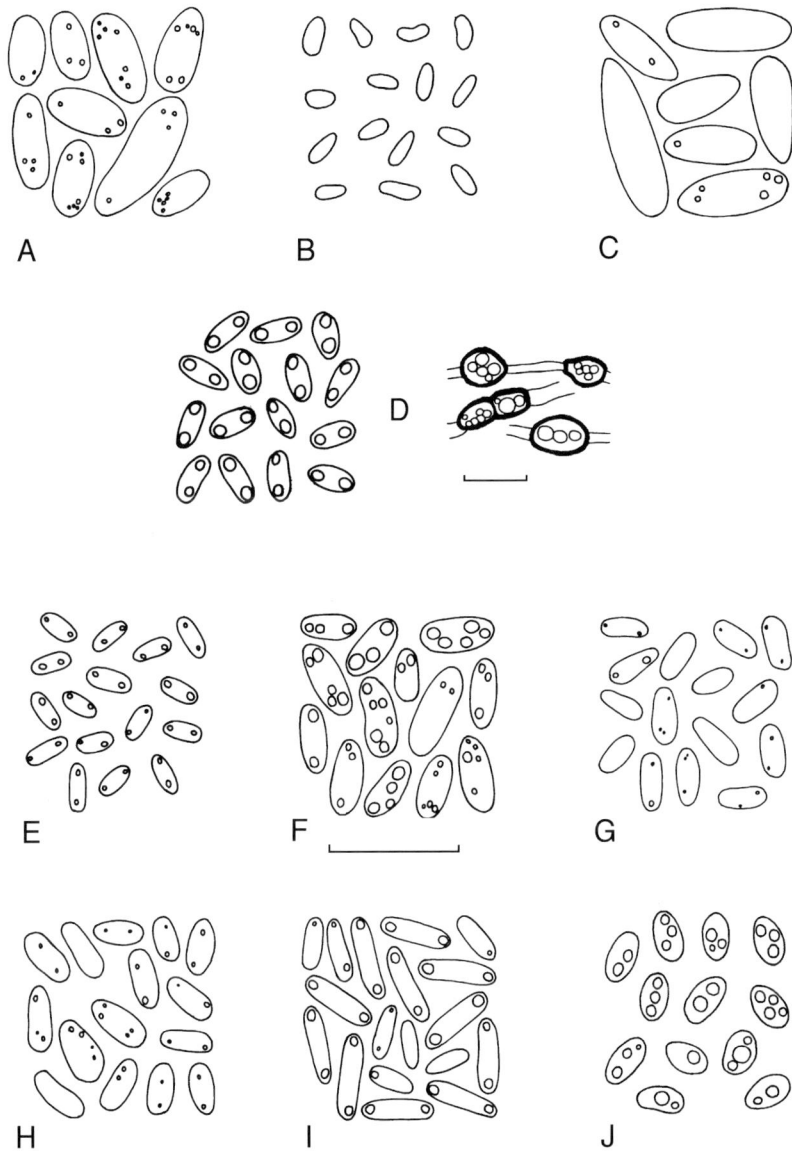

Fig. 12. Variation of conidia and some hyphal chlamydospores. **A** *Phoma obtusa*. **B** *Phoma opuntiae*. **C** *Phoma paspali*. **D** *Phoma pereupyrena*. **E** *Phoma piperis*. **F** *Phoma plurivora*. **G** *Phoma poolensis* var. *poolensis*. **H** *Phoma poolensis* var. *verbascicola*. **I** *Phoma pratorum*. **J** *Phoma putaminum*. Bars 10 μm.

V H ≡ *Phoma opuntiicola* Boerema, Gruyter & Noordel. apud de Gruyter & Noordeloos *in* Persoonia **15**(1): 77, 78. 1992 [as '*opunticola*']; not *Phoma opuntiicola* Speg. *in* An. Mus. nac. Hist. nat. B. Aires **6** [= IIa, **3**] [Fg. Arg. novi v. crit.]: 316. 1899.

Diagn. data: de Gruyter & Noordeloos (1992) (fig. 3: shape of conidia, reproduced in present Fig. 12B; cultural descriptions, partly adopted below),
Boerema *et al.* (1995) (nomenclature).

Description in vitro

Pycnidia globose or depressed globose with long neck, 40–150 μm diam.,
glabrous or somewhat hairy, mostly solitary. Conidial matrix whitish to pale olivaceous grey.

Conidia irregular, ellipsoidal to subcylindrical, eguttulate, 2.5–3.5 ×
1–1.5 μm.

Colonies on OA *c.* 3 cm diam. after 7 days [on MA *c.* 4 cm], regular, without aerial mycelium, olivaceous grey to olivaceous greenish or citrine with paler
sectors; reverse similar. On MA felted white aerial mycelium, colony greenish
olivaceous. Pycnidia mostly in concentric rings and associated with radially
hyphal strands, on and partly in the agar. Addition of a drop of NaOH usually
causes a slightly sienna discoloration that slowly turns bluish green.
Representative culture CBS 376.91.

Ecology and distribution

A pathogen of *Opuntia* spp. (Cactaceae) causing dead spots on the leaf-like
stems (cladodes): necrotic spot. The fungus is indigenous to South America
(e.g. Peru), but probably also widespread distributed in other areas where the
hosts are growing. It may easily be confused with the *Asteromella*-spermatial
state of *Mycosphaerella opuntiae* (Ellis & Everh.) Dearn., commonly known as
Phyllosticta concava Seaver, see van der Aa & Vanev (2002).

Phoma paspali P.R. Johnst., Fig. 12C

Phoma paspali P.R. Johnst. *in* N.Z. Jl Bot. **19**: 181. 1981.

Diagn. data: Johnston (1981) (tab. 1: cultural and microscopic characteristics;
fig. 10: cultures on OA and MA; original description *in vitro*, New Zealand), de
Gruyter *et al.* (1998) (fig. 8, conidial shape, reproduced in present Fig. 12C;
cultural descriptions, partly adopted below).

Description in vitro

Pycnidia globose to irregularly shaped, with 1 to many papillate ostioles, later
often developing an elongated neck, 100–140 μm diam., glabrous, solitary or
confluent. Conidial matrix white to straw-coloured.

Conidia obclavate-ovoid to ellipsoidal, mostly eguttulate, but sometimes
with 1–3 small polar guttules, 5.5–8.5(–11) × 2.5–4 μm.

Colonies very fast growing on OA and MA, completely filling a Petri dish in
7 days, on OA 4–6 cm after 5 days, on MA 6–7 cm after 5 days. Growth on

OA regular to irregular, colourless with faint green tinge to grey olivaceous, with scanty to abundant floccose, white to olivaceous grey aerial mycelium; reverse dark herbage green to grey olivaceous. On MA aerial mycelium woolly and colony colourless to olivaceous buff. Pycnidia scattered, sometimes on the agar, but mostly partly submerged in the agar. The fungus showed a positive reaction with the sodium hydroxide test: on application of a drop of NaOH green → red (E$^+$ reaction). Representative culture CBS 560.81.

Ecology and distribution

Isolated from soil and various grasses (*Paspalum*, *Dactylis* and *Lolium* spp.) in New Zealand. Probably widespread on Gramineae in Australasia. Necrophyte or weak wound parasite.

Phoma pereupyrena Gruyter *et al.,* Fig. 12D

> *Phoma pereupyrena* Gruyter, Noordel. & Boerema *in* Persoonia **15**(3): 390. 1993.
>> ≡ *Polyopeus pomi* A.S. Horne *in* J. Bot., Lond. **58**: 240. 1920; not *Phoma pomi* Schulzer & Sacc. *in* Hedwigia **23**: 109. 1884 [= *Phoma macrostoma* Mont. var. *macrostoma*, sect. *Phyllostictoides*, **E**]; not *Phoma pomi* Pass. *in* Atti R. Accad. naz. Lincei Rc. **4**(2): 96. 1888 [= *Asteromella mali* (Briard) Boerema apud Boerema & Dorenbosch *in* Versl. Meded. plziektenk. Dienst Wageningen **142** (Jaarb. 1964): 149. 1965].

Diagn. data: de Gruyter *et al.* (1993) (figs 19, 31: shape of conidia and chlamydospores, reproduced in present Fig. 12D; cultural descriptions, partly adopted below) (1998) (errata).

Description in vitro

Pycnidia globose to irregular with 1–3 ostioles, non-papillate or slightly papillate, but occasionally with long tubular outgrowths, 50–210 μm diam., glabrous or 'hairy' (semi-pilose), solitary or confluent. Conidial matrix white to straw-coloured.

Conidia ellipsoidal, with 2 large polar guttules, mostly 3.5–5 × 1.5–2 μm (occasionally a small percentage larger conidia may be produced, up to 9 × 3 μm).

Chlamydospores unicellular, intercalary, solitary or in chains, subglobose to ellipsoidal, olivaceous with greenish guttules, mostly 6–9 μm diam.

Colonies on OA *c.* 7 cm diam. after 7 days, regular, with felted to downy olivaceous grey aerial mycelium, on the agar distinct radiating mycelial strands; reverse olivaceous grey, iron grey or grey olivaceous. Pycnidia abundantly formed on or partly in the agar. Representative culture CBS 267.92.

Ecology and distribution

In Eurasia this fungus has been isolated from above-ground parts (fruits and leaves) of quite different woody and herbaceous plants. It seems to be a plurivorous opportunistic parasite. The fungus *in vitro* shows much resemblance to the ubiquitous soil fungus *Phoma eupyrena* Sacc. (this section), but can be recognized by its larger conidial diversity and different cultural behaviour. Its original classification in a separate genus, *Polyopeus* A.S. Horne, is based on the fact that under certain conditions the pycnidia in culture develop several long tubular outgrowths or 'necks', compare Kidd & Beaumont (1924).

Phoma piperis (Tassi) Aa & Boerema, Fig. 12E

> *Phoma piperis* (Tassi) Aa & Boerema apud de Gruyter, Noordeloos & Boerema *in* Persoonia **15**(3): 398. 1993.
> > ≡*Phyllosticta piperis* Tassi *in* Boll. R. Orto bot. Siena **3**(2): 28. 1900 ['1899'].

Diagn. data: de Gruyter *et al.* (1993) (fig. 27: conidial shape, reproduced in present Fig. 12E; cultural descriptions, partly adopted below).

Description in vitro

Pycnidia globose-papillate or flask-shaped with 1–2 ostioles, 75–160 μm diam., glabrous or with hyphal outgrowths, solitary or confluent. Conidial matrix buff.
   Conidia oblong, with two indistinct polar guttules, mostly 3–5 × 1.5 μm.
   Colonies on OA *c.* 5.5 cm diam. after 7 days, regular, colourless with only a few tufts of white aerial mycelium; reverse colourless to pale greyish buff [on MA similar growth rate with fine granulose smoke grey aerial mycelium]. On all media fine needle-like crystals are formed. Production of pycnidia, mainly on, but sometimes also (partly) in the agar. Representative culture CBS 268.93.

Ecology and distribution

This fungus is described and well known as causal organism of irregular leaf spots on shrubs of *Piper longus* (Indian long pepper) and other *Piper* spp. (Piperaceae). Members of the genus *Peperomia* (of South American origin) may also be attacked: leaf spot.

Phoma plurivora P.R. Johnst., Fig. 12F

> *Phoma plurivora* P.R. Johnst. *in* N.Z. Jl Bot. **19**: 181. 1981.

Diagn. data: Johnston (1981) (tab. 1: cultural and microscopic characteristics; fig. 11: cultures on OA and MA; original description *in vitro*, New Zealand), de Gruyter *et al.* (1998) (fig. 17: conidial shape, reproduced in present Fig. 12F; cultural descriptions, partly adopted below).

Description in vitro

Pycnidia globose to irregular with 1 papillate ostiole, 80–260 μm diam., glabrous, solitary to confluent. Conidial matrix buff to pale saffron.

Conidia broadly ellipsoidal to oblong, usually with several, distinct large or small polar guttules, 3.5–6(–8) × 1.5–2.5(–3) μm.

Colonies on OA 4–5 cm after 7 days, regular, colourless to grey olivaceous, with or without sparse grey olivaceous aerial mycelium; reverse also colourless to grey olivaceous. Growth rate on MA also 4–5 cm; olivaceous buff/greenish olivaceous to olivaceous; reverse similar. Pycnidia abundant, both on and (partly) in the agar. Representative cultures CBS 558.81, CBS 284.93.

Ecology and distribution

Typically a plurivorous fungus, saprophyte or opportunistic parasite, probably indigenous to Australasia. In New Zealand it has been isolated from various dicotyledonous and monocotyledonous herbaceous plants, as well as from trees and shrubs.

Phoma poolensis Taubenh. var. *poolensis,* Fig. 12G

> *Phoma poolensis* Taubenh., Dis. Greenhouse Crops 203. 1919, var. *poolensis*
>> = *Phoma oleracea* var. *antirrhini* Sacc. *in* Sylloge Fung. **3**: 135. 1884 [no priority: another rank, Art. 11.3].
>> = *Phyllosticta antirrhini* P. Syd. *in* Beibl. Hedwigia **38**: 134. 1899 [sometimes wrongly quoted as '*Phoma*'].
>> = *Phoma antirrhini* Dzhalag. *in* Nov. Sist. Nizsh. Rost. **1965**: 157. 1965.

Diagn. data: Smiley (1920) (figs 1–3: disease symptoms on leaves and stems of snapdragon sub *Phyllosticta antirrhini*; fig. 4: drawing of germtube entering through stomate and penetrating epidermal cells; figs 5, 6: drawings of pycnidia *in vivo* and conidia swelling at germination; fig. 7: cultures on potato and cornmeal agar), Guba & Anderson (1919) (figs 1–3: disease symptoms; figs 4–8: drawings of mycelium, pycnidia and conidia), Marchionatto (1948) (figs 1, 2, pls I, II; pathogenicity to seedlings; colour plate of cultural characteristics; drawing of pycnidia and conidia), Cejp (1965) (variability of conidia *in vivo*), Neergaard (1977) (occurrence on seed), Maiello (1978) (figs 1–3; influence ultraviolet light and nutrition on pycnidium production), Boerema (1978) (tab. 3: synonymy; classification), de Gruyter *et al.* (1993) (fig. 13: conidial shape, reproduced in present Fig. 12G; cultural descriptions, partly adopted below).

Description in vitro

Pycnidia globose to subglobose, sometimes shortly papillate, but an ostiole often fails to come, so that fully mature pycnidia are still non-ostiolate (for release of the conidia the pycnidial wall then has to burst open), mostly 60–170 μm diam., glabrous, solitary or confluent. Conidial matrix buff or whitish.

Conidia ellipsoidal, eguttulate or with (1)2(3) inconspicuous polar guttules, mostly 3.5–5 × 1.5–2 μm.

Colonies on OA *c.* 7 cm diam. after 7 days, regular, almost colourless but somewhat olivaceous tinged, with appressed floccose, white aerial mycelium; reverse colourless with olivaceous tinge or greenish olivaceous. Pycnidia usually most numerous in the centre, on and in the agar. All strains studied showed on MA a positive reaction with the sodium hydroxide test: on application of a drop of NaOH green → red (E$^+$ reaction). Representative culture CBS 116.93.

Ecology and distribution

A worldwide recorded pathogen of cultivated varieties of snapdragon, *Antirrhinum majus* (Scrophulariaceae): leaf spot and (basal) stem rot. It is frequently encountered on seed of this host: damping-off in seedlings; sometimes apparently quite destructive. The fungus also has been isolated from other Scrophulariaceae. Comparative inoculation experiments demonstrated the occurrence of different host forms, which may have been promoted by the common seed transmission. Isolates from snapdragon and figwort (*Scrophularia nodosa*) showed to be predisposed to the phenomenon of producing *in vitro* mainly non-ostiolate pycnidia (compare the cultural experiments by Rajak & Rai, 1983b, 1984; copied in Rai, 1998). This abnormality is not observed in isolates of the variety *verbascicola*, treated below.

Phoma poolensis var. *verbascicola* (Ellis & Kellerm.) Aa & Boerema, Fig. 12H

> *Phoma poolensis* var. *verbascicola* (Ellis & Kellerm.) Aa & Boerema apud de Gruyter, Noordeloos & Boerema *in* Persoonia **15**(3): 385. 1993.
> > ≡ *Phyllosticta verbascicola* Ellis & Kellerm. *in* Bull. Torrey bot. Club **11**: 115. 1884.
> > [= 'Phoma verbascicola' sensu e.g. Allesch. *in* Rabenh. Krypt.-Flora [ed. 2], Pilze **6** [Lief. 64]: 327. 1899 [vol. dated '1901'] and Grove, Br. Coelomycetes **1**: 112–113. 1935; see Note.]

Diagn. data: de Gruyter *et al.* (1993) (fig. 14: conidial shape, reproduced in present Fig. 12H; cultural descriptions, partly adopted below), Boerema *et al.* (1996) (documentation of misidentifications).

Description in vitro

Pycnidia resembling those of the type var. *poolensis*, but always with 1–2(–5) distinct papillate ostioles, 70–225 μm diam., glabrous, solitary or confluent. Conidial matrix colourless or whitish.

Conidia ellipsoidal, usually with 2(–4) small polar guttules, often somewhat longer and/or broader than those of the type var.: 3.5–5.5 × 1.5–2.5 μm.

Colonies on OA show about the same growth rate and cultural characteristics as var. *poolensis*. Usually abundant production of ostiolate pycnidia, on and in the agar. Cultures on MA are characterized by rather compact woolly to floccose white aerial mycelium; reverse conspicuous saffron to cinnamon discoloured. On application of a drop of NaOH all strains tested showed on MA the E$^+$ reaction: green → red. Representative culture CBS 127.93.

Ecology and distribution

This fungus is in Europe and North America frequently reported and collected from dead stems and seed capsules of various *Verbascum* spp. (Scrophulariaceae); always under the misapplied name *Phoma verbascicola* (Schwein.) Cooke. The type material of the basionym of the latter, *Sphaeria verbascicola* Schweinitz (PH, duplicate BPI; Philadelphia and Beltsville, USA) contains only immature ascomata, probably belonging to *Pleospora scrophulariae* (Desm.) Höhn. [but also interpreted as referring to a species of *Mycosphaerella*: *M. verbascicola* (Schwein.) Fairman, Proc. Rochester Acad. Sci. **4**: 176. 1905].

Inoculation experiments have shown that *Ph. poolensis* var. *verbascicola* may cause damping-off of young plants and necrotic lesions on stems and leaves. The fungus did not cause disease symptoms on snapdragon, *Antirrhinum majus*, the principle host of the type variety *Ph. poolensis* var. *poolensis*.

Phoma pratorum P.R. Johnst. & Boerema, Fig. 12I

Phoma pratorum P.R. Johnst. & Boerema *in* N.Z. Jl Bot. **19**: 395. 1981.

Diagn. data: Johnston & Boerema (1981) (figs 2, 4: cultures on OA and MA, drawing of conidia; original description *in vitro*, New Zealand), de Gruyter *et al.* (1998) (fig. 2, conidial shape, reproduced in present Fig. 12I; cultural descriptions, partly adopted below).

Description in vitro

Pycnidia globose with 1–3 non-papillate or slightly papillate ostioles, 90–250 μm diam., glabrous, solitary or confluent. Conidial matrix white.

Conidia cylindrical with (1–)2 distinct small or relatively large polar guttules, 3.5–7(–8.5) × 1–2 μm.

Colonies on OA 2–2.5 cm diam. after 7 days (4–5 cm after 14 days), dark herbage green with felted to finely floccose, white to olivaceous grey aerial mycelium; reverse dull green to olivaceous. Pycnidia abundant, mainly immersed in the agar. Representative cultures CBS 445.81, CBS 286.93.

Ecology and distribution

Frequently isolated from the surface of living leaves of various grasses in New Zealand (e.g. *Dactylis, Lolium* and *Paspalum* spp.). This epiphytic saprophyte of Gramineae is probably also common in Australia and recently also recorded in Europe.

Phoma putaminum Speg., Fig. 12J

> *Phoma putaminum* Speg. *in* Atti Soc. crittogam. ital. **3**: 66. 1881; not *Phoma putaminum* Hollós *in* Növén Közlen **6**, Extr. 3. 1907.
>
> > ≡ *Aposphaeria putaminum* (Speg.) Sacc. *in* Sylloge Fung. **3**: 177. 1884.
> > ≡ *Coniothyrium putaminum* (Speg.) Kuntze, Revisio Gen. Pl. **3**(2): 459. 1898.
> > = *Phoma radicicola* McAlpine, Fung. Dis. Stone-fruit Trees Melb.: 126. 1902; not *Phoma radicicola* Maubl. *in* Bull. Soc. mycol. Fr. **21**: 90. 1905 [= *Phoma anserina* Marchal, this section].
> > = *Phoma dunorum* ten Houten, Kiemplziekt. Conif. [Thesis Univ. Utrecht]: 88. 1939.

Diagn. data: Boerema & Dorenbosch (1973) (pl. 3a, b: cultures on OA; tab. 1: synoptic table on characters *in vitro* with drawing of conidia; tab. 2: isolate sources), ten Houten (1939) (fig. 1: pycnidial and conidial shape; cultural characteristics), Sutton (1980) (fig. 229B: drawing of conidia; specimens in CMI collection), de Gruyter & Noordeloos (1992) (fig. 5: conidial shape, reproduced in present Fig. 12J; cultural descriptions, partly adopted below).

Description in vitro

Pycnidia globose, mostly with a short neck, 60–300 μm diam., covered with hyphal strands (hairy), mostly solitary but also confluent. Conidial matrix whitish/salmon.

Conidia broadly ellipsoidal, mostly with (1)2 or 3(4) conspicuous greenish guttules, 3–4 × 2–2.5 μm.

Colonies on OA 4–6 cm diam. after 7 days, regular, without aerial mycelium, colourless but exuding a pigment which causes a honey discoloration of the agar; reverse honey with a greenish olivaceous tinge. On MA sparse aerial mycelium of more or less erect hairs, later becoming floccose; colony colour between ochraceous and umber with slight olivaceous tinge; reverse umber with honey margin. Pycnidia scattered, esp. towards margin of colony, also in concentric zones, immersed in the agar, sometimes entirely in the agar. Representative cultures CBS 130.69, CBS 372.91.

Ecology and distribution

A typical soil-borne species which has been isolated from the subterranean parts of various herbaceous and woody plants. It is generally regarded as a saprophyte, but may act as opportunistic parasite on roots, etc. Most isolates of

the fungus are from Europe (Germany, the UK, The Netherlands) and North
America (Alaska, Canada, USA), but the fungus is also recorded in Australia,
Africa (Nigeria) and South-east Asia (Myanmar, India, Malaysia) and South
America (Venezuela).

Phoma selaginellicola Gruyter *et al.*, Fig. 13A

> *Phoma selaginellicola* Gruyter, Noordel. & Boerema *in* Persoonia **15**(3):
> 399. 1993.
>
> > ≡ *Phyllosticta selaginellae* Sacc. *in* Malpighia **11**: 304. 1897;
> > not *Phoma selaginellae* Cooke & Massee *in* Grevillea **16**:
> > 102. 1888.

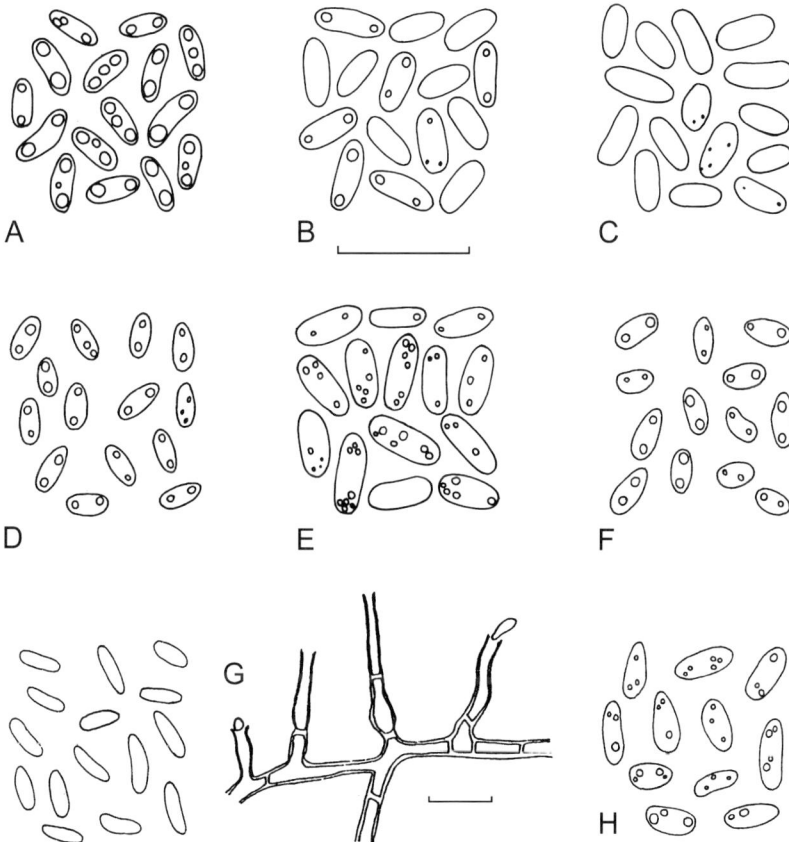

Fig. 13. Conidial variation and some hyphal conidiophores. **A** *Phoma selaginellicola.* **B** *Phoma
senecionis.* **C** *Phoma subherbarum.* **D** *Phoma tropica.* **E** *Phoma urticicola.* **F** *Phoma valerianae.*
G *Phoma vasinfecta*; synanam. *Phialophora chrysanthemi.* **H** *Phoma viburnicola.* Bars 10 μm.

Diagn. data: de Gruyter *et al.* (1993) (fig. 28: conidial shape, reproduced in present Fig. 13A; cultural descriptions, partly adopted below).

Description in vitro

Pycnidia globose with 1 distinct, non-papillate ostiole, 125–330 µm diam., glabrous, solitary. Conidial matrix straw coloured.

Conidia ellipsoidal, sometimes curved, with 2 large polar, and 1 central, usually smaller, guttules, mostly 3–4.5 × 1.5 µm.

Colonies on OA *c.* 5–5.5 cm diam. after 7 days, regular, colourless to (oli-vaceous) buff, with fine velvety olivaceous grey aerial mycelium; reverse (oliva-ceous) buff with grey (olivaceous) concentric zones. Pycnidia are produced abundantly on the agar, partly also in the aerial mycelium. Representative cul-ture CBS 122.93.

Ecology and distribution

This seems to be an opportunistic parasite of *Selaginella* spp. (Selaginellaceae). In Italy the fungus is repeatedly found on wilting leaves of wild plants of *S. hel-vetica*. In a Dutch nursery it occurred in association with leaf necroses of a selected cultivar of the same species. The conidial size is variable *in vivo* and probably dependent on the age and ecological situation of the host.

Phoma senecionis P. Syd., Fig. 13B

> *Phoma senecionis* P. Syd. in Beibl. Hedwigia **38**: 136. 1899.
> = *Phyllosticta albobrunnea* Bubák & Wróbl. apud Bubák *in* Hedwigia **57**: 330. 1916.

Diagn. data: de Gruyter *et al.* (1993) (fig. 6: conidial shape, reproduced in pre-sent Fig. 13B; cultural descriptions, partly adopted below).

Description in vitro

Pycnidia subglobose with 1–2 distinct ostioles, non-papillate or slightly papil-late, 160–250 µm diam., glabrous, wall yellow brown, solitary or confluent. Frequently also micropycnidia occur, 50–90 µm diam. Conidial matrix salmon.

Conidia oblong to ellipsoidal, eguttulate or with 2(3) small polar guttules, mostly 4–6.5 × 1.5–2.5 µm.

Colonies on OA *c.* 4.5–5 cm diam. after 7 days, regular, with sparse floc-cose, white aerial mycelium; reverse colourless or pale yellowish. On MA the yellow staining of the agar appears to be caused by a pigment (complex) which on addition of a drop of NaOH becomes red. The pycnidia develop mainly in the aerial mycelium. In the agar conspicuous dense strands of dark-coloured short-celled hyphae occur. Representative culture CBS 160.78.

Ecology and distribution

This fungus has been found on and isolated from dead tissue of *Senecio jacobea* and other *Senecio* spp. (Compositae). It probably occurs wherever the hosts are growing. The records so far are from Europe and New Zealand. It seems to be a necrophyte with a specific host relation.

Phoma subherbarum Gruyter *et al.*, Fig. 13C

> *Phoma subherbarum* Gruyter, Noordel. & Boerema *in* Persoonia **15**(3): 387. 1993.

Diagn. data: de Gruyter *et al.* (1993) (fig. 15: conidial shape, reproduced in present Fig. 13C; cultural descriptions, partly adopted below).

Description in vitro

Pycnidia globose to subglobose or irregularly shaped, with 1(–3) papillate ostioles, mostly 90–200 μm diam., glabrous or with short hyphal outgrowths, solitary or confluent. Fertile micropycnidia, 30–70 μm diam., frequently also occur. Conidial matrix white to rosy buff.
   Conidia oblong to ellipsoidal, mostly eguttulate, occasionally with a few very small guttules, 4–5(–6.5) × 1.5–2 μm.
   Colonies on OA *c.* 8 cm diam. after 7 days, regular, without aerial mycelium, colourless with saffron-flesh and olivaceous zones; reverse similar. On MA also *c.* 8 cm diam., with felted or floccose olivaceous grey or smoke grey aerial mycelium, reverse rust to chestnut obvious due to the release of a pigment (not changing with NaOH). Pycnidia usually develop abundantly on and in the agar. Representative culture CBS 250.92.

Ecology and distribution

This saprophyte, resembling in many respects the ubiquitous *Phoma herbarum* Westend. (this section), seems to be a fungus from (South?) American origin. The isolates studied were obtained from necrotic leaves of wild potatoes (Solanaceae) in Peru (Andes, alt. 2500 m) and the surface of overwintered seed of maize (Gramineae), collected in Canada (Ottawa: 'most common fungus isolated from discoloured seed').

Phoma tropica R. Schneid. & Boerema, Fig. 13D

> *Phoma tropica* R. Schneid. & Boerema *in* Phytopath. Z. **83**: 361[–365]. 1975.
> > Θ = '*Sphaeronaema coloratum* P.N. Mathur & Thirum.' [in 1965 deposited in CBS-collection, but never published: 'collection-name'].

Diagn. data: Schneider & Boerema (1975b) (figs 1, 2: cultures on OA and MA; figs 3–6: photographs of pycnidia, mycelia, pycnidial wall with ostiole and conidia *in vitro*; tab. 1: isolate sources; original description), Sutton (1980) (fig. 229E: drawing of conidia; specimens in IMI collection), de Gruyter & Noordeloos (1992) (fig. 16: conidial shape, reproduced in present Fig. 13D; cultural descriptions, partly adopted below), Rai & Rajak (1993) (cultural characteristics), Rai (1998) (descriptions of several isolates from India; effect of different physical and nutritional conditions on morphology and cultural characters).

Description in vitro

Pycnidia subglobose, with 1–5 conspicuous dark circumvalated ostioles, 100–300(–400) μm diam., glabrous, solitary. Conidial matrix white yellowish.
  Conidia ellipsoidal, with 2(3) distinct guttules, (2–)3–4(–6) × 1–2(–2.5) μm.
  Colonies on OA, 5–5.5 cm diam. after 7 days, regular, colourless to greenish olivaceous or dull green, aerial mycelium usually poorly developed, dark olivaceous; reverse greenish olivaceous, outer margin light green. Pycnidia scattered over colony or in concentric rings on the agar. Representative cultures CBS 436.75, CBS 498.91.

Ecology and distribution

This thermotolerant saprophyte (still growth at 35°C) is in western Europe a common occupant of greenhouses, where it has been found on necrotic tissue of a wide variety of ornamental plants. The fungus probably originates from South America. Recently it has been repeatedly recorded from India [apart from the common form with transparent light-brown, nearly neck-less pycnidia frequently also a form with dark flask-shaped pycnidia (like 'Sphaeronaema coloratum'; see also Rai, 1998)].

Phoma urticicola Aa & Boerema, Fig. 13E

Teleomorph: **Didymella urticicola Aa & Boerema**

Phoma urticicola Aa & Boerema apud Boerema *in* Trans. Br. mycol. Soc. **67**: 303[–306]. 1976.

Diagn. data: Boerema (1976) (figs 2: drawings of pycnidia and pseudothecia, discrete and aggregated in a stroma, asci and ascospores; fig. 11: cultures on OA; tab. 9: synoptic table on cultural characters with drawings of conidia; detailed descriptions of both morphs *in vitro*, partly adopted below), de Gruyter et al. (1998) (fig. 14: conidial shape, reproduced in present Fig. 13E).

Description in vitro

Pycnidia globose to bottle-shaped (pyriform) with 1–2 papillate ostioles, 110–500 μm diam., glabrous, solitary or confluent. Conidial matrix pale buff to saffron. [In fresh cultures soon also pseudothecia develop, see below.]

Conidia irregular ellipsoidal, usually with 2 or many small guttules, (3–)4–6.5(–8.5) × (1.5–)2–3(–3.5) μm.

Pseudothecia (intermixed with pycnidia, discrete or in a small number aggregated in a stroma) subglobose to pyriform. Asci 40–65 × 8–12 μm, 8-spored. Ascospores 12–18 × 5.5–7.5 μm, 2-celled (on ageing rarely 3–4-celled), the lower cell usually slightly tapering to the base, the upper cell widest near the septum, tapering gradually to a broadly rounded apex.

Colonies on OA 5–6 cm diam. after 7 days, regular, colourless, but with grey tinge due to development of pycnidia, sometimes with dark herbage green sectors, often fluffy whitish aerial mycelium which discolours in the light to saffron-salmon; reverse colourless to grey, sometimes with green sectors. Pycnidia abundant, on and in the agar; in fresh cultures soon followed by the development of pseudothecia. Representative culture CBS 121.75.

Ecology and distribution

Common in Europe on dead stems of nettle, especially *Urtica dioica* (Urticaceae), usually with simultaneous production of anamorph and teleomorph. In the past, i.e. before 1976, this fungus was always confused with other species of *Phoma* and *Didymella* occurring on *Urtica* spp.

Phoma valerianae Henn., Fig. 13F

Phoma valerianae Henn. *in* Nyt Mag. Naturvid. **42**: 29. 1904.
= *Phyllosticta valerianae-tripteris* f. *minor* Unamuno *in* Mems R. Soc. esp. Hist. nat. **15**: 348. 1929.

Diagn. data: de Gruyter & Noordeloos (1992) (fig. 8: conidial shape, reproduced in present Fig. 13F; cultural descriptions, partly adopted below).

Description in vitro

Pycnidia globose with distinct neck, 120–340 μm diam., glabrous, mostly solitary. Conidial matrix amber.

Conidia ellipsoidal, with 2 distinct, small or relatively large polar guttules, (2.5–)3–4 × (1.5–)2 μm.

Colonies on OA *c.* 6.5 cm diam. after 7 days, regular, greenish grey, flat with finely woolly aerial mycelium; reverse zonate, grey olivaceous black. On MA relatively slow growing, *c.* 3.5 cm in diam., with very fine compact velvety greenish olivaceous grey aerial mycelium. Pycnidia scattered on and in the agar. Representative cultures CBS 630.68, CBS 499.91.

Ecology and distribution

In Europe frequently occurring on dead stems of *Valeriana* spp. (Valerianaceae), and apparently seed-borne. The fungus also has been isolated from other, not related, herbaceous plants. There are no data on pathogenicity;

records of disease symptoms on Valerianaceae associated with *Phoma* usually refer to *Ph. valerianellae* Gindrat *et al.*, sect. *Phyllostictoides* (**E**).

Phoma vasinfecta Boerema *et al.*, Fig. 13G

Hyphomycetous synanamorph: ***Phialophora chrysanthemi* (Zachos *et al.*) W. Gams (p.p.)**

> *Phoma vasinfecta* Boerema, Gruyter & Kesteren *in* Persoonia **15**(4): 484. 1994.
>> = *Phoma tracheiphila* f. sp. *chrysanthemi* K.F. Baker, L.H. Davis, S. Wilhelm & W.C. Snyder *in* Can. J. Bot. **63**: 1733. 1985.

Diagn. data: Baker *et al.* (1949a) (first record on florists' chrysanthemum in the USA, as 'Deuterophoma sp.'), Taylor (1962) (occurrence on chrysanthemum varieties in Australia), Robertson (1967) (description of isolate from pyrethrums in England, morphology of pycnidia, conidia and hyphomycetous synanamorph, 'Cephalosporium sp.', *in vitro*; pathogenicity; cultural characteristics in comparison with similar Hyphomycete isolates from chrysanthemum), Baker *et al.* (1985) (figs 1, 2: disease symptoms on chrysanthemum plants; morphology of pycnidia and conidia on stems of florists' chrysanthemum, similarity with *Phoma tracheiphila* on citrus; cultural characters: production of pycnidia and hyphomycetous synanamorph; pathogenicity tests on chrysanthemum and citrus), Boerema *et al.* (1994) (*in vitro* and *in vivo* descriptions, partly adopted below), Boerema *et al.* (1996: 157) (former interpretations as 'Deuterophoma' and 'Cephalosporium').

Description in vitro

Pycnidia subglobose, usually with a short neck, but often without visible ostiole, mostly 55–160 μm diam., glabrous, solitary, rarely in groups. Conidial matrix milky-cream.

Conidia subcylindrical or slightly curved, eguttulate, mostly 2–3(–3.5) × (0.5–)1–1.5 μm (*in vivo* usually somewhat larger, 2.5–4 × 1–1.5 μm).

Colonies variable in growth rate, depending on strain and agar medium, on OA mostly between 1 and 3.5 cm diam. after 7 days, sparse yellow-green aerial mycelium, colony margin irregular and whitish, compact. Pigment production variable, depending on strain, whitish pink becoming reddish brown grey, or greenish grey becoming dark grey-green to black; reverse also variable in colours, ranging from distinctly pink or purple/reddish brown to dark greenish grey or almost black. Pycnidia production variable, superficially on the agar. On application of a drop of NaOH the reddish diffusible pigments turn blue.

Hyphal conidia arising from free conidiogenous cells on the aerial mycelium (representing the *Phialophora*-anamorph) are variable in size, mostly 2–5.5 × 1–2(–2.5) μm.

Ecology and distribution

A vascular-inhabiting pathogen of florists' chrysanthemum (Compositae), *Dendranthema*-Grandiflorum hybrids (formerly known as e.g. *Chrysanthemum morifolium* and *C. indicum*): slow wilt, *Phoma* decline disease. Occasionally recorded in Europe, North America and Australia. As vascular pathogen with a *Phialophora*-synanamorph the fungus resembles *Phoma tracheiphila* (Petri) L.A. Kantsch. & Gikaschvili, the cause of 'mal secco disease' of citrus trees in the Mediterranean and Black Sea areas. However, the latter fungus, type species of *Deuterophoma* Petri, produces scleroplectenchymatous pycnidia and belongs to *Phoma* sect. *Plenodomus* (**G**). The *Phialophora*-anamorph of *Ph. vasinfecta* is indistinguishable from *Phialophora chrysanthemi* (Zachos *et al.*) W. Gams ('*Cephalosporium*-wilt' of *Chrysanthemum* spp.; see Zachos *et al.*, 1960). The latter also may be interpreted as representing the hyphomycetous synanamorph of *Ph. vasinfecta*, but developing separately and independently.

Phoma viburnicola Oudem., Fig. 13H

> *Phoma viburnicola* Oudem. *in* Versl. gewone Vergad. wis- en natuurk. Afd.
> K. Akad. Wet. Amst. **9**: 298. 1900; *in* Ned. kruidk. Archf III, **2**(1): 247. 1900.
> = *Phyllosticta opuli* Sacc. *in* Michelia **1**(2): 146. 1878; *in*
> Sylloge Fung. **3**: 16–17. 1884; not *Phoma opuli* Thüm. *in*
> Instituto, Coimbra **28** sub Contr. Fl. myc. Lusit. III n. 564.
> 1881 ['1880–81']; quoted *in* Hedwigia **2**: 24. 1882.

Diagn. data: Boerema & Griffin (1974) (tabs 1, 2: isolates from *Viburnum*; differential criteria), de Gruyter & Noordeloos (1992) (fig. 14: conidial shape, reproduced in present Fig. 13H; cultural descriptions, partly adopted below).

Description in vitro

Pycnidia globose to elongated-flask shaped, with 1 or 2–3 ostioles, initially papillated, later neck-like elongated, 100–300 µm diam., glabrous, solitary or in clusters of 2–3 pycnidia. Conidial matrix whitish.

Conidia ellipsoidal, with 2–4 small, mainly polar guttules, 3.5–5.5 × 1.5–2 µm.

Colonies on OA, between 4.5 and 5 cm diam. after 7 days, regular, pale olivaceous grey to grey olivaceous, with some greyish aerial mycelium at centre; reverse grey olivaceous. On MA same growth rate, with dense woolly white or olivaceous aerial mycelium. Pycnidia scattered or in concentrical rings both in and on the agar as well as in the aerial mycelium. With addition of a drop of NaOH a slightly sienna discoloration of the agar may occur. Representative cultures CBS 523.73, CBS 500.91.

Ecology and distribution

In Europe a widespread occurring opportunistic parasite of woody plants. The fungus is originally described from leaf spots and stem lesions on *Viburnum* spp. (Caprifoliaceae), but serious disease symptoms on those shrubs are usually caused by *Phoma exigua* var. *viburni* (Roum. & Sacc.) Boerema (sect. *Phyllostictoides*, **E**: producing always some septate conidia), see Boerema & Griffin (1974).

9

B *Phoma* sect. *Heterospora*

Phoma sect. *Heterospora* Boerema *et al.*

> Phoma sect. Heterospora Boerema, Gruyter & Noordel. in Persoonia 16(3): 336. 1997.
> Type: *Phoma heteromorphospora* Aa & Kesteren
> Synanamorph *Stagonosporopsis* Died. *in* Annls mycol. **10**: 142. 1912.

The section *Heterospora* refers to *Phoma* species producing distinctly large conidia in addition to relatively small 'phomoid' conidia. The small conidia are mainly aseptate, generally measuring 3–11 × (1–)1.5–4(–5) µm. The dimensions of the large conidia vary *in vitro* usually in the range (15–)20–25(–28) × 3.5–6(–7) µm. *In vivo* their dimensions are still more variable, (8–)11–30 (occ. 45–62) × (2.5–)3–8 (occ. 12–15) µm. The large conidia may remain continuous ('macrosporoid'), but often become two- or more-celled by secondary septation ('ascochytoid/stagonosporoid'). They break (split) easily at the septa: conidial fragmentation.[7] Pycnidia like those in sect. *Phoma*, simple or complex, thin-walled, pseudoparenchymatous and ostiolate, glabrous, but sometimes with hyphal outgrowths.

The type species of this section, *Phoma heteromorphospora* Aa & Kesteren, always produces *in vivo* small and large conidia in the same pycnidium, but *in vitro* the pycnidia contain mainly small conidia with only a few large conidia (Fig. 14A). The closely related *Phoma dimorphospora* (Speg.) Aa & Kesteren produces only small conidia *in vitro* (Fig. 14B). Some other species always show a mixture of small and large conidia both *in vivo* and *in vitro* (Fig. 14C). However, most species of the section *Heterospora* are characterized by mainly small conidia *in vitro*, whereas *in vivo* the pycnidia contain either mainly small or mainly large conidia.

---

[7] A phenomenon never observed in conidia of true species of *Ascochyta* or *Stagonospora* characterized by wall-thickening septation (see Introduction).

Dominant 'ascochytoid/stagonosporoid' phenotypes *in vivo*, may be differentiated as separate synanamorphs belonging to the genus *Stagonosporopsis* Died. (1912: 141–142). The latter was originally separated from *Ascochyta* on the basis of occasional multiseptate conidia ('*Stagonospora*'-like). Diedicke (1912: 141) indicated that seven species belonged to this genus. The first species combination described, *Stagonosporopsis actaeae* (Allesch.) Died. (Diedicke, 1912: 144), interpreted by many authors as the type species, represents the 'ascochytoid/stagonosporoid' phenotype of *Phoma actaeae* in this section. The combination *Stagonosporopsis boltshauseri* (Sacc.) Died., also published in 1912 (Diedicke, 1912–15: 400), was chosen as lectotype by Clements & Shear (1931). It represents the 'ascochytoid/stagonosporoid' phenotype of *Phoma subboltshauseri* in this section.

Most species of the section are pathogens specific to particular hosts. The two types of conidia may play different roles in the life cycle. Some species produce small conidia especially on dead host material, whereas pycnidia with large conidia only develop in association with disease symptoms. The large conidia generally develop with fluctuating humidity and with desiccation, whereas small conidia are often formed in humid conditions. This may occur in the same pycnidium and appears to be reversible. Temperature may also be important. Large conidia usually germinate more quickly than the smaller ones.

Only one species of this section is known as a plurivorous necrophyte. None has been experimentally connected with a teleomorph, but in one case a single identity with a species of *Didymella* Sacc. ex Sacc. was suggested. Chlamydospores are absent, but dark swollen cells occasionally occur.

Notes: A comparable *Phoma/Stagonospora*-like conidial dimorphism is recorded in anamorphs of some species of the Ascomycete genera *Leptosphaeria* Ces. & De Not. and *Phaeosphaeria* I. Miyake (Sivanesan, 1984; Leuchtmann, 1984). However, in those anamorphs the relatively large septate conidial phenotype commonly dominates, not only *in vivo* but also *in vitro*. Therefore they are usually only referred to *Stagonospora* (Sacc.) Sacc. char. emend. Leuchtmann (1984). Most of the associated *Phoma*-like microconidial forms (Leuchtmann, 1984) do not have a specific name, see Chapter 17 'Miscellaneous'.

Finally it should be noted that some species in the sections *Plenodomus* (**G**) and *Sclerophomella* (**F**) occasionally also produce some relatively large ascochytoid conidia *in vivo*.

Diagn. lit.: Laskaris (1950) (study of conidial variability under different conditions: *Phoma* in artificial culture, '*Diplodina delphinii*' on *Delphinium* hybrids: see this section under *Ph. aquilegiicola*), van der Aa & van Kesteren (1979, 1980) (description and synonymy of *Ph. heteromorphospora*, type species of the section *Heterospora*), Boerema *et al.* (1997) (precursory contribution to this chapter; documentation of the concept of this section), Boerema *et al.* (1999) (addition of *Ph. schneiderae*, synanam. *Stagonosporopsis lupini*).

Synopsis of the Section Characteristics

- Pycnidia simple or complex, usually thin-walled, pseudoparenchymatous and ostiolate, glabrous, but sometimes with hyphal outgrowths.

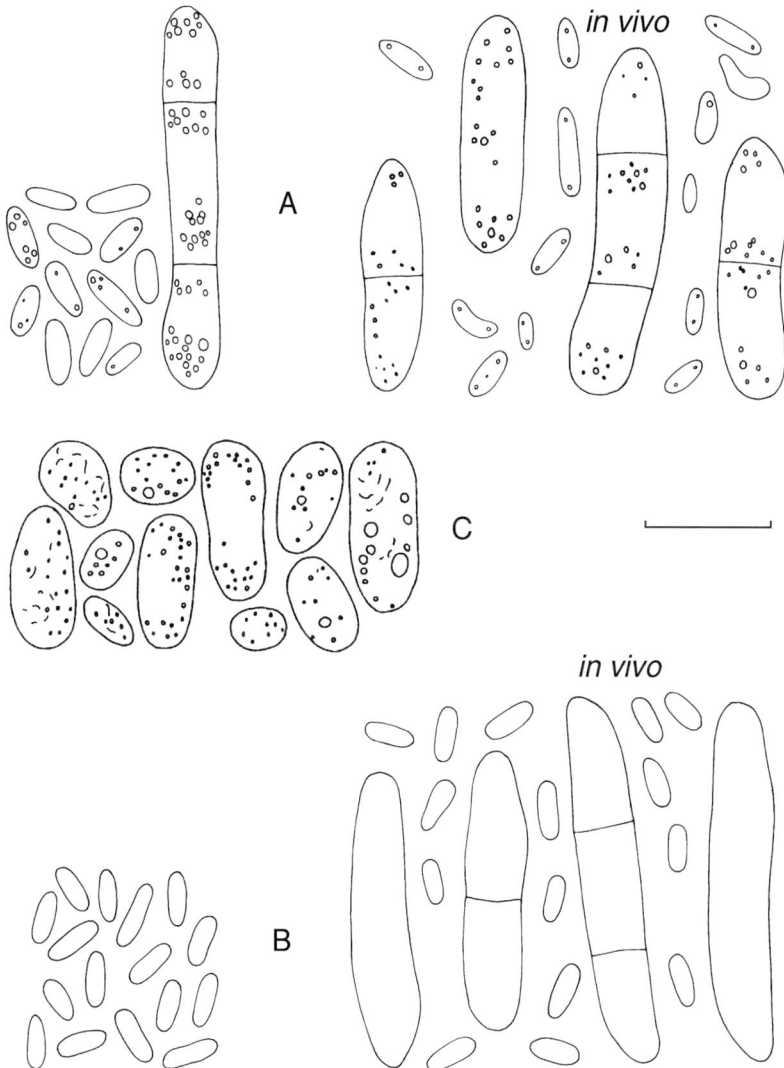

Fig. 14. A *Phoma heteromorphospora*, type species of *Phoma* sect. *Heterospora*. Conidial shape and size. *In vitro* the pycnidia contain mainly small aseptate conidia, but also a few large and mostly 1–2-septate conidia are usually present. *In vivo* the pycnidia are always heterosporous; the macroconidia vary in shape, size and septation. **B** *Phoma dimorphospora*, closely related to the type species. Conidia generally eguttulate. *In vitro* the pycnidia usually contain only small aseptate conidia. **C** *Phoma andigena*. Conidia always dimorphic, *in vitro* and *in vivo*, usually aseptate and variable in shape, with numerous guttules. Bar 10 μm.

- Conidia usually mainly relatively small and aseptate *in vitro*, but often also with some distinctly large conidia, which may become 1(–2)-septate and easily break at the septa; the latter conidial phenotype may be dominant *in vivo*: *Stagonosporopsis*-synanamorphs.
- Chlamydospores absent, but dark swollen cells occur in some species.
- Teleomorph usually absent (if present probably belonging to *Didymella*).

Key to the Species and Varieties Treated of Section *Heterospora*

Differentiation based on the characteristics *in vitro*

This includes notes on the conidial characters *in vivo* (sometimes as *Stagonosporopsis*-synanamorph).

The distinguishing character of members of the section *Heterospora*, the large sized conidial dimorph, is most conspicuous under natural growth conditions. *In vitro* the conidia are often mostly small and aseptate; large conidia may be wanting, especially in old isolates. Direct identification of the *Heterospora* species *in vitro* and their differentiation from members of other sections of *Phoma* is therefore often difficult. Most species of sect. *Heterospora* are pathogens with a restricted host range and/or distribution. Thus the host-fungus index on p. 125, including a code indicating conidial variability *in vivo*, may be very helpful in identification of species.

1. a.   Conidia hyaline, isolates associated with disease symptoms **2**
    b.   Conidia with a typical yellow tinge, they are usually somewhat curved and attenuated at one end, aseptate, 4–8.5 × 1.5–3 µm, or 1(–2)-septate, 7–16 × 2–3 µm ....*Ph. samarorum*
         [Common saprophyte in Eurasia and North America. *In vivo*, conidia much more variable, small-aseptate or large and 1–3-septate, sometimes up to 25 × 3.5 µm, synanam. *Stag. fraxini*.]

2. a.   Large conidial dimorph always present, aseptate ('macrosporoid'), c. half mixed with small conidia (phomoid), on Solanaceae **3**
    b.   Large conidial dimorph, if present, becoming septate ('asochytoid-stagonosporoid'). Most conidia always small-aseptate (phomoid) **4**

3. a.   Large conidia 8.5–12.5 × 5–6.5 µm, broadly cylindrical, sometimes median constricted. Small conidia 4–7.5 × 2–5 µm, subglobose to ovoid. Both types with numerous guttules...*Ph. andigena*
         [Pathogen of wild and cultivated tuber-bearing *Solanum* spp. in the Andes of South America. *In vivo* conidia also dimorph, the large type often 14–22 × 5–7 µm, occasionally 1-septate, the small type mostly 6–8 × 2–2.5 µm.]
    b.   Large conidia 8.5–19.5(–25.5) × 3–5.5 µm, elongate ellipsoidal or subcylindrical, often constricted in the middle, with numerous small, polar guttules. Small conidia 5–8 × 2–3 µm, elongate ellipsoidal or ovoid, eguttulate. Characteristic production of whitish, dichotomous branched crystal strands ....*Ph. crystalliniformis*
         [Pathogen of tomato and wild species of potato in the Andes of South America. *In vivo* also dimorph and conidia similar in shape and size to those *in vitro*.]

4. a. Growth rate slow, < 3.5 cm after 7 days, on *Chenopodium* spp. **5**
    b. Growth rate moderate to fast, > 3.5 cm after 7 days **6**

5. a. Conidia usually of two different types; mainly small, aseptate, usually
        4–7 × 1.5–2 μm, but some much larger, mostly 1–2-septate, 12.5–26.5 ×
        3–5 μm..*Ph. heteromorphospora*
        [Common pathogen of *Chenopodium* spp. in Europe. *In vivo*, also het-
        erosporous; the large sized conidia mostly aseptate, but sometimes 1(–3)-
        septate, up to 27 × 7 μm.]
    b. Conidia always aseptate phomoid, 3–5.5 × 1.5–2 μm.........*Ph. dimorphospora*
        [Common pathogen of *Chenopodium* spp. in North and South America. *In
        vivo* conidia always dimorph, partly small-aseptate, partly large-aseptate or
        1(–2)-septate, up to 25 × 7 μm.]

6. a. Colonies producing multicellular chlamydospores, commonly known as
        dictyochlamydospores (compare sect. *Peyronellaea*, **D**). Unicellular
        chlamydospores may also be present **17**
    b. Dictyochlamydospores absent, but unicellular chlamydospores or unicellular
        chlamydospore-like structures may be present **7**

7. a. Colonies producing a diffusible pigment, staining the agar yellowish to ochre **8**
    b. Colonies greenish olivaceous to olivaceous, not producing a diffusible
        pigment **11**

8. a. Yellow/green crystals are formed on MA, NaOH reaction reddish (not an E$^+$
        reaction), phomoid conidia (sub)cylindrical to ellipsoidal **9**
    b. No crystals are formed, NaOH reaction yellow/green, gradually changing
        to red (E$^+$ reaction), phomoid conidia ellipsoidal to more or less obclavate-
        fusiform **10**

9. a. On OA diffusible pigment staining the agar pale luteous to amber/ochraceous,
        often with a distinct yellow pigmentation around the pycnidia, conidia of two
        types, phomoid aseptate, 4.5–9.5 × 2–3 μm, or ascochytoid 1-septate,
        14–28.5 × 4–7 μm, guttulate ....*Ph. actaeae*
        [Pathogen of *Actaea* and *Cimicifuga* spp. Recorded from Europe and North
        America. *In vivo*, conidia sometimes small-aseptate, but usually large and
        1(–2)-septate, up to 28.5 × 7 μm, synanam. *Stag. actaeae.*]
    b. On OA diffusible pigment staining the agar primrose to olivaceous buff, no distinct
        yellow pigmentation around the pycnidia, conidia usually aseptate phomoid,
        4–6.5 × 1.5–2 μm, without or with a few small guttules, occasionally large and 1-
        septate, ascochytoid, 14.5–24 × 4–7 μm ....*Ph. dennisii* var. *dennisii*
        [Recorded from *Solidago* spp. in Europe and North America. *In vivo*, conidia
        small-aseptate or mainly large and 1–2-septate, up to 28 × 6 μm, synanam.
        Stag. dennisii.]
        Note: A similar fungus, but lacking the diffusible pigment and with
        somewhat smaller 1-septate conidia, has once been isolated from a human
        cornea, USA ....*Ph. dennisii* var. *oculo-hominis*

10. a. Growth rate on OA *c.* 8 cm after 7 days, diffusible pigment staining the agar
        saffron to fulvous, conidia usually aseptate, 6–10.5 × 2–4 μm, occasionally 1-
        septate up to 13 × 5 μm...*Ph. glaucii*

[Reported from various wild *Papaveraceae* in Europe. *In vivo* conidia similar or larger, occasionally aseptate, mostly 1(–2)-septate, sometimes distinctly large, up to 23 × 6 μm, synanam. *Stag. chelidonii.*]

    b. Growth rate on OA 5.5–7.5 cm after 7 days, diffusible pigment staining the agar rosy buff to honey, conidia usually aseptate, 5–8 × 2–2.5 μm, occasionally 1-septate up to 15 × 5 μm ....*Ph. aquilegiicola* [Worldwide recorded pathogen of wild and cultivated *Aquilegia* and perennial *Aconitum* and *Delphinium* spp.: dark leaf and stem lesions, crown rot. Occasionally also on other Ranunculaceae. *In vivo* conidia similar or mostly larger and mainly 1(–2)-septate, up to 25 × 5.5 μm, synanam. *Stag. aquilegiae.*]

11. a. Conidia usually of two types, mainly phomoid and aseptate with some distinctly large septate ascochytoid-stagonosporoid conidia **12**
    b. Conidia always of phomoid type, i.e. usually aseptate, but occasionally elongated and 1-septate [they may reach double the length of aseptate conidia, but not three times the length as with the ascochytoid-stagonosporoid conidia *in vivo*] **13**

12. a. Phomoid aseptate conidia usually small-cylindrical, 5.5–9 × 1.5–2 μm; occasionally some larger 1-septate conidia, mostly 9–15 × 2–4 μm (in fresh cultures distinctly larger and often more-celled, stagonosporoid) ....*Ph. nigripycnidia* [Specific pathogen of *Vicia cracca* and other *Vicia* spp. in south-eastern Europe: leaf and stem spot. *In vivo*, conidia sometimes small-aseptate, but usually very large and 1–2(–4)-septate, up to 45 × 12 μm, synanam. *Stag. nigripycnidiicola.*]
    b. Phomoid aseptate conidia variable, 4–15 × 1.5–5 μm, always some large 1-septate ascochytoid conidia 15.5–22 × 4–5 μm, usually guttulate, ellipsoidal to fusiform-allantoid ....*Ph. delphinii* [Specific pathogen of *Delphinium* spp., but also found on *Aconitum* spp. All records are from Europe. *In vivo*, conidia may be small-aseptate, but also mainly large and 1(–2)-septate, often 15–22 × 4–5 μm, synanam. *Stag. delphinii.*]

13. a. NaOH test positive, yellow/green, later red (E$^+$ reaction), conidia ellipsoidal to more or less obclavate-fusiform, on Ranunculaceae **10**
    b. NaOH test negative or not specific (E$^-$) **14**

14. a. Pycnidia initially honey coloured, later black. Conspicuous ostioles **15**
    b. Pycnidia already initially black, ostiole inconspicuous **12a**

15. a. Colonies in addition to pycnidia producing unicellular chlamydospores in short chains or clusters. Aerial mycelium scarce on OA, pycnidia non-papillate or papillate, conidia mainly aseptate, 5.5–13.5 × 2.5–3.5 μm, 1-septate conidia 9.5–15(–21) × 2.5–4 μm...*Ph. schneiderae* [Specific pathogen of lupins, *Lupinus* spp., indigenous to North and South America. Occasionally also found in Europe: leaf spot and blight. *In vivo* conidia similar or mostly larger, 1–3-septate, 15–30 × 5 μm, synanam. *Stag. lupini.*]
    b. Chlamydospores absent **16**

16. a. Colony with coarsely floccose aerial mycelium on OA, pycnidia globose to subglobose, conidia usually aseptate, 3.5–9 × 1.5–2.5 μm, occasionally 1-septate, up to 11 × 3.5 μm...*Ph. subboltshauseri*

[Worldwide recorded pathogen of bean, *Phaseolus vulgaris*: leaf spot disease. Recently also found on cowpea, *Vigna unguiculata. In vivo* conidia always predominantly large, 1–3(–5)-septate, up to 34 × 9 µm, synanam. *Stag. hortensis.*]

    b.   Colony with floccose aerial mycelium on OA, usually sparse after 14 days, pycnidia globose to papillate, sometimes with an elongated neck, conidia usually aseptate, 4–8.5 × 1.5–3 µm, occasionally 1-septate, 7–16 × 2–3.5 µm ....*Ph. trachelii*

[Common seed-borne pathogen of *Campanula* and *Trachelium* spp. in Eurasia and North America: leaf, stem and flower spot. *In vivo*, conidia sometimes all small and aseptate, or larger and mainly 1(–2)-septate, up to 23 × 6 µm, synanam. *Stag. bohemica.*]

17. a.   Growth rate moderate, 4.5–5.5 cm on OA after 7 days, colony colourless to olivaceous grey, needle-like crystals may be formed, dictyochlamydospores usually intercalary, conidia usually 4–8.5 × 2–3 µm, frequently also 1-septate, up to 13 × 4 µm...*Ph. clematidina*

[Isolated from leaf spots and stem lesions on naturally wilting cultivars and hybrids of *Clematis* spp. In Australia, Eurasia and North America. Apparently an opportunistic parasite. *In vivo*, also distinctly large 1-septate ascochytoid conidia may be produced, up to 28 × 6.5 µm.]

    b.   Growth rate fast, 8–8.5 cm on OA after 7 days, colony grey olivaceous, or olivaceous grey to dull green, crystals absent, dictyochlamydospores usually intercalary and solitary, conidia mostly 4–7.5 × 2.5–3.5 µm, occasionally septate, up to 15 × 5.5 µm ....*Ph. narcissi*

[Worldwide recorded pathogen of Amaryllidaceae, esp. *Narcissus* and *Hippeastrum* spp.: leaf spot, neck rot, red spot disease, red leaf spot. *In vivo*, distinctly large, mainly 3-septate stagonosporoid conidia are often produced, up to 28 × 8 µm, synanam. *Stag. curtisii.*]

Distribution and Facultative Host Relations in Section *Heterospora*

The conidial variability found *in vivo* in different pycnidia (,) or/and in the same pycnidium (+) is coded as follows:

o   =   aseptate relatively small phomoid conidia
θ   =   secondarily 1-septate phomoid conidia
O   =   aseptate distinctly large conidia
Ө   =   secondarily 1(–2)-septate large conidia ('ascochytoid')
Ө   =   secondarily 1–3(–5)-septate large conidia ('stagonosporoid')

Plurivorous   Eurasia and North America, esp. temperate regions:
                                                Ph. samarorum
                                               (synanam. *Stag. fraxini*)
                                               [In necrotic tissue of herbaceous, gramineous and woody plants; conidia o,+ (O+) Ө (+Ө).]

With specific or preferred host

Amaryllidaceae
esp. *Narcissus* and
Hippeastrum spp.

Ph. narcissi
(synanam. *Stag. curtisii*)
[Worldwide known as causal organism of red spots on leaves, bulb scales and flowers; conidia o(+θ), (O+)Θ + ⊖.]

Campanulaceae
Campanula and
Trachelium spp.

Ph. trachelii
(synanam. *Stag. bohemica*)
[In Eurasia, North and South America frequently recorded as causal organism of leaf, stem and flower spot; conidia o,+ (O+)Θ.]

Chenopodiaceae
Chenopodium spp.

Ph. dimorphospora
[In North and South America recorded in association with leaf spots and stem lesions; conidia o + O + Θ(+⊖).]
Ph. heteromorphospora
[In Europe frequently found in association with leaf spots and stem lesions; conidia o + O + Θ + ⊖.]

Compositae
Solidago spp.

Ph. dennisii var. *dennisii*
(synanam. *Stag. dennisii*)
[*Solidago* spp. are possibly also the natural source of the var. *oculo-hominis.*]
[In Europe and North America found on, and isolated from last year's dead stems; so far no data on pathogenicity; conidia o, (O+)Θ(+⊖).]

Leguminosae
Lupinus spp. (esp.
American species)

Ph. schneiderae
(synanam. *Stag. lupini*)
[In North and South America well known as causal organism of spots of leaves, stems and pods; occasionally also recorded in Europe; leaf spot and blight; conidia o + Θ + (O+)Θ + ⊖.]

Phaseolus vulgaris; occ.
also on *Vigna unguiculata*

Ph. subboltshauseri
(synanam. *Stag. hortensis*)
[Worldwide known as causal organism of stunting and red-brown blotches on stems, leaves and pods; leaf spot disease; conidia o + (O+)Θ + ⊖.]

Vicia spp., esp. *V. cracca*

Ph. nigripycnidia
(synanam. *Stag. nigripycnidiicola*)
[In south-eastern Europe frequently recorded as causal organism of leaf and stem spots: leaf spot, stem lesions; conidia o, Θ + ⊖.]

Papaveraceae,
e.g. *Chelidonium,*   *Ph. glaucii*
Corydalis, Dicentra and   (synanam. *Stag. chelidonii*)
Glaucium spp.   [In Europe frequently recorded and isolated from wild
                                            species; conidia o(+θ), (O+)Θ(+⊖).]

Ranunculaceae
esp. *Aconitum, Aquilegia,*   *Ph. aquilegiicola*
but also on perennial   (synanam. *Stag. aquilegiae*)
Delphinium spp.   [Worldwide recorded pathogen, found in association with
                                            leaf spots, stem lesions and crown rot; conidia o(+θ),
                                            (O+)Θ(+⊖).]

Actaea and   *Ph. actaeae*
Cimicifuga spp.   (synanam. *Stag. actaeae*)
                                            [In Europe and North America recorded on living and wilt-
                                            ing leaves of wild and cultivated plants; conidia o,
                                            (O+)Θ(+⊖).]

Clematis spp.   *Ph. clematidina*
                                            [In Australasia, Eurasia and North America frequently
                                            recorded on cultivated species in association with lesions
                                            on stems and leaves; conidia o(+θ),+ (O+)Θ(+⊖).]

esp. *Delphinium* spp.,   *Ph. delphinii*
but also recorded on   (synanam. *Stag. delphinii*)
Aconitum spp.   [Only recorded in Europe; found in association with leaf
                                            spots and stem rot; conidia o(+θ),+ (O+)Θ(+⊖).]

Solanaceae
Lycopersicon esculentum   *Ph. crystalliniformis*
                                            [In the Andes of South America (esp. in Colombia and
                                            Venezuela) recorded pathogen, causing necrotic spots on
                                            all aerial parts of tomato, followed by total plant necrosis
                                            and mummification of the fruits: local name 'Carate'; coni-
                                            dia o + O.]

Solanum spp.   *Ph. andigena*
Serie *Tuberosa*   [In the Andes of South America (esp. in Bolivia and Peru)
                                            recorded pathogen of wild and cultivated species of potato,
                                            causing numerous small leaf spots, which later coalesce:
                                            black potato blight and leaf spot; conidia o + O(+Θ).]
                                            Ph. crystalliniformis
                                            [In the Andes region of Colombia and Venezuela this nox-
                                            ious pathogen of tomato (see above) is also found in asso-
                                            ciation with leaf spots on *S. tuberosa* subsp. *andigena*;
                                            conidia o + O.]

Seeds and fruits

Soil   *Ph. samarorum*
                                            (synanam. *Stag. fraxini*)
                                            [See above under plurivorous; conidia o,+ (O+)⊖(+Θ).]

Substrata of mammal (incl. human) origin
                    Ph. dennisii var. *oculo-hominis*
              [Isolated from a human cornea (ulcer) in the USA.]

Descriptions of the Taxa

Differentiation based on study in culture. Characteristics *in vitro* and *in vivo*.

Phoma actaeae Boerema *et al.,* Fig. 15A

Pycnidial dimorph with large conidia, 1(occ. 2)-septate, ascochytoid (-stagonosporoid): ***Stagonosporopsis actaeae*** **(Allesch.) Died.**

> *Phoma actaeae* Boerema, Gruyter & Noordel. *in* Persoonia **16**(3): 347. 1997.
> synanamorph:
> > *Stagonosporopsis actaeae* (Allesch.) Died. *in* Annls mycol. **10**: 141. 1912.
> > > ≡ *Actinonema actaeae* Allesch. *in* Ber. bayer. bot. Ges. **5**: 7. 1897.
> > > = *Marssonina actaeae* Bres. *in* Hedwigia **32**: 33. 1893.
> > > > ≡ *Ascochyta actaeae* (Bres.) Davis *in* Trans. Wis. Acad. Sci. Arts Lett. **19**: 656. 1919.

Diagn. data: Diedicke (1912) (disease symptoms may explain original classification in *Actinonema*: cuticle sometimes wrinkled in dendritic lines; first indication as a species of *Stagonosporopsis in vivo*), (1912–15) (fig. 14a, b: morphology of pycnidium and conidia *in vivo*; description as *Stagonosporopsis actaeae*), Boerema *et al.* (1997) (fig. 2A: drawings of conidia *in vitro* and *in vivo*, reproduced in present Fig. 15A; *in vitro* and *in vivo* descriptions, partly adopted below; formal diagnosis of *Ph. actaeae in vitro*).

Description in vitro

Pycnidia irregular globose, with 1(–2) non-papillate or papillate ostioles, 80–250 μm diam., glabrous or with some hyphal outgrowths, solitary or confluent, on and in the agar and also in aerial mycelium. Micropycnidia frequently occur, 40–80 μm diam. Conidial matrix white to salmon.
      Conidia always mainly phomoid, aseptate, (sub-)cylindrical to ellipsoidal, with several small, more or less polar guttules, variable in dimensions, 4.5–10 × 2–3 μm; occasionally some much larger and 1-septate (ascochytoid), usually with several large guttules, 14–28.5 × 4–7 μm.
      Colonies on OA 6.5–8 cm diam. after 7 days, regular to somewhat irregular, pale luteous to amber/ochraceous due to a diffusible pigment which on application of a drop of NaOH changes to scarlet/vinaceous, finely floccose to coarsely floccose/woolly, pale olivaceous aerial mycelium; reverse pale luteous to ochraceous with citrine due to the production of crystals [citrine green and needle-like but also small, yellowish/brownish; the needle-like crystals develop abundantly in colonies on MA]. Representative cultures CBS 105.96, CBS 106.96.

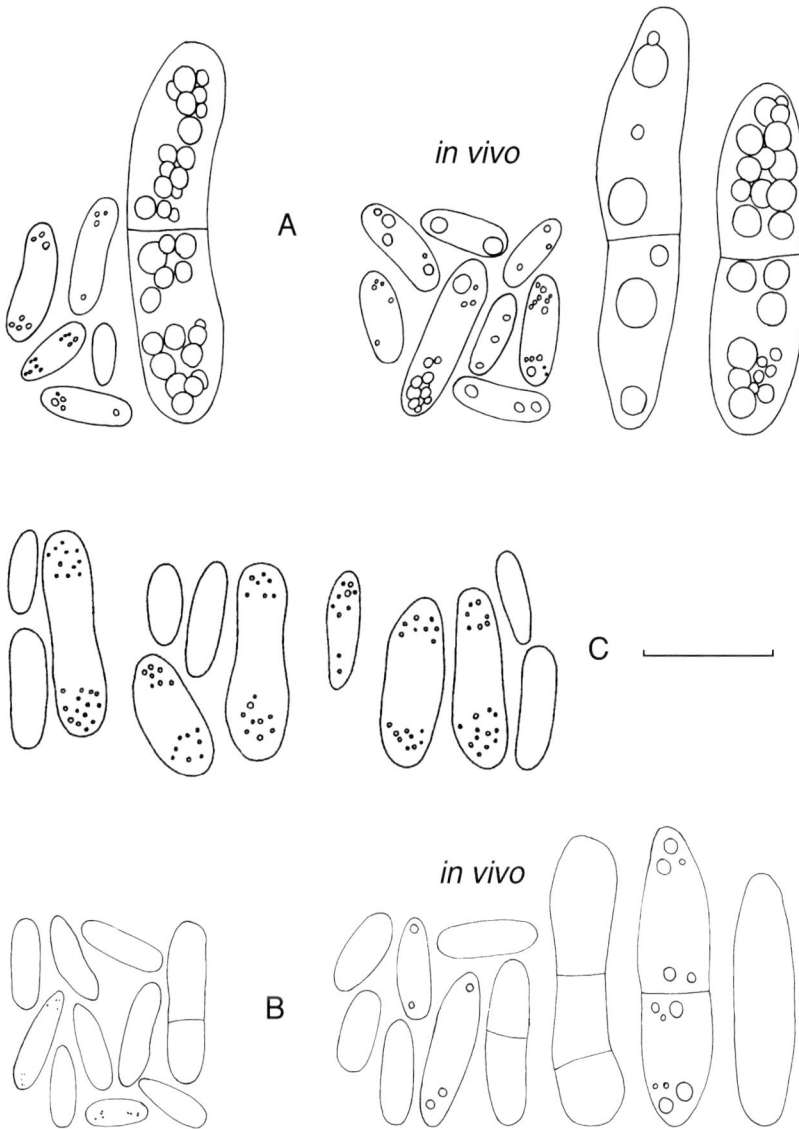

Fig. 15. A *Phoma actaeae. In vitro* conidia variable, mainly relatively small and aseptate, occasionally large and 1-septate. *In vivo* the pycnidia sometimes contain only small phomoid conidia, but usually only large 1(–2)-septate conidia, the *Stagonosporopsis* synanamorph: *S. actaeae.* **B** *Phoma aquilegiicola.* Conidia phomoid *in vitro,* usually aseptate, but some larger ones 1-septate. Pycnidia *in vivo* may contain similar conidia, aseptate and occasionally 1-septate, but mostly much larger conidia, usually 1(–2)-septate, synanamorph *Stagonosporopsis aquilegiae.* **C** *Phoma crystalliniformis.* Conidia always dimorphic, *in vitro* and *in vivo,* aseptate and variable in shape; the small conidia eguttulate, the larger with numerous small polar guttules. Bar 10 μm.

In vivo *(especially on* Actaea spicata*)*

Pycnidia (on indefinite blackened areas of the leaves, epiphyllous, scattered) mostly 100–130 µm diam., subglobose with 1(–2) more or less papillate ostioles.

Conidia sometimes phomoid and aseptate, but usually large and 1(–2)-septate (ascochytoid-stagonosporoid), characteristic for the synanamorph *Stagonosporopsis actaeae*: cylindrical, straight or somewhat curved, usually with several guttules, mostly 17–24 × 5–6 µm, hyaline but becoming some-what olivaceous with age.

Ecology and distribution

In Europe and North America (USA) the *Stagonosporopsis* synanamorph is recorded on living and wilting leaves of wild and cultivated *Actaea* and *Cimicifuga* species. The fungus seems to be a specific pathogen of these peren-nial plants belonging to the Ranunculaceae.

Phoma andigena Turkenst., Fig. 14C

Conidia dimorphic, *c.* half small and half large in same pycnidium [*in vitro* and *in vivo*].

> *Phoma andigena* Turkenst. apud Boerema, de Gruyter & Noordel. *in* Persoonia **16**(1): 131. 1995.
>> H ≡ *Phoma andina* Turkenst. *in* Fitopatología **13**: 67–68. 1978; not *Phoma andina* Sacc. & P. Syd. *in* Annls mycol. **2**: 170. 1904.

Diagn. data: Turkensteen (1978) (original Latin description), Turkensteen in Hooker (1980) (fig. 49F: shape of 'infective' conidia; description *in vitro*, pig-ment production), EPPO (1984) (fig. 1: disease symptoms; notes on biology and economic importance, conidial morphology and cultural characters), Noordeloos *et al.* (1993) (fig. 2: conidial shape, reproduced in present Fig. 14C; detailed description, partly adopted below; production of crystals and pig-ment of 'radicinin').

Description *in vitro*

Pycnidia pale at first, then black, globose with late developing opening (pore), 55–160 µm diam., glabrous, scattered, but often more or less concentrical arranged, on the agar and in the aerial mycelium.

Conidia always dimorphic, usually unicellular and variable in shape, with numerous guttules; one type relatively small, 4–7.5 × 2–5 µm (mean 5 × 4 µm), subglobose to ovoid, the second type 8.5–12.5 × 5–6.5 µm (mean 11 × 5.6 µm), broadly cylindrical, sometimes with median constriction. In old cul-tures occasionally considerably larger, 1-septate conidia have been found.

Colonies on OA 3–3.5(–4) cm diam. after 7 days, olivaceous greenish to olivaceous black, often with darker concentric zones and/or sectors, radically fil-amentous; aerial mycelium sparse, white, floccose, especially at centre; reverse similar, a slightly more grey. Crystals (radicinin) may occur within 2–3 weeks, but often poorly developed, the substance usually diffuses as yellow pigment into the agar. Under certain conditions complexes or chains of darker coloured swollen cells may develop. Representative culture CBS 101.80.

In vivo *(tuber-bearing* Solanum *spp.)*

Pycnidia (embedded in small leaf spots, at the upper side only, later also in elongated lesions on stems and petioles) resembling those *in vitro*, but usually somewhat larger, 125–200 μm diam.

Conidia also dimorphic, but somewhat longer than those *in vitro*; the small type mostly 6–8 × 2–2.5 μm, the large type often 14–22 × 5–7 μm, usually aseptate, but occasionally 1-septate, pluriguttulate.

Ecology and distribution

A pathogen of wild and cultivated species of potato, *Solanum* spp. series *Tuberosa* (Solanaceae) in the Andes of South America (Bolivia and Peru) at altitudes above 2000 m, causing numerous small leaf spots which later may coalesce so that leaves turn blackish and appear scorched: black potato blight and leaf spot. The fungus survives in soil on plant debris; infection takes place in cool, moist weather with temperatures below 15°C. The small conidia should be not infective and should not germinate on artificial media. *Ph. andigena* (*andina*) is included in the list of quarantine organisms of the European Plant Protection Organization (EPPO, 1984).

Phoma aquilegiicola M. Petrov, Fig. 15B

Pycnidial dimorph with large conidia, occ. aseptate, mostly 1(–2)-septate, ascochytoid: *Stagonosporopsis aquilegiae* (Rabenh.) Boerema *et al.*

> *Phoma aquilegiicola* M. Petrov *in* Acta Inst. Bot. Acad. Sci. USSR Pl. crypt. [Trudÿ bot. Inst. Akad. Nauk SSSR] Fasc. 1: 281. 1933.
>> V = *Phyllosticta aquilegiicola* Brunaud *in* Act. Soc. linn. Bordeaux **54**: 244. 1890 [as 'aquilegicola'].
>> = *Sclerophomella aconiticola* Petr. *in* Hedwigia **65**: 308. 1925.
>> Θ ≡ *Phoma aconiticola* Petr. *in* Fungi polon. exs. No. 643. 1921 [without description].
>> = *Phoma aconiticola* Nagai, Shishido & Tsuyama *in* J. Fac. Agric. Iwata Univ. **10**(1): 23. 1970.
> synanamorph:
> *Stagonosporopsis aquilegiae* (Rabenh.) Boerema, Gruyter & Noordel. *in* Persoonia **16**(3): 354. 1997.
>> ≡ *Depazea aquilegiae* Rabenh. *in* Klotzschii Herb. mycol. Cent. 17, No. 165. 1852.

H = *Ascochyta aquilegiae* (Rabenh.) Höhn. *in* Annls mycol. **3**: 406. 1905 [later homonym, see below].

= *Phyllosticta aquilegiae* Roum. & Pat. *in* Revue mycol. **5**: 28. 1883.

≡ *Ascochyta aquilegiae* (Roum. & Pat.) Sacc. *in* Sylloge Fung. **3**: 396. 1884.

≡ *Ascochytella aquilegiae* (Roum. & Pat.) Tassi *in* Boll. R. Orto bot. Siena **5**: 27. 1902.

≡ *Actinonema aquilegiae* (Roum. & Pat.) Grove *in* J. Bot., Lond. **56**: 343. 1918.

= *Diplodina delphinii* Laskaris *in* Phytopathology **40**: 620. 1950 [July].

≡ *Ascochyta laskarisii* Melnik *in* Nov. Sist. niz. Rast. **8**: 211. 1971; *in* Mikol. Fitopatol. **7**(2): 142. 1973 [not *Ascochyta delphinii* Melnik *in* Nov. Sist. niz. Rast. **1968**: 173. 1968 = *Phoma delphinii* (Rabenh.) Cooke, this section].

H = *Diplodina delphinii* Golovin *in* Cent. Asian Univ. Studies II [N.S.], **14**(5): 34. 1950 [December].

Diagn. data: Laskaris (1950) (figs 1, 2; disease symptoms on *Delphinium* cultivars; agens *in vitro* initially classified as *Phoma* sp.; description of large sized conidial dimorph *in vivo* as *Diplodina delphinii*; drawings of conidia, both *in vivo* and *in vitro*, and chlamydospore-like swollen cells and pycnidial primordia *in vitro*), Cejp (1965) (disease symptoms on *Aquilegia* cultivars; description of *Phoma*-phenotype *in vivo* as *Phyllosticta aquilegiicola*), Nagai *et al.* (1970) (fig. 1; description of *Phoma*-phenotype on *Aconitum* cultivars as *Phoma aconiticola*; drawings of pycnidia and conidia *in vivo*), Buchanan (1987) (fig. 1; description and drawing of large sized conidial dimorph as *Ascochyta aquilegiae*; *Phoma* phenotype *in vivo* considered to be a different fungus), Boerema *et al.* (1997) (fig. 3A: different conidial shape *in vitro* and *in vivo*, reproduced in present Fig. 15B; synonymy and *in vitro* and *in vivo* descriptions, partly adopted below; similarity with pycnidial anamorph of *Didymella inaequalis* Corbaz), de Gruyter & Boerema (1998) (note on the taxonomic and nomenclatural confusion of *Phoma* spp. on *Delphinium*).

Description in vitro

Pycnidia globose to subglobose, with usually 1 non-papillate or slightly papillate ostiole, 120–300 μm diam., glabrous or with mycelial outgrowths, solitary, on and in the agar, and in aerial mycelium. Micropycnidia usually present, 30–60 μm diam. Conidial matrix white, occasionally rosy buff.

Conidia always mainly phomoid, aseptate, ellipsoidal to more or less obclavate-fusiform, eguttulate or with some small polar guttules, mostly 5–8 × 2–2.5 μm, occasionally some large 1-septate conidia are present, up to 15 × 4 μm.

Colonies on OA 5.5–7.5 cm diam. after 7 days, regular, usually colourless, with finely woolly/floccose, white to pale olivaceous grey aerial mycelium; some isolates staining the agar rosy buff/saffron with a diffusible pigment;

reverse similar. All strains tested showed E^+ reaction on application of a drop of NaOH (green → red). True chlamydospores were absent, but chains and clusters of dark swollen cells may form. Representative cultures CBS 108.96, CBS 107.96.

In vivo *(especially on* Aquilegia vulgaris*)*

Pycnidia (subepidermal on leaf lesions which are usually marginal, grey to brown with a dark brown border and sometimes with radiating mycelial fibrils; also on carpels and old stems) mostly 80–170 μm diam., relatively thick-walled, globose to subglobose with 1 distinct ostiole.

Conidia of pycnidia on leaf lesions mostly large and septate (ascochytoid/stagonosporoid), typical for the synanamorph *Stagonosporopsis aquilegiae*: cylindrical or somewhat irregular, guttulate or eguttulate, 1(–2)-septate and occasionally broken at the septa, but sometimes aseptate, mostly (10–)13–20 × (3–)4–5(–5.5) μm. In pycnidia on withered leaves, stems and carpels phomoid conidia often predominate, mostly aseptate but occasionally 1-septate, similar to conidia produced *in vitro*, but they are usually larger and more variable in shape, eguttulate or guttulate, (5–)6–9(–14) × (2–)2.5–3.5(–5) μm.

Ecology and distribution

In Eurasia frequently found in association with dark leaf spots and stem lesions on wild and cultivated *Aquilegia* spp. (Ranunculaceae; perennials), especially *A. vulgaris*. However, it occurs also on other Ranunculaceae, e.g. perennial *Aconitum* and *Delphinium* spp. In the USA it has been an important cause of crown rot of cultivated delphiniums. In Japan the fungus has recently caused severe damage on cultivated aconitums. The *Stagonosporopsis* synanamorph is also recorded in Australasia (New Zealand).

Phoma clematidina (Thüm.) Boerema, Fig. 16A

Conidial dimorph large, 1-septate, asochytoid [not obtained *in vitro*].

> *Phoma clematidina* (Thüm.) Boerema apud Boerema & Dorenbosch *in* Versl. Meded. plziektenk. Dienst Wageningen **153** (Jaarb. 1978): 17[–18]. 1979.
>> ≡ *Ascochyta clematidina* Thüm. *in* Bull. Soc. imp. Nat. Moscou **55**: 98. 1880.
>> = *Phyllosticta clematidis* Brunaud *in* Annls Soc. Sci. nat. Charente-Infér. **26**: 9. 1889.
>> ‡ ≡ *Phyllosticta clematidicola* Brunaud *in* Act. Soc. linn. Bordeaux **53**: 9. 1898 [superfluous name, replacing an older homonym, Art. 52].
>> = *Ascochyta vitalbae* Briard & Har. apud Briard *in* Revue mycol. **13**: 17. 1891.
>>> ≡ *Diplodina vitalbae* (Briard & Har.) Allesch. *in* Rabenh. Krypt.-Flora [ed. 2] Pilze **6** [Lief. 69]: 683. 1900 [vol. dated '1901'].

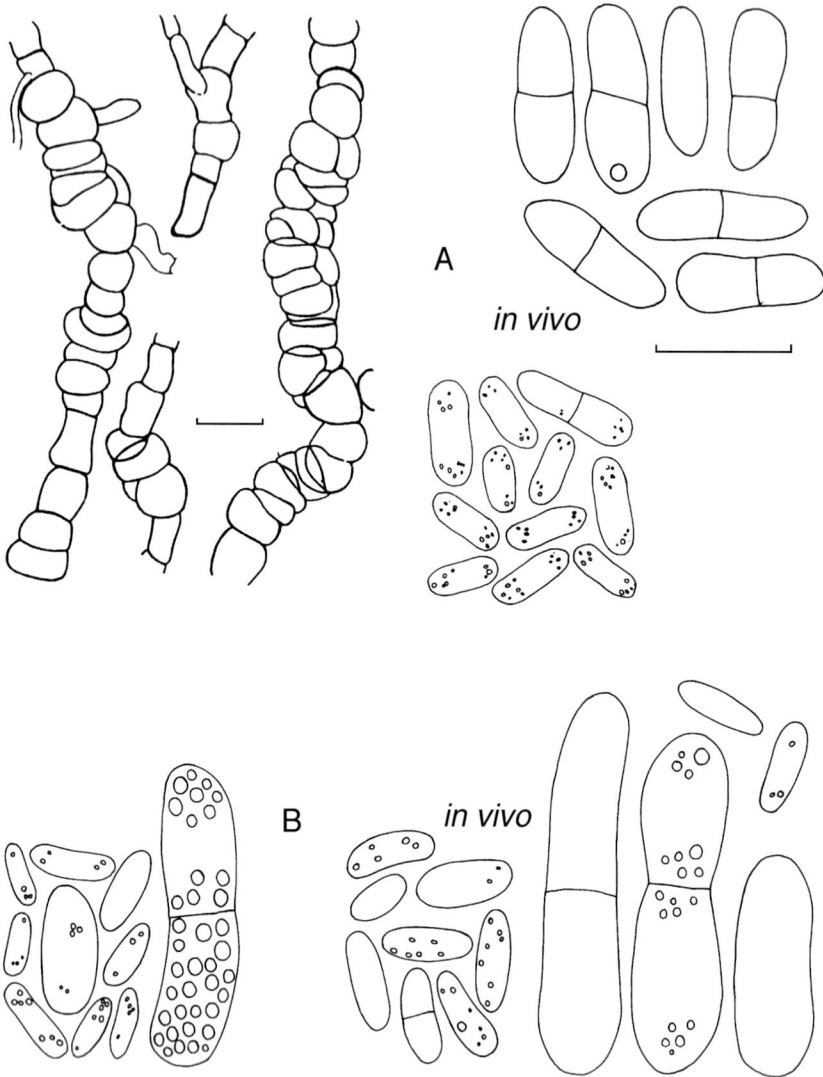

Fig. 16. A *Phoma clematidina*. Conidia *in vitro* mainly aseptate, but some frequently also 1-septate. Conidia *in vivo* aseptate and 1-septate, but larger than those *in vitro*; sometimes distinctly large and mainly 1-septate (ascochytoid). Multicellular chlamydospores intercalary, often somewhat botryoid and in combination with unicellular chlamydospores. **B** *Phoma delphinii*. Conidia *in vitro* mostly aseptate and notably variable in shape and size, but always including a number of large 1-septate conidia. Pycnidia *in vivo* may contain mainly aseptate conidia, but these are usually less variable than *in vitro*; most conidia, however, may also be large and 1(–2)-septate; synanamorph *Stagonosporopsis delphinii*. Bars 10 μm (the smaller referring to the chlamydospores).

> = *Diplodina clematidina* Fautrey & Roum. apud Roumeguère *in* Revue mycol. **14**: 105. 1892.
> H = *Phyllosticta clematidis* Ellis & Dearn. *in* Can. Rec. Sci. **5**: 268. 1893; not *Phyllosticta clematidis* Brunaud, see above.
> = *Ascochyta indusiata* Bres. *in* Hedwigia **35**: 199. 1896.
> = *Ascochyta davidiana* Kabát & Bubák *in* Öst. bot. Z. **54**: 25. 1904.

Diagn. data: Gloyer (1915) (pls L, LI, LII fig. 1, LIII: disease symptoms; pl. LII figs 2, 3: drawing of chlamydospores and conidia; pl. IV figs 1, 2: photographs of pycnidium and culture on agar), Ebben & Last (1966) (figs 1, 2: disease symptoms; figs 3, 4: cultures on OA and MA, and leafpatch with pycnidia), Boerema & Dorenbosch (1979) (taxonomy), Boerema (1993) (fig. 6A: drawings of pycnidia, the conidial phenotypes and chlamydospores reproduced in present Fig. 16A; description, adopted below), Boerema *et al.* (1997) (discussion of conidial dimorphism), van der Aa & Vanev (2002) (*Phyllosticta* synonyms).

Description in vitro

Pycnidia usually subglobose with dark circumvallated ostioles, 110–120 μm diam., mostly with some hyphae around the ostioles, mostly solitary. Pycnidial wall composed of isodiametric or somewhat elongated cells. Conidial matrix honey-coloured or pinkish.

Conidia most variable in shape and dimensions, subellipsoidal to cylindrical, on OA usually (3.5–)4–8.5(–9) × 2–3(–3.5) μm, occasionally larger and 1-septate, 9–13 × 3–4 μm (*in vivo* usually much larger and mainly 1-septate, ascochytoid, see below).

Chlamydospores usually scanty, uni- or multicellular, where unicellular usually intercalary in short chains, guttulate, thick-walled, green/brown, 8–10 μm diam., where multicellular, irregular dictyo/phragmosporous, often somewhat botryoid and in combination with unicellular chlamydospores, tan to dark brown 3–50 × 12–25 μm.

Colonies on OA 4.5–5.5 cm diam. after 7 days (on MA 2.5–4 cm diam.). Aerial mycelium whitish to olivaceous grey. Hyphae often conspicuously curved, particularly on MA. Reverse buff/yellowish, often with needle-like crystals (anthraquinone pigments). Representative culture CBS 108.79.

In vivo *(Clematis spp.)*

Pycnidia (subepidermal in leaf spots, also in stem lesions and on dead stubs) resembling those *in vitro*, also relatively small, *c.* 120 μm diam.

Conidia different in size, generally phomoid, but larger than those *in vitro*, aseptate or 1-septate, (6–)8–10(–13) × 3–4 μm (av. 9.5 × 3.2 μm). Sometimes, however, the septate conidia are distinctly large (ascochytoid), (10–)12–22(–28) × 4.5–6(–6.5) μm (av. 18.5 × 5.8 μm). In old pycnidia the conidia become dark coloured and occasionally 2-septate.

Ecology and distribution

This fungus is in Eurasia, Australasia and North America frequently isolated from leaf spots and stem lesions on naturally wilting cultivars and hybrids of various *Clematis* spp. (Ranunculaceae). It may be regarded as an opportunistic parasite. The above noted wide variability of the conidia *in vivo* and *in vitro* suggests that the fungus includes a mixture of strains from different geographical origin, i.e. American and Eurasian sources. On account of the production of dictyochlamy- dospores it has also been included in the key of *Phoma* sect. *Peyronellaea* (**D**).

Phoma crystalliniformis (Loer. *et al.*) Noordel. & Gruyter, Fig. 15C, Colour Plate I-F

Conidia dimorphic, about half small and half large in same pycnidium [*in vitro* and *in vivo*].

> *Phoma crystalliniformis* (Loer. *et al.*) Noordel. & Gruyter apud Noordeloos, de Gruyter, van Eijk & Roeijmans *in* Mycol. Res. **97**: 1344. 1993.
>> ≡ *Phoma andina* var. *crystalliniformis* Loer., R. Navarro, Lobo & Turkenst. *in* Fitopatología **21**(2): 100. 1986.

Diagn. data: Loerakker *et al.* (1986) (fig. 1AB: drawings of pycnidia, conidio- genous cells and conidia; fig. 1CD: cultures on OA and MA with production of crystals in the media), Loerakker & Boerema (1987) (fig. 14: disease symp- toms), Noordeloos *et al.* (1993) (fig. 4: conidial shape, reproduced in present Fig. 15C; detailed description, partly adopted below; production of dendritic crystals of 'brefeldin A').

Description in vitro

Pycnidia hyaline at first, then yellow brown, finally blackish, (sub)globose with late developing of a wide opening (pore), initially papillate, later on often fun- nel-shaped, 150–300 µm diam., glabrous, often aggregated to complex struc- tures, in agar and aerial mycelium. Conidial matrix white to pale cinnamon.
        Conidia always dimorphic, aseptate and variable in shape; one type rela- tively small, 5–8 × 2–3 µm (mean 6.6 × 2.5 µm), elongate ellipsoidal or ovoid, without guttules, the second type larger, 8.5–19.5(–25.5) × 3–5.5 µm (mean 13.6 × 4.8 µm), elongate ellipsoidal or subcylindrical, often constricted in the middle, with numerous small polar guttules. Intermediate forms may occur.
        Colonies on OA (5–)5.5–6(–7) cm diam. after 7 days, pallid with darker olivaceous patches or sectors, often zonate with alternating olivaceous buff or greenish zones, aerial mycelium white to pale salmon, sparse, often hardly developed; reverse similar.
        Crystals (brefeldin A) readily formed in a stellate or dendritic pattern of whitish, dichotomous branched strands. Representative culture CBS 713.85.

In vivo *(Lycopersicon esculentum, Solanum tuberosum subsp.* andigena*)*

Pycnidia (superficially occurring on necrotic spots) resembling usually in shape and size those described *in vitro.*
        Conidia also dimorphic and similar in size and shape to those *in vitro.*

Ecology and distribution

A Solanaceae-pathogen recorded on tomato, *Lycopersicon esculentum*, and a wild species of potato, *Solanum tuberosum* subsp. *andigena*, in the Andes of South America (Colombia, Venezuela) at altitudes between 1500 and 3700 m. The fungus causes necrotic spots on all aerial parts of the tomato, followed by total plant necrosis and mummification of the fruits. The tomato disease (crop losses up to 100%) is locally known as 'Carate'. On the wild potato the fungus is frequently found in association with leaf spots.

Phoma delphinii (Rabenh.) Cooke, Fig. 16B

Pycnidial dimorph with large conidia, mostly 1(–2)-septate, ascochytoid: ***Stagonosporopsis delphinii*** **Lebedeva**

> *Phoma delphinii* (Rabenh.) Cooke *in* Grevillea **20**: 113. 1892 [misapplied, see Grove, Br. Coelomycetes **1**: 80. 1935].
>> ≡ *Sphaeria delphinii* Rabenh. *in* Klotzschii Herb. mycol. ed. 2, Cent. 8, No. 747. 1845 [Fiedler's Exs.].
>> H = *Phyllosticta delphinii* Lobik *in* Bolez. Rast. **17** [1928] (3–4) [Morbi Plant **17**]: 167. 1929; not *Phyllosticta delphinii* Clem. ex Seaver *in* N. Am. Flora **6**(1): 64. 1922 [second impression 1961] [= *Asteromella* sp. cf. van der Aa & Vanev, 2002].
>> = *Ascochyta delphinii* Melnik *in* Nov. Sist. niz. Rast. **1968**: 173. 1968.

synanamorph:
> *Stagonosporopsis delphinii* Lebedeva *in* Notul. syst. Inst. cryptog. Horti bot. petropol. **1**(8): 156. 1922.

Diagn. data: Boerema *et al.* (1997) (fig. 4A: conidial dimorphism *in vitro* and *in vivo*, reproduced in present Fig. 16B; synonymy and differentiation against other *Phoma* species on Ranunculaceae; *in vitro* and *in vivo* descriptions, partly adopted below), de Gruyter & Boerema (1998) (note on the taxonomic and nomenclatural confusion of *Phoma* spp. on *Delphinium*).

Description in vitro

Pycnidia globose to subglobose, with 1–2 distinct ostioles, 90–300 µm diam., glabrous, solitary or aggregated, abundant on and in the agar, and in aerial mycelium. Micropycnidia usually present, 30–70 µm diam. Conidial matrix rosy buff to salmon/saffron.

Conidia mostly phomoid, aseptate, ellipsoidal to fusiform-allantoid, usually with several guttules, notably variable in shape and size, 4–15 × 1.5–5 µm; but always including a number of large 1-septate conidia (ascochytoid), 15.5–22 × 4–5 µm.

Colonies on OA 3.5–4.5 cm diam. after 7 days, regular, colourless to grey olivaceous with scarce, finely woolly, pale olivaceous grey aerial mycelium;

reverse olivaceous buff to grey olivaceous. On application of a drop of NaOH a weak greenish bluish discoloration may occur, but does not change to red (E-negative). Chlamydospores absent, but clusters of dark swollen cells may be present. Representative culture CBS 134.96.

In vivo *(especially on* Delphinium consolida*)*

Pycnidia (subepidermal scattered in dark brown leaf spots, also superficial on capsules and old stems), mostly 120–270 µm diam., globose to subglobose with 1(–3) distinct non-papillate ostioles.

Conidia sometimes all aseptate phomoid, (3–)4–7.5 × 1.5–2.5 µm, but often mixed with some larger, ultimately septate conidia, 15–22(–25) × 4–5(–5.5) µm. These large, 1(–2)-septate ascochytoid conidia may predominate, synanamorph: *Stagonosporopsis delphinii*.

Ecology and distribution

Most records of this fungus are on species of *Delphinium* (Ranunculaceae; inclusive *Consolida*) in Europe. In Russia it has also been reported from a species of *Aconitum* (same tribus as *Delphinium*). The most common host seems to be *D. consolida* (*Consolida regalis*). The fungus causes angular, often marginal, dark leaf spots and appears as a necrophyte on capsules and old stems. It closely resembles *Phoma aquilegiicola* of this section (synanam. *Stagonosporopsis aquilegiae*), but can be easily differentiated *in vitro* by its slower growth rate, the negative E reaction and different shape and size of the conidia. The phomoid conidial phenotype is sometimes confused with *Phoma ajacis* (Thüm.) Aa & Boerema (sect. *Phoma*, **A**) and the *Stagonosporopsis* synanamorph with *Phoma xanthina* Sacc. (sect. *Macrospora*, **H**).

Phoma dennisii Boerema var. *dennisii*, Fig. 17A

Pycnidial dimorph large, mainly 1(–2)-septate, ascochytoid: **Stagono-sporopsis dennisii Boerema et al.**

> *Phoma dennisii* Boerema *in* Trans. Br. mycol. Soc. **67**: 307. 1976, var. *dennisii*.
>
> > ≡ *Phoma oleracea* var. *solidaginis* Sacc. *in* Sylloge Fung. **3**: 135. 1884; not *Phoma solidaginis* Cooke *in* Grevillea **13**: 95. 1885.
>
> synanamorph:
> > *Stagonosporopsis dennisii* Boerema, Gruyter & Noordel. *in* Persoonia **16**(3): 350. 1997.

Diagn. data: Boerema & Bollen (1975) (fig. 6B: production of phomoid and ascochytoid conidia *in vitro*; possible relation with conidiogenesis), Boerema (1976) (fig. 13: cultures on OA; tab. 9: synoptic table on cultural characters with drawing of conidia), Boerema *et al.* (1997) (fig. 2B: conidial dimorphism

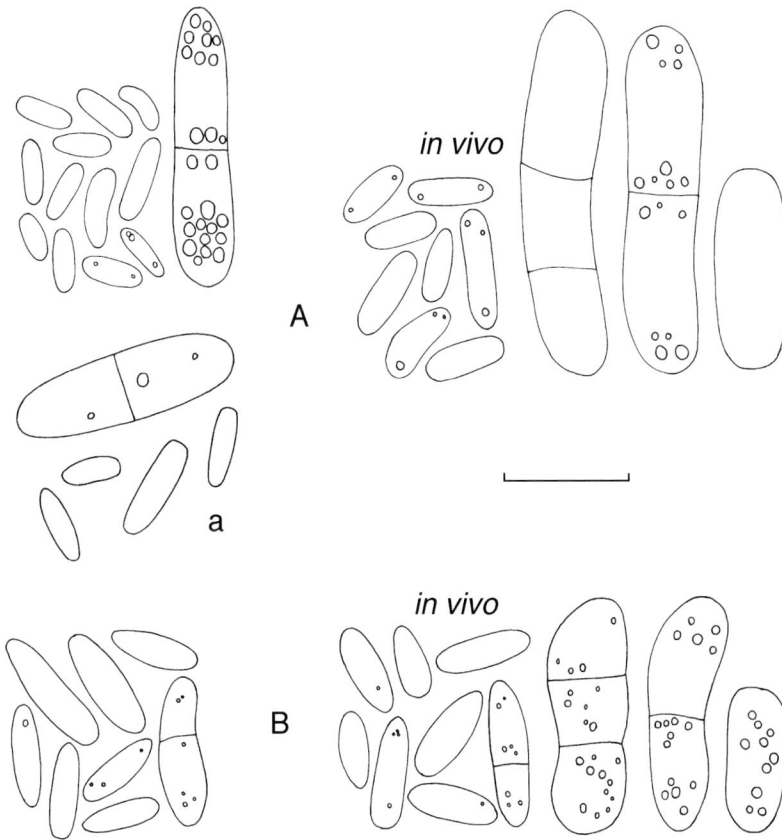

Fig. 17. A *Phoma dennisii.* *In vitro* conidia mainly small and aseptate, but occasionally also some larger 1-septate conidia are present. *In vivo* pycnidia usually contain only small aseptate conidia, but also pycnidia with large, mainly 1–2-septate conidia may occur, synanamorph *Stagonosporopsis dennisii* [**Aa** conidia from original isolate of the var. *oculo-hominis*]. **B** *Phoma glaucii.* *In vitro* conidia variable phomoid, mostly aseptate, occasionally 1-septate. *In vivo* pycnidia may contain similar phomoid conidia, but also larger 0–2-septate conidia, synanamorph *Stagonosporopsis chelidonii.* Bar 10 μm.

in vitro and also *in vivo*, reproduced in present Fig. 17A: formal diagnosis of *Stag. dennisii*; *in vitro* and *in vivo* descriptions, partly adopted below).

Description in vitro

Pycnidia globose to irregular, with 1–2 non-papillate or papillate ostioles, often developing a long neck, 80–260 μm diam., glabrous, confluent on the agar, sometimes formed in the agar. Conidial matrix white to buff.

Conidia always mainly phomoid, aseptate, ellipsoidal to cylindrical, eguttulate or with a few small guttules, 4–6.5 × 1.5–3 μm; occasionally some large 1-septate conidia (ascochytoid) are present, 14.5–24 × 4–7 μm.

Colonies on OA 7–8 cm diam. after 7 days, regular with coarsely floccose, white to smoke grey aerial mycelium, colourless to primrose/olivaceous buff, due to a diffusible pigment, often with dull green to olivaceous sectors (with salmon appearance by conidial exudate); reverse similar [on MA reverse partly citrine green due to crystals of the 'foveata-type', see *Phoma foveata* Foister (sect. *Phyllostictoides*, **E**)]. With addition of a drop of NaOH a coral discoloration occurs [most conspicuous on MA]. Representative culture CBS 631.68.

In vivo *(especially on* Solidago canadensis*)*

Pycnidia (superficial on dead stems) mostly 100–150 μm diam. and globose ostiolate.

Conidia. In most collections the pycnidia contain only aseptate conidia resembling those *in vitro*, (4–)5–6(–7) × 1.5–2(–2.5) μm. Sometimes, however, pycnidia with large, mainly 1–2-septate conidia (ascochytoid/stagonosporoid) occur, characteristic for the synanamorph *Stagonosporopsis dennisii*: subcylindrical-ellipsoidal, usually with several guttules, 15.5–28 × 4–6 μm.

Ecology and distribution

In Europe and North America (Canada) found on and isolated from last year's dead stems of golden rod, *Solidago* species (Compositae) of both Eurasian and North American origin. The fungus is especially frequently recorded on *S. canadensis*. So far no data on pathogenicity.

Phoma dennisii var. *oculo-hominis* (Punith.) Boerema *et al.*, Fig. 17Aa

Conidial dimorph large, mainly 1-septate, ascochytoid [only known *in vitro*].

> *Phoma dennisii* var. *oculo-hominis* (Punith.) Boerema, Gruyter & Noordel. *in* Persoonia **16**(3): 351. 1997.
> > ≡ *Phoma oculo-hominis* Punith., Trans. Br. mycol. Soc. **67**: 142–143. 1976.

Diagn. data: Punithalingam (1976) (fig. 1A, B, C; description of original isolate on potato dextrose agar (PDA), illustrated with sections of pycnidium and pycnidial wall, and picture of conidial variability, reproduced in present Fig. 17Aa), Boerema *et al.* (1997) (characteristics *in vitro*, adopted below).

Description in vitro

Pycnidia similar to those of the type var. *dennisii*.

Conidia variable in size and shape; the 1-septate conidia are in comparison with those of the type variety relatively small, mostly 9–16 × 4.5 μm.

Colonies: the growth rate and cultural characteristics are about the same as those of the type var. [crystals of the 'foveata-type' may be formed abundantly on MA].

However, on OA and MA it is differing in the absence of a diffusible pigment and there is no coral discoloration with addition of a drop of NaOH. Original culture IMI 193307 = CBS 634.92.

Ecology and distribution

Once isolated from a human cornea (ulcer) in the USA (Tennessee). The cultural characteristics of this fungus suggest a natural relation with the 'golden rod fungus', *Ph. dennisii* (var. *dennisii*, e.g. common on dead stems of *Solidago canadensis*).

Phoma dimorphospora (Speg.) Aa & Kesteren, Fig. 14B

Conidial dimorph large, aseptate and 1(occ. 2)-septate, ascochytoid(-stagonosporoid) [not obtained *in vitro*].

> *Phoma dimorphospora* (Speg.) Aa & Kesteren *in* Persoonia **10**(2): 269[–270]. 1979.
> > ≡ *Phyllosticta dimorphospora* Speg. *in* An. Mus. nac. Buenos Aires III, **20**: 334. 1910.
> > = *Stagonospora chenopodii* Peck *in* Rep. N.Y. St. Mus. nat. Hist. **40**: 60. 1887 [sometimes erroneously listed as *Stag. chenopodii* 'House']; not *Phoma chenopodii* S. Ahmad *in* Sydowia **2**: 79. 1948 (sect. *Macrospora*, **H**).

Diagn. data: van der Aa & van Kesteren (1979) (fig. 2: 'macro- and microconidia' *in vivo*; documentation of synonymy; description *in vivo* and cultural characters), Boerema *et al.* (1997) (fig. 1B: conidial dimorphism *in vitro* and *in vivo*, reproduced in present Fig. 14B; *in vitro* and *in vivo* descriptions, partly adopted below).

Description in vitro

Pycnidia globose to subglobose, with 1(–8) papillate ostioles, 110–470 μm diam., glabrous or with mycelial outgrowths, solitary or aggregated on the agar. Micropycnidia frequently occur, 60–100 μm diam. Conidial matrix sordid white to buff or saffron.

Conidia always phomoid, aseptate, cylindrical to ellipsoidal, eguttulate, 3–5.5 × 1.5–2 μm.

Colonies on OA slow growing, 2–2.5 cm diam. after 7 days (3–3.5 cm after 14 days), regular, colourless to pale luteous with finely felted white aerial mycelium; reverse pale luteous to cinnamon [on MA growth rate faster, 2–3 cm in 7 days (4.5–5 cm in 14 days); colony citrine green, reverse pale luteous to citrine green, later ochraceous to isabelline]. Representative culture CBS 165.78.

In vivo *(especially on* Chenopodium quinoa*)*

Pycnidia (in very pale brown leaf spots or in eye-shaped lesions on stems) resembling those *in vitro*, but seldom over 300 μm diam. and only with rather flat papillae.

Conidia always dimorphic, some small, mostly 4–5 × 2–2.5 μm, aseptate, short cylindrical or ellipsoidal, eguttulate; some much larger, 16–22.5(–25) × 4–4.5(–7) μm, mostly aseptate, but also 1-, or seldom 2-septate (ascochytoid-stagonosporoid), straight or slightly curved, mostly without guttules.

Ecology and distribution

A common pathogen on species of *Chenopodium* (Chenopodiaceae) in North and South America. Leaf and stem spot. Closely allied to the European *Phoma heteromorphospora* Aa & Kesteren (this section), but easy to distinguish *in vitro* by the absence of large conidia.

Phoma glaucii Brunaud, Fig. 17B

Pycnidial dimorph with large conidia, occ. aseptate, mostly 1(–2)-septate, ascochytoid-stagonosporoid: ***Stagonosporopsis chelidonii* (Bres.) Died.**

V   *Phoma glaucii* Brunaud *in* Annls Soc. Sci. nat. La Rochelle **1892**: 97. 1892 [as 'glauci']; not *Phoma glaucii* Therry *in* Revue mycol. **13**: 10. 1891.

= *Diplodina glauci* Cooke & Massee *in* Grevillea **17**: 79. 1889.

≡ *Ascochyta glauci* (Cooke & Massee) Died. *in* Krypt.-Flora Mark Brandenb. 9, Pilze **7** [Heft 2]: 383. 1912 [vol. dated '1915'].

= *Ascochyta dicentra* Oudem. *in* Ned. kruidk. Archf III, **2**(3): 721. 1902.

= *Diplodina chelidonii* Naumov *in* Bull. Soc. oural Amat. Sci. nat. [= Zap. ural'. Obshch. Lyub. Estest.] **35**: 32 [extrait]. 1915.

H = *Diplodina chelidonii* Ade *in* Mitt. bad. Landesver. Naturk. II [N.F.], **1**: 332. 1924 [later homonym, see above].

= *Phyllosticta corydalina* Picbauer *in* Sb. vys. Sk. zeměd. Brné Fak. **18**: 20. 1931.

= *Ascochyta papaveris* var. *dicentrae* Grove, Br. Coelomycetes **1**: 301. 1935.

= *Phoma chelidonii* I.E. Brezhnev *in* Bot. Mater. Otd. spor. Rast. Bot. Inst. Akad. Nauk SSSR **7**: 190. 1951.

synanamorph:

Stagonosporopsis chelidonii (Bres.) Died. *in* Krypt.-Flora Mark Brandenb. 9, Pilze **7** [Heft 2]: 398. 1912 [vol. dated '1915'].

≡ *Phyllosticta chelidonii* Bres. *in* Hedwigia **35**: 199. 1896 [for convenience by Diedicke *in* Annls mycol. **10**: 141. 1912 indicated as '*Ascochyta chelidonii* (Bres.) Died.'].

H = *Ascochyta chelidonii* Kabát & Bubák *in* Hedwigia **46**: 290. 1907; not *Ascochyta chelidonii* Lib. *in* Pl. cryptog. Ard., Fasc. 3, No. 204. 1834 [≡ *Septoria chelidonii* (Lib.) Desm. *in* Annls Sci. nat. (Bot.) II, **17**: 110. 1842].

≡ *Ascochyta chelidoniicola* Melnik *in* Nov. Sist. niz. Rast. **12**: 204. 1975.

Diagn. data: Diedicke (1912–15) (disease symptoms; description *in vivo* as *Stagonosporopsis chelidonii*; change of conidial production from 'phomoid' as described by Bresadola to 'stagonosporoid' interpreted as a matter of ripening), Boerema *et al.* (1997) (fig. 2C: conidial shape *in vitro* and *in vivo*, reproduced in present Fig. 17B; *in vitro* and *in vivo* descriptions, partly adopted below; observations on the influence of humidity on conidial development, whether phomoid or ascochytoid-stagonosporoid).

Description in vitro

Pycnidia globose to irregular, with 1 or 2(–4) papillate ostioles, 90–300 μm diam., glabrous or with mycelial outgrowths, solitary or confluent, abundant on and in the agar, and in aerial mycelium. Micropycnidia also present, 60–90 μm diam. Conidial matrix rosy buff to salmon/saffron.

Conidia always mainly phomoid, aseptate subcylindrical-ellipsoidal to more or less obclavate fusiform, eguttulate or with some small guttules, 6–10.5 × 2–4 μm; occasionally some larger 1-septate conidia occur, up to 13 × 5 μm.

Colonies on OA *c.* 8 cm diam. after 7 days, regular, mostly colourless with coarsely floccose, white aerial mycelium, sometimes the agar at margin staining saffron to fulvous due to a diffusible pigment; reverse similar [on MA the agar staining luteous to amber in stellate pattern]. Application of a drop of NaOH resulted in a green spot turning red: E$^+$ reaction. Representative culture CBS 114.96.

In vivo *(especially on* Chelidonium majus*)*

Pycnidia (subepidermal on irregular yellowish brown leaf spots with darker border; also in long stretches on dead stems, on dried seed capsules and on fading or dead leaves) mostly 100–200 μm diam., usually depressed globose to ellipsoidal, with 1 small inconspicuous ostiole.

Conidia usually display about the same range of variability as those *in vitro*, subcylindrical to fusiform, often aseptate, mostly 7–8(–8.5) × (2.5–)3(–3.5) μm, but frequently becoming 1-septate, (7–)10–13(–15) × (2.5–)3–4 μm; they are often microguttulate. Sometimes, especially under dry circumstances, the pycnidia may contain distinctly larger conidia, representing the synanamorph *Stagonosporopsis chelidonii*, (13.5–)15–21(–23) ×

4–5(–6.5) μm, subcylindrical to ellipsoidal and rounded at the ends, mostly 1-septate, but occasionally also aseptate or 2-septate; they may be very pale brown tinged and always have small guttules.

Ecology and distribution

This fungus has been isolated and reported from quite different wild Papaveraceae throughout Europe, e.g. species of *Chelidonium*, *Corydalis*, *Dicentra* and *Glaucium*. The *Stagonosporopsis* phenotype of the fungus is known as pathogen, causing leaf spot, but most records refer to phomoid phenotypes (aseptate and 1-septate conidia) colonizing fading leaves and dead stems.

Phoma heteromorphospora Aa & Kesteren, Fig. 14A

Conidial dimorph large, aseptate and 1(2–3)-septate, ascochytoid-stagonosporoid.

> *Phoma heteromorphospora* Aa & Kesteren *in* Persoonia **10**(4): 542. 1980.
>> *H* ≡ *Phoma variospora* Aa & Kesteren *in* Persoonia **10**(2): 268 (Nov.) 1979; not *Phoma variospora* Shreem. *in* Indian J. Mycol. Pl. Path. **8**: 221. (July) 1979 ['1978'].
>> ≡ *Phyllosticta chenopodii* Westend. *in* Bull. Acad. r. Sci. Lett. Beaux-Arts Belg. II, **2**: 567. 1857; not *Phyllosticta chenopodii* Sacc. *in* Sylloge Fung. **3**: 55. 1884 = *Phoma exigua* Desm. var. *exigua* (sect. *Phyllostictoides*, **E**); not *Phoma chenopodii* S. Ahmad *in* Sydowia **2**: 79. 1948 (sect. *Macrospora*, **H**).
>> ≡ *Septoria westendorpii* G. Winter *in* Hedwigia **26**: 26. 1887; not *Phoma westendorpii* Tosquinet apud Westend. *in* Bull. Acad. r. Sci. Lett. Beaux-Arts Belg. II, **2**: 564. 1857.

Diagn. data: van der Aa & van Kesteren (1979) (fig. 1; description *in vivo* as *Ph. variospora*; documentation of synonymy; cultural characters; illustration of 'macro- and microconidia' *in vivo*), (1980) (name change in *Ph. heteromorphospora*), Boerema *et al.* (1997) (fig. 1A: conidial dimorphism *in vitro* and *in vivo*, reproduced in present Fig. 14A; *in vitro* and *in vivo* descriptions, partly adopted below).

Description in vitro

Pycnidia globose to subglobose, with 1–8 papillate ostioles, 120–350 μm diam., glabrous or with mycelial outgrowths, solitary or aggregated on the agar. Micropycnidia frequently occur, 50–100 μm diam. Conidial matrix white to primrose.

Conidia always dimorphic; mainly small and aseptate, (3–)4–7 × (1–)1.5–2(–2.5) μm, subcylindrical to ellipsoidal, without or with inconspicuous guttules; but also much larger, mostly 1–2-septate, 12.5–26.5 × 3–5 μm, subcylindrical with abundant guttules.

Colonies slow growing on OA [and on MA], 1–1.5 cm diam. after 7 days (3 cm in 14 days), somewhat irregular, pale luteous with velvety white aerial mycelium; reverse also pale luteous. Representative culture CBS 448.68.

In vivo (*especially on* Chenopodium album*)*

Pycnidia (in pale yellowish brown or whitish leaf spots with narrow purplish-brown border) similar to those *in vitro*, but up to 550 µm diam. and usually distinctly papillate.

Conidia always heterosporous; partly small, narrow, 3–6 × 1–1.5 µm, aseptate, subcylindrical to ellipsoidal, sometimes curved, minutely biguttulate; some clearly larger, but very variable in dimensions, (8–)15–20(–27) × (3–)3.5–4.5(–7) µm, mostly aseptate, but sometimes 1-septate or, rarely 2–3-septate, ellipsoidal-cylindrical or somewhat irregular in shape, irregularly multiguttulate.

Ecology and distribution

A very common pathogen on various species of *Chenopodium* (Chenopodiaceae) in Europe. Leaf spot. Very similar and closely related to the American *Phoma dimorphospora* (Speg.) Aa & Kesteren (this section). *In vitro*, they can be distinguished easily by the fact that the latter does not produce large conidia on artificial media. On account of the septate macroconidia *Ph. heteromorphospora* is sometimes confused with *Ascochyta caulina* (P. Karst.) Aa & Kesteren, the conidial anamorph of *Pleospora calvescens* (Fr.) Tul. & C. Tul. (Boerema *et al.* 1987).

Phoma narcissi (Aderh.) Boerema *et al.*, Fig. 18A

Pycnidial dimorph with large conidia, mainly 3-septate, stagonosporoid: ***Stagonosporopsis curtisii* (Berk.) Boerema**

Phoma narcissi (Aderh.) Boerema, Gruyter & Noordel. *in* Persoonia **15**(2): 215. 1993.

≡ *Phyllosticta narcissi* Aderh. *in* Centbl. Bakt. ParasitKde Abt. 2, **6**: 632[–633]. 1900 [May].

H = *Phyllosticta narcissi* Oudem. *in* Ned. kruidk. Archf III, **2**(1): 227. 1900 [August].

≡ *Phyllosticta oudemansii* Sacc. & P. Syd. *in* Sylloge Fung. **16**: 849. 1902.

= *Phyllosticta hymenocallidis* Seaver *in* N. Am. Flora **6**(1): 12. 1922 [second impression 1961].

= *Phoma amaryllidis* Kotthoff & Friedrichs *in* Obst- u. Gartenbau Ztg **18**: 32. 1929.

= *Phyllosticta gemmipara* Zondag *in* Tijdschr. PlZiekt. **35**: [97–]106. 1929.

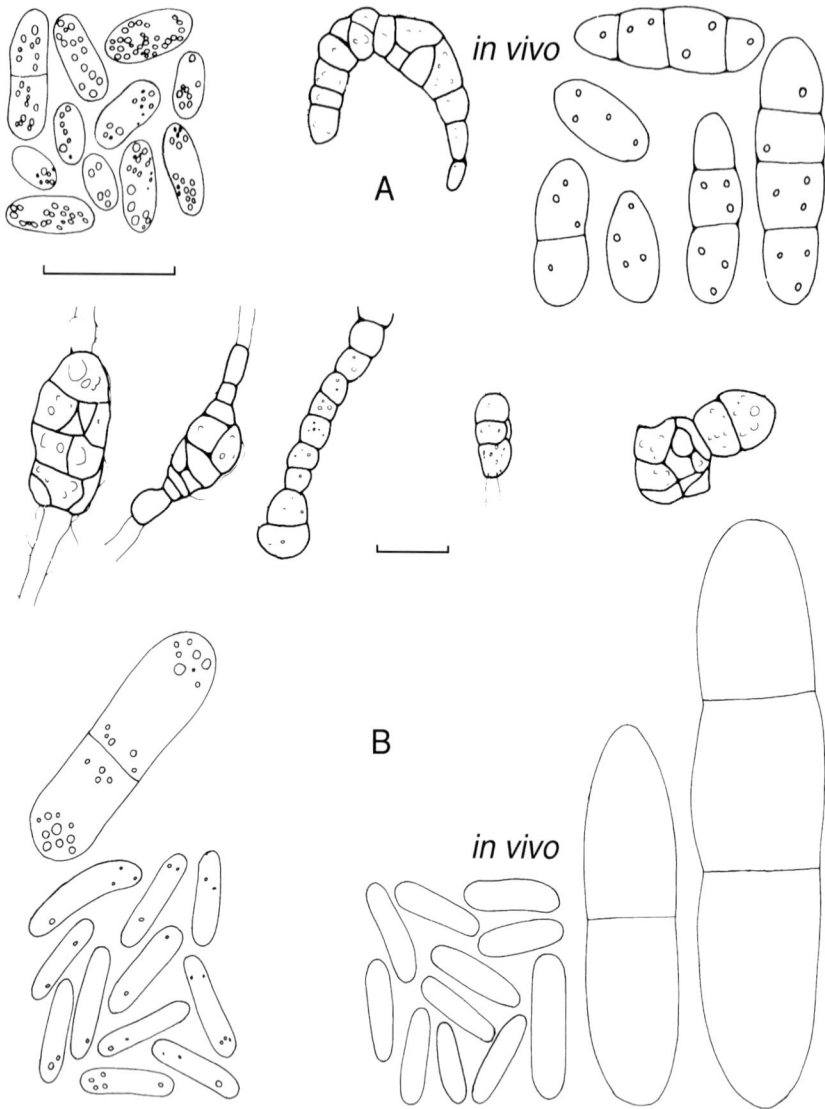

Fig. 18. A *Phoma narcissi*. Conidia *in vitro* often nearly all aseptate. *In vivo* they may be similar, but also much larger and in majority 1- or 3-septate, synanamorph *Stagonosporopsis curtisii*. Multicellular chlamydospores mostly intercalary, irregular botryoid-dictyosporous, usually solitary; often also short chains of unicellular chlamydospores. **B** *Phoma nigripycnidia*. Conidia usually small and aseptate *in vitro* but some large 1-septate conidia may be present, especially in fresh cultures. Pycnidia sometimes contain only small aseptate conidia *in vivo*, but usually only very large 1–2(–4)-septate conidia are present, synanamorph *Stagonosporopsis nigripycnidiicola*. Bar conidia 10 μm, bar chlamydospores 20 μm (as in sect. *Peyronellaea*, **D**).

synanamorph:

    Stagonosporopsis curtisii (Berk.) Boerema apud Boerema & Dorenbosch *in* Versl. Meded. plziektenk. Dienst Wageningen **157** (Jaarb. 1980): 20. 1981.

        ≡ *Hendersonia curtisii* Berk. apud Cooke *in* Nuovo G. bot. ital. **10**: 19. 1878 ['Berk. in herb. Curt.']; not *Phoma curtisii* Sacc. *in* Sylloge Fung. **3**: 860. 1884 [as new name for *Phoma maculans* (Berk. & M.A. Curtis) Sacc. *in* Sylloge Fung. **3**: 116. 1884; not *Phoma maculans* (Lév.) Sacc. *in* Sylloge Fung. **3**: 103. 1884].

        ≡ *Stagonospora curtisii* (Berk.) Sacc. *in* Sylloge Fung. **3**: 451. 1884.

        = *Stagonospora narcissi* Hollós *in* Annls hist.-nat. Mus. natn. hung. **5**: 354. 1906.

        = *Stagonospora crini* Bubák & Kabát *in* Hedwigia **47**: 361. 1908.

Diagn. data: Feekes (1931) (fig. 10, pl. III; characteristics *in vivo* and *in vitro*; pathogenicity and disease symptoms; synonymy), Creager (1933) (figs 1–8; pathogenicity and disease symptoms; pycnidial origin; conidial variability *in vivo* and *in vitro*), Smith (1935) (fig. 1, tab. I; host range; synonymy), Pag (1965) (figs 1–4: conidial shape *in vivo* and *in vitro*), Boerema & Dorenbosch (1981) (fig. 5: conidial shape *in vivo* and *in vitro*; taxonomy, classification in *Stagonosporopsis*), Boerema & Hamers (1989) (nomenclature; disease symptoms and literature references), Boerema (1993) (fig. 5A: drawings of pycnidia, the conidial phenotypes and dictyochlamydospores reproduced in present Fig. 18A; classification in *Phoma*; description *in vitro* adopted below), Boerema *et al.* (1997) (nomenclatural differentiation in two synanamorphs as listed above; characters *in vivo*, adopted below).

Description in vitro

Pycnidia globose, usually somewhat compressed, with a definite ostiole, 110–275 µm diam., often covered by hyphae. Conidial matrix rosy buff.

    Conidia broadly ellipsoidal, finely guttulate, 4–7.5(–8) × (2–)2.5–3.5(–4) µm, occasionally larger and 1-septate, 8–15 × 3–5.5 µm (*in vivo* often large and multiseptate, see below).

    Chlamydospores uni- and multicellular, where unicellular usually intercalary in short chains, dark brown, 8–15 µm diam., where multicellular mostly intercalary, occasionally terminal, usually solitary, sometimes in series of 2–3 elements, irregular botryoid-dictyosporous, mostly somewhat curvate and very dark brown, 8–35 µm diam.

    Colonies on OA fast-growing, 8–8.5 cm diam. after 7 days, mycelium compact woolly-fluffy, smoky grey; reverse grey olivaceous to olivaceous black. Hyphae and chlamydospores often bearing more or less hemispherical or flattened droplet-like deposits, which when becoming darker give an impression of ornamentation. Representative culture CBS 251.92.

In vivo *(Amaryllidaceae)*

Pycnidia (subepidermal in dead leaf tips and in spots on leaves and scales) similar to those *in vitro*, but more variable in size (up to 350 μm diam.).

Conidia often mainly aseptate and only a few 1-septate ones, 4.5–8(–10) × 2.5–4(–5) μm (av. 6.8–7.5 × 3–3.8 μm), thus resembling the conidia *in vitro*. However, the pycnidia in the field sometimes contain larger conidia which are mostly 3-septate, 13.4–28 × 4.8–8.0 μm (av. 21 × 6.5 μm): synanamorph *Stagonosporopsis curtisii*. In this case aseptate conidia are relatively rare, but 1-septate ones may also be present, 8–16 × 3.2–6.4 μm (av. 11.5 × 4.5 μm).

Ecology and distribution

A worldwide pathogen of *Narcissus*, *Hippeastrum* and various other Amaryllidaceae, causing leaf scorch, neck rot, red spot disease, red leaf spot. The synonymy reflects, apart from its plurivorous character, the extreme variability of its conidia *in vivo*. In phytopathological literature this pathogen is commonly known under the synanamorphic synonyms *Stagonospora curtisii* or *Stagonosporopsis curtisii*.

On account of the occurrence of multicellular chlamydospores this fungus is also incorporated in the key of *Phoma* sect. *Peyronellaea* (**D**).

Phoma nigripycnidia Boerema et al., Fig. 18B

Pycnidial dimorph with large conidia, 1–2(occ. 3–4)-septate, stagonosporoid: ***Stagonosporopsis nigripycnidiicola* (Ondřej) Boerema et al.**

Phoma nigripycnidia Boerema, Gruyter & Noordel. *in* Persoonia **16**(3): 356. 1997.
synanamorph:
Stagonosporopsis nigripycnidiicola (Ondřej) Boerema, Gruyter & Noordel. *in* Persoonia **16**(3): 356. 1997.
≡ *Ascochyta nigripycnidiicola* Ondřej *in* Biológia, Bratisl. **23**: 816. 1968.

Diagn. data: Ondřej (1968) (fig. 6: culture on 'Czapek-Dox' agar; fig. 9: drawing of conidia and disease symptoms on *Vicia cracca*), Ondřej (1970) (tabs 5, 6: conidial dimensions *in vivo* and *in vitro*), Boerema & Bollen (1975) (fig. 6A: conidial shape *in vivo* and *in vitro*), Boerema et al. (1997) (fig. 3B: conidial dimorphism *in vitro* and *in vivo*, reproduced in present Fig. 18B; *in vitro* and *in vivo* descriptions, partly adopted below), Punithalingam & Spooner (2002) (fig. 12: photographs of stagonosporoid conidia *in vivo*, fig. 13: drawing of conidial variation *in vivo* and fragmented septate conidium).

Description in vitro

Pycnidia irregularly globose, sometimes papillate, but ostioles inconspicuous, 220–360 μm diam., glabrous, usually solitary on the agar and in the aerial mycelium.

Conidia usually relatively small and aseptate, 5.5–9 × 1.5–2 μm, cylindrical to allantoid with some small guttules; occasionally also some 1-septate conidia occur, 9–15 × 2–4 μm. [In fresh cultures also much larger, more-celled conidia, resembling those *in vivo*, have been recorded.]

Colonies on OA 6–6.5 cm diam. after 7 days, regular, colourless to dull green and pale luteous/citrine, aerial mycelium floccose, white to olivaceous grey; reverse becoming pale luteous/citrine. Representative culture CBS 116.96.

In vivo *(especially on* Vicia cracca*)*

Pycnidia (subepidermal, scattered in ochraceous brown circular leaf spots and elongated stem lesions) 120–250 μm diam., subglobose-papillate, dark with inconspicuous ostiole.

Conidia usually very large, mostly 20–45 × 7–12 μm, 1–2(occasionally 3–4)-septate, i.e. stagonosporoid, typical for the synanamorph *Stagonosporopsis nigripycnidiicola*. Sometimes, however, the dark papillate pycnidia in the leaf spots contain only small aseptate conidia, similar to those of *Ph. nigripycnidia in vitro*. The mature stagonosporoid conidia easily break into parts (see Punithalingam & Spooner, 2002; difference with conidia in true species of *Ascochyta* or *Stagonospora*).

Ecology and distribution

This fungus is in south-eastern Europe (esp. the Czech Republic and Slovakia) a common pathogen of *Vicia cracca* (Leguminosae): leaf and stem spot; recently it is also recorded on *V. cracca* in Surrey, England. The fungus has been found occasionally on *Vicia sepium* and *Vicia sativa* (susceptibility proved by inoculation).

Phoma samarorum Desm., Fig. 19A

Pycnidial dimorph with large conidia, 1–3-septate, stagonosporoid: ***Stagonosporopsis fraxini* (Allesch.) Died.**

> *Phoma samarorum* Desm. *in* Pl. cryptog. N. France [ed. 1] Fasc. 7, No. 349. 1828.
>> ≡ *Phomopsis samarorum* (Desm.) Höhn. *in* Hedwigia **62**: 87. 1921 [misapplied].
>> *V* ≡ *Septoria samarorum* (Desm.) Wollenw. & Hochapfel *in* Z. ParasitKde **8**: 604. 1936 [as '*samararum*'; based on secondary collection with many stagonosporoid conidia].
>> *V* ≡ *Diplodina samarorum* (Desm.) Nevod. *in* Fungi USSR Fasc. 1, No. 14. 1952 [as '*samararum*'].
>> ≡ *Stagonospora samarorum* (Desm.) Boerema *in* Persoonia **6**(1): 25. 1970 [with reference to collections with stagonosporoid conidia].

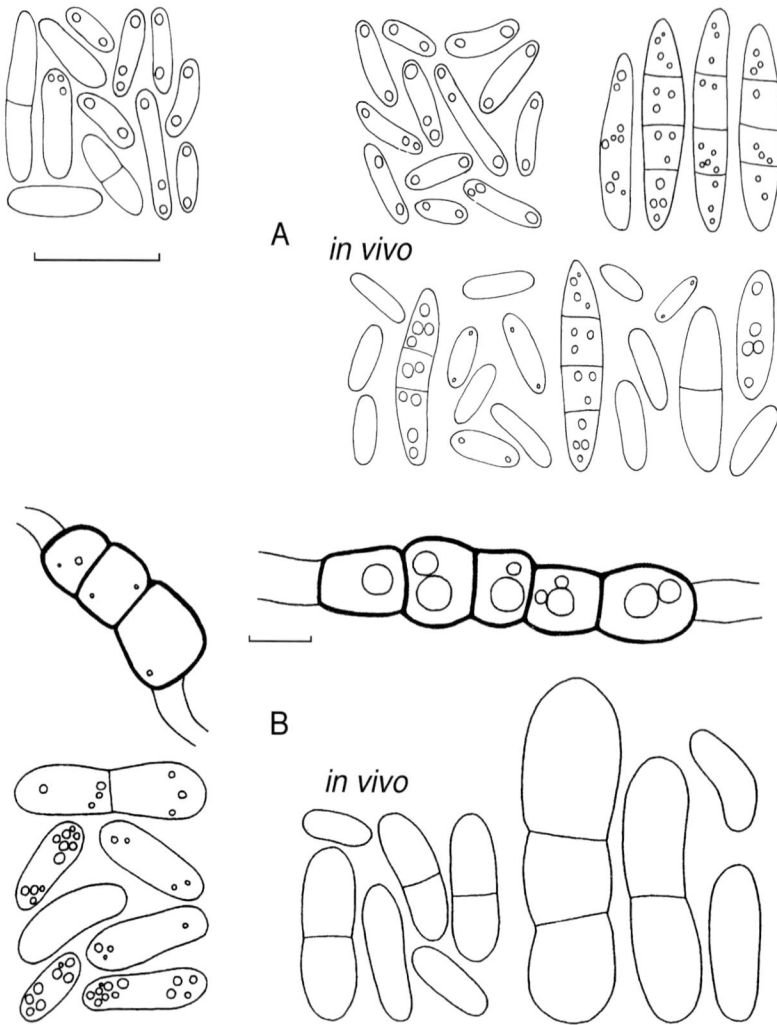

Fig. 19. A *Phoma samarorum.* Conidia *in vitro*, phomoid and variable, mainly aseptate, but the larger ones often 1(–2)-septate. Conidia *in vivo*, sometimes mostly aseptate phomoid and sometimes mainly larger, fusoid in shape and 1–3-septate; synanamorph *Stagonosporopsis fraxini*. However, mixtures with various intermediate conidia forms may also occur. **B** *Phoma schneiderae.* Conidia *in vitro* phomoid, mainly aseptate, occasionally elongated and 1-septate. Chlamydospores usually intercalary in short chains (but sometimes clustered and terminal). *In vivo* large conidia predominate, 1–3-septate, synanamorph *Stagonosporopsis lupini*, but some small phomoid conidia are normally present as well (occasionally mainly small phomoid conidia are produced). Bars 10 μm (the smaller referring to the chlamydospores).

= *Ascochyta orni* Sacc. & Speg. *in* Michelia **1**(2): 167. 1878 [apparently based on a collection with many ascochytoid conidia].

= *Ascochyta sambuci* Sacc. *in* Michelia **1**(2): 170. 1878 [based on a collection with many fusoid ascochytoid conidia].

= *Phoma herbarum* f. *salicariae* Sacc. *in* Michelia **2**(1): 93. 1880.

Θ = *Phoma herbarum* f. *ansoniae-salicifoliae* Berl. & Roum. *in* Revue mycol. **9**: 178 ['162']. 1887.

H = *Ascochyta fraxini* Bubák & Kabát *in* Hedwigia **52**: 346. 1912 [based on a collection with many ascochytoid conidia]; not *Ascochyta fraxini* Lib. *in* Pl. cryptog. Ard. Fasc. 1, No. 48. 1830.

= *Phyllosticta avenae* Lobik *in* Bolez. Rast. **17**(3–4): 166. 1928.

= *Phoma fusispora* Wehm. *in* Mycologia **38**: 320. 1946.

synanamorph:

Stagonosporopsis fraxini (Allesch.) Died. *in* Krypt.-Flora Mark Brandenb. 9, Pilze **7** [Heft 2]: 399. 1912 [vol. dated '1915'] [as '(Oud.) Died.'].

≡ *Diplodina fraxini* Allesch. *in* Rabenh. Krypt.-Flora [ed. 2], Pilze **6** [Lief. 69]: 687. 1899 [vol. dated '1901'] [as '(Oud.)', but based on an illegitimate name (see after this), which means (Art. 58.1) that it has to be treated as a new name dating from 1899].

H ≡ *Ascochyta fraxini* Oudem. *in* Ned. kruidk. Archf II, **5**: 497. 1889 [illegitimate as later homonym of *Ascochyta fraxini* Lib., 1830, see above under *A. fraxini* Bubák & Kabát].

≡ *Ascochytella fraxini* (Allesch.) Petr. apud Sydow & Sydow *in* Annls mycol. **22**: 266. 1924 [as '(Oud.)'].

≡ *Pseudodiplodia fraxini* (Allesch.) Petr. *in* Sydowia **7**: 304. 1953 [as '(Oud.)'].

= *Stagonospora oryzae* Hara, Dis. Rice Plant (Japan) 184. 1918.

Diagn. data: Boerema (1970) (conidial characteristics of two herbarium specimens, filed as forms of *Ph. herbarum*; provisional assignation of the fungus to *Stagonospora*; cultural characters and sources of isolates), Boerema & Dorenbosch (1973) (tab. 9: isolate sources sub *Stagonospora samarorum*; fig. 1: drawing of conidia *in vitro*; synonym), Boerema et al. (1997) (fig. 5: conidial shape *in vitro* and *in vivo*, reproduced in present Fig. 19A; differentiating of the *Stagonosporopsis*-synanamorph; *in vitro* and *in vivo* descriptions, partly adopted below).

Description in vitro

Pycnidia globose to irregular, with 1–3 usually papillate ostioles, sometimes developing long necks, 120–270 μm diam., with mycelial outgrowths, solitary on the agar, or confluent to large pycnidial structures up to 900 μm diam. Conidial matrix ochraceous.

Conidia mainly aseptate, 4–8.5 × 1.5–3 μm, sometimes larger and often 1(–2)-septate, 7–16 × 2–3 μm, slightly yellowish, ellipsoidal to cylindrical, usually attenuate at one end, with or without distinct polar guttules.

Colonies on OA 2–2.5 cm in diam. after 7 days, regular with floccose-woolly, white aerial mycelium, colourless to primrose or rosy buff to salmon due to a diffusible pigment [not changing with NaOH], later with olivaceous grey tinges; reverse similar. Representative cultures CBS 138.96, CBS 139.96.

In vivo

Pycnidia (in necrotic host tissue) resemble those *in vitro*, usually solitary, often with interwoven hyphae and obvious ostioles.

Conidia always show the characteristic yellow tinge as noted above. However, they are much more variable in size and shape than those *in vitro*, sometimes being fusoid and depending on the humidity changing from phomoid to ascochytoid-stagonosporoid. The pycnidia may contain only aseptate conidia, 5.5–10 × 1.5–3 μm, only 1–3-septate conidia, up to 17 × 3.5 μm, or a mixture of both types of conidia. Sometimes the conidia are explicitly large, 17–25 × 2.5–3.5 μm. They may be distinctly guttulate, but also eguttulate.

Ecology and distribution

The sources of the isolates indicate that it is a common saprophytic soil fungus in the whole of Eurasia. The fungus also has been isolated and described from different substrata in North America (USA). It has been found on necrotic tissue of quite different herbaceous and gramineous plants as well as deciduous trees and shrubs.

Phoma schneiderae Boerema et al., Fig. 19B

Pycnidial dimorph with large conidia, 1–3-septate, stagonosporoid: **Stagonosporopsis lupini (Boerema & R. Schneid.) Boerema et al.**

> *Phoma schneiderae* Boerema, Gruyter & P. Graaf *in* Persoonia **17**(2): 282[–283]. 1999; **18**(2): 159. 2003 [info type specimen].
> synanamorph:
> *Stagonosporopsis lupini* (Boerema & R. Schneid.) Boerema, Gruyter & P. Graaf *in* Persoonia **17**(2): 283. 1999.
> > ≡ *Ascochyta lupini* Boerema & R. Schneid. *in* Versl. Meded. plziektenk. Dienst Wageningen **162** (Jaarb. 1983): 28. 1984 [introduced with the intention to cover both phenotypes, but holotype stagonosporoid].

Θ = *Ascochyta pisi* var. *lupini* Sacc. in Fungi Columb. [Ed. Bartholomew] No. 4506. 1915 [without description].

‡ = *Ascochyta caulicola* var. *lupini* Grove, Br. Coelomycetes **1**: 305. 1935 [not validly published, no Latin description, Art. 36].

Diagn. data: Boerema (1984a) (Latin description, covering both phenotypes; fig. 9: conidial variability *in vivo* and *in vitro*, drawings from R. Schneider; fig. 10: disease symptoms, reproduced from Frey & Yabar, 1983), Boerema, *et al.* (1999) (differentiation of the phomoid phenotype as *Phoma schneiderae*, based on characteristics *in vitro*; figs 1, 2: conidial shape *in vitro* and *in vivo* and chlamydospores *in vitro*, reproduced in present Fig. 19B; *in vitro* and *in vivo* descriptions, partly adopted below).

Description in vitro

Pycnidia globose to subglobose, glabrous, with 1(–3) non-papillated or papillated ostioles, 80–230 μm diam., solitary or confluent. Conidial matrix white to buff.

Conidia mainly aseptate, 5.5–13.5 × 2.5–3.5 μm, sometimes larger and 1-septate (9.5–)11–15(–21) × 2.5–4 μm, ellipsoidal to more or less obclavate, eguttulate or with several small, scattered guttules.

Chlamydospores globose to oblong, 7–18 μm diam., olivaceous with greenish guttules, in short chains or clustered, intercalary or terminal.

Colonies on OA 5.5–6.5 cm in diam. after 7 days, regular to slightly irregular with scarce, felty olivaceous grey aerial mycelium, colourless to grey olivaceous/olivaceous grey or greenish grey/greenish black; reverse similar. Representative culture CBS 101.494.

In vivo *(Lupinus spp.)*

Pycnidia (subepidermal in concentric rings on reddish/brown leaf spots with dark edges, or scattered on brown lesions on stems or pods), variable in diameter, mostly 200–250 μm, globose to subglobose with one distinct non-papillate ostiole.

Conidia usually for the most part large, 1–3-septate (stagonosporoid), typical of the synanamorph *Stagonosporopsis lupini*: cylindrical with obtuse ends, mostly 15–30 × 5–9 μm. Some smaller, aseptate, phomoid conidia, 8–14 × 3–5 μm, are normally present as well. Sometimes only 0–1-septate conidia are formed, which resemble those of *Phoma schneiderae in vitro*.

Ecology and distribution

This fungus is a specific, seed-borne pathogen of lupins, *Lupinus* spp. (Leguminosae), causing spots on leaves, stems and pods: leaf spot and blight. In South America it is a serious problem in the cultivation of *L. mutabilis*, while in North America it has been regularly found on the wild *L. perennis*. In

Europe, this fungus used to be known only from occasional findings on *L. arboreus* in England, but it has recently been isolated from the economically more important *L. albus* in Poland and England.

Phoma subboltshauseri Boerema *et al.*, Fig. 20A

Pycnidial dimorph with large conidia, 1–3(–5)-septate, stagonosporoid: **Stagonosporopsis hortensis (Sacc. & Malbr.) Petr.**

> *Phoma subboltshauseri* Boerema, Gruyter & Noordel. *in* Persoonia **16**(3): 360. 1997.
>
> synanamorph:
> *Stagonosporopsis hortensis* (Sacc. & Malbr.) Petr. *in* Annls mycol. **19**: 21. 1921.
>
> > ≡ *Hendersonia hortensis* Sacc. & Malbr. *in* Michelia **2**(3): 629. 1882 [as '*Hendersonia (Stagonospora) hortensis*']; not *Phoma hortensis* Brunaud *in* Sylloge Fung. **11**: 485. 1895 [as '*hortense*'].

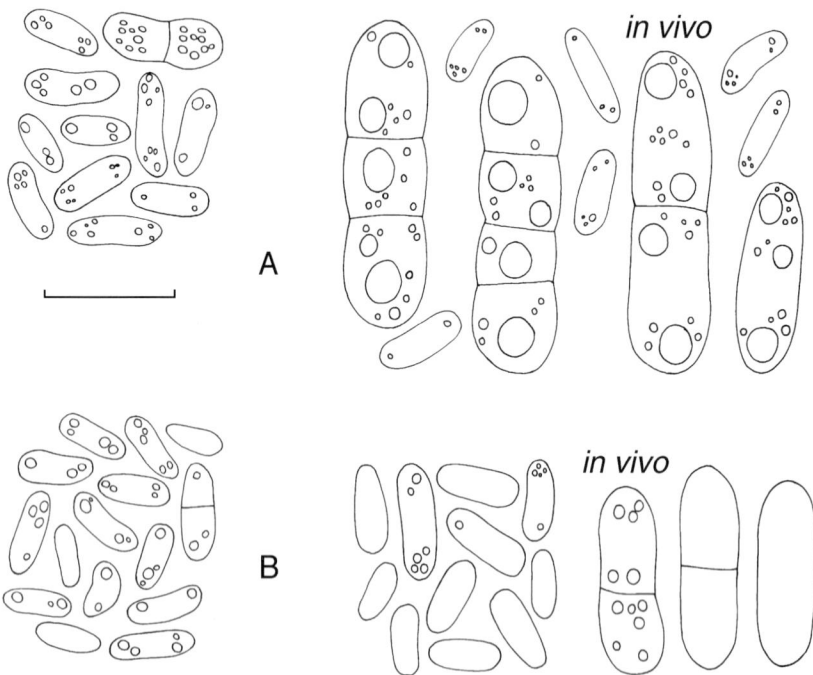

Fig. 20. A *Phoma subboltshauseri*. Conidia *in vitro* phomoid, mostly aseptate, occasionally 1-septate. Conidia *in vivo*, predominantly large and 1–3(–5)-septate, synanamorph *Stagonosporopsis hortensis*; but the pycnidia usually also contain some small aseptate conidia. **B** *Phoma trachelii*. Conidia *in vitro*, phomoid, usually aseptate, occasionally the larger ones 1-septate. Conidia *in vivo*, sometimes all relatively small and aseptate (phomoid), but usually much larger, 0–1(–2)-septate; synanamorph *Stagonosporopsis bohemica*. Bar 10 μm.

≡ *Stagonospora hortensis* (Sacc. & Malbr.) Sacc. & Malbr. *in*
Sylloge Fung. **3**: 446. 1884.

H ≡ *Ascochyta hortensis* (Sacc. & Malbr.) Jörstad [Jørstad] *in*
Meld. St. plpatol. Inst. **1**: 74. 1945; not *Ascochyta hortensis*
Kabát & Bubák *in* Hedwigia **44**: 353. 1905 [≡ *Phoma
funkiae-albomarginatae* Punith. *in* Mycol. Pap. **159**: 189.
1988].

= *Ascochyta boltshauseri* Sacc. apud Boltshauser *in* Z.
PflKrankh. **1**: 136. 1891 [Midsummer]; *in* Sylloge Fung.
10: 303. 1892.

≡ *Stagonosporopsis boltshauseri* (Sacc.) Died. *in* Annls
mycol. **10**: 141–142. 1912; *in* Krypt.-Flora Mark
Brandenb. 9, Pilze **7** [Heft 2]: 400. 1912 [vol. dated
'1915'].

≡ *Stagonospora boltshauseri* (Sacc.) Grigoriu *in* Annls
Inst. phytopath. Benaki II, **11**: 113. 1975.

≡ *Phoma boltshauseri* (Sacc.) Boerema, Pieters & Hamers
in Neth. J. Pl. Path. **99**, Suppl. **1**: 17. 1993.

= *Stagonopsis phaseoli* Erikss. *in* Bot. Centbl. **12** [**47**]: 298.
1891 [Sept.].

Diagn. data: Sprague (1935) (tab. 1, 2; first report of *Ascochyta boltshauseri* on
Phaseolus beans in the USA; cross-inoculation experiments; synonymy), Petrak
(1943) (synonymy of *Ascochyta boltshauseri* with *Stagonospora hortensis*; discussion on classification and nomenclature), Grigoriu (1975) (figs 3–9, tabs
I–VI; first observation in Greece; disease symptoms on beans; illustrations and
descriptions of the morphology of the fungus *in vivo* and *in vitro*; proposed
classification in *Stagonospora*; influence of temperature, humidity, pH and C/N
of media on growth and germination of conidia), Boerema & Verhoeven
(1979) (classification in *Stagonosporopsis*), Boerema *et al.* (1981a) (tab. 3:
diagnostic characters in comparison with vars of *Phoma exigua* on *Phaseolus*
beans), Boerema *et al.* (1993b) (classification in *Phoma*), Boerema *et al.*
(1997) (fig. 4B: conidial shape *in vitro* and *in vivo*, reproduced in present Fig.
20A; introduction of the binomium *Phoma subboltshauseri* for the phomoid
phenotype *in vitro*; *in vitro* and *in vivo* descriptions, partly adopted below).

Description in vitro

Pycnidia globose to subglobose, with 1 non-papillate ostiole, 90–230 μm diam.,
glabrous, solitary, sometimes confluent on and in the agar and in aerial mycelium.
Micropycnidia usually occur, 30–70 μm diam. Conidial matrix white to buff.

Conidia usually aseptate 3.5–9 × 1.5–2.5 μm, occasionally somewhat
larger and 1-septate, up to 11 × 3.5 μm, cylindrical to ellipsoidal with 1–several
small guttules at each end.

Colonies on OA 4.5–5.5 cm in diam. after 7 days, regular, greenish olivaceous, grey olivaceous or dull green, margin usually more like citrine green,
aerial mycelium coarsely floccose, white to olivaceous grey; reverse olivaceous

at centre, towards margin greenish olivaceous, grey olivaceous to citrine green. Representative cultures CBS 104.42, CBS 572.85.

In vivo *(on* Phaseolus vulgaris*)*

Pycnidia (subepidermal in concentric rings on reddish/brown leaf spots; occasionally scattered in sunken dark reddish/brown lesions on stems and pods) mostly 100–200 μm diam., globose to subglobose with 1 distinct ostiole.

Conidia always for the most part large and 1–3(–5)-septate (stagonosporoid), typical of the synanamorph *Stagonosporopsis hortensis*: cylindrical with obtuse ends, constricted at the septa and usually with four large and several small guttules, mostly (16–)18–22(–34) × (4–)5–8(–9) μm. Usually the pycnidia also contain some relatively small aseptate conidia, 5–10 × 1.5–2.5 μm, resembling those of *Phoma subboltshauseri in vitro*. Fragmentation of the large septate conidia may occur.

Ecology and distribution

A worldwide pathogen of beans, *Phaseolus vulgaris* (Leguminosae), causing stunting and red-brown blotches on stems, leaves and pods: leaf spot disease. Recently the fungus has also been found on cowpea, *Vigna unguiculata*. This is not surprising because the African genus *Vigna* and the American genus *Phaseolus* are closely related and susceptible to their mutual pathogens.

Phoma trachelii Allesch., Fig. 20B

Pycnidial dimorph with large conidia, 1(–2)-septate, ascochytoid: **Stagonosporopsis bohemica (Kabát & Bubák) Boerema *et al*.**

> *Phoma trachelii* Allesch. in Fungi bavar. Exs. [Ed. Allesch. & Schnabl] Cent. 4, No. 360. 1894; in Allg. bot. Z. **1**: 26. 1895.
>> = *Phyllosticta alliariaefoliae* Allesch. in Rabenh. Krypt.-Flora [ed. 2] Pilze **6** [Lief. 60]: 109. 1898 [vol. dated '1901'].
>> H ≡ *Phyllosticta fallax* Allesch. in Beibl. Hedwigia **36**: 159. 1897; not *Phyllosticta fallax* Sacc. & Roum. in Michelia **2**(3): 620. 1882; not *Phoma fallax* Berk. in Hooker, Fl. Nov.-Zeland. **2** [Flowerless Pl.]: 193. 1855.
>> = *Phyllosticta veraltiana* C. Massal. ex Sacc. in Annls mycol. **9**: 25. 1911.
>>> ≡ *Phoma veraltiana* (C. Massal. ex Sacc.) Aa in van der Aa & Vanev, Revision Phyllosticta, CBS: 474. 2002.

> synanamorph:
> *Stagonosporopsis bohemica* (Kabát & Bubák) Boerema, Gruyter & Noordel. in Persoonia **16**(3): 361. 1997.
>> ≡ *Ascochyta bohemica* Kabát & Bubák apud Bubák & Kabát in Hedwigia **44**: 352. 1905.
>>> ≡ *Stagonospora bohemica* (Kabát & Bubák) Tobisch in Öst. bot. Z. **83**: 142. 1934.

Diagn. data: Sauthoff (1962) (figs 1–5, 6, 7: disease symptoms on *Campanula* spp.; fig. 5: drawing of 1-septate and aseptate conidia *in vivo*, with conidial germination and fragmentation; tabs 1–3: pycnidial and conidial variability *in vivo* and *in vitro*, host range, disease symptoms), Brewer & Boerema (1965) (pl. 4, transm. electr. micrographs of conidia *in vivo*), Boerema & Bollen (1975) (fig. 6C: characteristics of conidia and conidiogenous cells *in vivo* and *in vitro*), Neergaard (1977) (occurrence on seed), Boerema *et al.* (1997) (fig. 4C: conidial shape *in vitro* and *in vivo*, reproduced in present Fig. 20B; synonymy and *in vitro* and *in vivo* descriptions, partly adopted below).

Description in vitro

Pycnidia globose or compressed to more or less bottle shaped, with 1–3 distinct papillate ostioles, sometimes with elongated necks, 110–300 μm diam., glabrous, solitary or confluent, abundant, mostly on, or partly in the agar. Conidial matrix dirty white to pale vinaceous/buff.

Conidia usually aseptate, 4–8.5 × 1.5–3 μm, occasionally larger and 1-septate (ascochytoid), 7–16 × 2–3.5 μm, ellipsoidal to cylindrical, sometimes eguttulate but usually with distinct polar guttules. The small phomoid conidia are sometimes produced on elongated cells which look like large aseptate ascochytoid conidia that have not seceded (Boerema & Bollen, 1975).

Colonies on OA 4–4.5 cm in diam. after 7 days, regular, initially with white floccose aerial mycelium, but without obvious aerial mycelium after 14 days, always rather colourless but with greenish olivaceous centre or sector; reverse similar. Representative cultures CBS 384.68, CBS 379.91.

In vivo *(especially on* Campanula isophylla*)*

Pycnidia (subepidermal, usually concentrically arranged in large grey brown to black leaf spots; also on roundish or elongated lesions on petals and stems) resembling those *in vitro*, mostly 75–255 μm diam., depressed globose to ellipsoidal, with 1, only slightly papillate dark bordered ostiole.

Conidia sometimes all relatively small and aseptate, similar to those *in vitro*. However, the pycnidia may also contain many large conidia, which mostly become 1(–2)-septate (ascochytoid), typical of the synanamorph *Stagonosporopsis bohemica*: cylindrical, rounded at both ends, often slightly curved and mostly eguttulate, (11–)13–23 × 4–6 μm. Together with these large conidia some small aseptate conidia usually occur. The mature 1-septate conidia easily break into two parts.

Ecology and distribution

A common seed-borne pathogen of wild and cultivated species of *Campanula* and *Trachelium* (Campanulaceae) in Eurasia and in North and South America. Found most frequently on *C. isophylla* 'Alba', the 'Star of Bethlehem'; causing leaf, stem and flower spot.

10 C *Phoma* sect. *Paraphoma*

Phoma sect. *Paraphoma* (Morgan-Jones & J.F. White) Boerema

> *Phoma* sect. *Paraphoma* (Morgan-Jones & J.F. White) Boerema apud van der Aa, Noordeloos & de Gruyter *in* Stud. Mycol. **32**: 7. 1990.
>
> > ≡ *Paraphoma* Morgan-Jones & J.F. White *in* Mycotaxon **18**(1): 59. 1983.
>
> Type: *Paraphoma radicina* (McAlpine) Morgan-Jones & J.F. White ≡ *Phoma radicina* (McAlpine) Boerema.

The species of section *Paraphoma* are characterized by a copious production of mainly septate setae on the surface of the relatively thick-walled, pseudo-parenchymatous and distinctly ostiolate pycnidia. The conidia are always aseptate, both *in vivo* and *in vitro*. The setae may be stiff or rather hyphal-like and either short or relatively long. They may be scattered over the entire surface of the pycnidium as shown in the type species of the section, *Phoma radicina* (Fig. 21A), but often they are most abundant around the ostiole (Fig. 21B). Pycnidia with setae mainly around the ostiole superficially resemble those of the genus *Pyrenochaeta* De Not. emend. R. Schneider (1979). That genus, however, is characterized by elongated, branched conidiophores instead of simple doliiform or ampulliform conidiogenous cells (see Fig. 5C).

Some species of this section produce single chlamydospores, solitary or in series and complexes. So far none of the species has been associated with a teleomorph.

It is curious that most species of *Phoma* sect. *Paraphoma* are typical soil fungi, often associated with monocotyledonous plants (Gramineae, Amaryllidaceae, Iridaceae, Liliaceae, Orchidaceae and Zingiberaceae).

Diagn. lit.: Boerema & Bollen (1975) (occurrence of setae in *Phoma* spp. and difference with species of *Pyrenochaeta* on account of conidiogenesis), Morgan-Jones & White (1983b) (introduction of a new genus *Paraphoma* 'to accomo-date *Phoma radicina* (McAlpine) Boerema, a species with setose pycnidia'),

Boerema (1985) (*Paraphoma* first treated as a section of *Phoma*), van der Aa *et al.* 1990 (formal introduction of *Phoma* sect. *Paraphoma*), de Gruyter & Boerema (2002) (precursory contribution to this chapter).

Synopsis of the Section Characteristics

- Pycnidia simple or complex, relatively thick-walled, pseudoparenchymatous and ostiolate, with hairy or rigid setae.

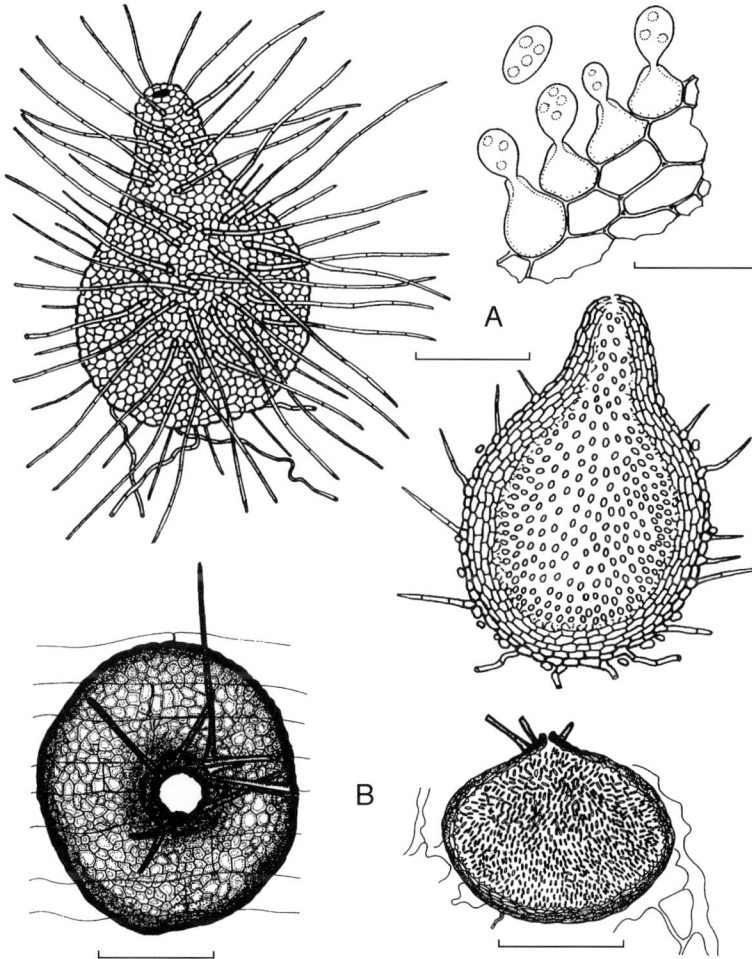

Fig. 21. A *Phoma radicina*, type species of *Phoma* sect. *Paraphoma*. Surface view and vertical section of pycnidium with setae scattered over the entire pycnidial wall; inner part of wall with conidiogenous cells. **B** *Phoma terrestris*. Surface view and vertical section of pycnidium with setae around the ostiole. Bar pycnidia 100 μm, bar conidiogenous cells 10 μm. Drawing A after Morgan-Jones & White (1983b; with permission), B after Punithalingam & Holliday (1973a; with permission).

- Conidia aseptate, both *in vitro* and *in vivo*.
- Chlamydospores, if present, simple, solitary or in series and clusters.
- Teleomorph unknown.

Key to the Species and Varieties Treated of Section *Paraphoma*

Differentiation based on the characteristics *in vitro*

1. a.   Chlamydospores absent **2**
    b.   Chlamydospores present **8**

2. a.   Characteristic fragmentation of hyphae occurs (Fig. 23A), colony greenish to rosy vinaceous/orange on OA, conidia 3.5–6 × 1.5–3 μm .... *Ph. septicidalis* [Common soil- and airborne saprophyte in Europe; also found in Africa.]
    b.   Fragmentation of hyphae absent **3**

3. a.   Conidia very small, subglobose, not exceeding 3.5 μm, colony greenish, often with coral pigmentation ....*Ph. carteri* [Isolated from discoloured bark and wood of oaks, *Quercus* spp., in North America and Europe.]
    b.   Conidia exceeding 3.5 μm **4**

4. a.   Colony distinct pale luteous on OA, due to a diffusible pigment production, conidia 3–6 × 1–3 μm... *Ph. radicina* [Common on roots of herbaceous and woody plants in Australia, Eurasia and North America. Also isolated from bulbs, cysts of nematodes and various soil samples. Harmless or saprophytic.]
    b.   Colony on OA greenish, greyish, brownish or vinaceous...............................**5**

5. a.   Colony vinaceous on OA, due to production of pigmented grains of exudate, conidia 4–6 × 2–2.5 μm ....*Ph. terrestris* [Causal organism of pink root of onions, *Allium cepa*. Apparently a common soil fungus in North and South America. Records from Europe, Africa and Australasia are generally associated with the cultivation of *Allium* crops. Frequently also isolated from roots of grasses and other herbaceous plants, but usually without disease symptoms.]
    b.   Colony on OA greenish, greyish or brownish **6**

6. a.   Conidia not exceeding 5.5 μm **7**
    b.   Conidia exceeding 5.5 μm **8**

7. a.   Colony greenish to greyish on OA, conidia 4–5.5 × 2–2.5 μm ....
         ....*Ph. leveillei* var. *leveillei* [Plurivorous worldwide recorded soil fungus. Usually as necrophyte, but the basal and underground parts of monocotyledonous plants may be affected.]
    b.   Colony greenish on OA, conidia 3.5–4.5 × 1.5–2 μm, probably also a cosmopolitan soil fungus ....*Ph. leveillei* var. *microspora* [In worldwide records and saprophytic behaviour resembling the type variety.]

8 a.   Conidia ellipsoidal to subglobose with several small guttules, colony white to
        greyish/greenish/brownish on OA, conidia 5–7.5 × 2–2.5 µm; sclerotial bodies
        covered by short setae present ....*Ph. glycinicola*
        [Specific pathogen of *Glycine* spp.: leaf spot. Widespread in Africa. Varieties of
        soybean, *Glycine max* (originally native of eastern Asia), appear to be very
        susceptible.]
   b.   Conidia cylindrical to allantoid with two or more distinct guttules, colony
        colourless to greenish/brownish on OA, conidia cylindrical to allantoid, 4.5–7 ×
        1–2 µm, sclerotial bodies absent ....*Ph. briardii*
        [Soil fungus, probably widespread occurring in the agricultural fields of western
        Europe. Specifically recorded from roots of *Monocotyledonae*. No data on
        pathogenicity.]

9. a.   Growth rate fast, 5–7 cm on OA after 7 days, conidia highly
        variable and relatively large, 3.5–10.5 × 1.5–4.5 µm, colony greenish
        on OA ....*Ph. gardeniae*
        [Probably common soil-borne fungus in India, but also reported from Curaçao.
        Opportunistic parasite of woody and herbaceous plants.]
   b.   Growth rate slow to moderate, 1–4.5 cm on OA after 7 days..........................**9**

10. a.   Growth rate slow, 1–1.5 cm on OA after 7 days, conidia 4–5.5 ×
        1.5–2.5 µm, colony brownish on OA and MA, NaOH reaction greenish
        (not an E⁺ reaction) ....*Ph. indica*
        [Thus far only known from leaf spots of sugarcane, *Saccharum officinarum*, in
        India. No data on pathogenicity.]
   b.   Growth rate slow to moderate, > 2.5 cm on OA after 7 days **11**

11. a.   Growth rate slow to moderate, 2.5–3 cm on OA after 7 days, conidia
        3–5 × 1.5–2 µm, colony greyish to greenish on OA, NaOH reaction
        negative ....*Ph. terricola*
        [Soil fungus, probably widespread occurring in the agricultural fields of western
        Europe, possibly parasitic on cysts of *Globodera pallida*.]
   b.   Growth rate moderate, 3.5–4.5 cm on OA after 7 days, conidia 4–6 ×
        2–2.5 µm, colony vinaceous on OA, NaOH reaction vinaceous/violet on OA
        (not an E⁺ reaction) **5a**

Possibilities of identification *in vivo*

Many species of sect. *Paraphoma* are plurivorous soil fungi and therefore difficult
to identify without isolation and comparative study *in vitro*. However, pycnidia
and conidia of some plurivorous species show specific characteristics which may
offer possibilities for identification *in vivo*. Species infecting a specific host may
cause recognizable disease symptoms, e.g. *Ph. terrestris* causing pink root of
onions. The same holds for *Ph. glycinicola* causing a serious leaf spot on soybean in Africa. The presence of *Ph. carteri* on oaks is so far only established by
disease symptoms; pycnidia have not yet been observed *in vivo*.

Table 2. Characteristics of setae in species of *Phoma* sect. *Paraphoma*.

	Setae short, up to 100 μm	Setae of moderate length	Setae long, exceeding 200 μm	Setae mainly around ostiole	Setae scattered over pycnidium
Ph. gardeniae	+			+	
Ph. indica	+			+	
Ph. setariae	+			+	
Ph. terricola	+			+	
Ph. leveillei var. *microspora*	+			+	+
Ph. carteri		+			+
Ph. glycinicola		+		+	+
Ph. briardii			+	+	+
Ph. leveillei var. *leveillei*			+		+
Ph. radicina			+		+
Ph. septicidalis			+		+
Ph. terrestris			+	+	

Several *Pyrenochaeta* spp. which, according to their type material (Schneider, 1979), should belong to sect. *Paraphoma*, could not be identified or related to species so far studied in culture. Most of those taxa are known only from single collections. However, Schneider (1979) noted that three differently named African/American collections from leaf spots on *Pennisetum*, *Saccharum* and *Setaria* spp. refer to the same species. An old isolate of that pathogen (CBS 333.39) from *Saccharum officinarum* in Brazil, exhibited specific cultural characters, but remained sterile. Nevertheless this still insufficiently known species has been included in this chapter, with annotations on its characteristics *in vivo*:

Pycnidia distinguished from those of the species described *in vitro* by uniform short continuous setae, 15–75 μm, mostly around the ostiole. Conidia 6–10(–12) × 2.5–4 μm, ellipsoidal, subcylindrical to fusoid or irregular, usually with 2 distinct guttules...*Ph. setariae* [Possibly a widely distributed weak pathogen of Gramineae, which only becomes noticeable in conditions favourable for spread. Recorded from Guinea, Nigeria, Brazil and North America: leaf necroses.]

Distribution and Facultative Host and Substratum Relations in Section *Paraphoma*

Plurivorous
(But with preference of roots
from monocotyledonous
plants, see also below)

America (North and South):
   Ph. terrestris
[Generally saprophytic, but roots of susceptible plants may be affected; as causal organism of pink root of onions worldwide known.]

<u>Europe (western), esp. agricultural fields:</u>
Ph. briardii
[Saprophytic; recorded from France, Belgium, The Netherlands and Germany.]
[Saprophytic, but possibly parasitic on cysts of *Globodera pallida*; recorded from Belgium and The Netherlands.]

<u>Worldwide (probably), but esp. in temperate regions:</u>
Ph. leveillei vars *leveillei* and *microspora*
[Generally saprophytic, but basal and underground parts of monocotyledonous plants may be affected.]
Ph. radicina
[Generally saprophytic, esp. on root surfaces; recorded from Australia, Eurasia and North America.]
Ph. septicidalis
[Saprophytic, 'pioneer flora'; recorded from western Europe and South America.]
[Opportunistic parasite of Gramineae: leaf necroses; recorded from Africa, North and South America.]

<u>Worldwide (probably) in subtropical regions:</u>
Ph. gardeniae
[Opportunistic parasite of woody and herbaceous plants; common in India.]

Hosts mentioned in the discussion of the species:

Chenopodiaceae
Beta vulgaris
(rhizosphere)

             Ph. terricola
             [Record from Europe.]

Fagaceae
Quercus spp.

             Ph. carteri
             [In Europe and North America isolated from discoloured bark and wood; although dieback has been attributed to it (USA: *Pyrenochaeta*-dieback), probably only an opportunistic pathogen.]

Leguminosae
Arachis hypogaea

             Ph. gardeniae
             [Record from India.]

Glycine spp., esp. *Gl. max*   *Ph. glycinicola*
             [Serious pathogen, widespread recorded in Africa: leaf spot.]

Glycine max (roots)   *Ph. septicidalis*
             [Record from Africa.]

MONOCOTYLEDONAE

Amarylidaceae
Narcissus sp. (roots)   *Ph. leveillei* var. *leveillei*
                                      [Record from Europe.]

Gramineae
Milium effusum   *Ph. briardii*
(stem debris)   [Record from Europe.]
Oryza sativa (roots)   *Ph. leveillei* var. *leveillei*
                                      Ph. terrestris
                                      [Records from Asia and North America.]
Pennisetum typhoides   *Ph. setariae*
                                      [Record from Africa; disease: leaf spot.]
Saccharum officinarum   *Ph. indica*
                                      [Recorded in India; disease: leaf spot.]
                                      Ph. setariae
                                      [Record from South America.]
Secale cereale (roots)   *Ph. radicina*
                                      Ph. leveillei var. *leveillei*
                                      [Both worldwide distributed.]
                                      Ph. briardii
                                      [So far only known from Europe.]
Setaria lutescens (roots)   *Ph. terrestris*
                                      [Record from North America.]
                                      Ph. setariae
                                      [Record from North America; disease: leaf spot.]
Triticum aestivum   *Ph. terricola*
(rhizosphere)   [Record from Europe.]
Zea mays (roots)   *Ph. terrestris*
                                      [Record from North America.]

Iridaceae
Iris spp. (roots)   *Ph. radicina*
                                      [Record from Europe.]

Liliaceae
Allium spp., esp. *A. cepa*   *Ph. terrestris*
                                      [Worldwide distributed; disease: pink root.]

Orchidaceae
Phalaenopsis sp.   *Ph. briardii*
                                      [Record from Europe.]

Zingiberaceae
Elettaria cardamomum   *Ph. leveillei* var. *microspora*
                                      [Record from Central America.]

Soil   *Ph. briardii*
                                      Ph. gardeniae

> Ph. *leveillei* vars *leveillei* and *microspora*
> Ph. *radicina*
> Ph. *septicidalis*
> Ph. *terrestris*
> Ph. *terricola*

Substrata of animal and inorganic origin
> Ph. *radicina*
> Ph. *terricola*
> [Both isolated from cysts of phytonematodes.]
> Ph. *gardeniae*
> Ph. *septicidalis*
> [Both isolated from air.]
> Ph. *leveillei* var. *microspora*
> [Isolated from water.]

Descriptions of the Taxa

Characteristics based on study in culture.

Phoma briardii Gruyter & Boerema, Fig. 22A

> *Phoma briardii* Gruyter & Boerema *in* Persoonia **17**(4): 555. 2002 ['2001'].
>> ≡ *Pyrenochaeta leptospora* Sacc. & Briard *in* Revue mycol. **11**: 16. 1889; not *Pyrenochaeta leptospora* Speg. *in* Ann. Mus. Buenos Aires **20**: 354. 1910; not *Phoma leptospora* Speg., Fungi Chilens. 145. 1910, nor *Phoma leptospora* Sacc. *in* Annls mycol. **11**: 553. 1913.
>> ‡ ≡ *Pyrenochaeta spegazziniana* Trotter *in* Sylloge Fungi. **25**: 190. 1931 [illegitimate: nomenclaturally superfluous name; Trotter probably intended to replace the later homonym *Pyr. leptospora* Speg. (l.c.) with *Pyr. spegazziniana* but listed the latter as a substitute for *Pyr. leptospora* Sacc. & Briard].

Diagn. data: de Gruyter & Boerema (2002) (fig. 9: conidial shape, reproduced in present Fig. 22A; cultural descriptions, partly adopted below).

Description in vitro

Pycnidia setose, globose to subglobose with usually 1 papillate or non-papillate ostiole, 70–265 µm diam., honey/citrine, later olivaceous to olivaceous black; setae relatively long, exceeding 200 µm, spread over the upper surface, more densely around the ostiole. Walls made up of 2–6(–10) layers of cells, outer layers pigmented.

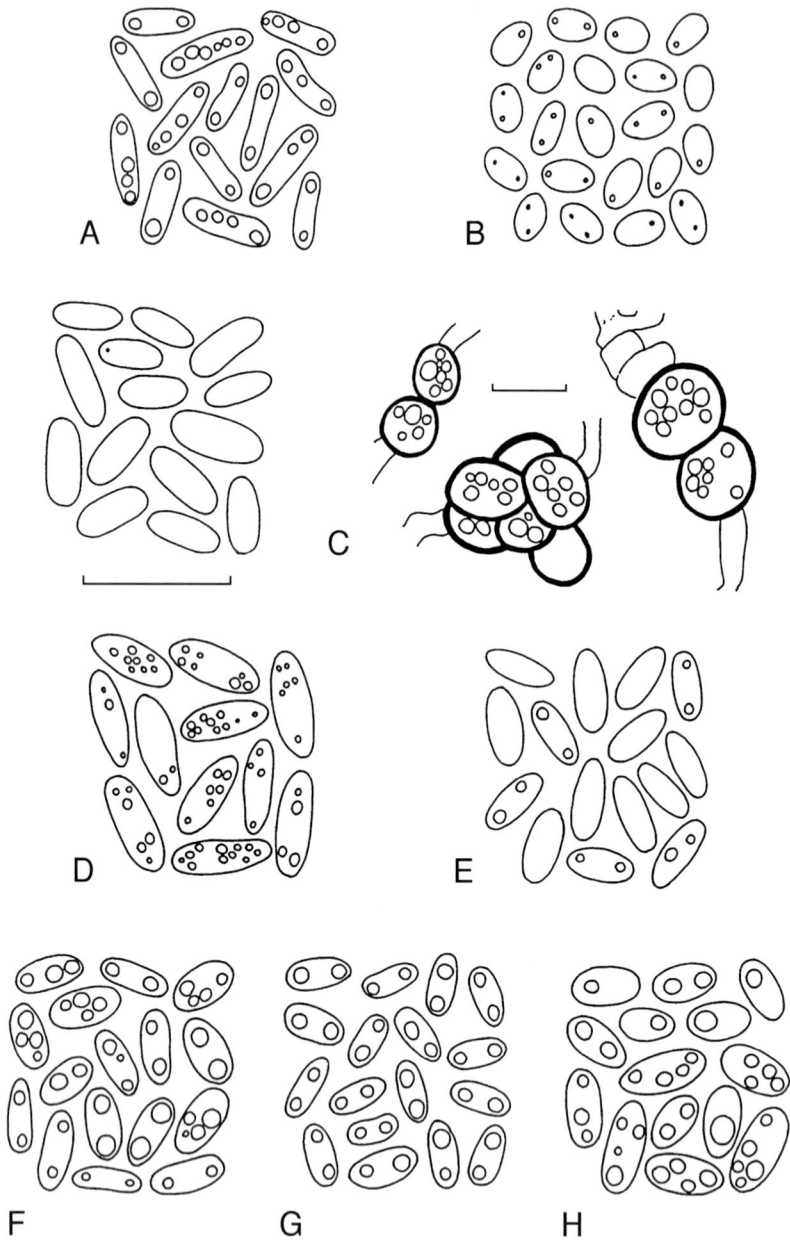

Fig. 22. Shape of conidia and some hyphal chlamydospores. **A** *Phoma briardii.* **B** *Phoma carteri.* **C** *Phoma gardeniae.* **D** *Phoma glycinicola.* **E** *Phoma indica.* **F** *Phoma leveillei* var. *leveillei.* **G** *Phoma leveillei* var. *microspora.* **H** *Phoma radicina.* Bars 10 μm.

Micropycnidia occur, mainly 20–50 μm diam. Conidial matrix whitish.

Conidia cylindrical to allantoid with 2 or more distinct guttules, 4.5–7 × (1–)1.5–2 μm.

Colonies on OA *c.* 2 cm diam. after 7 days (*c.* 4.5 cm after 14 days), regular, colourless to grey olivaceous/olivaceous, with finely velvety pale olivaceous grey aerial mycelium. Pycnidia scattered, both on and in the agar. Representative culture CBS 101635.

Ecology and distribution

This fungus has been repeatedly isolated from soil in agricultural fields in Europe (especially Belgium, Germany and The Netherlands). The French type collection was on stem debris of millet (*Milium effusum*; Gramineae). Other records refer specifically to roots of Monocotyledonae, Gramineae (e.g. *Secale cereale*) and Orchidaceae (*Phalaenopsis* sp.). So far there are no data on pathogenicity.

Phoma carteri Gruyter & Boerema, Fig. 22B

> *Phoma carteri* Gruyter & Boerema *in* Persoonia **17**(4): 547. 2002 ['2001'].
> ≡ *Pyrenochaeta minuta* J.C. Carter *in* Bull. Ill. nat. Hist. Surv. **21**: 214. 1941; not *Phoma minuta* Wehm. *in* Mycologia **38**: 318. 1946, nor *Phoma minuta* Alcalde *in* An. Jard. bot. Madr. **10**: 233. 1952.

Diagn. data: Carter (1941) (figs 35, 36; original description on cornmeal agar, with photographs of pycnidia and mycelial mat; disease symptoms on oaks), de Gruyter & Boerema (2002) (fig. 3: conidial shape, reproduced in present Fig. 22B; cultural descriptions, partly adopted below).

Description in vitro

Pycnidia setose, globose with 1(–2) non-papillate ostioles, 80–230 μm diam., greenish olivaceous/olivaceous, later olivaceous black; setae of moderate length, up to 200 μm, spread over the upper surface. Walls made up of 2–6 layers of cells, sometimes partly thicker due to protuding cells into the pycnidial cavity, outer layers pigmented. Conidial matrix buff/rosy buff.

Conidia subglobose with 1(–2) guttules, 2.5–3.5 × 2–2.5 μm.

Colonies on OA *c.* 2.5 cm diam. after 7 days (4.5–5.5 cm after 14 days), regular to somewhat irregular, olivaceous buff/greenish olivaceous to grey olivaceous, often below the colony a coral pigmentation; aerial mycelium finely floccose/finely woolly, (pale) olivaceous grey. With addition of a drop of NaOH the coral pigmentation discolours to violet. Pycnidia scattered on the agar and in aerial mycelium. Representative culture CBS 101633.

Ecology and distribution

Isolated from discoloured bark and wood of different species of oaks (*Quercus alba, Q. palustris* and *Q. suber*; Fagaceae) in North America (USA, Illinois) and Europe (The Netherlands, Spain). Although dieback has been attributed to this fungus (USA), it is probably only an opportunistic pathogen.

Phoma gardeniae (S. Chandra & Tandon) Boerema, Fig. 22C

> Phoma gardeniae (S. Chandra & Tandon) Boerema apud Boerema & Dorenbosch *in* Versl. Meded. plziektenk. Dienst Wageningen **156** (Jaarb. 1979): 27. 1980.
>> ≡ *Pyrenochaeta gardeniae* S. Chandra & Tandon *in* Mycopath. Mycol. appl. **29**: 274–275. 1966.

Diagn. data: Chandra & Tandon (1966) (fig. 2; original description), Boerema & Dorenbosch (1980) (classification, occurrence in Curaçao), Sharma *et al.* (1990) (cultural characteristics sub 'Phoma leveillei'), de Gruyter & Boerema (2002) (figs 10, 15: shape of conidia and chlamydospores, reproduced in present Fig. 22C; cultural descriptions, partly adopted below).

Description in vitro

Pycnidia setose, globose to irregular with usually 1 slightly papillate or non-papillate ostiole, 50–180 μm diam., olivaceous to olivaceous black; setae relatively short, up to 100 μm, concentrated around ostiole. Walls made up of 3–8 layers of cells, or filling nearly the entire cavity, outer layers pigmented. Conidial matrix white or flesh.

Conidia ellipsoidal to ovoid, with or without small guttules, (3.5–)5–8.5(–10.5) × (1.5–)2–3.5(–4.5) μm.

Chlamydospores globose to subglobose, usually intercalary, solitary or aggregated, 6–15 μm diam., ochraceous to olivaceous with greenish guttules.

Colonies on OA 5–7 cm diam. after 7 days, regular, grey olivaceous to greenish olivaceous towards margin or colourless with grey olivaceous to dull green sectors (reverse olivaceous grey to leaden grey/leaden black, or with grey olivaceous to olivaceous grey sectors), with finely floccose, grey olivaceous to olivaceous grey aerial mycelium. With addition of a drop of NaOH a weak reddish/brownish discolouring may occur, but this is not specific. Representative culture CBS 626.68.

Ecology and distribution

The original Indian isolate of this species was made from leaf spots of the cape jasmine, *Gardenia jasminoides*. In India it seems to be a common soil- and air-borne fungus, which may act as an opportunistic pathogen of woody as well as herbaceous plants (once isolated from *Arachis hypogaea*). The fungus is also reported from Curaçao. It has been confused with *Phoma leveillei* Boerema & G.J. Bollen var. *leveillei* (this section), but can be easily distinguished from the latter by its highly variable, relatively large, often eguttulate conidia and the production of chlamydospores.

Phoma glycinicola Gruyter & Boerema, Fig. 22D

> *Phoma glycinicola* Gruyter & Boerema *in* Persoonia **17**(4): 554. 2002 ['2001'].
>> ≡ *Pyrenochaeta glycines* R.B. Stewart *in* Mycologia **49**: 115. 1957; not *Phoma glycines* Sawada *in* Spec. Publ., Coll. Agric., Nat. Taiwan Univ. **8**: 129. 1959 [‡ not published with Latin diagnosis].

Diagn. data: Stewart (1957) (fig. 1; original description, disease symptoms with illustrations), de Gruyter & Boerema (2002) (fig. 8: conidial shape, reproduced in present Fig. 22D; cultural descriptions, partly adopted below).

Description in vitro

Pycnidia setose, globose to irregular with 1(–3) non-papillate or papillate ostioles, 70–240 μm diam., honey/olivaceous, later olivaceous black; setae of moderate length, up to 200 μm, spread over the upper surface, more densely around the ostiole. Walls made up of of 4–11 layers of cells, outer layers pigmented. Conidia matrix rosy buff to salmon/saffron.

Conidia ellipsoidal to subglobose, with several small guttules, 5–7.5 × 2–2.5 μm.

Sclerotia setose, globose to subglobose, up to 600 μm diam., develop together with pycnidia; setae very short, up to 10 μm, spread over the surface. The cell structure of these sclerotia resemble those of the pseudoparenchymatous 'pycnosclerotia' found in some species of *Phoma* sect. *Sclerophomella* (**F**).

Colonies on OA 2–2.5 cm diam. after 7 days (3.5–5 cm after 14 days), regular, white to pale olivaceous grey/pale dull green to brick, with scarce, finely felty, white aerial mycelium. Dark red mycelial fragments occur, due to crystallization of the pigments. The brick pigments turn to greenish blue with addition of a drop of NaOH. In the agar often reddish pigmented grains of exudate, resembling small crystals. Pycnidia abundant, often accompanied by sclerotia, mainly in concentric rings both on and in the agar, and in aerial mycelium as well. Representative culture IMI 294986.

Ecology and distribution

Recorded as serious pathogen of *Glycine* spp. (Leguminosae): leaf spot in Africa (Ethiopia, Zambia, Zimbabwe). The primary indigenous host is probably *Glycine javanica*. Varieties of soybean, *Glycine max* (originally native of eastern Asia) appear to be very susceptible. The leaf spots, at first small, reddish-brown, soon become necrotic and may fall out, giving the plants a very ragged appearance. In susceptible varieties of soybean leaf abscission is the most damaging aspect of the disease.

Phoma indica (T.S. Viswan.) Gruyter & Boerema, Fig. 22E

> *Phoma indica* (T.S. Viswan.) Gruyter & Boerema *in* Persoonia **17**(4): 556. 2002 ['2001'].
>
> > ≡ *Pyrenochaeta indica* T.S. Viswan. *in* Curr. Sci. **26**: 118. 1957.

Diagn. data: de Gruyter & Boerema (2002) (fig. 11: shape of conidia, reproduced in present Fig. 22E; cultural descriptions, partly adopted below).

Description in vitro

Pycnidia setose, globose to subglobose, with 1–2 papillate ostioles, 55–240 μm diam., citrine/honey, later sienna to olivaceous/olivaceous black; setae relatively short, up to 100 μm, mainly concentrated around the ostiole. Walls made up of 3–12 layers of cells, outer layer(s) pigmented. Conidial matrix pale luteous.

Conidia ellipsoidal, usually with 2 guttules, 4–5.5 × 1.5–2.5 μm.

Chlamydospores globose to subglobose, mostly terminal, solitary or in short chains, occasionally clustered, 5–11 μm diam., olivaceous with greenish guttules.

Colonies on OA slow growing, c. 1 cm diam. after 7 days (c. 3 cm after 14 days), regular, olivaceous with sparse, felty, pale olivaceous grey aerial mycelium. With addition of a drop of NaOH an herbage green discoloration occurs (*not* E$^+$ reaction). Pycnidia scattered, both on and in the agar. Representative culture IMI 062569.

Ecology and distribution

Found on the whitish centre of fusiform dirty brownish leaf spots of sugarcane, *Saccharum officinarum* (Gramineae), in India. There are no data on pathogenicity. On the spots a species of *Melanospora* was also found. *Ph. indica* produces significantly smaller conidia than *Ph. setariae* (H.C. Greene) de Gruyter & Boerema (this section) recorded in Brazil from leaf spots of sugarcane.

Phoma leveillei Boerema & G.J. Bollen var. *leveillei*, Fig. 22F

> *Phoma leveillei* Boerema & G.J. Bollen *in* Persoonia **8**(2): 115. 1975, var. *leveillei*.
>
> > ≡ *Vermicularia acicola* Moug. & Lév. apud Léveillé *in* Annls Sci. nat. (Bot.) III, **9**: 259. 1848 [as 'Moug. Lév.']; not *Phoma acicola* (Lév.) Sacc. *in* Michelia **2**(2): 272. 1881 [= *Sclerophoma pythiophila* (Corda) Höhn., see Sutton, 1980 : 487].

≡ *Pyrenochaeta acicola* (Moug. & Lév.) Sacc. *in* Sylloge Fung. **3**: 220. 1884 [as '(Lév.) Sacc.'].

= *Pyrenochaeta phlogis* Massee *in* Bull. misc. Inf. R. bot. Gdns Kew **1907**: 241. 1907; not *Phoma phlogis* Roum. *in* Revue mycol. **6**: 160. 1884 [= *Phoma acuta* (Hoffm. : Fr.) Fuckel subsp. *acuta* f. sp. *phlogis*, sect. *Plenodomus*, **G**].

= *Pyrenochaeta oryzae* Shirai ex I. Miyake *in* J. Coll. Agric. imp. Univ. Tokyo **2**(4): 255. 1910; not *Phoma oryzae* Catt. *in* Arch. Bot. crittog. Pavia **2–3**: 118. 1877 (1879?); not *Phoma oryzae* Cooke & Massee *in* Grevillea **16**: 15. 1887 [= *Phoma minutispora* P.N. Mathur, sect. *Phoma*, **A**]; not *Phoma oryzae* Hori, 'Nosakubutsu-Biyogatu' 111–113. 1903 [in Japanese].

= *Pyrenochaeta lupini* Sibilia *in* Annali Bot. **18**: 284. 1930; not *Phoma lupini* Ellis & Everh. *in* Bull. Washburn [Coll.] Lab. nat. Hist. **1**: 6. 1884 [sect. *Phoma*, **A**]; not *Phoma lupini* N.F. Buchw. in Møller, Fungi Faeröes **2**: 153. 1958.

= *Pyrenochaeta calligoni* Kravtzev apud Schwarzman & Kravtzev *in* Trudÿ Inst. Bot., Alma Ata **9**: 45. 1961; not *Phoma calligoni* Murashk. *in* Trans. Agric. Forest. Omsk **9**: 6. 1928.

= *Pyrenochaeta spinaciae* Verona & Negru apud Negru & Verona *in* Mycopath. Mycol. appl. **30**: 309. 1966; not *Phoma spinaciae* Bubák & K. Krieg. apud Bubák *in* Annls mycol. **10**: 47. 1912 [= *Phoma betae* A.B. Frank, sect. *Pilosa*, **I**].

= *Pyrenochaeta anthyllidis* Manoliu & Mítítíuc *in* Reprium nov. Spec. Regni veg. [Feddes Reprium] **87**: 142. 1976.

Diagn. data: Dorenbosch (1970) (pls 3 figs 1, 2, 5 figs 3, 4, tab. I, 1; characteristics *in vitro* sub *Pyrenochaeta acicola* with photographs of pycnidia and mycelial mat on OA and MA, isolate sources, history and designation of neotype; relation with the genus *Phoma*), Domsch *et al.* (1980) (fig. 283: pycnidia, conidia and conidiogenous cells; occurrence in soil, literature references), Sutton (1980) (fig. 228C; specimens in CMI collection), Boerema & Hamers (1989) (synonymy, concept as collective species, association with disease symptoms on bulbs and roots), de Gruyter & Boerema (2002) (fig. 6: shape of conidia, reproduced in present Fig. 22F; cultural descriptions, partly adopted below).

Description in vitro

Pycnidia setose, subglobose with usually 1 non-papillate or slightly papillate ostiole, 180–270 μm diam., olivaceous to olivaceous black; setae relatively long, exceeding 200 μm, spread over the upper surface. Walls made up of 2–7 layers of cells, outer layers pigmented. Conidial matrix white to buff.

Conidia subglobose to ellipsoidal with 2 or more distinct guttules, 4–5.5 × 2–2.5 μm.

Colonies on OA 2–2.5 cm diam. after 7 days (4–5 cm after 14 days), regular to somewhat irregular, grey olivaceous/olivaceous grey to dull green, with woolly, (pale) olivaceous grey aerial mycelium. Luteous to ochraceous pigmented grains of exudate, resembling small crystals, may be produced in the agar.

Chlamydospores absent, but hyphal swollen cells may occur. Pycnidia abundant, scattered or in concentric rings, on the agar as well as in aerial mycelium. Representative cultures CBS 260.65, CBS 101634.

Ecology and distribution

A worldwide soil fungus (Eurasia, North America, Africa, Australia), regarded as a collective species with much variability in morphological and physiological characters. Generally it behaves like a saprophyte; all listed synonyms being associated with necrotic plant tissue. However, the basal and underground parts of monocotyledonous plants such as Gramineae and Amaryllidaceae may be affected by it (reported from e.g. *Oryza sativa*, *Secale cereale* and *Narcissus* spp.). The fungus has been confused with morphologically very similar soil fungi of this section: *Phoma terrestris* H.N. Hansen (characterized by the production of a red pigment) and *Phoma terricola* Boerema (distinguished by abundant production of chlamydospores).

Phoma leveillei var. *microspora* Gruyter & Boerema, Fig. 22G

Phoma leveillei var. *microspora* Gruyter & Boerema *in* Persoonia **17**(4): 553. 2002 ['2001']; **18**(2): 159. 2003 [info type specimen].

Diagn. data: de Gruyter & Boerema (2002) (fig. 7: shape of conidia, reproduced in present Fig. 22G; original description, partly adopted below).

Description in vitro

Pycnidia setose, globose to subglobose, with 1(–2) non-papillate ostioles, (20–)80–270 μm diam., greenish olivaceous/olivaceous, later olivaceous black; setae relatively short, up to 100 μm, spread over the upper surface, more densely around the ostiole. Walls made up of 2–7 layers of cells, outer layers pigmented. Conidial matrix off-white.

Conidia ellipsoidal to oblong, with 2 distinct guttules, 3.5–4.5 × 1.5–2 μm.

Colonies on OA *c.* 2.5 cm diam. after 7 days (5 cm after 14 days), regular, grey olivaceous to dull green/dark herbage green, with finely woolly, pale olivaceous grey aerial mycelium. With addition of a drop of NaOH a weak greenish discolouring may occur, but this is not specific. Pycnidia scattered or in concentric rings, mainly on the agar. Representative culture CBS 102876.

Ecology and distribution

In saprophytic behaviour and apparently worldwide distribution (south-east Europe, soil and water, and Central America, fruits of *Elettaria cardamomum*,

Zingiberaceae) this variety resembles the very variable and ubiquitous *Phoma leveillei* Boerema & G.J. Bollen (var. *leveillei*). Morphologically, however, it is distinguished by significantly smaller conidia and shorter setae.

Phoma radicina (McAlpine) Boerema, Figs 21A, 22H

> *Phoma radicina* (McAlpine) Boerema apud Boerema & Dorenbosch *in* Versl. Meded. plziektenk. Dienst Wageningen **153** (Jaarb. 1978): 20. 1979.
> > ≡ *Pyrenochaeta radicina* McAlpine, Fung. Dis. Stone-fruit Austr. 127. 1902.
> > ≡ *Paraphoma radicina* (McAlpine) Morgan-Jones & J.F. White *in* Mycotaxon **18**: 60. 1983.

Diagn. data: Boerema & Dorenbosch (1979) (classification in *Phoma*; sources of isolates), Morgan-Jones & White (1983b) (fig. 1A–D: drawings of pycnidia, conidiogenous cells, conidia and setae; pls 1A–F, 2A–F: photographs of agar plate cultures, pycnidia, conidia, setae and wall cells; history, indication as type of *Paraphoma*, cultural and morphological description; occurrence on cyst of *Heterodera glycines*), de Gruyter & Boerema (2002) (figs 1A, 4: characteristics of pycnidia, conidiogenous cells and conidia, reproduced in present Figs 21A, 22H; cultural descriptions, partly adopted below).

Description in vitro

Pycnidia setose, globose to subglobose, with 1(–2) non-papillate or papillate ostioles, 180–450 µm diam., honey/olivaceous, later olivaceous black; setae relatively long, exceeding 200 µm, spread over the upper surface. Walls made up of 3–7 layers of cells, outer layers pigmented. Conidial matrix whitish to buff.
   Conidia ellipsoidal to subglobose, usually with several guttules, (3–)4–6 × (1–)2–3 µm.
   Colonies on OA *c.* 3 cm diam. after 7 days (*c.* 5 cm after 14 days), regular, with pale olivaceous grey aerial mycelium; colony colour pale luteous due to production of a diffusable pigment, with coral concentric zones; reverse pale luteous to amber. Red pigmented grains of exudate, resembling small crystals, are produced in the agar. A greenish discolouring may occur with the NaOH spot test. Pycnidia abundant, solitary or confluent, mainly on the agar. Representative cultures CBS 111.79, CBS 102875.

Ecology and distribution

Recorded from a wide variety of woody and herbaceous plants in Australia, Eurasia and North America. Very often isolated from root surfaces of mono-cotyledonous plants such as Iridaceae and Gramineae, e.g. bulbs of *Iris* spp. and roots of *Secale cereale*. Also from cysts of phytonematodes and various soil samples. The fungus may be regarded as harmless or saprophytic. It repre-sents the type of the section *Paraphoma*.

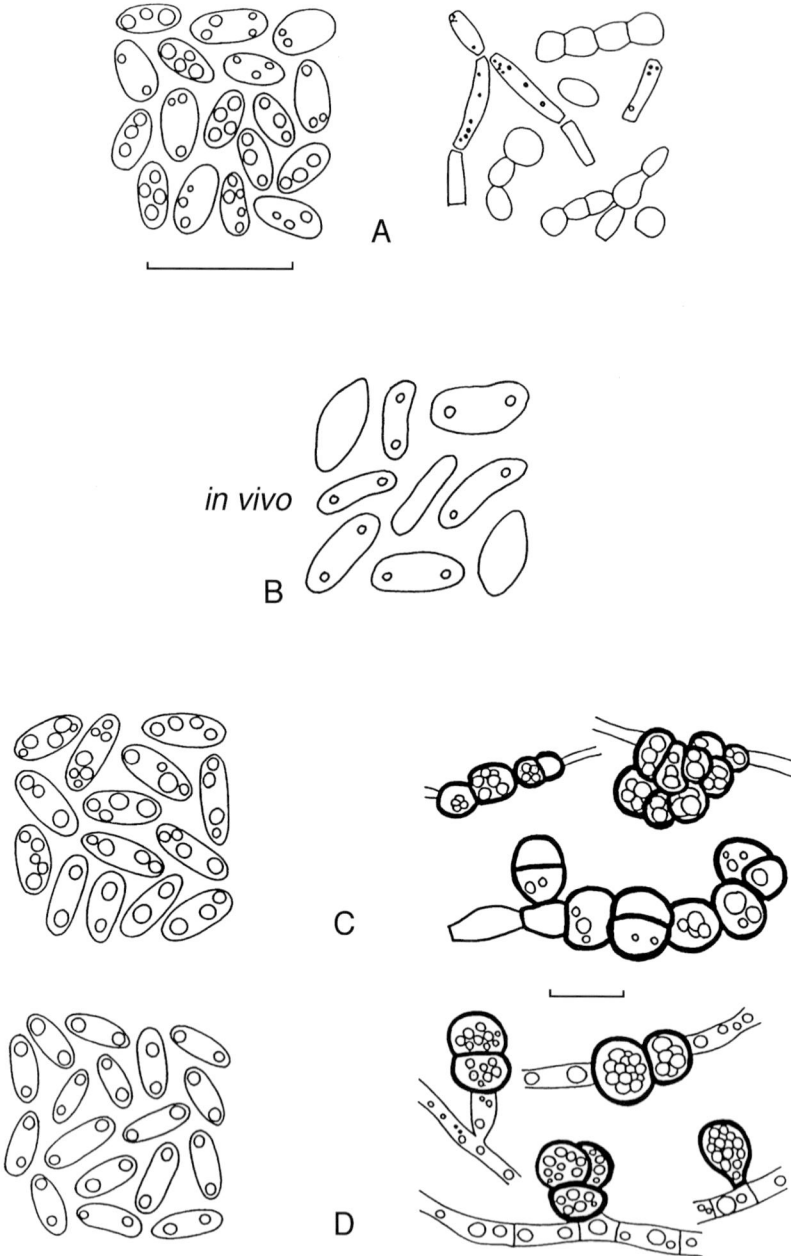

Fig. 23. Shape of conidia and some hyphal characteristics, fragmentation and chlamydospores. **A** *Phoma septicidalis.* **B** *Phoma setariae.* **C** *Phoma terrestris.* **D** *Phoma terricola.* Bars 10 μm.

Phoma septicidalis Boerema, Fig. 23A

> *Phoma septicidalis* Boerema apud Boerema & Dorenbosch *in* Versl.
> Meded. plziektenk. Dienst Wageningen **153** (Jaarb. 1978): 20. 1979.
> > ≡ *Pyrenochaeta telephii* Allesch. *in* Ber. bayer. bot. Ges. **4**:
> > 33. 1896; not *Phoma telephii* (Vestergr.) Kesteren *in* Neth.
> > J. Pl. path. **78**: 117. 1972 (sect. *Phyllostictoides*, **E**).

Diagn. data: Boerema & Dorenbosch (1979) (classification in *Phoma*, isolate sources), de Gruyter & Boerema (2002) (figs 2, 13: shape of conidia, fragmentation of hyphae, reproduced in present Fig. 23A; cultural descriptions, partly adopted below).

Description in vitro

Pycnidia setose, globose to subglobose, with 1(–2) non-papillate ostioles, 70–170 μm diam., honey to olivaceous, later olivaceous black; setae relatively long, exceeding 200 μm, spread over the upper surface. Walls made up of 2–8 layers of cells, sometimes partly thicker due to protuding of cells into the pycnidial cavity, outer layers pigmented.

Micropycnidia also occur, 25–50 μm diam. Conidial matrix whitish.

Conidia subglobose to ellipsoidal with several small or large guttules, 3.5–5(–6) × 1.5–3 μm.

Colonies on OA 2–4 cm diam. after 7 days (5–7.5 cm after 14 days), regular to slightly irregular, citrine/greenish olivaceous to dull green, rosy vinaceous to orange pigmented towards margin, with felty, (pale) olivaceous grey/grey olivaceous aerial mycelium. With addition of a drop of NaOH the rosy vinaceous margin may discolour to livid violet/purple. Fragmentation of hyphae frequently occurs. Pycnidia abundant, mainly on the agar. Representative cultures CBS 112.79, CBS 101636.

Ecology and distribution

In western Europe a widespread soil- and airborne saprophyte ('pioneer flora'). The fungus is also found in South Africa (isolated from roots of *Glycine max*, Leguminosae) and may be equally common in other parts of the world. The epithet '*septicidalis*' refers to the easy fragmentation of the hyphae *in vitro*.

Phoma setariae (H.C. Greene) Gruyter & Boerema, Fig. 23B

> *Phoma setariae* (H.C. Greene) Gruyter & Boerema *in* Persoonia **17**(4): 559. 2002 ['2001'].
> > ≡ *Pyrenochaeta setariae* H.C. Greene *in* Trans. Wis. Acad. Sci. Arts Lett. **53**: 211. 1964.
> > = *Pyrenochaeta sacchari* Bitanc. *in* Arquivos Inst. biol., S Paulo **9**: 301. 1938; not *Phoma sacchari* (Cooke) Sacc. *in* Sylloge

Fung. **3**: 166. 1884, not *Phoma sacchari* Gutner apud
Bondartseva-Monteverde, Gutner & Novoselova *in* Acta Inst.
bot. Acad. Sci. URSS, Ser. II, Fasc. 3: 789. 1936 [= *Phoma
gutneri* N. Pons *in* Fitopat. Venezolana **3**(2): 40. 1990].
= *Pyrenochaeta penniseti* J. Kranz *in* Sydowia **22**: 360. 1968.

Diagn. data: Bitancourt (1938) (first description of the fungus; figs A–L: leaf
spot symptoms on sugarcane, photographs of pycnidia with setae, conidia and
culture of isolate, = CBS 333.39, on potato dextrose agar), Schneider (1979)
(study types of the *Pyrenochaeta*-synonyms: *P. setariae* 'ähnelt *P. sacchari* und
P. penniseti und gehört zu *Phoma* Sacc. emend. Boerema & Bollen'), de
Gruyter & Boerema (2002) (classification in sect. *Paraphoma*).

Description in vivo

(Fresh cultures were not at our disposal; CBS 333.39 was sterile.)

Pycnidia (initially epiphyllous, later also amphigenous in oval-fusiform, often
confluent spots) subglobose with usually 1 papillate ostiole, 100–150 µm
diam., brownish, lighter at the base and darker towards the ostiole; setae very
short, continuous, 15–75 µm, mainly concentrated around the ostiole. Walls
made up of 2–5 layers of cells, outer layer(s) pigmented.
        Conidia variable, broadly ellipsoidal, subcylindrical to subfusoid or irregu-
lar with usually 2 distinct guttules, 6–10(–12) × 2.5–4 µm.

Ecology and distribution

Possibly a widely distributed weak pathogen of Gramineae, which only
becomes noticeable in conditions favourable for spread. The records refer to
Pennisetum typhoides in Africa (Guinea, Nigeria), *Saccharum officinarum* in
Brazil and *Setaria lutescens* in North America: leaf necroses. As the infection
progresses the first infected leaves may die back completely.

Phoma terrestris H.N. Hansen, Figs 21B, 23C

Phoma terrestris H.N. Hansen *in* Phytopathology **19**: 699. 1929.
        ≡ *Pyrenochaeta terrestris* (H.N. Hansen) Gorenz, J.C. Walker
        & Larson *in* Phytopathology **38**: 838. 1948; not *Phoma
        terrestris* R.K. Saksena, Nand & A.K. Sarbhoy *in*
        Mycopath. Mycol. appl. **29**: 86. 1966 [= *Phoma multiros-
        trata* var. *macrospora* Boerema, sect. *Phoma*, **A**].

Diagn. data: Hansen (1929) (figs 3–5; study of the pink-root disease of onions
in California, formerly ascribed to a species of *Fusarium*; description of the
causal 'pinkish' *Phoma* species, which showed to be a common soil inhabitant
in the USA; illustrations of pycnidial primordia with chlamydospores and cultural characters on cornmeal agar), Gorenz *et al.* (1948) (figs 1–3: photographs

of pycnidia on roots and seedlings; transfer of the fungus to *Pyrenochaeta* on account of the occurrence of setae around the ostiole), Punithalingam & Holliday (1973a) (figs A–D; CMI description with sketches of pycnidia, part of pycnidial wall, conidiogenous cells and conidia; phytopathological notes), Boerema (1985) (fig. 9, tab. 2; classification in *Phoma* sect. *Paraphoma*), de Gruyter & Boerema (2002) (figs 1B, 5, 14: characteristics of pycnidia, conidia and chlamydospores, reproduced in present Figs 21B, 23C; cultural descriptions, partly adopted below).

Description in vitro

Pycnidia setose, globose to subglobose with 1(–3) usually papillate ostioles, 120–370 µm diam., honey, later olivaceous to olivaceous black; setae relatively long, exceeding 200 µm, mainly concentrated around the ostiole. Walls made up of 4–11 layers of cells, outer layer(s) pigmented. Conidial matrix whitish.

Conidia ellipsoidal with several distinct guttules, 4–6 × 2–2.5 µm.

Chlamydospores globose to subglobose, terminal or intercalary, solitary or aggregated, 6–12 µm diam., ochraceous to olivaceous, with greenish guttules.

Colonies on OA 3.5–4.5 cm after 7 days (on MA 1.5–4 cm), regular with felty to finely woolly, pale olivaceous grey aerial mycelium. Colony colour brick to vinaceous, due to a red pigment in the mycelium, sometimes partly primrose, with fulvous to rust patches. Vinaceous or amber pigmented grains, resembling small crystals, may be produced in the agar. With appplication of a drop of NaOH, the brick to vinaceous pigments become vinaceous/violet. Pycnidia scattered or in concentric rings, mostly on the agar. Representative cultures CBS 377.52, CBS 335.87.

Ecology and distribution

This well-known causal organism of pink root of onion, *Allium cepa* (Liliaceae), is apparently a widely distributed soil fungus in North America (USA and Canada) and probably also a common soil inhabitant in some regions of South America (Argentina, Brazil, Venezuela). Records from Europe, Africa and Australia are generally associated with the cultivation of onions or other crops of *Allium* (leek, shallot, garlic and chive). The fungus is frequently isolated from the roots of Gramineae (e.g. *Setaria lutescens*) and other herbaceous plants, but usually without any disease symptoms. However, the roots of maize plants (*Zea mays*) and rice (*Oryza sativa*) may also be affected. The fungus is characterized by a red pigment in the mycelium and this easily distinguishes it from the morphologically very similar *Phoma leveillei* Boerema & Bollen var. *leveillei* and from *Phoma terricola* Boerema.

Phoma terricola Boerema, Fig. 23D

> *Phoma terricola* Boerema *in* Versl. Meded. plziektenk. Dienst Wageningen **163** (Jaarb. 1984): 38. 1985; not *Phoma terricola* 'Agnihothrudu' *in* Soil

Sci. **91**: 135. 1961 [a nomen nudum erroneously adopted in Mathur, Coelom. India: 185. 1979].

= *Pyrenochaeta decipiens* Marchal, Champ. copr. **6**: 8. 1891; not *Phoma decipiens* Montagne, Fl. chil. cell. **7**: 488. 1852.

Diagn. data: Boerema (1985) (fig. 10B; classification in *Phoma*, sources of isolates, drawings of pycnidia, chlamydospores and conidia), de Gruyter & Boerema (2002) (figs 12, 16: shape of conidia and chlamydospores, reproduced in present Fig. 23D; cultural descriptions, partly adopted below).

Description in vitro

Pycnidia setose, globose with usually 1, non-papillate or slightly papillate ostiole, 130–250 µm diam., honey to citrine, later olivaceous black; setae relatively short, up to 100 µm, mainly concentrated around the ostiole. Walls made up of 2–5 layers of cells, outer layer(s) pigmented. Conidial matrix whitish.

Conidia ellipsoidal to subcylindrical with usually 2 guttules, 3–5(–5.5) × 1.5–2 µm.

Chlamydospores globose to subglobose, intercalary or terminal, solitary or aggregated, 6–10 µm diam., ochraceous to olivaceous with greenish guttules.

Colonies on OA 2.5–3 cm after 7 days (4.5–5 cm after 14 days), regular to somewhat irregular, pale olivaceous grey/grey olivaceous to dark herbage green/dull green (reverse dull green/olivaceous to leaden grey/leaden black) with finely floccose pale olivaceous grey to pale greenish grey aerial mycelium. Pycnidia scattered or in concentric rings, both on and in the agar. Representative culture CBS 343.85.

Ecology and distribution

A soil fungus recorded in western Europe (Belgium, The Netherlands). Probably widespread occurring in agricultural fields. For example, it has been found in the rhizosphere of wheat, *Triticum aestivum* (Gramineae) and beet, *Beta vulgaris* (Chenopodiaceae), and isolated from cysts of the golden nematode of potatoes (*Globodera pallida*). The pycnidia of this fungus are very similar to *Phoma leveillei* Boerema & Bollen var. *leveillei*, but can be easily distinguished by an abundant production of chlamydospores. It also morphologically resembles *Phoma terrestris* H.N. Hansen, but does not produce red pigment.

11 D *Phoma* sect. *Peyronellaea*

Phoma sect. *Peyronellaea* (Goid. ex Togliani) Boerema

> *Phoma* sect. *Peyronellaea* (Goid. ex Togliani) Boerema apud van der Aa,
> Noordeloos & de Gruyter *in* Stud. Mycol. **32**: 6. 1990.
> > ≡ *Peyronellaea* Goid. *in* Atti Acad. naz. Lincei Rc. VIII, **1**:
> > 455. 1946 [without Latin diagnosis] ex Togliani *in* Annali
> > Sper. agr. II, **6**: 93. 1952.
> Type: *Peyronellaea glomerata* (Corda) Goid. ex Togliani ≡ *Phoma glomerata* (Corda) Wollenw. & Hochapfel

The section *Peyronellaea* is characterized by possession of multicellular chlamydosporal structures, which may occur in combination with unicellular chlamydospores. The present concept of the section includes: (i) species producing dictyochlamydospores resembling certain types of conidia in the dematiaceous Hyphomycete genus *Alternaria* Nees: Fr., typical alternarioid to irregular botryoid configurations (Figs 24, 25C and 26–28); (ii) species producing multicellular chlamydospores indistinguishable from the conidia in the dematiaceous Hyphomycete genus *Epicoccum* Link, epicoccoid shape (Fig. 25A); and (iii) species producing aggregates of unicellular chlamydospores looking like pseudosclerotia: pseudosclerotioid masses (Figs 25B, 28B).

The pycnidia and the chlamydospores occur as two different asexual forms (anamorphs) adapted to the conditions of growth. The carbon:nitrogen (C/N) ratio of the medium proved to be a determining factor; at low values there is greater production of pycnidia, at higher values the development of chlamydospores usually increases. However, genetic differences are also involved, sometimes the chlamydosporal anamorph may also develop separately and independently. It is important to note that it has been repeatedly shown that strains of typical species of sect. *Peyronellaea* may lose the ability to form multicellular chlamydospores (Boerema *et al.*, 1965a: 48, 49; Dorenbosch, 1970: 10).

Most species of the section occasionally produce micropycnidia developing from a hyphal cell or from a single chlamydospore cell.

The conidia are always mainly one-celled *in vitro*, but some species show secondary septation of long conidia. *In vivo* the septate condition is often more prominent. In old pycnidia the conidia usually become light-brown and occasionally also (extra) septated.

It should be noted that botryoid-alternarioid multicellular chlamydospores are found in cultures of two species classified in sect. *Heterospora* (**B**), namely *Ph. clematidina* (Thüm.) Boerema, and *Ph. narcissi* (Aderh.) Boerema *et al.* (both also incorporated in key and host list of this section). Pycnidia and conidia in section *Peyronellaea* resemble those found in some other sections of *Phoma*. For instance most species of this chlamydosporic section produce thin-walled ostiolate pycnidia containing only one-celled conidia, indistinguishable from those of sect. *Phoma* (**A**). The pycnidia of species also producing occasionally some two-celled conidia resemble the pycnidia of members of sect. *Phyllostictoides* (**E**). Finally the species with an epicoccoid synanamorph produces thick-walled poroid pycnidia resembling those of sect. *Sclerophomella* (**F**).

None of the species of this section has so far been associated with a teleomorph.

Diagn. lit.: Boerema *et al.* (1965a, 1968, 1971, 1973, 1977) (history, synonymy and misapplications of the species 'ascribed to *Peyronellaea*'), White & Morgan-Jones (1983) (species belonging to 'a group characterized by possession of phaeodictyochlamydospores'), Boerema (1993) (precursory contribution to this chapter; documentation of the concept of the present section).

Synopsis of the Section Characteristics

- Pycnidia simple or complex, mostly thin-walled, pseudoparenchymatous, distinctly ostiolate and glabrous, but sometimes with hyphal outgrowths; one species in this section produces thick-walled poroid pycnidia.
- Conidia mostly unicellular *in vitro* but some sometimes also two-celled.
- Multicellular chlamydosporal structures always present (alternarioid, botryoid-alternarioid, pseudosclerotioid or epicoccoid), sometimes unicellular chlamydospores also present.
- Teleomorph unknown.

Key to the Species and Varieties Treated of Section *Peyronellaea*

Differentiation based on the characteristics *in vitro*

1. a. Colonies in addition to pycnidia producing multicellular chlamydospores resembling the conidia in *Alternaria*: typical alternarioid to irregular-botryoid;

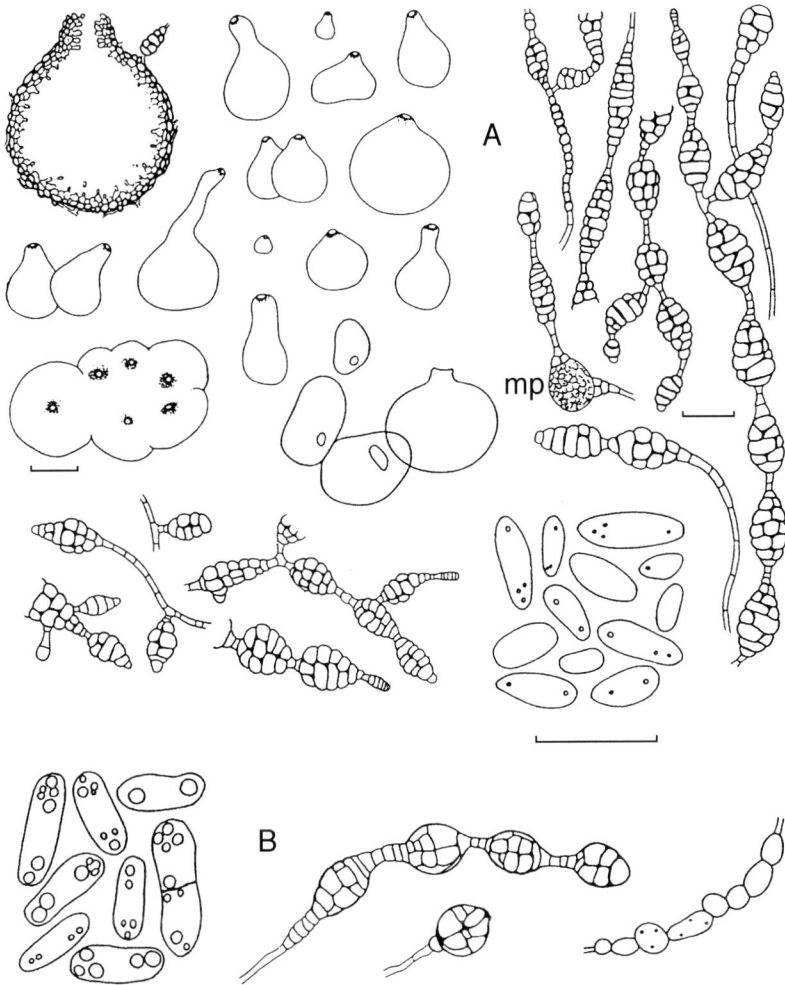

Fig. 24. A *Phoma glomerata*, type species of *Phoma* sect. *Peyronellaea*. Structure of pycnidia, conidia and aerial alternarioid chlamydospores as found in cultures of different strains of the fungus. The multicellular-dictyosporous chlamydospores of this genetically heterogeneous species are highly variable in shape and dimensions and produced in chains resembling the conidial chains of *Alternaria alternata* (Fr. : Fr.) Keissl., mp = micropycnidium. **B** *Phoma subglomerata*, e.g. characterized by short chains of unicellular and alternarioid multicellular chlamydospores. Conidia relatively long and occasionally two-celled: secondary septation. Bar pycnidia 100 μm, bar chlamydospores 20 μm, bar conidia 10 μm.

   sometimes looking like pseudosclerotia: pseudosclerotioid.; often also unicellular chlamydospores occur (Figs 24–28) **2**
b. Colonies in addition to pycnidia producing multicellular chlamydospores indistinguishable from the conidia of *Epicoccum nigrum* Link (Fig. 25A); pycnidia
   subglobose, stromatic, intermixed with sclerotia; conidia variable, mostly 3–7 ×
   1.5–3 μm ....*Ph. epicoccina*

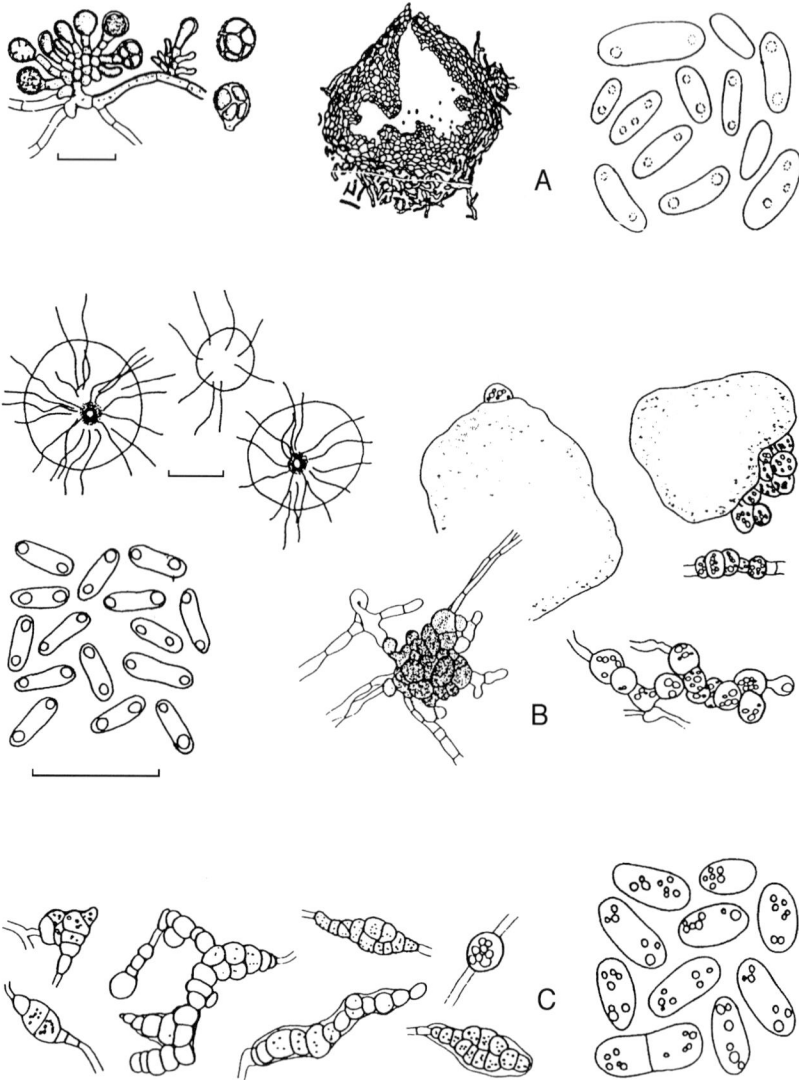

Fig. 25. A *Phoma epicoccina,* produces more or less stromatic pycnidia, highly variable conidia and clusters of multicellular-phragmosporous epicoccoid chlamydospores indistinguishable from the sporodochia of *Epicoccum nigrum* Link. **B** *Phoma chysanthemicola,* characterized by aggregates of unicellular chlamydospores forming large irregularly shaped pseudosclerioid masses in the aerial mycelium. Pycnidia often 'hairy': semi-pilose. **C** *Phoma americana,* shape of unicellular and alternarioid multicellular chlamydospores found in a culture of this species. Conidia occasionally two-celled. Bar pycnidia 100 μm, bar chlamydospores 20 μm, bar conidia 10 μm. Drawings A partly after Punithalingam *et al.* (1972), B partly after Sutton (1980), with permission.

[Plurivorous and worldwide distributed. Saprophytic. Especially common on dead seed coats.]

2. a. Pseudosclerotioid chlamydospores absent..**3**
   b. Pseudosclerotioid chlamydospores present (Figs 25B, 28B); pycnidia to varying degrees covered by hyphae (semi-pilose)..**16**

3. a. Colonies conspicuously dark cyan blue; chlamydospores and pycnidia also cyan blue; conidia mostly 5–7 × 2–3 μm..*Ph. cyanea*
      [Thus far only known from wheat field debris in South Africa.]
   b. Colonies not blue pigmented **4**

4. a. Pycnidia glabrous **5**
   b. Pycnidia to varying degrees covered by hyphae; occasionally conidia two-celled **12**

5. a. Multicellular chlamydospores typical alternarioid-dictyosporous or phragmo-sporous, mostly terminal (catenate or solitary), but sometimes also intercalary (Figs 24, 25C, 26, 27A) **6**
   b. Multicellular chlamydospores more irregular botryoid-alternarioid in shape, intercalary or terminal, mostly solitary (Figs 27B, 28) **13**

6. a. Multicellular chlamydospores frequently catenate and explicitly dictyosporous (solitary dictyosporous chlamydospores also occur) **7**
   b. Multicellular chlamydospores mostly solitary, dictyosporous or phragmosporous (some catenation may occur)..**8**

7. a. Abundant production of chains of alternarioid chlamydospores; no unicellular chlamydospores; pycnidia variable; conidia variable one-celled, mostly 4–8.5 × 1.5–3 μm ....*Ph. glomerata*
      [Plurivorous, also on animal (human) and inorganic substrata. Worldwide dis-tributed, often on dead seed coats; most records are from temperate regions. Soil fungus. Opportunistic parasite: blight, leaf spots, fruit rot.]
   b. Apart from short chains of alternarioid chlamydospores also chains of unicellu-lar chlamydospores; pycnidia usually subglobose; conidia relatively large, 7–12 × 2–3.5 μm, occasionally two-celled, 12–17 × 3–4 μm.........*Ph. subglomerata*
      [Occasionally found in southern Europe, Central America and South Africa. Opportunistic parasite: leaf spots. Repeatedly confused with *Ph. glomerata*.]

8. a. Apart from alternarioid chlamydospores (mainly solitary) always many unicel-lular chlamydospores, relatively large with conspicuous guttules **9**
   b. Solitary alternarioid chlamydospores in sympodial arrangement; unicellular chlamydospores, if present, relatively small; pycnidia usually ampulliform **10**

9. a. Pycnidia variable, often globose-divided; conidia variable, mostly 5–7 × 1.5–2.5 μm; no growth at 30°C.............................*Ph. pomorum* var. *pomorum*
      [Plurivorous and worldwide distributed, but most records are from temperate regions. Soil fungus. Opportunistic parasite, esp. common on Pomoideae: leaf spots, fruit rot.]
   b. Variety adapted to relatively high temperatures: good growth at 30°C; pycnidia not divided; conidia relatively large, mostly 5–8.5 × 2–3 μm.............................
      ....*Ph. pomorum* var. *calorpreferens*
      [Occasionally found in Europe and North America.]

10. a. Pycnidia abundantly produced at 22°C; conidia variable in shape and
        dimensions **11**
    b. Pycnidia occur only at temperature ranges of 28–30°C; at room
        temperature only alternarioid chlamydospores, often with a kind of halo;
        conidia consistent in shape, ellipsoidal-obovoid, mostly 4–5.5 ×
        2.5–3 μm ....*Ph. jolyana* var. *sahariensis*
        [Recorded in northern Africa.]

11. a. Conidia biguttulate and eguttulate, mostly 4–7 × 2–4 μm ....
        ....*Ph. jolyana* var. *jolyana*
        [Plurivorous and common in tropical and subtropical regions. Apparently
        saprophytic.]
    b. Variety adapted to cold climate: at room temperature abundant sympodial clus-
        ters of alternarioid chlamydospores; conidia often with several polar guttules,
        relatively long, mostly 5–9 × 2–3.5 μm ....*Ph. jolyana* var. *circinata*
        [Recorded in southern Siberia.]

12. a. Multicellular chlamydospores, alternarioid, intercalary and terminal (Fig. 3A),
        solitary; unicellular chlamydospores usually single and with conspicuous gut-
        tules; pycnidia globose, often 'hairy' and confluent; conidia mostly 5–8 ×
        2–3 μm, occasionally two-celled, 8–12 × 3–3.5 μm ....*Ph. americana*
        [Plurivorous; so far only known from warm regions in North America. The
        unicellular chlamydospores resemble very much those of *Ph. pomorum*, this
        key no. 9.]
    b. Multicellular chlamydospores irregular botryoid-alternarioid, generally inter-
        calary, solitary or in complexes with series of unicellular chlamydospores:
        Characteristic for two pathogenic species which may produce *in vivo* apart
        from relatively small, mainly aseptate conidia, also distinctly larger conidia
        which frequently become septate (ascochytoid/stagonosporoid). On account of
        this conidial dimorphism, both pathogens, i.e. *Ph. clematidina*[8] and *Ph.
        narcissi*[9] are treated in..sect. *Heterospora*, **B**

13. a. Colonies extremely variable, reverse usually with reddish-lilac or pinkish discol-
        oration; botryoid-alternarioid chlamydospores intercalary or terminal, often
        with discrete individual cellular elements, mainly in subtropical regions **14**
    b. Colonies rather uniform, felted, reverse greyish to black; extremely irregular
        botryoid-alternarioid chlamydosporal configurations **15**

14. a. Abundant production of intercalary botryoid-alternarioid chlamydospores and
        series of unicellular chlamydospores; pycnidia papillate-rostrate; conidia vari-

[8] Growth rate 4.5–5.5 cm on OA after 7 days. Conidia *in vitro* usually 4–8.5 × 2–3 μm,
frequently 1-septate, up to 13 × 4 μm ....*Ph. clematidina*
[Opportunistic parasite of *Clematis* spp. *In vivo* distinctly large 1-septate ascochytoid conidia
may be produced, up to 28 × 6.5 μm.]
[9] Growth rate 8–8.5 cm on OA after 7 days. Conidia *in vitro* mostly 4–7.5 × 2.5–3.5 μm, occasionally septate, up to 15 × 5.5 μm..*Ph. narcissi*
[Pathogen of Amaryllidaceae, esp. *Narcissus* and *Hippeastrum* spp. *In vivo* distinctly large,
mainly 3-septate stagonosporoid conidia are often produced, up to 28 × 8 μm, synanam.
Stagonosporopsis curtisii.]

able, mostly 4.5–7 × 2–3 μm; often a reddish or yellowish discoloration below the colony..*Ph. sorghina*
[Plurivorous and common in tropical and subtropical regions; esp. on Gramineae: spots on leaves and stems, damping off. Occasionally also recorded in temperate regions (glasshouses) and areas with continental climate.]
  b. Production of botryoid-alternarioid chlamydospores usually scarce, intercalary and terminal; pycnidia papillate-rostrate; conidia relatively broad, mostly 6–7 × 3.5–4 μm ....*Ph. pimprina*
[Thus far only known from soil in India. The dictyochlamydospores resemble very much those of *Ph. sorghina*.]

15. a. Apart from botryoid-alternarioid chlamydosporal configurations also pseudo-sclerotioid structures present **16**
   b. No pseudosclerotioid structures present; botryoid-alternarioid chlamydospores mostly intercalary and solitary; pycnidia subglobose; conidia eguttulate, variable in dimensions, mostly 4–7 × 2.5–3.5 μm ....*Ph. zantedeschiae*
[Specific pathogen of the arum or calla lilly, *Zantedeschia aethiopica*: leaf blotch. Recorded from South Africa, western Europe, North and South America.]

16. a. Production of irregular botryoid-alternarioid chlamydospores, unicellular chlamydospores and pseudosclerotioid masses; pycnidia subglobose, papillate or rostrate, often hairy; conidia relatively large, biguttulate, mostly 9–10 × 2–3 μm..*Ph. violicola*
[Recorded on various *Viola* spp. in Europe. Apparently an opportunistic parasite: leaf spots.]
   b. Usually abundant production of irregular pseudosclerotioid masses of chlamydospores; often a reddish or yellowish discoloration below the colony; pycnidia subglobose, sometimes confluent, often hairy; conidia mostly 4–5.5 × 1.5–2 μm...*Ph. chrysanthemicola*
[Common saprophytic soil fungus in most temperate regions. A specific pathogenic form on florists' chrysanthemums is known from western Europe and North America: root rot and basal stem rot.]

Note

All the species once classified in the synonymous genus *Peyronellaea* have been treated in a synoptic table by Boerema *et al.* (1977: tab. 1). This concerns:
Ph. glomerata
Ph. jolyana var. *jolyana*
Ph. jolyana var. *circinata*
Ph. pomorum (sensu lato)
Ph. sorghina

Possibilities of identification *in vivo*

Exact identification of representatives of *Phoma* sect. *Peyronellaea* is usually possible only by study *in vitro*. However, observations *in vivo* sometimes indicate that a *Phoma* species producing multicellular chlamydospores is involved.

Old pycnidia of typical members of sect. *Peyronellaea* often contain brownish conidia which may become (extra)septated. Furthermore, the multicellular chlamydospores are sometimes produced on the pycnidial wall, near the ostiole. This holds especially for *Ph. glomerata* and *Ph. pomorum*; pycnidia of the latter on leaves are often associated with chains of chlamydospores.

The connection between the carbon:nitrogen ratio of the substratum and the development of pycnidia and/or chlamydospores explains why the pycnidia of species of *Phoma* sect. *Peyronellaea* are never associated with abundant production of multicellular chlamydospores *in vivo*. Also with *Phoma epicoccina* there are no records of simultaneous abundant production of pycnidia and *Epicoccum*-spores.

The opportunistic plant pathogens in sect. *Peyronellaea* are mainly plurivorous, but the host specific pathogens may be recognized by their association with typical disease symptoms. Finally *Ph. cyanea* is easy to identify by the exceptional cyan blue colour of the pycnidia.

Distribution and Facultative Host and Substratum Relations in Section *Peyronellaea*

Plurivorous
(But often with special
host or substratum
relations)

America, esp. subtropical regions of south-east USA:
Ph. americana

Worldwide, but esp. in temperate regions:
Ph. chrysanthemicola
Ph. epicoccina, synanam. *Epicoccum nigrum*
Ph. glomerata
Ph. pomorum var. *pomorum*
adapted to higher temperatures:
var. *calorpreferens*

Worldwide, but esp. in tropical and subtropical regions:
Ph. sorghina
Ph. subglomerata
Ph. jolyana var. *jolyana*
adapted to desert conditions:
var. *sahariensis*
adapted to cool continental climate:
var. *circinata*

With special host and/or substratum relation:

Amaryllidaceae
esp. *Narcissus* and
Hippeastrum spp.

Ph. narcissi (on account of the occurrence of distinctly large, mainly 3-septate conidia *in vivo*, synanam. *Stagonosporopsis curtisii*, treated under sect. *Heterospora*, **B**)
[Worldwide known as causal organism of red spots on leaves, bulb scales and flowers.]

Araceae
Zantedeschia aethiopica   *Ph. zantedeschiae*
                                   [Worldwide found in association with brown blotches on
                                   leaves and spathes: leaf blotch; probably indigenous to
                                   southern Africa.]

Compositae
Dendranthema–Grandiflorum   *Ph. chrysanthemicola*
hybrids (formerly known as,   f. sp. *chrysanthemicola*
e.g. *Chrysanthemum*
morifolium and *C. indicum*)   [In western Europe and North America known as causal
                                   organism of root rot and basal stem rot.]
Scorzonera hispanica   *Ph. chrysanthemicola*
                                   [In Belgium frequently found in association with stunted
                                   roots.]

Gramineae,
e.g. *Oryza, Saccharum,*   *Ph. americana*
Sorghum, Triticum and   *Ph. sorghina*
Zea spp.   *Ph. subglomerata*
                                   [In tropical and subtropical regions occurring in association
                                   with infections of seeds and lesions on leaves (leaf spot)
                                   and stems.]

Ranunculaceae
Clematis spp.   *Ph. clematidina* (on account of the occurrence of dis-
                                   tinctly large, ascochytoid 1-septate conidia *in vivo*,
                                   treated under sect. *Heterospora*, **B**)
                                   [In Australasia, Eurasia and North America frequently
                                   recorded on cultivated plants in association with lesions on
                                   stems and leaves.]

Rosaceae
Fragaria (×) *ananassa*   *Ph. pomorum* var. *pomorum*
                                   [In Europe frequently associated with the black root rot
                                   complex.]
Malus, Prunus and   *Ph. pomorum* var. *pomorum*
Pyrus spp.(Pomoideae)   [+ probably var. *calorpreferens*]
                                   [In Eurasia and North America frequently found in associa-
                                   tion with leaf spots and fruit rot.]

Violaceae
Viola spp.   *Ph. violicola*
                                   [In Europe repeatedly recorded in association with leaf spots.]

Vitaceae
Vitis vinifera   *Ph. glomerata*
                                   [In southern Europe known as causal organism of blight of
                                   vine flowers and grapes.]

Seeds and fruits

> *Ph. epicoccina*, synanam. *Epicoccum nigrum*
> *Ph. glomerata*
> *Ph. jolyana* var. *jolyana*
> *Ph. pomorum* var. *pomorum*

[All saprophytic on dead seed coats; occasionally associated with fruit rot.]

> *Ph. sorghina*

[May cause reduction in seed germination and post-emergence death of seedlings of gramineous hosts; occasionally found in association with fruit rot.]

Soil

> *Ph. chrysanthemicola*
> *Ph. cyanea*
> *Ph. glomerata*
> *Ph. pomorum* var. *pomorum*

[All isolated from soil in temperate regions.]

> *Ph. jolyana* var. *jolyana*
> *Ph. jolyana* var. *sahariensis*
> *Ph. pimprina*
> *Ph. pomorum* var. *calorpreferens*
> *Ph. sorghina*

[All isolated from soil in tropical and subtropical regions.]

Substrata of animal (incl. human) and inorganic origin

> *Ph. americana*

[Repeatedly isolated from cysts of *Heterodera glycines* in the USA.]

> *Ph. epicoccina*, synanam. *Epicoccum nigrum*

[In northern Europe once isolated from a human toenail.]

> *Ph. glomerata*

[In southern Europe frequently recorded in association with human mycosis, but the fungus seems to be not a causal but an aggravating factor. Other sources of isolates were chemical products, paint, wool fibres and butter.]

> *Ph. pomorum* var. *pomorum*

[In Japan isolated from earthen pots.]

> *Ph. sorghina*

[In India isolated from lesions on human skin. Also isolated from insects and paper.]

Descriptions of the Taxa

Characteristics based on study in culture.

Phoma americana Morgan-Jones & J.F. White, Fig. 25C

Chlamydospore-anamorph uni- and multicellular, alternarioid.

Phoma americana Morgan-Jones & J.F. White *in* Mycotaxon **16**(2): 406[–412]. 1983.

Diagn. data: Morgan-Jones & White (1983a) (figs 1–2: drawings of pycnidia, wall cells around ostiole, conidiogenous cells, conidia, chlamydospores and dictyochlamydospores; pls 1A–H, 2A–F, 3A–H: photographs of agar plate cultures, pycnidia, wall cells, chlamydospores and dictyochlamydospores; cultural and morphological description; isolate sources), Rai & Rajak (1993) (cultural characteristics), Boerema (1993) (fig. 3B: shape of chlamydospores and conidia reproduced in present Fig. 25C; cultural description adopted below).

Description in vitro

Pycnidia subglobose, 100–220 μm diam., papillate or with a short cylindrical neck, often multiostiolate and confluent reaching up to 850 μm diam. Pycnidial wall composed of isodiametric or somewhat elongated cells. Conidial matrix salmony in colour.

Conidia irregular cylindrical-ellipsoidal, mostly 5–8(–8.5) × 2–3(–3.5) μm, occasionally 1-septate, 8–12(–13.5) × 3–3.5(–4) μm.

Chlamydospores very variable, terminal or intercalary, solitary or in chains, uni- or multicellular, when septate phragmosporous or dictyosporous, somewhat irregular-botryoid, smooth or roughened, pale brown to brown, occasionally heavily melanized and roughened as in *Epicoccum*-conidia, mostly 15–25 μm diam. Unicellular chlamydospores mostly 7–18 μm diam. and with conspicuous guttules.

Colonies on OA moderately fast growing at room temperature, 5–6 cm diam. after 7 days; at 30°C also fast growing, even up to 6.5 cm diam.; on MA always rather slow growing, mostly 3.5 cm diam. Aerial mycelium tenous, particularly in a wide marginal zone, underground greenish olivaceous, reverse also greenish olivaceous; on MA darker, olivaceous black. Representative culture CBS 185.85.

Ecology and distribution

This plurivorous soil fungus is so far only known from isolates obtained in the south-east of North America, mainly in regions with a subtropical climate. This explains its ability to grow fast at 30°C. The substratum records refer to leaves of wheat, *Triticum aestivum*, roots of maize, *Zea mays* (both Gramineae), and cysts of *Heterodera glycines*. The fungus resembles in some respects *Ph. pomorum* var. *calorpreferens* Boerema *et al.*, but can easily be differentiated by the papillated or rostrated pycnidia and its slow growth on MA.

Phoma chrysanthemicola Hollós [sensu lato],[10] Fig. 25B

Chlamydospore-anamorph uni- and multicellular, pseudosclerotioid.

> *Phoma chrysanthemicola* Hollós in Annls hist.-nat. Mus. natn hung. **5**: 456. 1907.
>> = *Phoma radicis-oxycocci* Ternetz in Jb. wiss. Bot. **44**: 365. 1907.

Diagn. data: Dorenbosch (1970) (tab. 1, pl. 1 fig. 1, pl. 3 figs 3, 6, pl. 6 figs 1, 2; characteristics *in vitro* of isolates from soil and members of the *Compositae*; photographs of pseudosclerotoid masses of chlamydospores and mycelial mat on OA and MA, colour plate of red discoloured medium; history and designation of neotype), Boerema (1975) (soil isolates not pathogenic to florists' chrysanthemums, i.e. host of type), Schneider & Boerema (1975a) (tab. 1; host range; introduction of forma specialis *chrysanthemicola*), Domsch *et al.* (1980) (occurrence in soil; literature references), Sutton (1980) (fig. 226D: drawings of pseudosclerotium and conidia; specimens in CMI collection), Johnston (1981) (tab. 1, figs 1, 13, cultural and microscopic characteristics; occurrence in New Zealand), Hauptmann & Schickedanz (1986) (figs 1–11; detailed description and illustration: scanning electron micrographs of conidia and chlamydospores), Rai & Rajak (1993) (cultural characteristics), Boerema (1993) (fig. 7A: shape of pycnidia, conidia and pseudosclerotoid masses of chlamydospores, reproduced in present Fig. 25B; cultural description adopted below), Rai (1998) (effect of different physical and nutritional conditions on the morphology and cultural characters).

Description in vitro

Pycnidia subglobose, 150–350 μm diam., with papillated ostiole or flask-shaped, solitary but often coalescing to large irregular fructifications with many ostioles. Fertile micropycnidia frequently occur, 40–100 μm diam. Conidial matrix yellowish or cream-coloured.

Conidia cylindrical with two guttules, often somewhat dumb-bell shaped, (3.5–)4–5.5(–6.5) × 1.5–2(–2.5) μm, hyaline.

Chlamydospores numerous, globose or subglobose, 5–11 μm diam., olivaceous, single or in long chains and often aggregated into large irregularly shaped olivaceous black pseudosclerotioid masses. These structures may be present in abundance in the aerial mycelium; however, it may take several weeks before they appear.

Colonies on OA rather slowly growing, 3–4 cm diam. after 7 days (on MA also 3–4 cm diam.). Variable in appearance, e.g. strains with abundant woolly

[10] So far this soil fungus has never been isolated from leaves; therefore, we have not adopted a synonymy with *Phyllosticta leucanthemi* Speg. as suggested by van der Aa & Vanev (2002).

aerial mycelium and strains with less conspicuous loose velvety mycelium, in both cases pale olivaceous grey; reverse similar, but in most strains after a few days a sienna discoloration of the agar occurs, associated with the formation of orange/red crystals. Occasionally the discolorations are more red/violet and sometimes yellow (apparently a complex of pigments is involved). The discoloration appears in closed or interrupted concentric zones; addition of NaOH causes the colour to fade. Representative culture CBS 522.66.

Ecology and distribution

In its widest sense this fungus appeared to be a common saprophytic soil fungus in most temperate regions. It has been isolated from roots and stems of various herbaceous plants, esp. Compositae, in central and western Europe, south-eastern Asia, North America and Australasia.

In Belgium the fungus is frequently found in association with stunted roots of scorzonera (*Scorzonera hispanica*). Finally it should be noted that occasionally strains of *Ph. chrysanthemicola* do not produce chlamydospores (see Dorenbosch, 1970: 10 and Ternetz, 1907: 365). In the key to species of *Phoma sect. Phoma* (**A**) these deviating strains have been incorporated.

f. sp. **chrysanthemicola** [R. Schneid. & Boerema *in* Phytopath. Z. **83**: 242. 1975].
The lectotype specimen of the fungus occurs on florists' chrysanthemum, *Dendranthema*–Grandiflorum hybrids (formerly known as, e.g. *Chrysanthemum morifolium* and *C. indicum*), and refers to a specialized pathogenic form of the fungus recorded in western Europe and North America (USA and Canada): root rot or basal stem rot.

Phoma cyanea Jooste & Papendorf, Fig. 26A

Chlamydospore-anamorph uni- and multicellular, alternarioid.

Phoma cyanea Jooste & Papendorf *in* Mycotaxon **12**(2): 444. 1981.

Diagn. data: Jooste & Papendorf (1981) (figs 1–7: photographs of pycnidium, conidia, hyphae and dictyochlamydospores; original cultural and morphological description), Boerema (1993) (fig. 1B: shape of conidia and chlamydospores, reproduced in present Fig. 26A; cultural description adopted below).

Description in vitro

Pycnidia conspicuous cyan blue, subglobose to globose, mostly 100–300 μm diam., usually with a short neck and a wide 'collaretted' ostiole, mostly solitary. Conidial matrix whitish.

Conidia oblong ellipsoidal or obovoid, sometimes slightly curved, occasionally clavate, (4–)5–7(–10) × 2–3(–4) μm.

Chlamydospores conspicuous cyan blue, variable and irregular, uni- or multicellular, where unicellular usually in short chains, intercalary or terminal, with somewhat thick walls encrusted in blue crystals, mostly 8–10 μm diam.; in older cultures often multicellular, variable-dictyosporous, solitary or in short

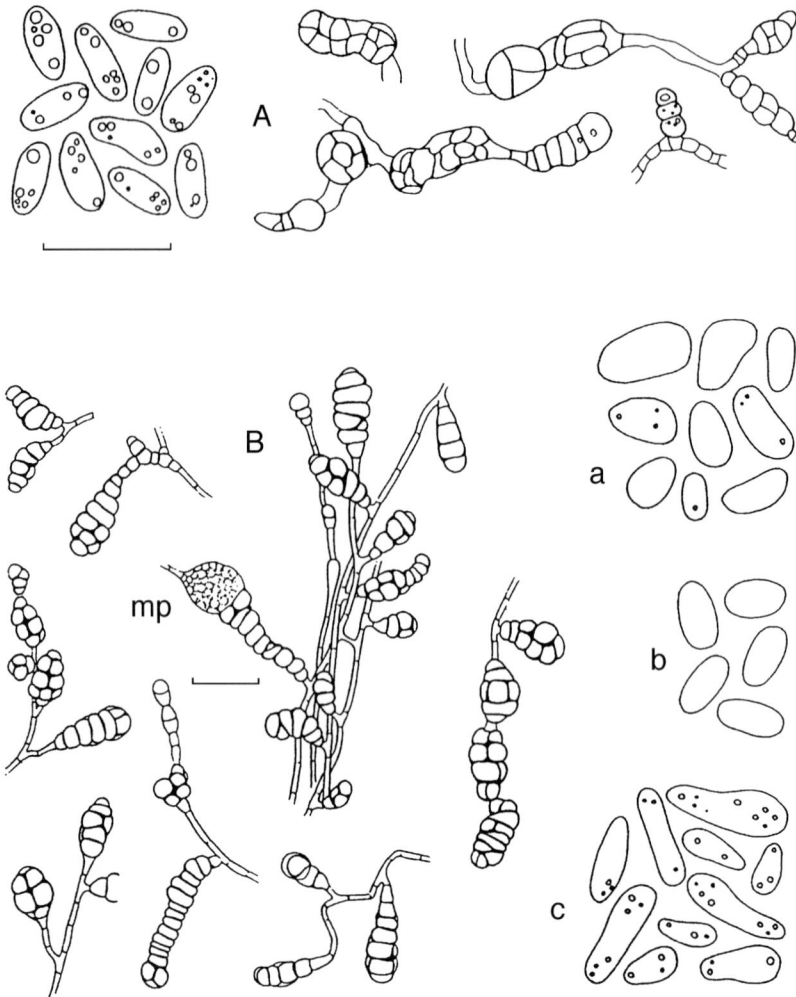

Fig. 26. A *Phoma cyanea,* structure of conidia and some unicellular and alternarioid multicellular chlamydospores. This species at once can be differentiated by the dark cyan blue colour of the pycnidia, hyphae and chlamydospores. **B** Varieties of *Phoma jolyana,* characterized by alternarioid multicellular chlamydospores becoming more or less lateral by continued growth of the hyphae. Note the alternating arrangement; mp = micropycnidium. Conidia (a) var. *jolyana,* (b) var. *sahariensis* and (c) var. *circinata.* Bar chlamydospores 20 μm, bar conidia 10 μm.

chains, intercalary or terminal on branched hyphae, often in combination with unicellular chlamydospores, with relatively thick walls encrusted in blue crystals, 10–50 × 10–20 μm.

Colonies on OA becoming conspicuous dark cyan blue coloured, moderately fast growing 5–6 cm diam. after 7 days (on MA also fast growing, 6–7 cm diam.), mycelium cottony, floccose, consisting of hyaline or light to dark cyan blue hyphae, occasionally encrusted in cyan blue crystals. The blue pigment is unique among *Phoma* species. Representative culture CBS 388.80.

Ecology and distribution

This fungus is so far only known from wheat field debris (*Triticum aestivum*; Gramineae) in South Africa. The exceptional blue colour of hyphae, pycnidia and chlamydospores was found to be a stable characteristic after several subcultures.

Phoma epicoccina Punith. *et al.*, Fig. 25A

Chlamydospore-anamorph multicellular, epicoccoid: ***Epicoccum nigrum* Link**

> *Phoma epicoccina* Punith., Tulloch & Leach *in* Trans. Br. mycol. Soc. **59**: 341. 1972.

synanamorph:
> *Epicoccum nigrum* Link *in* Mag. Ges. naturf. Fr. Berl. **7**: 32. 1816 (p.p.).

Diagn. data: Punithalingam *et al.* (1972) (figs 1A–D, 2A–C: drawings of pycnidium, conidiogenous cells, conidia, pycnosclerotium and epicoccoid chlamydospores; original description; similarity of epicoccoid anamorph with *Epicoccum nigrum*), Sutton (1980) (fig. 227A: conidia and *Epicoccum* conidium; specimens in CMI collection), Punithalingam (1982c) (figs A–D; CMI description), Monte *et al.* (1990) (physiological and biochemical study of six isolates), Boerema (1993) (fig. 7B; cultural description adopted below), Arenal *et al.* (2000) (fig. 1; internally transcribed spacer (ITS) sequencing support for *Phoma epicoccina* and *Epicoccum nigrum* being the same biological species).

Description in vitro

Pycnidia subglobose to globose, 120–200 μm diam., solitary or confluent, intermixed among pycnosclerotia, 200–400 μm diam. Pycnidial wall and sclerotia composed of compressed brownish cells, heavily pigmented and thick-walled on the outer side. Conidial matrix whitish cream.
    Conidia most variable in shape and dimensions, usually shortly cylindrical, sometimes slightly curved, eguttulate or with 2–3 polar guttules, 3–7(–10) × 1.5–3(–3.5) μm.
    Chlamydospores, produced in sporodochia, representing the *Epicoccum*-anamorph, multicellular-phragmosporous, but the septa being obscured by the dark-brown to black verrucose outer wall, subglobose-pyriform, often with a paler basal cell, variable in dimensions, but mostly 15–35 μm diam., arising in gradually growing clusters as solitary, terminal elements of mycelial branches, from a more or less globose pseudoparenchymatous stroma. The sporodochia may be present in abundance, scattered or aggregated.
    Colonies on OA moderately fast growing, 6–7 cm diam. after 7 days. Aerial mycelium floccose, extremely variable in pigmentation, mostly yellowish to bright yellow or pink to purple red. The pigment also diffuses into the agar, but the colour of the mycelium and the agar are not always the same; reverse also variable in colour, but mostly grey to almost black in the centre. Representative culture CBS 173.73.

Ecology and distribution

A worldwide recorded seed-, air- and soil-borne saprophyte. It appeared to be a common contaminant of grass seeds, but is also recorded from different parts of various other plants. The fungus has been isolated once from a human toe-nail. Information on pathogenicity to plants is unclear. The *Epicoccum*-anamorph is indistinguishable from *Epicoccum nigrum* Link, which shows a similar variable pigmentation, but develops separately and independently *in vivo* and *in vitro*. However, sequencing of the ITS regions of ribosomal DNA from isolates of *Ph. epicoccina* and *Epicoccum nigrum* (Arenal *et al.* 2000) indicate that both anamorphs represent the same biological species.

Phoma glomerata (Corda) Wollenw. & Hochapfel, Fig. 24A

Chlamydospore-anamorph multicellular, alternarioid; three times described in *Alternaria*.

> *Phoma glomerata* (Corda) Wollenw. & Hochapfel *in* Z. ParasitKde **8**: 592. 1936.
>
> > ≡ *Coniothyrium glomeratum* Corda, Icon. Fung. **4**: 39. 1840.
> > ≡ *Aposphaeria glomerata* (Corda) Sacc. *in* Sylloge Fung. **3**: 175. 1884.
> > ≡ *Peyronellaea glomerata* (Corda) Goid. *in* Atti Accad. nac. Lincei Rc. VIII, **1**: 455. 1946 [‡ not validly published: name of the genus published without a Latin diagnosis and desig-nation of type species, Art. 10, 36.1, 43] ex Togliani *in* Annali Sper. agr. II, **6**: 93. 1952 [with formal Latin diagno-sis and designating *Coniothyrium glomeratum* Corda sensu Wollenweber & Hochapfel l.c. as type species].
> > = *Phoma fibricola* Berk. *in* Hook. J. Bot. **5**: 41. 1853.
> > > ≡ *Aposphaeria fibricola* (Berk.) Sacc. *in* Sylloge Fung. **3**: 176. 1884.
> > > ≡ *Coniothyrium fibricola* (Berk.) Kuntze, Revisio Gen. Pl. **3**(2): 459. 1898.
> > ‡ ≡ *Peyronellaea fibricola* (Berk.) Goid. *in* Atti Accad. nac. Lincei Rc. VIII, **1**: 455. 1946 [not validly published, see *Peyr. glomerata*].
> > H = *Phyllosticta berberidis* Westend. *in* Bull. Acad. r. Sci. Lett. Beaux-Arts Belg. II, **2**: 567. 1857; not *Phyllosticta berberidis* Rabenh. *in* Klotzschii Herb. mycol. [Ed. Rabenh.] Cent. 19, No. 1865. 1854 [= *Phoma macros-toma* Mont. var. *macrostoma*, sect. *Phyllostictoides*, **E**].
> > > ≡ *Phyllosticta westendorpii* Thüm. *in* Byull. mosk. Obshch Ispýt. Prir. **55**: 229 [= Pilzflora Sibiriens: 304]. 1880.
> > = *Phoma olivarum* Thüm. *in* Boll. Soc. Adriat. Sci. Nat. **8**: 241. 1883.
> > = *Aposphaeria consors* Schulzer & Sacc. *in* Hedwigia **23**: 109. 1884.

≡ *Coniothyrium consors* (Schulzer & Sacc.) Kuntze, Revisio Gen. Pl. **3**(2): 459. 1898 (as '*concors*'].

‡ ≡ *Peyronellaea consors* (Schulzer & Sacc.) Goid. *in* Atti Accad. nac. Lincei Rc. VIII, **1**: 455. 1946 [not validly published, see *Peyr. glomerata*].

= *Plenodomus oleae* Cavara *in* Revue Mycol. **10**: 1888.

≡ *Phoma oleae* (Cavara) Sacc. *in* Sylloge Fung. **10**: 146. 1892.

= *Phoma herbarum* var. *euphorbiae-gayonianae* Pat., Cat. rais. Pl. cell. Tun. 116. 1897.

= *Phyllosticta auerswaldii* Allesch. *in* Rabenh. Krypt.-Flora [ed. 2], Pilze **6** [Lief. 59]: 25. 1898 [vol. dated '1901'].

= *Phoma herbarum* f. *chrysanthemi-corymbosi* Allesch. *in* Rabenh. Krypt.-Flora [ed. 2], Pilze **6** [Lief. 64]: 330. 1899 [vol. dated '1901'].

‡ ≡ *Peyronellaea herbarum* f. *chrysanthemi-corymbosi* (Allesch.) Goid. *in* Atti Accad. nac. Lincei Rc. VIII, **1**: 455. 1946 [not validly published, see *Peyr. glomerata*].

= *Phoma radicis-andromedae* Ternetz *in* Jb. wiss. Bot. **46**: 366–367. 1907.

= *Phoma radicis-vaccinii* Ternetz *in* Jb. wiss. Bot. **46**: 366–367. 1907.

= *Phoma richardiae* Mercer *in* Mycol. Centbl. [Mykol. Zentbl.] **2**: 244, 297, 326. 1913.

≡ *Coniothecium richardiae* (Mercer) Jauch *in* An. Soc. cient. argent. **144**: 456. 1947 [misapplied, see discussion under *Phoma zantedeschiae* Dippen. (this section), and Boerema & Hamers (1990)].

≡ *Peyronellaea richardiae* (Mercer) Goid. *in* Atti Accad. nac. Lincei Rc. VIII, **1**: 454. 1946 [not validly published, see *Peyr. glomerata*].

= *Phoma conidiogena* Schnegg *in* Zentbl. Bakt. ParasitKde Abt. 2, **43**: 326[–364]. 1915.

‡ ≡ *Peyronellaea conidiogena* (Schnegg) Goid. *in* Atti Accad. nac. Lincei Rc. VIII, **1**: 455. 1946 [not validly published, see *Peyr. glomerata*].

= *Phoma alternariaceum* F.T. Brooks & Searle *in* Trans. Br. mycol. Soc. **7**: 193. 1921.

‡ ≡ *Peyronellaea alternariaceum* (F.T. Brooks & Searle) Goid. *in* Atti Accad. nac. Lincei Rc. VIII, **1**: 455. 1946 [not validly published, see *Peyr. glomerata*].

= *Phoma monocytogenetica* Curzi apud Curzi & Barbaini *in* Atti Ist. bot. Univ. Lab. crittogam. Pavia III, **3**: 169. 1927.

= *Phoma fumaginoides* Peyronel apud Filippopulos *in* Boll. Staz. Patol. veg. Roma II, **7**: 332[–336]. 1927 [simultaneously introduced with *Alternaria fumaginoides* Peyronel, see below at chlamydospore-anamorph].

‡ ≡ *Peyronellaea fumaginoides* (Peyronel) Goid. *in* Atti
Accad. nac. Lincei Rc. VIII, **1**: 452. 1946 [not validly
published, *see Peyr. glomerata*] ex Leduc *in* Revue gén.
Bot. **65**: 542. 1958.

= *Phoma hominis* A. Agostini & Tredici apud Pollacci *in* Atti
Ist. bot. Univ. Lab. crittog. Pavia IV, **6**: 154. 1935 [‡ not
validly published: not accompanied by a Latin description,
Art. 36] ex A. Agostini & Tredici *in* Atti Ist. bot. Univ. Lab.
crittog. Pavia IVa, **9**: 187. 1937 [simultaneously introduced
with *Alternaria hominis* A. Agostini & Tredici, *see* below at
chlamydospore-anamorph].

‡ ≡ *Peyronellaea hominis* (A. Agostini & Tredici) Goid. *in*
Atti Accad nac. Lincei Rc. VIII, **1**: 455. 1946 [not validly
published, *see Peyr. glomerata*].

= *Phoma polymorpha* Verona *in* Cellulosa **1939**: 27. 1939
[as 'n. nom.' for *Alternaria polymorpha* L. Planchon, *see*
below at chlamydospore-anamorph].

‡ ≡ *Peyronellaea polymorpha* (L. Planch.) Goid. *in* Atti
Accad. nac. Lincei Rc. VIII, **1**: 455. 1946 [not validly
published, *see Peyr. glomerata*].

‡ = *Peyronellaea veronensis* Goid. *in* Atti Accad. nac. Lincei
Rc. VIII, **1**: 451 [455, 458]. 1946 [not validly published:
without a Latin description, Art. 36, and genus not valid,
see Peyr. glomerata].

= *Peyronellaea stipae* Lacoste *in* C.r. hebd. Séanc. Acad. Sci.,
Paris **241**: 818. 1955 [‡ not validly published: not accom-
panied by a Latin description, Art. 36.1] ex Lacoste *in*
Revue Mycol. **22** [Suppl. Colon. 1]: 14. 1957.

= *Phoma saprophytica* Eveleigh *in* Trans. Br. mycol. Soc. **44**:
582. 1961.

= *Phyllosticta gymnosporiicola* V.G. Rao *in* Mycopath. Mycol.
appl. **22**: 161. 1964.

= *Peyronelleae ruptilis* Kusnezowa *in* Nov. Sist. Niz. Rast. **8**:
192. 1971.

= *Peyronellaea sibirica* Kusnezowa *in* Nov. Sist. Niz. Rast. **8**:
193. 1971.

= *Phyllosticta rubicola* var. *macrospora* Cejp, Fassatiová &
Zavřel *in* Zprávy Vlastiv. Úst. Olomouci, Cislo **153**: 7.
1971.

= *Peyronellaea sibirica* var. *allii* Kusnezowa *in* Nov. Sist. Niz.
Rast. **8**: 196. 1971.

V = *Peyronellaea zhdanovii* Kusnezowa *in* Nov. Sist. Niz. Rast.
8: 198. 1971 [as 'zhdanovi'].

Chlamydospore-anamorph three times described in *Alternaria*:
Alternaria polymorpha L. Planchon *in* Annls Sci. nat. (Bot.) VIII, **11**: 48[–89].
1900 [in 1939 renamed in *Phoma*, see above].

= *Alternaria fumaginoides* Peyronel apud Filippopulos *in* Boll. Staz. Patol. veg. Roma II, **7**: 332. 1927 [simultaneously introduced with *Phoma fumaginoides* Peyronel, see above].

= *Alternaria hominis* A. Agostini & Tredici *in* Atti Ist. bot. Univ. Lab. crittogam. Pavia IVa, **9**: 187. 1937 [simultaneously introduced with *Phoma hominis* A. Agostini & Tredici, see above].

Diagn. data: Boerema *et al.* (1965a) (fig. 2, pl. 1 figs 1–8, pl. 2 figs 9–12; description and illustrations; synonymy and variability; habitat), (1968) (synonymy and variability), (1971) (tab. I; synonymy and host range), (1973) (misidentifications), (1977) (full synonymy; synoptic table on characters *in vitro*), Morgan-Jones (1967) (figs A–F; CMI description with drawings of pycnidia, conidia, wall cells, hyphae and dictyochlamydospores), Dorenbosch (1970) (pl. 6 fig. 5, pl. 2, figs 5–6, tab. I; characteristics *in vitro* with photographs of dictyochlamydospores and mycelial mat on OA and MA; isolates from soil), Boerema & Dorenbosch (1973) (pl. 3 figs e–h, tab. I; synonymy; sources of isolates; differentiating diagnostic characters), Punithalingam (1979c) (fig. 7A–D, pl. 7 figs 19–21; mycotic diseases of humans), Domsch *et al.* (1980) (fig. 282a–c; occurrence in soil; literature references), Sutton (1980) (fig. 227F; drawings of conidia and dictyochlamydospores; specimens in CMI collection), Johnston (1981) (tab. 1, figs 5, 14, photographs of agar plate cultures and dictyochlamydospores; occurrence in New Zealand), White & Morgan-Jones (1987b) (pls 1A–F, 2A–D, 3A–H: photographs of agar plate cultures, pycnidia, conidia and dictyochlamydospores; history; cultural and morphological descriptions), Rai & Rajak (1993) (cultural characteristics), Boerema (1993) (fig. 1A: shape of pycnidia, conidia and dictyochlamydospores, reproduced in present Fig. 24A; cultural description adopted below), Rai (1998) (effect of different physical and nutritional conditions on the morphology and cultural characters), van der Aa & Vanev (2002) (*Phyllosticta* synonyms).

Description in vitro

Pycnidia subglobose to obpyriform, 100–300 µm diam., papillate or with necks of various length, usually solitary but sometimes coalescing. Pycnidial wall three to five layers thick; the outer portion composed of more or less isodiametric but rounded and sometimes inflated cells with dark extracellular deposits. In aerial mycelium and arising from a single dictyochlamydospore cell frequently fertile micropycnidia occur, 20–50 µm diam. Conidial matrix at first rosy buff to salmon, later darkening and becoming olivaceous-brown.

Conidia variable in shape and dimensions, mostly ovoid-ellipsoidal, sometimes slightly curved, (3.5–)4–8.5(–10) × 1.5–3(–3.5) µm, hyaline but with age becoming pale olive-brown and minutely roughened.

Chlamydospores highly variable in shape and dimensions, generally multicellular-dictyosporous, occasionally solitary-terminal, but usually in branched or unbranched chains of 2–20 or more elements, alternarioid, smooth at first, later roughened, dark brown to black, (18–)30–65(–80) × (12–)15–25(–35) µm.

Colonies on OA relatively slow to fast growing, 3.5–7 cm diam. after 7 days; at 30°C always very slow growing; on MA at room temperature always fast growing 6.5–7.5 cm diam. Most variable in appearance, strains (sectors) with rather sparse aerial mycelium and abundant aerial mycelium, dense and woolly in places, olivaceous, greenish olivaceous, olivaceous buff or dull green; reverse dark olivaceous to blackish beneath sectors with dense mycelium, paler elsewhere. Representative culture CBS 528.66.

Ecology and distribution

A worldwide distributed soil fungus, which has been isolated from various kinds of plants (belonging to about one hundred different host plant genera) as well as from animal (human) and inorganic material. It occurs with special frequency on dead seed coats. Most records are from temperate regions, but the fungus appears to be particularly common in the subtropics. It is frequently found in association with symptoms of blight, leaf spots and fruit rot (e.g. blight of vine flowers and grapes, *Vitis vinifera*, Vitaceae). Generally it is considered to be a secondary invader or opportunistic pathogen; this holds also for the various records of the fungus in association with mycotic diseases of humans (Punithalingam, 1979c). Other sources of isolates were chemical products, paint, wool fibres, wood and butter. The fungus shows extreme variability in the shape and dimensions of the characteristic alternarioid dictyochlamydospores, which explains its many *Peyronellaea*-synonyms.

Phoma jolyana Piroz. & Morgan-Jones var. *jolyana*, Fig. 26Ba

Chlamydospore-anamorph multicellular, alternarioid.

> *Phoma jolyana* Piroz. & Morgan-Jones *in* Trans. Br. mycol. Soc. **51**: 200. (June) 1968, var. *jolyana*.
>> ≡ *Peyronellaea musae* P. Joly *in* Revue mycol. **26**: 97. (July) 1961.
>> ≡ *Phoma musae* (P. Joly) Boerema, Dorenb. & Kesteren *in* Persoonia **4**(1): 63. 1965; not *Phoma musae* (Cooke) Sacc. *in* Sylloge Fung. **3**: 163. 1884; not *Phoma musae* C.W. Carp. *in* Rep. Hawaii agric. Exp. Stn **1918**: 39. 1919.
>> ≡ *Phoma jolyi* M. Morelet *in* Bull. Soc. Sci. nat. Archéol. Toulon Var **177**: 9. (July) 1968.
>> = *Peyronellaea nainensis* Tandon & Bilgrami *in* Curr. Sci. **30**: 344. (Dec.) 1961; not *Phoma nainensis* Bilgrami *in* Curr. Sci. **32**: 175. 1963.

Diagn. data: Boerema *et al.* (1965a) (fig. 4, pl. 4 figs 22–27; description and illustrations sub syn. *Phoma musae*) (1968) (nomenclature), (1971) (host range), (1973) (nomenclature), (1977) (varietal differentiation; synoptic table on characters *in vitro*), Sutton (1980) (fig. 228A: drawings of conidia and dictyochlamydospores; specimens in CMI collection), Rai (1985) (isolation from air,

India), Morgan-Jones & Burch (1987b) (fig. 1A–E: drawings of pycnidium, conidia, wall cells, chlamydospores and dictyochlamydospores; history; cultural and morphological description sensu lato), van der Aa *et al.* (1990) (varietal differentiation), Rai & Rajak (1993) (cultural characteristics), Boerema (1993) (fig. 3A–a: shape of conidia and dictyochlamydospores, reproduced in present Fig. 26Ba; cultural description adopted below), Rai (1998) (effect of different physical and nutritional conditions on the morphology and cultural characters).

Description in vitro

Pycnidia subglobose to obpyriform, 150–200 μm diam., papillate, mostly uniostiolate and solitary. Pycnidial wall with a characteristic inner layer of thin-walled subhyaline distinctly angular cells. In aerial mycelium frequently fertile micropycnidia occur. Conidial matrix usually salmony in colour.

Conidia variable in shape and dimensions, mostly broad, oblong-ellipsoidal to obovoid or somewhat allantoid, (3.5–)4–7(–8.5) × 2–4 μm.

Chlamydospores generally multicellular-dictyo/phragmosporous, occasionally intercalary, but frequently as terminal elements of short lateral branches, sometimes becoming lateral through continued growth of a constituent cell, usually solitary, smooth or irregular roughened, tan to dark brown, 13–45(–50) × 7–20(–25) μm.

Colonies on OA moderately fast growing at room temperature, 5–5.5 cm diam. after 7 days. Optimum growth at *c.* 25°C; no growth at 30°C. Aerial mycelium felted, blackish, greenish olivaceous to dull black; reverse dark olivaceous buff to amber. Representative culture CBS 463.69.

Ecology and distribution

A common soil-borne fungus in subtropical regions of Eurasia and Africa. Occasionally also found in regions with continental climate. In Siberia and in the Sahara adapted varieties of the fungus occur, see below (var. *circinata* and var. *sahariensis*). Although *Ph. jolyana* occasionally has been recorded as a plant pathogen this fungus is probably always only a secondary invader of dead or weakened plant tissue.

Phoma jolyana var. *circinata* (Kuznez.) Boerema *et al.,* Fig. 26Bc

Chlamydospore-anamorph multicellular, alternarioid.

> *Phoma jolyana* var. *circinata* (Kusnez.) Boerema, Dorenb. & Kesteren *in* Kew Bull. **31**(3): 535. 1977 ['1976'].
>> ≡ *Peyronellaea circinata* Kusnezowa *in* Nov. Sist. Nas. Rast **8**: 189. 1971.
>> = *Peyronellaea nigricans* Kusnezowa *in* Nov. Sist. Nas. Rast **8**: 191. 1971.

Diagn. data: Boerema *et al.* (1977) (pl. 20A–D; differential characteristics and illustrations; synoptic table on characters *in vitro*), Morgan-Jones & Burch

(1987b) (fig. 2: drawing of hyphae, dictyochlamydospores and conidia; pl. 1A–I: photographs of agar plate cultures, pycnidial wall cells, hyphae and dictyochlamydospores), Boerema (1993) (fig. 3A–c: shape of conidia, reproduced in present Fig. 26Bc; cultural description adopted below).

Description in vitro

Pycnidia similar to those of the type var. *jolyana.*
     Conidia generally somewhat longer, smaller and more irregular in shape than those of the type var.: (3.5–)5–9 × 2–3.5 μm.
     Chlamydospores multicellular-dictyo/phragmosporous, similar to those of the type var., but much more abundant, forming large irregular clusters.
     Colonies on OA at room temperature show about the same rate of growth as var. *jolyana,* but differing in the powdery appearance (clusters of dictyochlamydospores) and honey discoloration below the colony (on MA citrine-olivaceous); no growth at 30°C. Representative culture CBS 285.76.

Ecology and distribution

This variety refers to Russian isolates made in Novosibirsk. The abundant production of thick-walled dictyochlamydospores may be interpreted as an adaptation to the cool continental climate.

Phoma jolyana var. *sahariensis* (Faurel & Schotter) Boerema *et al.*, Fig. 26Bb

Chlamydospore-anamorph multicellular, alternarioid.

> *Phoma jolyana* var. *sahariensis* (Faurel & Schotter) Boerema, Dorenb. & Aa apud Boerema *in* Versl. Meded. plziektenk. Dienst Wageningen **159** (Jaarb. 1982): 27. 1983.
> > ≡ *Sphaeronaema sahariense* Faurel & Schotter *in* Revue mycol. **30**: 156. 1965; not *Phoma sahariensis* Faurel & Schotter *in* Revue mycol. **30**: 154. 1965.

Diagn. data: Boerema (1983b) (fig. 9; differential characteristics and illustration), Boerema (1993) (fig. 3A–b: drawings of dictyochlamydospores, a pycnidium and conidia, the latter reproduced in Fig. 26Bb; cultural description adopted below).

Description in vitro

Pycnidia occur only at temperature ranges of 28–30°C, but do not differ essentially from those of the type var. *jolyana;* however, they may have a pronounced neck.
     Conidia are somewhat shorter and much more consistent in shape than those of the type var.: ellipsoidal-obovoid, 4–5.5(–6) × (2–)2.5–3 μm.
     Chlamydospores multicellular-dictyo/phragmosporous, similar to those of the type var., but often with a kind of halo.

Colonies on OA at room temperature show about the same rate of growth as var. *jolyana*, but are distinguished by conspicuous yellow olivaceous aerial mycelium, yellow discoloration below the colony and absence of pycnidia. Optimum growth and pycnidia production at *c.* 28–30°C. Representative culture CBS 448.83.

Ecology and distribution

This variety is recorded from hare droppings in Centr. Sahara, desert soil in Egypt and seed of *Cucumis sativus* (Cucurbitaceae) of European origin. Apparently a variety adapted to relatively high temperatures. The frequently occurring pronounced neck of the pycnidia explains the originally classification in the genus *Sphaeronaema*.

Phoma pimprina P.N. Mathur *et al.,* Fig. 27B

Chlamydospore-anamorph uni- and multicellular, botryoid-alternarioid.

> *Phoma pimprina* P.N. Mathur, S.K. Menon & Thirum. apud Mathur & Thirumalachar *in* Sydowia **13**: 146a–147. 1959.

Diagn. data: Mathur & Thirumalachar (1959) (description on Czapek agar), Boerema (1993) (fig. 4B: shape of conidia and dictyochlamydospores, reproduced in present Fig. 27B; cultural description adopted below).

Description in vitro

Pycnidia subglobose to globose, mostly 115–230 μm diam., usually with pronounced necks and wide ostioles, mostly solitary and not confluent. Conidial matrix salmony.

Conidia broad, oblong-ovate, usually fine guttulate, (4–)6–7(–8.5) × (3–)3.5–4(–4.5) μm.

Chlamydospores usually scanty, variable and irregular, uni- or multicellular, intercalary or terminal, mostly solitary, when septate usually dictyosporous, often somewhat botryoid, subhyaline to brown, 8–35 μm diam., when nonseptate usually in chains, 5–15 μm diam.

Colonies on OA rather slow growing, 4–5.5 cm diam. after 7 days. Flat, with scarce aerial mycelium and abundant production of pycnidia; reverse often with lilac/pinkish discoloration. Representative culture CBS 246.60.

Ecology and distribution

This fungus is so far only known from soil in India (various isolates). The fungus in many respects resembles *Ph. sorghina* (Sacc.) Boerema *et al.* (this section), a common soil fungus in the tropics and subtropics. However, it can be differentiated very easily by the consistent broad oblong-ovate conidia. Rai (1998) erroneously treated this species as identical with *Phoma capitulum* V.H. Pawar *et al.* (sect. *Phoma*, **A**).

Phoma pomorum Thüm. var. *pomorum*, Fig. 27Aa

Chlamydospore-anamorph uni- and multicellular, alternarioid.

Phoma pomorum Thüm., Fungi pomicoli 105. 1879, var. *pomorum*.
Θ = *Depazea prunicola* Opiz in Malá Encyclop. Nauk. Náklad cesk. Mus. **10**: 120. 1852 [without description].

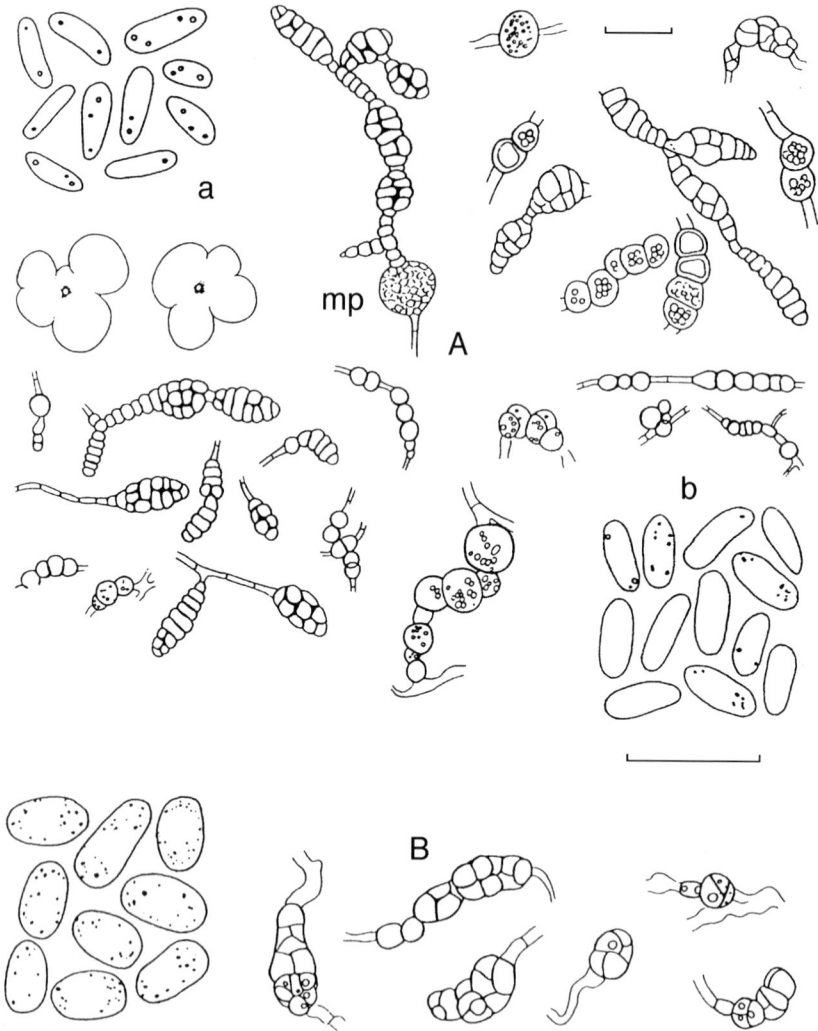

Fig. 27. **A** Varieties of *Phoma pomorum*, characterized by similar complex unicellular and alternarioid multicellular chlamydospores; mp = micropycnidium. The conidia of type var. *pomorum* (a) are highly variable in shape and dimensions, and generally smaller than those of the southern var. *calorpreferens* (b). **B** *Phoma pimprina* produces botryoid-alternarioid multicellular chlamydospores similar to those of *Ph. sorghina* (Fig. 28A), but easily differentiated by the broad conidia. Bar pycnidia 100 μm, bar chlamydospores 20 μm, bar conidia 10 μm.

= *Phyllosticta pyrina* Sacc. *in* Michelia **1**(2): 134. 1878; not *Phoma pyrina* (Fr. : Fr.) Cooke *in* Grevillea **20**: 86. 1892 ≡ *Phoma pyrina* (Fr. : Fr.) Ellis *in* Proc. Acad. nat. Hist. Philad. **1895**: 28. 1895 [as '(Schw.)'; refers to a species of *Myxofusicoccum* or *Paradiscula*, see Boerema & Dorenbosch (1973: 16) and Sutton (1977: 166)].
≡ *Coniothyrium pyrinum* (Sacc.) J. Sheld. *in* Torreya **7**: 142. 1907 [misapplied].
‡ V ≡ *Peyronellaea pyrina* (Sacc.) Goid. *in* Atti Accad. nac. Lincei Rc. VIII, **1**: 455. 1946 [as '*pirina*'; not validly published: name of the genus published without a Latin diagnosis and designation of type species, Arts 10, 36.1, 43].
= *Phyllosticta persicae* Sacc. *in* Michelia **1**(2): 147. 1878; not *Phoma persicae* Sacc. *in* Michelia **1**(5): 526. 1879 [= *Phomopsis* sp.].
= *Phyllosticta prunicola* Sacc. *in* Michelia **1**(2): 157. 1878 [as '(Opiz?) Sacc.'].
H ≡ *Phoma prunicola* (Sacc.) Wollenw. & Hochapfel *in* Z. ParasitKde **8**: 595. 1936 [as '(Opiz) n.c.']; not *Phoma prunicola* Schwein. *in* Trans. Am. phil. Soc. II, **4**: 249. 1832 ['1834'; = Synopsis Fung. Am. bor.].
V ≡ *Coniothyrium prunicola* (Sacc.) Husz *in* Magy. kertéz. szölész. Föisk. Közl. **5**: 23. 1939 [as '*prunicolum*'].
‡ ≡ *Peyronellaea prunicola* (Sacc.) Goid. *in* Atti Accad. Nac. Lincei Rc. VIII, **1**: 455. 1946 [as '(Opiz) comb. nov.'; not validly published, see *Peyr. pyrina*].
≡ *Sphaceloma prunicola* (Sacc.) Jenkins *in* Arg. Inst. Biol. S. Paulo **39**: 233. 1971 [misapplied].
= *Phyllosticta quernea* Thüm. *in* Mycoth. univ. Cent. 18, No. 1787. 1880.
= *Phoma herbarum* f. *capparidis* Sacc. *in* Michelia **2**(1): 93. 1880.
= *Phoma succedanea* Pass. *in* Erb. crittog. ital. II, No. 1378. 1884.
≡ *Phyllosticta succedanea* (Pass.) Allesch. *in* Rabenh. Krypt.-Flora [ed. 2], Pilze **6** [Lief. 60]: 97. 1898 [vol. dated '1901'].
= *Phoma arisari* Bres. *in* Bolm Soc. broteriana **9**: 34. 1891.
V = *Phyllosticta lycopi* Ellis & Everh. *in* Proc. Acad. nat. Sci. Philad. **1891**: 76. 1891 [as '*lycopodis*'].
= *Phyllosticta pruni-spinosae* Allesch. *in* Ber. bot. Ver. Landshut **12**: 10. 1892.
≡ *Phyllosticta prunicola* var. *pruni-spinosae* (Allesch.) Allesch. *in* Rabenh. Krypt.-Flora [ed. 2], Pilze **6** [Lief. 60]: 70. 1898 [vol. dated '1901'].
= *Phyllosticta pruni-avium* Allesch. *in* Ber. bot. Ver. Landshut **12**: 15. 1892.

‡   ≡ *Peyronellaea pruni-avium* (Allesch.) Goid. *in* Atti Accad.
    nac. Lincei Rc. VIII, **1**: 455. 1946 [not validly published,
    see *Peyr. pyrina*].

= *Phoma fictilis* Delacr. *in* Bull. Soc. mycol. Fr. **9**: 186. 1893.

‡   ≡ *Peyronellaea fictilis* (Delacr.) Goid. *in* Atti Accad. nac.
    Lincei Rc. VIII, **1**: 455. 1946 [not validly published, *see*
    Peyr. pyrina].

= *Phyllosticta perforans* Ellis & Everh. *in* Proc. Acad. nat. Sci.
Philad. **1893**: 157. 1893.

    ≡ *Phyllostictella perforans* (Ellis & Everh.) Tassi *in* Boll. R.
    Orto bot. Siena **4**: 5. 1901.

= *Phoma herbarum* var. *tulostomatis* Pat., Cat. rais. Pl. cell.
Tun. 116. 1897.

= *Phyllosticta alni-glutinosae* P. Syd. *in* Beibl. Hedwigia **36**:
157. 1897.

= *Phyllosticta cardamines* Allesch. apud Allescher &
Hennings *in* Bibltheca bot. **42**: 49. 1897.

V = *Phyllosticta cydoniicola* Allesch. apud Sydow *in* Beibl.
Hedwigia **36**: 158. 1897 [as 'cydoniaecola'].

V   ≡ *Phyllosticta cydoniae* var. *cydoniicola* (Allesch.) Cif. *in*
    Annls mycol. **20**: 38. 1922 [as 'cydoniaecola'].

= *Phyllosticta persicicola* Oudem. *in* Beibl. Hedwigia **37** :
313. 1898.

= *Phoma pruni-japonicae* P. Syd. *in* Beibl. Hedwigia **38**: 136.
1899.

‡ V ≡ *Peyronellaea pruni-japonicae* (P. Syd.) Goid. *in* Atti
    Accad. nac. Lincei Rc. VIII, **1**: 455. 1946 [as 'pruni-
    japonica'; not validly published, see *Peyr. pyrina*].

= *Phyllosticta zonata* Ellis & Everh. *in* Bull. Torrey bot. Club
27: 54. 1900.

= *Phyllosticta tirolensis* Bubák apud Bubák & Kabát *in* Ost.
bot. Z. **54**: 181. 1904.

= *Phyllosticta leucosticta* C. Massal. *in* Oss. fitol. Madonna
Verona **2**: 11 [extr.]. 1908.

= *Phyllosticta iserana* Kabát & Bubák apud Bubák & Kabát
in Hedwigia **47**: 354. 1908.

= *Phyllosticta albomaculans* Kabát & Bubák *in* Hedwigia **47**:
354. 1908.

= *Phyllosticta persicophila* Traverso & Miglardi, Flora micol.
prov. Venezia: 11. 1911.

= *Phyllosticta deutziicola* Petr. *in* Annls mycol. **12**: 471. 1914.

= *Phyllosticta epignomonia* Bubák & Vleugel apud Vleugel *in*
Svensk bot. Tidskr. **11**: 320. 1917.

= *Phyllosticta pruni-nanae* Săvul. & Sandu *in* Hedwigia **75** :
185. 1936.

= *Phyllosticta evodiae* Săvul. & Negru *in* Buletin şti Acad.
Repub. pop. rom. **5**: 417. 1953.

= *Peyronellaea nicotiae* Parg.-Leduc *in* Revue gén. Bot. **65**: 545. 1958.

= *Phyllosticta bucklandii* Shreem. *in* Indian J. Mycol. Pl. Path. **3**: 113. 1974 ['1973'].

H V = *Phyllosticta indica* Shreem. *in* Indian J. Mycol. Pl. Path. **3**: 112. 1974 ['1973'] [as 'indicii']; not *Phyllosticta indica* Roum. & P. Karst. apud Karsten, Roumeguère & Hariot *in* Revue mycol. **12**: 79. 1890 [= *Asteromella* sp.].

= *Phoma cyperi* R.K. Upadhyay, Strobel & W.M. Hess *in* Can. J. Bot. **68**: 2063. 1990.

‡ *H* = *Phoma platensis* Cordo & Merlo *in* Invest. Agr.: Prod. veg. **6**(3): 415. 1991 [not validly published: nomenclatural type not indicated, Art. 37]; not *Phoma platensis* Speg. *in* An. Soc. cient. argent. **10**: 29. 1880 [= *Phomopsis platensis* (Speg.) Sousa da Câmara *in* Agron. lusit. **11**: 60. 1949]. [A somewhat deviating strain of *Ph. pomorum* cf. IMI 317307, dried culture in 1987 made by C.A. Cordo.]

Diagn. data: Boerema *et al.* (1965a) (fig. 3, pl. 3 figs 13–21; description and illustrations sub syn. *Phoma prunicola*; synonymy and misapplications; habitat), (1968) (synonymy), (1971) (tab. II; nomenclature and host range), (1973) (misidentifications), (1977) (full synonymy; misinterpretations; synoptic table on characters *in vitro*), Boerema & Dorenbosch (1965) (occurrence on apple trees), Morgan-Jones (1967) (figs A–G: drawings of pycnidia, wall cells, conidia, hyphae, chlamydospores and dictyochlamydospores; CMI description sub *Phoma prunicola*), Dorenbosch (1970) (pl. 3 figs 5, 6, pl. 6 fig. 6, tab. 1 sub *Phoma prunicola*; characteristics *in vitro* with photographs of chlamydospores, dictyochlamydospores and mycelial mat on OA and MA; isolates from soil), Boerema & Dorenbosch (1973) (pl. 2 fig. e, pl. 4 figs d–g, tab. 1; synonymy; sources of isolates; differentiating diagnostic characters), Jones (1976) (figs 1–11; ultrastructure of conidium ontogeny), Domsch *et al.* (1980) (fig. 285a–c; occurrence in soil; literature references), Sutton (1980) (fig. 229F: drawings of conidia and chlamydospores; specimens in CMI collection), Johnston (1981) (tab. 1, figs 12, 16: photographs of agar plate cultures and chlamydospores; records in New Zealand), Tuset & Portilla (1985) (tabs 1–2: dimensions of pycnidia, conidia and chlamydospores *in vitro*, figs 6–7: photographs and drawings of chlamydospores; occurrence on stone fruit trees in Spain), White & Morgan-Jones (1986) (fig. 1A–D: drawings of pycnidium, conidiogenous cells, conidia, chlamydospores and dictyochlamydospores; pl. 1A–H: photographs of agar plate cultures, pycnidia, conidia and chlamydospores; history; detailed cultural and morphological description), Rai & Rajak (1993) (cultural characteristics), Boerema (1993) (fig. 2A–a: shape of pycnidia, conidia, chlamydospores and dictyochlamydospores, reproduced in present Fig. 27Aa; varietal differentiation; cultural description adopted below), Rai (1998) (effect of different physical and nutritional conditions on the morphology and cultural characters), van der Aa & Vanev (2002) (*Phyllosticta* synonyms).

Description in vitro

Pycnidia usually subglobose with a distinct non-papillate ostiole, and often fur-rowed surface, solitary mostly 100–200 μm diam., but frequently confluent in groups up to 1000 μm diam. Pycnidial wall composed of a few layers of elon-gate cells. Fertile micropycnidia frequently occur. Conidial matrix usually whitish to cream, often later darkening to olivaceous brown.

Conidia, variable in shape and dimensions, mostly ovoid-ellipsoidal, (4–)5–7(–8) × 1.5–2.5(–3) μm, hyaline, but with age becoming light brown.

Chlamydospores highly variable in shape, unicellular and multicellular, where unicellular often in long chains, usually guttulate, thick-walled, smooth or roughened, pale brown to brown, 8–10 μm diam., where multicellular usu-ally dictyosporous, mostly terminal on mycelial branches, occasionally inter-calary in combination with chains of unicellular chlamydospores, smooth later roughened, brown to black, mostly 18–60 × 12–30 μm.

Colonies on OA moderately fast growing at room temperature, 4.5–6 cm diam. after 7 days; no growth at 30°C; on MA at room temperature 5.5–7.5 cm diam. Variable in appearance, strains with abundant olivaceous aerial mycelium and scattered pycnidia, others with flat colonies and pycnidia in sec-tors; reverse brownish to blackish beneath dense mycelium, cream-coloured elsewhere. Representative culture CBS 539.66.

Ecology and distribution

A worldwide occurring soil- and seed-borne opportunistic parasite, in older lit-erature commonly treated under the synonym *Phoma prunicola*. It appeared to be more common in regions with temperate climates; records from the trop-ics may concern the warmth preferring variety *calorpreferens*, treated below. The phytopathological literature on this fungus refers in most cases to its occurrence in association with leaf spots on Rosaceae, as apple, *Malus pumila*, pear, *Pyrus communis*, and species of *Prunus*. On apple the fungus has often been confused with *Ph. bismarckii* Kidd & Beaumont (sect. *Phoma*, **A**) and varieties of *Ph. macrostoma* Mont. (sect. *Phyllostictoides*, **E**), neither of which produce any chlamydospores. In Europe *Ph. pomorum* is also known as a 'common' opportunistic pathogen of Ericaceae and strawberry plants, *Fragaria* × *ananassa*; Rosaceae ('black root rot' complex). Further isolations have indicated that it has a rather wide range of substrata upon which it can grow (e.g. earthenware).

Phoma pomorum var. *calorpreferens* Boerema et al., Fig. 27Ab

Chlamydospore-anamorph uni- and multicellular, alternarioid.

Phoma pomorum var. *calorpreferens* Boerema, Gruyter & Noordel. apud Boerema *in Persoonia* **15**(2): 207. 1993.

Diagn. data: Boerema (1993) (fig. 2B–b: shape of conidia, reproduced in pre-sent Fig. 27Ab; cultural description adopted below).

Description in vitro

Pycnidia resemble those of the type var. *pomorum*, subglobose non-papillate, but usually smooth and not furrowed. Conidial matrix usually pinkish and not whitish-cream as in the type var.

Conidia are on average larger than those of the type var., and often oblong-ellipsoidal, (4–)5–8.5(–12) × 2–3(–3.5) μm.

Chlamydospores like those of var. *pomorum*, highly variable, unicellular and multicellular-dictyosporous; the unicellular chlamydospores may be very dark and sometimes extremely large, up to 25 μm diam.

Colonies on OA resemble very much those of var. *pomorum*, they have about the same growth rate at room temperature, but, in contrast with the type var., also grow very well at 30°C. Often abundant production of pycnidia and tufts of whitish aerial mycelium; reverse usually with yellowish (citrine) tinges. Representative culture CBS 109.92.

Ecology and distribution

This warmth-preferring variety of *Ph. pomorum* has been found in Europe as well as in North America. The isolate sources indicate a plurivorous behaviour, corresponding to that of var. *pomorum*. Various records of the latter may refer in fact to var. *calorpreferens*.

Phoma sorghina (Sacc.) Boerema *et al.*, Fig. 28A

Chlamydospore-anamorph uni- and multicellular, botryoid-alternarioid.

Phoma sorghina (Sacc.) Boerema, Dorenb. & Kesteren *in* Persoonia **7**(2): 139. 1973.

≡ *Phyllosticta sorghina* Sacc. *in* Michelia **1**(2): 140. 1878.
= *Phyllosticta sacchari* Speg. *in* Revta Fac. Agron. Univ. naz. La Plata **2**: 239. 1896.
= *Phoma insidiosa* Tassi *in* Boll. R. Orto Bot. Siena **1** ['1897']: 8. 1898.
= *Phyllosticta setariae* Ferraris *in* Malpighia **16**: 18. 1902.
= *Phyllosticta glumarum-setariae* Henn. *in* Annls Mus. r. Congo belge Sér. 4to, Bot. V, **2**: 101. 1907.
= *Phyllosticta glumarum-sorghi* Henn. *in* Annls Mus. r. Congo belge Sér. 4to, Bot. V, **2**: 101. 1907.
= *Phoma glumicola* Speg. *in* Revta Mus. La Plata **15: 36.** 1908.
≡ *Phyllosticta glumicola* (Speg.) Hara, Dis. Rice Plant 164. 1918.
= *Phyllosticta phari* Speg. *in* An. Mus. nac. Hist. nat. B. Aires III, **13**(= 20): 337. 1910 [preprint; vol. dated '1911'].
= *Phyllosticta penicillariae* Speg. *in* An. Mus. nac. Hist nat. B. Aires **26**: 129. 1914 [preprint; vol dated '1915'].

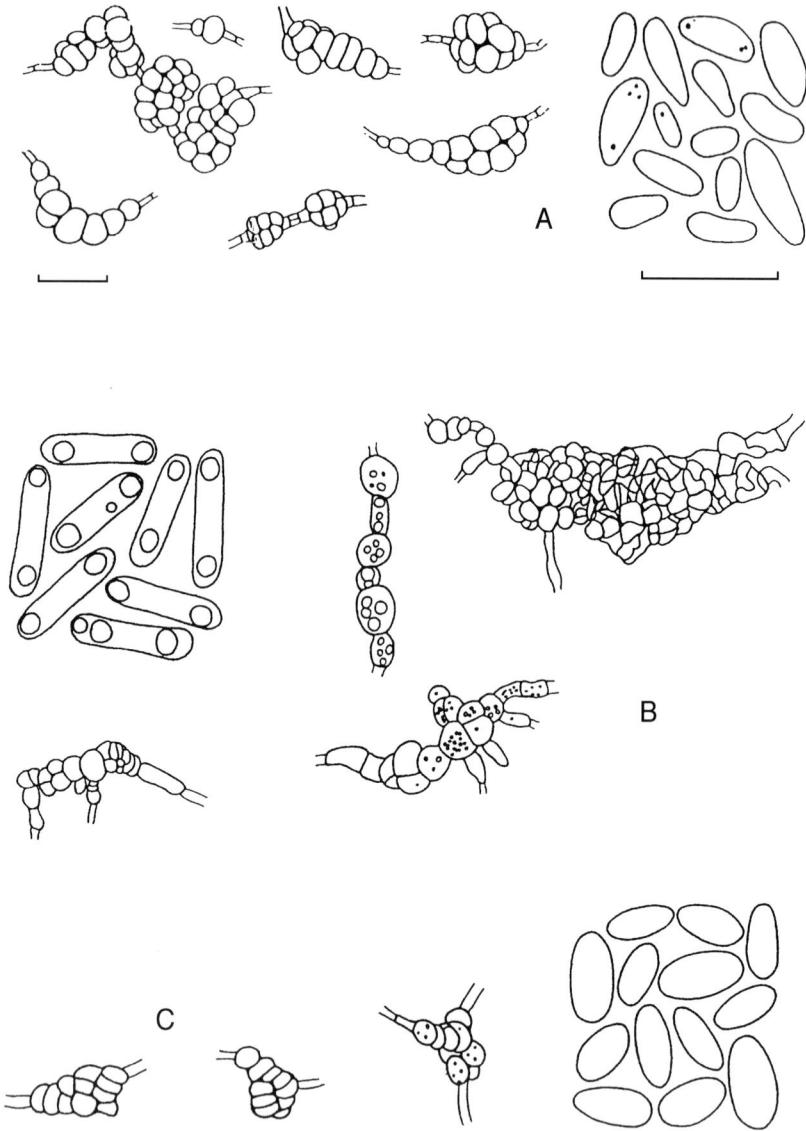

Fig. 28. **A** *Phoma sorghina*, characterized by intercalary botryoid-alternarioid multicellular chlamydospores and intermediate stages between unicellular and multicellular chlamydospores. Conidia highly variable in shape and dimensions. **B** *Phoma violicola*, characterized by various intermediate stages between irregular botryoid-alternarioid and pseudosclerotioid chlamydosporal structures. Conidia relatively large, with two conspicuous guttules. **C** *Phoma zantedeschiae*, characterized by irregular botryoid-alternarioid multicellular chlamydospores. Conidia eguttulate, unicellular and variable in dimensions. Bar chlamydospores 20 μm, bar conidia 10 μm.

= *Phoma depressitheca* Bubák *in* Annln naturh. Mus. Wien **28**: 203. 1914.

H = *Phyllosticta glumarum* Sacc. *in* Nuovo G. bot. ital. II, **23**: 207. 1916; not *Phyllosticta glumarum* (Ellis & Tracy) I. Miyake *in* J. Coll. Agric. imp. Univ. Tokyo **2**: 552. 1910 [≡ *Microsphaeropsis glumarum* (Ellis & Tracy) Boerema].

≡ *Phyllosticta oryzina* Padwick, Manual Rice Dis.: 163. 1950 [as '(Sacc.) Padw. nom. nov.'].

= *Phoma saccharina* Syd. & P. Syd. apud Sydow, Sydow & Butler *in* Annls mycol. **14**: 187. 1916 [cf. study of holotype by Pons, 1990].

V = *Phyllosticta hawaiiensis* Caum *in* Hawaii Plrs' Rec. **20**: 278. 1919 [as '*hawaicensis*'].

= *Ascochyta arachidis* Woron. *in* Notul. syst. Inst. crytog. Horti bot. petropol. **3**: 31. 1924 [cf. Mel'nik, 2000 translated from Mel'nik, 1977: 'identical with *Phyllosticta arachidis* Khokhr.', see below].

= *Phyllosticta arachidis* Khokhr. *in* Bolez. Vredit. maslich. Kultur [= diseases and pests of oil-yielding plants] **1**(2): 32. 1934 [cf. isol. and comp. fresh mat.].

= *Phoma chartae* Verona *in* Cellulosa **1939**: 27. 1939.

= *Phoma saccharicola* S. Ahmad *in* Biologia, Lahore **6**: 131. 1960 [cf. isol. and orig. description, compare Pons, 1990].

= *Peyronellaea indianensis* K.B. Deshp. & Mantri *in* Mycopath. Mycol. appl. **30**: 341. 1966.

≡ *Phoma indianensis* (K.B. Deshp. & Mantri) Boerema, Dorenb. & Kesteren *in* Persoonia **5**(2): 203. 1968.

= *Peyronellaea stemphylioides* Kusnezowa *in* Nov. Sist. Niz. Rast **8**: 199. 1971.

= *Phoma aspidioticola* Narendra & V.G. Rao *in* Mycopath. Mycol. appl. **54**: 137. 1974.

Diagn. data: Boerema *et al.* (1968) (fig. 1; pl. 10 figs 1–10, pl. 11 figs 11–14; description and illustrations sub syn. *Phoma indianensis*; isolates from non-gramineous hosts), (1973) (synonymy and occurrence on gramineous plants), (1977) (full synonymy; pigment production; toxicity to mammals, birds and insects; synoptic table on characters *in vitro*), Punithalingam & Holliday (1972c) (figs A–D: drawings of pycnidia, conidiogenous cells, conidia; CMI description sub *Phoma insidiosa*), Neergaard (1977) (occurrence on seed of rice), Sutton (1980) (fig. 227C: drawings of conidia; 'represented in CMI by more than 500 collections'), White & Morgan-Jones (1983) (figs 1A–E, 2: drawings of pycnidia, conidiogenous cells, conidia, chlamydospores and dictyochlamydospores; pls 1A–F, 2A–H: photographs of agar plate cultures, pycnidia, wall cells, chlamydospores and dictyochlamydospores; history; cultural and morphological description), Punithalingam (1985) (figs A–L: photographs of pycnidia, conidia and chlamydospores; CMI description with phytopathological notes), Rai (1985) (isolation from air, India), Rai (1989) (fig. 1; infection in human being), Monte *et al.* (1990) (physiological and biochemical study of 5

isolates: forming a distinct cluster), Pons (1990) (figs 1–2; specimens on sugarcane), Rai & Rajak (1993) (cultural characteristics), Boerema (1993) (fig. 4A:
shape of pycnidia, conidia and dictyochlamydospores, partly reproduced in
present Fig. 28A; cultural description adopted below), Rai (1998) (effect of different physical and nutritional conditions on morphology and cultural characters; 'red discoloration constant and unaffected by different factors').

Description in vitro

Pycnidia subglobose, 50–200 μm diam., usually with a distinct straight or
somewhat curved neck up to 80 μm long, occasionally touching but usually
not confluent. The outer elongated cells of the pycnidial wall are often characteristically inflated. In aerial mycelium occasionally aberrant small non-ostiolate
pycnidia occur, 5–23 μm diam. Conidial matrix usually salmony in colour.

Conidia variable in shape and dimensions, mostly ovoid-ellipsoidal, sometimes curved, (4–)4.5–7(–8.5) × (1.5–)2–3(–3.5) μm, hyaline, or sometimes
very pale brown.

Chlamydospores highly variable and irregular, uni- or multicellular, mostly
intercalary, sometimes terminal-lateral, solitary or in chains, when septate usually dictyosporous, often with a botryoid configuration, smooth, verrucose or,
rarely, tuberculate, subhyaline to brown, 8–35 μm diam., non-septate chlamydospores 5–15 μm diam.

Colonies on OA rather fast growing at room temperature, 5–7 cm diam.
after 7 days, at 30°C even faster growing; on MA also fast growing at room
temperature, 5–8 cm diam. Aerial mycelium fluffy, sometimes compact with
greyish green or whitish/salmon pink tinges and occasionally reddish exudate
droplets; reverse often with reddish discoloration and occasionally needle-like
crystals (anthraquinone pigments; yellow in acid conditions). About 50% of the
strains showed a positive reaction with the sodium hydroxide test: on application of a drop of NaOH green → red (E$^+$ reaction). Representative culture CBS
284.77.

Ecology and distribution

Worldwide occurring soil- and seed-borne fungus; particularly in the tropics and
subtropics, but occasionally also in temperate regions – especially on plants in
glasshouses – and in regions with a continental climate. In older phytopathological literature it is usually treated under the synonyms *Phoma insidiosa*,
Phyllosticta glumicola and *Phyllosticta sorghina*. The fungus is most frequently
associated with Gramineae (e.g. varieties of sorghum, *Sorghum vulgare*, rice,
Oryza sativa, and sugarcane, *Saccharum officinarum*): spots on leaves, glumes
and seed; root rot and dying-off. Usually, however, the fungus only behaves like
a weak parasite and secondary invader of diseased or weakened plants. It has
been isolated from about 80 different host plant genera, cattle and poultry feed
and several other substrates. Some strains of the fungus can produce a metabolite which is toxic to rats, chickens and insects. The fungus is also reported as an
opportunistic pathogen from man and mammal (erethematous lesions on the

skin). It has been suggested that *Ph. sorghina* belongs to *Mycosphaerella holci* Tehon, but this is implausible and not based on cultural experiments.

Phoma subglomerata Boerema *et al.,* Fig. 24B

Chlamydospore-anamorph uni- and multicellular, alternarioid.

> *Phoma subglomerata* Boerema, Gruyter & Noordel. apud Boerema *in* Persoonia **15**(2): 204. 1993.
>> ≡ *Ascochyta trachelospermi* Fabric. *in* Annali Sper. agr. II, **5**: 1445. 1951; not *Phoma trachelospermi* Tassi *in* Boll. R. Orto bot. Siena **3**(2): 30. 1900 ['1899'].

Diagn. data: Fabricatore (1951) (figs 1–12; original description and illustrations *in vitro*; photographs of agar plate cultures, pycnidia, dictyochlamydospores), Hosford (1975) (as 'Ph. glomerata'; figs 1, 2A–B, 3A–B, 4, 5, 6A–B, 7, 8A–B, 9, 10: photographs of pycnidium, conidia, chlamydospores and dictyochlamy-dospores; description *in vivo* and *in vitro*), Boerema (1993) (fig. 2A: shape of conidia and dictyochlamydospores, reproduced in present Fig. 24B; cultural description adopted below).

Description in vitro

Pycnidia subglobose to obpyriform, 135–225 μm diam., papillate, usually soli-tary but sometimes coalescing. Fertile micropycnidia often also occur. Conidial matrix usually salmony in colour.

Conidia variable in shape and dimensions, mostly oblong ellipsoidal and continuous, (5–)7–12(–15) × 2–3.5(–4) μm, but frequently longer and becom-ing 1-septated and constricted at the septum, (8.5–)12–17 × 3–4(–4.5) μm.

Chlamydospores mostly multicellular-dictyosporous, partly in short branched and unbranched chains of alternarioid elements, partly solitary on hyphal branches and lateral from hyphal strands; dark brown to black, mostly measuring 30–65 × 15–35 μm. In addition chains of unicellular chlamy-dospores and series of irregular short, olivaceous cells may occur.

Colonies on OA moderately fast growing at room temperature, 5–6 cm diam. after 7 days; slow growing at 30°C. Usually only sparse greyish green aerial mycelium and abundant production of pycnidia. Representative culture CBS 110.92.

Ecology and distribution

This fungus is only recently recognized as distinctly different from the cos-mopolitan *Ph. glomerata*. The few records so far known, are from warm regions in southern Europe (Italy), Central America (Mexico) and South Africa. The fungus appeared to be plurivorous and behaves like an opportunistic para-site in wet periods, especially on Gramineae; cf. records on triticale, *Triticale hexaploide*, wheat, *Triticum aestivum*, and maize, *Zea mays*: leaf spot.

Phoma violicola P. Syd., **Fig. 28B**

Chlamydospore-anamorph uni- and multicellular, irregular botryoid-alternarioid and pseudosclerotioid.

> *Phoma violicola* P. Syd. in Beibl. Hedwigia **38**: 137. 1899.
>> = *Phyllosticta violae* f. *violae-hirtae* Allesch. in Rabenh. Krypt.-Flora [ed. 2] Pilze **6** [Lief. 61]: 156. 1898 [vol. dated '1901'].
>> = *Phyllosticta violae* f. *violae-sylvaticae* Gonz. Frag. in Trab. Mus. nac. Cienc. nat. (Bot.), Madr. **7**: 35. 1914.
>> = *Phoma violae-tricoloris* Died. in Annls mycol. **2**: 179. 1904.

Diagn. data: Boerema (1993) (fig. 6B: shape of conidia and intermediate stages between chlamydospores and pseudosclerotioid structures, reproduced in present Fig. 28B; cultural description adopted below).

Description in vitro

Pycnidia usually subglobose, 125–250 μm diam., mostly uni-ostiolate, sometimes papillate or with a cylindrical neck of variable length, mostly solitary but sometimes aggregated. In the mycelium micropycnidia may occur, 60–100 μm diam. Conidial matrix whitish.
　　Conidia cylindrical, usually biguttulate (7.5–)9–10(–11) × 2–3 μm.
　　Chlamydospores highly variable and irregular, mostly intercalary but sometimes terminal, where unicellular usually in short chains, olivaceous-brown, 5–11 μm diam., where multicellular forming dictyosporous-botryoid configurations or pseudosclerotioid structures, olivaceous-brown in colour and very different in size and shape.
　　Colonies on OA relatively slow growing, *c.* 2 cm diam. after 7 days. Aerial mycelium felted, whitish to pale olivaceous grey, underground grey olivaceous to dull green, with olivaceous black concentric zones of pycnidia. Reverse grey olivaceous to greenish grey with olivaceous black concentric rings. Representative culture CBS 306.68.

Ecology and distribution

In Europe this fungus is repeatedly found in association with leaf spots on various wild species of *Viola* (Violaceae). The fungus seems to be an opportunistic parasite which may also affect cultivated *Viola* spp. In some reports the fungus has been identified as *Phyllosticta violae* Desm. s.s., but that species probably refers to immature ascomata of a *Mycosphaerella* sp., see van der Aa & Vanev (2002).

Phoma zantedeschiae Dippen., **Fig. 28C**

Chlamydospore-anamorph multicellular, irregular botryoid-alternarioid.

> *Phoma zantedeschiae* Dippen. in S. Afr. J. Sci. **28**: 284. 1931.
>> Θ = *Phyllosticta richardiae* Halst. in Rep. New Jers. agric. Coll. Exp. Stn **6** [= Rep. New Jers. St. agric. Exp. Stn **14** (**1893**)]: 400. 1894 [without description].
>> = *Phyllosticta richardiae* F.T. Brooks in Ann. appl. Biol. **19**: 18. 1932; not *Phoma richardiae* Mercer in Mycol. Centbl.

2: 244 [297, 326]. 1913 [= *Phoma glomerata* (Corda) Wollenw. & Hochapfel, this section].

Diagn. data: Dippenaar (1931) (original description of pycnidia), Brooks (1932) (sub *Phyllosticta richardiae* n.sp., pl. 1, figs 1–3: disease symptoms; description of pycnidia and cultural characteristics, 'old hyphae forming thick walled, much septated, irregular aggregations'), Jauch (1947) (figs 1–8: disease symptoms, drawings of conidia, photographs of dictyochlamydospores and agar plate cultures; tab. 1: cultural characteristics on different media), Boerema & Hamers (1990) (history), Boerema (1993) (fig. 5B: shape of conidia and dictyochlamydospores, reproduced in present Fig. 28C; cultural description adopted below).

Description in vitro

Pycnidia subglobose or depressed, 90–180 μm diam., usually uni-ostiolate, mostly solitary, sometimes compound. Conidial matrix greyish.

Conidia oval or ellipsoidal, often pointed at one end, variable in dimensions, (3–)4–7(–8) × (2–)2.5–3.5(–4) μm, eguttulate.

Chlamydospores variable-multicellular, mostly intercalary, occasional terminal, usually solitary, irregular botryoid-dictyosporous, dark brown, 15–40 μm diam.

Colonies on OA 7–8 cm in 7 days at 20–22°C, aerial mycelium fairly abundant, white when young, but rapidly becoming greyish-brown; reverse grey to almost black in the centre. Representative culture CBS 267.31.

Ecology and distribution

This fungus may cause large brown blotches on the leaves and 'flowers' (i.e. spathes) of the arum or calla lily, *Zantedeschia aethiopica* (Araceae; formerly *Richardia africana*). The disease, known as leaf blotch, is recorded from South Africa (centre of diversity for *Zantedeschia* spp.), western Europe, North and South America. The fungus has been confused with the ubiquitous *Phoma glomerata* (this section); compare Boerema & Hamers (1990).

It should be noted that occasionally strains of *Ph. zantedeschiae* do not produce dictyochlamydospores; in the key to species of *Phoma* sect. *Phoma* (**A**) these deviating strains have been incorporated.

Species Excluded from *Phoma* sect. *Peyronellaea*

(Misapplications in the synonymous genus *Peyronellaea*)

Formerly some fungi have been placed erroneously in the genus *Peyronellaea*; these misapplications are listed below. For documentation, identification and discussion see Boerema *et al.* (1977). Two misapplications appeared to refer to *Phoma exigua* Desm., the type species of the section *Phyllostictoides* (**E**).

Peyronellaea asteris (Bres.) Goid. [= *Phoma exigua*], *Peyronellaea chomatospora* (Corda) Goid., *Peyronellaea cincta* (Berk. & M.A. Curtis) Goid., *Peyronellaea destructiva* (Desm.) Goid. [= *Phoma exigua*], *Peyronellaea glomerata* '(Berk. & Broome)' Goid., and *Peyronellaea scabra* (McAlpine) Goid.

12 E *Phoma* sect. *Phyllostictoides*

Phoma sect. *Phyllostictoides* Žerbele ex Boerema

> *Phoma* sect. *Phyllostictoides* Žerbele ex Boerema *in* Mycotaxon **64**: 331. 1997.
>
> > ‡ ≡ *Ascochyta* sect. *Phyllostictoides* Žerbele *in* Trudỹ vses. Inst. Zashch. Rast. **29**: 20. 1971 [as a provisional name of a 'group-like section' with mainly aseptate conidia *in vitro*].
> >
> > ‡ ≡ *Phoma* sect. *Phyllostictoides* (Žerbele) Boerema apud van der Aa, Noordeloos & de Gruyter *in* Stud. Mycol. **32**: 6. 1990 [both names published without a Latin description].
>
> Type: *Ascochyta altheina* Sacc. & Bizz. = *Phoma exigua* Desm. var. *exigua*.

The pycnidia of species in section *Phyllostictoides* are similar to those of *Phoma* sect. *Phoma*: thin-walled, pseudoparenchymatous, glabrous but some-times with hyphal outgrowths. However, the species of this section are differen-tiated and characterized by the fact that *in vivo* the larger conidia often become two or even more celled by secondary septation (see Fig 29C–D). The percent-age of septate conidia depends on the environmental conditions and may vary *in vivo* between 5 and 95%. Under normal laboratory conditions the majority of conidia always remain aseptate, but usually some two- or more-celled coni-dia also occur. In most species the pycnidia have a predetermined opening or ostiole, but sometimes the pycnidia remain closed for a long time with final for-mation of a pore.

The section includes species with and without chlamydospores; if present they are unicellular, solitary or formed in series or complexes.

The section *Phyllostictoides* includes many species associated with leaf spots and leaf necroses. Under the Saccardoan system for anamorph genera, most collections of these species were arranged under '*Phyllosticta*'. The fre-quent occurrence of two-celled conidia *in vivo* explains why specimens on leaves were also classified in '*Ascochyta*', and in '*Diplodina*' when associated with stems.

Many species of sect. *Phyllostictoides* are anamorphs of species of *Didymella* Sacc. ex Sacc.

Diagn. lit.: Žerbele (1971) (foundation of the concept of the present section by its separation from true *Ascochyta* species by the production in culture of mainly aseptate '*Phyllosticta*' (= *Phoma*)-like conidia), van der Aa & van Kesteren (1971) (tabs 1, 2, variability in conidial dimensions and percentage of septate conidia as displayed by the type species of the section), Boerema & Bollen (1975) (pls 22, 23A–C, fig. 2H–J, electron microscopy of conidial septation in *Phoma* spp.; 'the secondary developing septa apparently attain their final thickness almost from the start', 'only a restricted zone of attachment to the lateral wall', 'septate conidia do not have thicker walls than one-celled conidia'), Boerema (1985) (introduction of '*Phoma* sect. *Phyllostictoides*'), van der Aa *et al.* (1990) (further documentation of the present section concept), Boerema (1997) (formal validation of the section), van der Aa *et al.* (2000) (characteristics and nomenclature of the type species).

Synopsis of the Section Characteristics

- Pycnidia simple or complex, relatively thin-walled, pseudoparenchymatous, distinct ostiolate or with a pore, glabrous, but sometimes with hyphal outgrowths (pycnidial wall usually thinner on leaves than on stems).
- Conidia mainly unicellular *in vitro*, but usually with some two- or more-celled conidia of 'common *Phoma* size'; 5–95% of the conidia may be septated *in vivo*.
- Chlamydospores, if present, unicellular; occasional occurrence of swollen cells.
- Teleomorph, if known, belonging to *Didymella* Sacc. ex Sacc.

Key to the Species and Varieties Treated of Section *Phyllostictoides*

Differentiation on the characteristics *in vitro*

1. a. Growth rate slow on OA, < 3.5 cm after 7 days **2**
     b. Growth rate moderate to fast on OA, > 3.5 cm after 7 days **6**

2. a. On OA and MA (dendritic) crystals are produced, chlamydospores present, conidia 4–14 × 3–5 µm, mainly aseptate, commonly 5–7 × 3–4 µm, 1-septate commonly 10–12 × 4–5 µm ....
....*Ph. arachidicola*, teleom. *Did. arachidicola*
[Specific pathogen of *Arachis hypogaea*, worldwide in groundnut-growing areas: net blotch, web blotch or leaf blotch.]
     b. Crystals or chlamydospores absent **3**

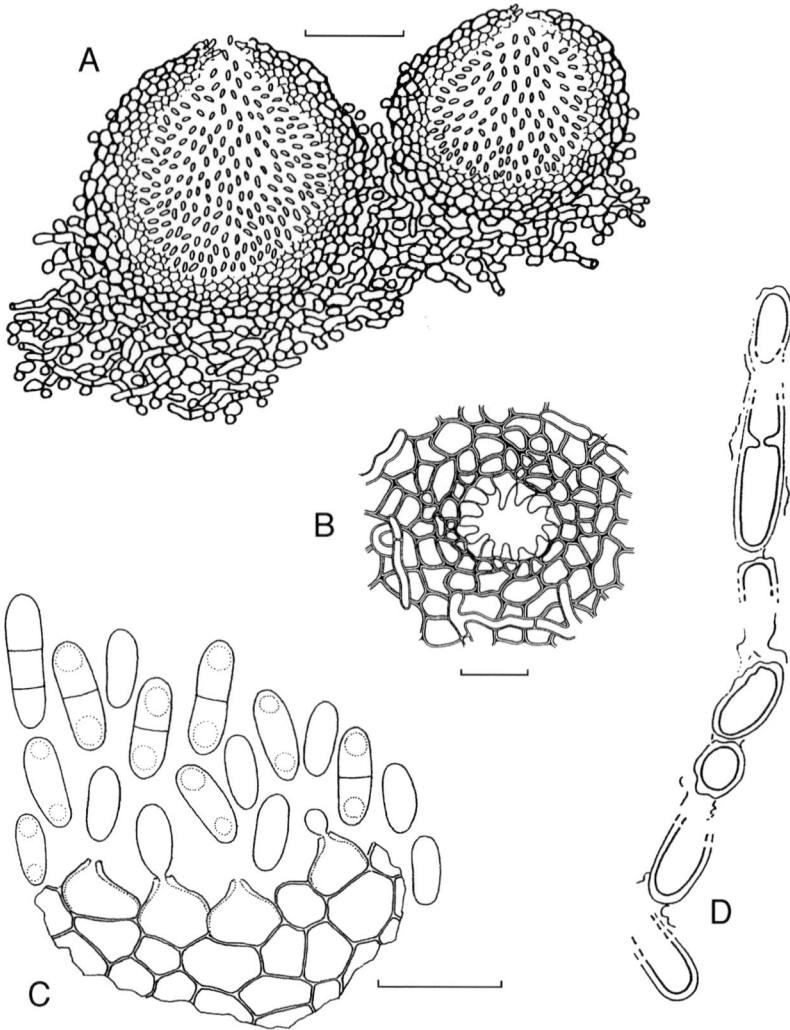

Fig. 29. *Phoma exigua* var. *exigua*, type species of *Phoma* sect. *Phyllostictoides.* **A** Vertical section of pycnidia and subtending mycelium, from 14-day-old colony. **B** Superficial view of an ostiolum, lined internally with papillate hyaline cells. **C** Conidiogenous cells and conidia. **D** Chain of conidia connected by a mucilaginous mass, showing one large, septate conidium between smaller, continuous ones. Drawings A and C after Morgan-Jones & Burch (1988d) with permission; the drawings B and D (cf. Boerema, 1965) are based on electron micrographs. Bar pycnidia 100 μm, bar ostiolum, conidiogenous cells and conidia 10 μm.

3. a. NaOH test on OA positive, greenish, then red (E⁺ reaction)............................**4**
    b. NaOH test on OA negative **5**

4. a. Growth variable, with irregular lobed or scalloped margin on OA and MA, (2.5–)5–8.5 cm after 7 days, i.e. sometimes slow, but mostly moderate to fast growing; colonies colourless or with various grey to greenish tinges, or olivaceous to olivaceous black, conidia very variable in shape and dimensions, one-

celled or becoming 1-septate, very occasionally 2-septate; aseptate conidia
2.5–12 × 2–3.5 μm, commonly 4–7 × 2–3.5 μm, 1(2)-septate conidia 5.5–13
× 2.5–5 μm, commonly 7–10 × 2.5–3.5 μm ....*Ph. exigua* var. *exigua*
[Worldwide distributed soil fungus; a plurivorous opportunistic plant parasite,
which may cause necroses on leaves and stems, and may produce a rot on
fleshy roots and tubers and at the bases of leaves and stems.]

> NB: A table to the host-specific varieties of this fungus is given on p. 246
> for *Ph. exigua* var. *diversispora*, in this key see..................................**35b**
> for *Ph. exigua* var. *forsythiae*..**40b**
> for *Ph. exigua* var. *heteromorpha* **8b**
> for *Ph. exigua* var. *lilacis*...**39a**
> for *Ph. exigua* var. *linicola* **4b**
> for *Ph. exigua* var. *noackiana* **44a**
> for *Ph. exigua* var. *populi* **16a**
> for *Ph. exigua* var. *viburni*..**40a**

b. Growth rate relatively slow on OA and MA, 2–4.5 cm after 7 days; colonies
compact, olivaceous grey to olivaceous black, conidia like those of the type
variety, see 4a ....*Ph. exigua* var. *linicola*
[Seed-borne pathogen of flax, *Linum usitatissimum*, recorded in Europe and
New Zealand: damping-off, foot rot.]

5. a. Colonies on OA irregular, grey olivaceous to olivaceous, citrine near margin,
with finely floccose to woolly, white aerial mycelium, aseptate conidia 6.5–11.5
× 2.5–3(–3.5) μm, conidiogenous cells relatively large, 5–13 × 6–12 μm.........
....*Ph. acetosellae*
[Common on ageing leaves of *Rumex acetosella* in Europe and North
America.]

   b. Colonies on OA irregular, olivaceous buff/pale luteous to citrine/olivaceous, with
very sparse, velvety, white aerial mycelium, aseptate conidia 4–7.5(–13) ×
2–4 μm, conidiogenous cells 4–7 × 4–7 μm...
....*Ph. argillacea*, teleom. *Did. applanata*
[Worldwide occurring pathogen of *Rubus idaeus*: cane blight or spur blight.]

6. a. Growth rate moderate on OA, 3.5–5 cm after 7 days....................................**7**
    b. Growth rate fast on OA, at least 5 cm after 7 days......................................**19**

7. a. NaOH test on OA positive, greenish, then red (E$^+$ reaction)..........................**8**
    b. NaOH test if positive, not an E$^+$ reaction...**9**

8. a. Growth rate variable on OA and MA, (2.5–)5–8.5 cm after one week, i.e.
mostly moderate to fast growing; colonies lobed or scalloped, colourless or
with various grey to greenish tinges, or olivaceous to olivaceous black...........**4a**
    b. Growth rate moderate on OA, 4–5 cm, relatively slow on MA, 2–2.5(–3) cm;
on OA colonies rather dark, grey olivaceous to olivaceous grey/olivaceous
black, with white to pale olivaceous grey/glaucus grey aerial mycelium, for
conidia see 4a ....*Ph. exigua* var. *heteromorpha*
[Specific pathogen of *Nerium oleander* in Europe and North America: dieback,
leaf necrosis.]

9. a.   Colonies on OA irregular due to recolonizing sectors, greenish olivaceous/citrine to grey olivaceous, olivaceous buff near margin, with sparse, finely floccose, white to pale grey olivaceous aerial mycelium, aseptate conidia (4–)5–7(–11.5) × 2.5–5 μm, septate conidia 9.5–14.5 × 2.5–5 μm ....
....*Ph. nepeticola*, teleom. *Did. catariae*
[Common pathogen of *Nepeta cataria*; also on other *Nepeta* spp. in Eurasia and North America: leaf spot, stem lesions.]

   b.   Colonies on OA different, conidia mainly aseptate...**10**

10. a.   Both *Phoma* anamorph and *Didymella* teleomorph are formed *in vitro*, colonies grey olivaceous to dull green, growth rate on OA 4.5–5 cm after 7 days, on MA 3.5–4 cm, conidia aseptate, (3.5–)5–8(–13.5) × 2–3(–4) μm, septate conidia up to 15 × 5 μm ....
..............*Ph. ligulicola* var. *inoxydabilis*, teleom. *Did. ligulicola* var. *inoxydabilis*
[Associated with necroses on leaves and stems of various wild and cultivated Compositae in Europe and Australia.]
   NB: The faster growing type variety of this fungus does not produce pseudothecia *in vitro* and only occasionally *in vivo*, see *Ph. ligulicola* var. *ligulicola*, teleom. *Did. ligulicola* var. *ligulicola*, this key..........................**28b**

   b.   Only a *Phoma* anamorph is formed *in vitro* **11**

11. a.   Colonies on OA peach/sienna to red/blood colour or dark vinaceous, due to the occurrence of a red pigment in the hyphae, with NaOH a violet discolouring may occur (not an E$^+$ reaction), conidia (4–)5–8(–11) × 2–3(–4) μm, septate conidia 8.5–14 × 2.5–4 μm..*Ph. macrostoma* var. *macrostoma*
[Opportunistic parasite, especially on woody plants; worldwide distributed.]
   NB: The fungus may lose the possibility to produce red pigment in the hyphae, see *Ph. macrostoma* var. *incolorata*, this key **18a**

   b.   Red pigment in hyphae absent, NaOH test on OA negative..........................**12**

12. a.   Especially on MA (dendritic) crystals are formed. In older cultures chlamydospores may be produced, conidia aseptate, 5–7(–10.5) × 1.5–4 μm, septate conidia sparse, similar size...............................*Ph. medicaginis* var. *medicaginis*
....*Ph. medicaginis* var. *macrospora*
[Pathogens of *Medicago sativa*: black stem disease; both worldwide distributed.]
   NB: The differentiation of these two varieties is based on conidial diversity *in vivo*, especially at low temperature. *In vitro* on agar media they are similar.

   b.   Crystals absent, chlamydospores absent **13**

13. a.   Colonies rather dark on OA, greenish olivaceous to grey olivaceous/olivaceous grey, or olivaceous to olivaceous black, aseptate conidia and septate conidia of similar size...**14**

   b.   Colonies on OA colourless to grey olivaceous to dull green/citrine green or rosy buff, aseptate conidia and septate conidia of similar size...............................**15**

14. a.   Colonies on OA greenish olivaceous/grey olivaceous to olivaceous, aseptate conidia (3.5–)5–7(–9.5) × 2.5–3.5 μm, aseptate conidia 6.5–13.5 × 3–4.5 μm
....*Ph. telephii*
[Pathogen of *Sedum telephium*, common in Europe: purple blotch disease.]

    b.   Growth rate moderate on OA, 4–5 cm, relatively slow on MA, 2–2.5(–3) cm; on OA colonies rather dark, grey olivaceous to olivaceous grey/olivaceous black, with white to pale olivaceous grey/glaucous grey aerial mycelium........**8b**

15. a.   Colonies on MA colourless to dull green, grey olivaceous/olivaceous grey or rosy buff **16**
    b.   Colonies on MA hazel/olivaceous or primrose/olivaceous buff/honey.............**18**

16. a.   Colonies on OA colourless to grey olivaceous, with pale olivaceous grey to glaucous grey aerial mycelium, for conidia see 4a ....*Ph. exigua* var. *populi*
    [Opportunistic pathogen on poplars, *Populus* spp. (occasionally on *Salix*) in Europe: necrotic black lesions.]
    b.   Colonies on OA colourless to (rosy) buff, dull green or olivaceous **17**

17. a.   Colonies on MA colourless to dull green, conidial exudate rosy buff to rosy vinaceous, aseptate conidia 4–7(–9) × 1.5–2.5(–3) μm, septate conidia 8.5–11.5 × 2–3.5 μm ....*Ph. destructiva* var. *diversispora*
    [Pathogen of *Lycopersicon esculentum* recorded in Europe; necroses on leaves, leaf stalks and stems: necrotic spot.]
        NB: The type variety of *Ph. destructiva* produces only aseptate conidia and therefore has been included in sect. *Phoma*, **A**.
    b.   Colonies on MA colourless to rosy buff, or pale olivaceous grey to dull green, conidial exudate buff, aseptate conidia (4–)5–7(–8.5) × (1.5–)2–3(–3.5) μm, septate conidia up to 10 μm..*Ph. digitalis*
    [Pathogen on *Digitalis* spp. in Europe and New Zealand: leaf spot.]

18. a.   Colonies on MA primrose/olivaceous buff, often with pale honey/olivaceous sectors, conidial exudate white to buff/rosy buff, pigmentless variety of *Ph. macrostoma* (see **11a**) ....*Ph. macrostoma* var. *incolorata*
    [Opportunistic parasite, especially on woody plants. Worldwide distributed.]
    b.   Colonies on MA hazel to olivaceous, conidial exudate off-white to primrose, aseptate conidia (5–)6–8.5(–11) × 2–3.5 μm, septate conidia up to 13.5 × 4 μm ....*Ph. laundoniae*
    [Recorded on fruits of *Prunus persica* in New Zealand; in that country possibly also occurring in association with other stone fruit trees.]

19. a.   Especially on MA crystals are formed, on OA and MA a diffusing pigment may be produced **20**
    b.   Crystals absent, on OA and MA no diffusible pigment production **26**

20. a.   Chlamydospores present **21**
    b.   Chlamydospores absent **23**

21. a.   Crystals needle-like, citrine green to yellow green. Chlamydospores only present when induced by bacteria **24a**
    b.   Crystals botryoid to dendritic, chlamydospores always produced...................**22**

22. a.   Colonies on OA greenish/yellowish olivaceous to olivaceous, (a) septate conidia 4–7.5 × 2–3.5 μm, chlamydospores 8–20 μm diam., crystals readily produced on MA after one week ....*Ph. pinodella*
    [Pathogen, especially on leguminous plants; diseases are known as black stem, foot rot and leaf spot. Worldwide distributed.]

b.   Colonies on OA colourless to pale olivaceous grey or greenish olivaceous/grey
     olivaceous, aseptate conidia 5–8 × 2–3.5 μm, septate conidia up to 12.5
     × 5 μm, chlamydospores 8–16 μm, crystals specific produced in fresh
     isolates on MA..*Ph. sojicola*
     [In Eurasia the most common species involved with the leaf and pod spot dis-
     ease of soybean, *Glycine max.*]

23. a.   On MA diffusible pigment crystallizes as yellow speckles. Growth rate on OA
         and MA extremely fast, > 8 cm after one week, aseptate conidia 5–10 ×
         2.5–4 μm, septate conidia up to 14 × 5 μm..........................*Ph. matteucciicola*
         [Pathogen of *Matteuccia struthiopteris* described in Canada: gangrene disease;
         also reported from ferns indigenous to Europe: *Dryopteris filix-mas* and
         Blechnum spicant: lethal effect on the gametophyte stage.]
    b.   On MA crystals are needle-like, citrine green to yellow green. Growth rate on
         OA and MA < 8 cm after 7 days **24**

24. a.   On MA growth rate fast, similar to those on OA, 6.5–8 cm after 7 days,
         colonies on OA buff/honey/amber due to the release of diffusible pigments, on
         MA crystals present, needle-like, citrine green to yellow-green, (a)septate coni-
         dia (3.5–)5–7(–11.5) × (1.5–)2–3(–3.5) μm....................................*Ph. foveata*
         [Pathogen of *Chenopodium quinoa* in South America: brown stalk rot; in
         Europe known as cause of tuber gangrene of potato, *Solanum tuberosum.*]
    b.   On MA growth rate slow to moderate, up to 5 cm after 7 days, also diffusible
         pigment on OA and needle-like crystals on MA **25**

25. a.   On MA growth rate 5 cm after 7 days, colonies on OA honey to pale luteous
         due to a diffusible pigment, on MA crystals needle-like, citrine green to yellow-
         green, aseptate conidia (5.5–)6.5–11 × 2–4 μm, septate conidia 9–14.5 ×
         3–5 μm..*Ph. rudbeckiae*
         [Pathogen of *Rudbeckia* spp. recorded in North America and Europe: leaf spot.]
    b.   On MA growth rate slow to moderate, 3–4 cm after 7 days, colonies on OA
         pale luteous to amber due to a diffusible pigment, on MA crystals needle-like,
         citrine green to yellow-green, aseptate conidia (3–)5–6.5(–8.5) × 1.5–3 μm......
         ....*Ph. artemisiicola*
         [On *Artemisia* spp. in Europe, should cause the premature death of the plants.]

26. a.   NaOH test positive, green, later red (E$^+$ reaction) **27**
    b.   NaOH test negative or if positive, not an E$^+$ reaction....................................**31**

27. a.   Growth on MA irregular, with a scalloped or lobed margin, growth rate on MA
         somewhat slower than those on OA **28**
    b.   Growth on MA regular to slightly irregular, growth rate on MA similar to those
         on OA **29**

28. a.   Growth rate variable on OA (2.5–)5–8.5 cm after 7 days, colonies colourless or
         with various grey to greenish tinges, or olivaceous to olivaceous black **4a**
    b.   Growth rate on OA *c.* 7 cm after 7 days, colonies colourless/greenish oliva-
         ceous to dull green/olivaceous, discolouring to sienna due to a diffusible pig-
         ment, conidia aseptate (3.5–)5–8(–13.5) × 2–3(–4) μm, septate conidia up to
         15 × 5 μm............*Ph. ligulicola* var. *ligulicola*, teleom. *Did. ligulicola* var. *ligulicola*

[Specific pathogen of *Dendranthema*-Grandiflorum hybrids (florists' chrysanthe-mum): ray (flower) blight). May affect all plant parts. Worldwide distributed.]
  NB: A slower growing E⁻ variety of this fungus produces also pseudothecia
in *vitro*: *Ph. ligulicola* var. *inoxydabilis*, teleom. *Did. ligulicola* var. *inoxyd-abilis*, this key...**10a**

29. a. Very fast growing on OA and MA, regular, completely filling the plates in 7
days, growth rate after 5 days already 6–7 cm, colonies colourless to olivaceous grey/grey olivaceous, conidial exudate buff to rosy buff/salmon, (a)sep-
tate conidia relatively small, 4.5–6.5 × 2–3 μm..............................*Ph. strasseri*
[Pathogen of *Mentha* spp.: recorded in Europe, Japan, New Zealand and
North America: rhizome and stem rot; occasionally also on other Labiatae.]
  b. Growth rate on OA and MA 6.5–8 cm after 7 days, regular to slightly irregular,
colonies grey olivaceous or dark green olivaceous, conidial exudate white to
off-white or off-white to buff, conidia more variable in size and often longer, up
to 9.5 or 10.5 μm..**30**

30. a. Pycnidia with 1–5 papillate ostioles developing to an elongate neck in a later
stage, conidia cylindrical-allantoid, mainly aseptate 4–9.5 × 1.5–2.5 μm, septate conidia up to 12 × 3.5 μm. Growth rate on OA and MA 6.5–7.5 cm after
7 days, colonies grey olivaceous/olivaceous grey ....*Ph. nemophilae*
[Pathogen of *Nemophila* spp. in Europe and North America, may cause damping-off of seedlings and decay of stems and leaves of older plants.]
  b. Pycnidia with or without papillate ostioles, conidia variable in shape, subglobose, ellipsoidal to oblong or allantoid, mainly aseptate (3.5–)5–8(–10.5) ×
2–3.5 μm, 1-septate conidia of similar size, sparse. Growth rate on OA and MA
7–8 cm after 7 days, colonies dark, greenish olivaceous to grey olivaceous/oli-
vaceous grey ....*Ph. sambuci-nigrae*
[Widespread on *Sambucus nigra* in Eurasia: leaf spot, shoot dieback.]

31. a. Growth rate on MA < 6 cm after 7 days..**32**
  b. Growth rate on MA > 6 cm after 7 days..**36**

32. a. Growth rate on MA up to 4 cm after 7 days, colonies on OA colourless to grey
olivaceous, with pale olivaceous grey to glaucous grey aerial mycelium **16a**
  b. Growth rate on MA exceeding 4 cm after 7 days **33**

33. a. Growth rate on OA < 6 cm after 7 days...**34**
  b. Growth rate on OA > 6 cm after 7 days **35**

34. a. Colonies on OA and MA with peach/sienna to red/blood colour or dark vinaceous, due to the occurrence of a red pigment in the hyphae, with NaOH a violet discolouring may occur (not an E⁺ reaction)...**11a**
  b. Colonies on OA colourless or with weak grey olivaceous/dull green sectors, on
MA primrose/olivaceous buff, often with weak honey/olivaceous sectors, NaOH
test negative **18a**

35. a. Colonies on MA dull green to citrine, (a)septate conidia 5–9(–15) × 2–5 μm,
chlamydospores absent..*Ph. rumicicola*
[Pathogen of *Rumex obtusifolius*, first recorded in New Zealand, probably
widespread.]

b.   Colonies on MA buff to grey olivaceous/olivaceous black, chlamydospores may
     be formed, 10–25 μm diam., for conidia see 4a......*Ph. exigua* var. *diversispora*
     [Pathogen of *Vigna unguiculata* and *Phaseolus vulgaris* in Africa and Europe:
     black node disease.]

36. a.   Conidia ellipsoidal to allantoid with several conspicuous, relatively large gut-
         tules, mainly aseptate, (5–)6–8(–10.5) × 1.5–3 μm, septate conidia up to
         13 × 3.5 μm. Colonies on OA colourless with an olivaceous/grey
         olivaceous to dull green stellate pattern.......................................*Ph. heliopsidis*
         [Pathogen of Compositae, e.g. *Heliopsis* spp. and *Ambrosia artemisiifolia*, in
         North America; mostly affecting the leaves, but also on stems and inflo-
         rescenses.]
    b.   Conidia different, more variable in shape, without or with small guttules **37**

37. a.   On woody plants **38**
    b.   On herbaceous plants...**41**

38. a.   Colonies relatively dark on OA and MA, olivaceous grey to grey olivaceous/dull
         green, with olivaceous black to leaden black in reverse, (a)septate conidia
         (3–)4–7(–9) × 2–3 μm, often eguttulate ....*Ph. tarda*
         [Pathogen of *Coffea arabica* known from Africa and South America: leaf blight
         and stem dieback.]
    b.   Colonies on OA colourless to greenish olivaceous/grey olivaceous to olivaceous
         grey, on MA similar, with leaden grey or olivaceous in reverse **39**

39. a.   On OA with abundant, compact tufted, white aerial mycelium, covering the
         entire greenish olivaceous colony; for conidia see 4a ....*Ph. exigua* var. *lilacis*
         [Specific pathogen of *Syringa vulgaris* (occasionally on *Forsythia*: damping-off,
         leaf necrosis, dieback; probably worldwide distributed.]
    b.   On OA sparse to abundant, velvety to finely floccose tufted, mainly (pale)
         olivaceous grey aerial mycelium; colony colourless to grey olivaceous/
         olivaceous grey..**40**

40. a.   On OA abundant velvety/finely floccose, tufted, mainly (pale) olivaceous grey
         aerial mycelium, for conidia see 4a ....*Ph. exigua* var. *viburni*
         [Pathogen of *Viburnum* spp. (occasionally on *Lonicera*): leaf spot, stem lesions
         (recorded in Europe and North America).]
    b.   On OA velvety to finely floccose/woolly, partly tufted, mainly (pale) olivaceous
         grey aerial mycelium, for conidia see 4a......................*Ph. exigua* var. *forsythiae*
         [Pathogenic on *Forsythia* spp. in Europe: shoot blight.]

41. a.   Aseptate conidia mainly cylindrical, width often < 2 μm, usually with 2–4 small
         polar guttules...**42**
    b.   Aseptate conidia variable, mainly ellipsoidal, width always > 2 μm, usually with
         several small guttules **43**

42. a.   Growth rate fast on OA and MA, 7–8 cm after 7 days, colonies on OA dull
         green, aseptate conidia 3.5–5.5(–7) × 1–2 μm, septate conidia up to 9 ×
         3 μm ....*Ph. valerianellae*
         [Seed-borne pathogen of Valerianaceae in Europe: damping-off.]

b. Growth rate on OA fast, *c.* 7 cm after 7 days, on MA 6–6.5 cm, colonies on OA olivaceous buff to greenish olivaceous/grey olivaceous, conidia (3.5–)5–8(–10.5) × 1.5–3 μm, septate conidia up to 18 × 3 μm............*Ph. rhei* [Pathogen of *Rheum* spp.: leaf spot. Worldwide distributed.]

43. a. (A)septate conidia relatively small, 4.5–6.5 × 2–3 μm, colonies on OA colourless to olivaceous grey/grey olivaceous **29a**
   b. (A)septate conidia variable in shape and size, 3.5–8(–10) × 2–3.5 μm, septate conidia of similar size or larger, up to 15.5 × 4.5 μm **44**

44. a. Growth rate on OA very fast, 7.5–8.5 cm after 7 days, on MA somewhat slower, 6.5–7.5 cm, colonies on OA olivaceous/iron grey or grey olivaceous/olivaceous, on MA greenish olivaceous to olivaceous, chlamydospores may be produced, for conidia see 4a ....*Ph. exigua* var. *noackiana* [Pathogen of *Phaseolus vulgaris*, known from South and Central America: black node disease.]
   b. Growth rate on OA and MA similar, also fast, chlamydospores absent **45**

45. a. Colonies on OA colourless/dull green to olivaceous/olivaceous grey, reverse buff to dull green/olivaceous, to leaden grey leaden black, aseptate conidia 4–8 × 2–3 μm, septate conidia up to 10 × 4.5 μm, pseudothecia of *Didymella* teleomorph may be produced.........*Ph. cucurbitacearum*, teleom. *Did. bryoniae* [Seed-borne pathogen of Cucurbitaceae: gummy stem blight, but it may affect all plant parts. Worldwide distributed.]
   b. Colonies on OA colourless/olivaceous buff to grey olivaceous, reverse grey olivaceous/olivaceous grey to olivaceous, olivaceous buff near margin, aseptate conidia (3.5–)5–8.5(–10) × 2–3.5(–4.5) μm, septate conidia up to 15.5 × 4.5 μm, in old cultures sterile, stilboid bodies may be formed...*Ph. lycopersici*, teleom. *Did. lycopersici* [Pathogen of *Lycopersicon esculentum* in Eurasia: canker, stem and fruit rot.]

Possibilities of identification *in vivo*

All species of sect. Phyllostictoides are characterized *in vivo* by the presence of a number of 1(–2)-septate conidia in thin-walled pycnidia.

The plurivorous opportunistic parasites in this section, *Ph. exigua* var. *exigua*, both varieties of *Ph. macrostoma*, as well as the plurivorous pathogen *Ph. pinodella* are often quite easy to recognize *in vivo* on account of the pycnidial characters (wide ostiole in *Ph. macrostoma*), their specific conidial variability (compare Figs 29, 31E, 32A), their ecology and host preferences (*Ph. exigua* on all kinds of herbaceous plants, see description *in vivo* on p. 244; *Ph. macrostoma* on various trees and shrubs and *Ph. pinodella* on leguminous plants).

The other species in this section and the pathogenic varieties of *Ph. exigua* have a restricted host range, which facilitates their identification *in vivo*. They are often associated with characteristic leaf spots (*Phyllosticta* synonyms) and lesions on fruits and stems. Therefore the phytopathological data are essential for their identification *in vivo*.

Following phytopathological studies we have included in sect. *Phyllostictoides* two important species of *Phoma* which could not yet be studied *in vitro*:

1.   Specific pathogen of *Cannabis sativa* (Cannabaceae), leaf and stem spot. On hemp in whole Eurasia and North America (McPartland, 1994) ....
....*Ph. cannabis*, teleom. *Did. cannabis*
Pycnidia *in vivo* globose-subglobose with slightly papillate ostiole, 65–180 µm diam., always associated with chlamydospores in host tissue. Conidia variable, oval to ellipsoidal, aseptate and 1-septate, 3–8 × 2–3 µm.

2.   Specific pathogen of *Carica papaya* (Caricaceae), fruit rot, trunk rot and leaf spot. Worldwide pathogen of papaw (Punithalingam, 1979b)......................
....*Ph. caricae-papayae*
Pycnidia *in vivo* subglobose with flat ostiole, 100–120 µm diam., conidia shortly cylindrical to ellipsoidal or slightly reniform, aseptate and 1-septate, mostly 8–10 × 2.5–3.5 µm.

Distribution and Facultative Host Relations in Section *Phyllostictoides*

Plurivorous
(But with special host
preference)

Cosmopolitan:
  Ph. exigua var. *exigua*
  [Opportunistic parasite, especially on herbaceous plants.]
    Ph. macrostoma var. *macrostoma*
    Ph. macrostoma var. *incolorata*
  [Opportunistic parasites, especially on woody plants.]
    Ph. pinodella
  [Pathogen, especially on leguminous plants; diseases are known as black stem, foot rot and leaf spot.]

With specific or preferred host:

Apocynaceae
Nerium oleander

Ph. exigua var. *heteromorpha*
[Recorded in Europe and North America: dieback, leaf necrosis.]

Vinca spp., esp. *V. minor*

Ph. exigua var. *heteromorpha* and other E⁻ strains of *Ph. exigua* ['var. *inoxydabilis*']
[Recorded in Europe and North America: stem blight and leaf spot.]

Cannabaceae
Cannabis sativa

Ph. cannabis
(teleom. *Did. cannabis*)
[Recorded in Eurasia and North America: leaf and stem spot.]

Caricaceae
Carica papaya   *Ph. caricae-papayae*
                                   [Worldwide recorded pathogen: fruit rot, trunk rot and leaf
                                   spot.]

Caprifoliaceae
Lonicera sp.   *Ph. exigua* var. *viburni*
                                   [Occasionally isolated in Europe; symptoms as with
                                   Viburnum spp.]
Sambucus nigra   *Ph. sambuci-nigrae*
                                   [Widespread in Eurasia: leaf spot, shoot dieback.]
Viburnum spp.   *Ph. exigua* var. *viburni*
                                   [Recorded in Eurasia and North America: leaf spot, stem
                                   lesions.]

Chenopodiaceae
Chenopodium quinoa   *Ph. foveata*
                                   [Recorded in Andes regions of South America: brown stalk
                                   rot; pathogen apparently via potatoes distributed in
                                   Europe.]

Compositae
Ambrosia artemisiifolia   *Ph. heliopsidis*
                                   [Host record from Canada; fungus in North America com-
                                   mon on *Heliopsis* spp.]
Artemisia spp.   *Ph. artemisiicola*
                                   [Recorded in southern Europe as cause of premature death
                                   of the plants.]
Cichorium intybus   *Ph. exigua* var. *exigua*
                                   [Frequently recorded as the cause of black root rot in
                                   Europe.]
Dahlia cultivars   *Ph. exigua* var. *exigua*
                                   [Frequently occurring in association with black rot of tubers.]
Dendranthema–Grandiflorum   *Ph. ligulicola* var. *ligulicola*
hybrids (formerly known as,   (teleom. *Did. ligulicola* var. *ligulicola*)
e.g. *Chrysanthemum*   [Worldwide recorded specific pathogen of florists' chrysan-
morifolium and *C. indicum*)   themum, indigenous to Japan. It may attack all plant parts;
                                   first recorded as causal organism of ray (flowers) blight.]
Heliopsis spp.   *Ph. heliopsidis*
                                   [Common pathogen in North America, affecting leaves,
                                   stems and inflorescences; occasionally also on other com-
                                   posites.]
Lactuca sativa   *Ph. exigua* var. *exigua*
                                   [Frequently occurring in association with foot rot.]
Rudbeckia spp., esp. *R. lacina* *Ph. rudbeckiae*
                                   [Known from North America and Europe: leaf spot.]
Tanacetum (*Chrysanthemum*/   *Ph. ligulicola* var. *inoxydabilis*
Pyrethrum) *cinerariifolium*,   (teleom. *Did. ligulicola* var. *inoxydabilis*)

T. parthenium, Zinnia
violacea (*elegans*)   [Recorded in Europe and Australia in association with
                                          necroses on leaves and stems of various composites; has
                                          been confused with the type variety.]

Crassulaceae
esp. *Sedum telephium*, also   *Ph. telephii*
on other species of *Sedum*   [Common pathogen in Europe: purple blotch disease.]

Cucurbitaceae
esp. *Cucumis sativus*,   *Ph. cucurbitacearum*
C. melo, Cucurbita pepo,   (teleom. *Did. bryoniae*)
Citrullus vulgaris   [Worldwide recorded seed-borne pathogen: gummy stem
                                          blight, including a variety of symptoms as stem rot, fruit rot,
                                          vine wilt and leaf spot.]

Hydrophyllaceae
Nemophila insignis and   *Ph. nemophilae*
N. atomaria   [Well-known seed-borne pathogen in Europe and North
                                          America: damping-off and decay of stems and leaves.]

Labiatae
Mentha spp., occasionally   *Ph. strasseri*
also on other *Labiatae*, as
Monarda didyma and
Stachys officinalis   [Recorded in Eurasia, North America and New Zealand:
                                          rhizome and stem rot.]
Nepeta cataria   *Ph. nepeticola*
                                          (teleom. *Did. catariae*)
                                     [Recorded in Eurasia and North America: leaf spot, stem
                                          lesions.]

Leguminosae
Plurivorous but with   *Ph. pinodella*
preference of leguminous
plants   [Worldwide distributed pathogen: black stem, foot rot, leaf
                                          spot.]
Arachis hypogaea   *Ph. arachidicola*
                                          (teleom. *Did. arachidicola*)
                                     [Known from Africa, Asia, North and South America as
                                          causal organism of net blotch, web blotch or leaf blotch.]
Glycine max   *Ph. sojicola*
                                     [In Eurasia the most common species involved with the leaf
                                          and pod spot.]
Medicago sativa   *Ph. medicaginis* var. *medicaginis*
                                          Ph. medicaginis var. *macrospora*
                                     [In Eurasia and North America both involved with the
                                          black stem disease; the var. *macrospora* is particularly
                                          widely distributed in the USA and Canada.]

Phaseolus vulgaris	*Ph. exigua* var. *exigua* [Worldwide recorded in association with brown specks on mature pods; also in older leaves and stems: speckle disease.] *Ph. exigua* var. *diversispora* [Recorded in Europe and East Africa as causal organism of black node disease; primary host *Vigna unguiculata*.] *Ph. exigua* var. *noackiana* [Known from South and Central America as agent of black node disease.]
Vigna unguiculata	*Ph. exigua* var. *diversispora* [In Africa commonly associated with symptoms of black node disease.]

Linaceae
Linum usitatissimum   *Ph. exigua* var. *linicola*
                                 [Well-known from Europe and New Zealand: damping-off and foot rot.]

Oleaceae
Forsythia hybrids   *Ph. exigua* var. *forsythiae*
                                 [Recorded in Europe in association with shoot blight.]
Syringa vulgaris   *Ph. exigua* var. *lilacis*
                                 [Known from Europe, North America and New Zealand: damping-off, leaf necrosis, dieback of shoots.]

Polygonaceae
Rheum spp.   *Ph. rhei*
                                 [Worldwide recorded pathogen: leaf spot.]
Rumex acetosella   *Ph. acetosellae*
                                 [In Europe and North America common on ageing leaves: leaf spot and stem necrosis.]
Rumex obtusifolius   *Ph. rumicicola*
                                 [Probably a worldwide occurring pathogen, but disease symptoms similar to those caused by a species of *Ramularia*: leaf spot.]

Polypodiaceae
Matteuccia struthiopteris   *Ph. matteucciicola*
                                 [Described from Canada: gangrene disease; also reported from ferns indigenous to Europe: *Dryopteris filix-mas* and *Blechnum spicant*: lethal effect on the gametophyte stage.]

Rosaceae
Prunus persica   *Ph. laundoniae*
                                 [Isolated from lesions on fruits in New Zealand; possibly also occurring in association with other stone fruit trees.]
Rubus idaeus   *Ph. argillacea*
                                 (teleom. *Did. applanata*)
                                 [Worldwide recorded under the teleomorphic name: cane blight or spur blight and irregular leaf necroses.]

Rubiaceae
Coffea arabica   *Ph. tarda*
                                    [Known from Africa and South America: leaf blight and
                                    stem dieback.]

Salicaceae
Populus spp., esp. *P. nigra*   *Ph. exigua* var. *populi*
and *P.* (×) *euramericana*   [Recorded in Europe as causal organism of necrotic black
                                    lesions.]
Salix sp.   *Ph. exigua* var. *populi*
                                    [Occasionally isolated in Europe, symptoms as with
                                    Populus spp.]

Scrophulariaceae
Digitalis spp., esp. *D.*   *Ph. digitalis*
 purpurea   [Recorded in Europe and New Zealand as seed-borne
                                    pathogen: leaf spot.]

Solanaceae
Capsicum annuum   '*Ph. exigua* var. *capsici*'
                                    [Recorded in China as seed-borne pathogen; invalidly pub-
                                    lished infraspecific taxon; identity doubtful, may refer to
                                    Phoma destructiva var. *diversispora*, see below under
                                    Lycopersicon esculentum.]
Solanum tuberosum   *Ph. exigua* var. *exigua*
                                    [Frequently in Eurasia occurring in association with potato
                                    gangrene.]
                                        Ph. foveata
                                    [In Europe occurring in association with potato gangrene,
                                    indigenous to Andes region of South America: primary host
                                    Chenopodium quinoa (Chenopodiaceae).]
Lycopersicon esculentum   *Ph. lycopersici*
                                    (teleom. *Did. lycopersici*)
                                    [Common pathogen in Eurasia and Africa: canker, stem
                                    and fruit rot.]
                                        Ph. destructiva var. *diversispora*
                                    [First recognized in The Netherlands: necrotic spot, on
                                    leaves, leaf stalks and stems, fruit rot; probably also else-
                                    where, see above with *Capsicum annuum*.]

Umbelliferae
Daucus carota   *Ph. exigua* var. *exigua*
                                    [Occasionally in Europe recorded in association with root
                                    rot.]

Valerianaceae
Valerianella locusta var.   *Ph. valerianellae*
oleracea, Valeriana spp.   [In Europe common seed-borne pathogen: damping-off.]

Descriptions of the Taxa

Characteristics based on study in culture. The type species and the species with a teleomorph are also described *in vivo*.

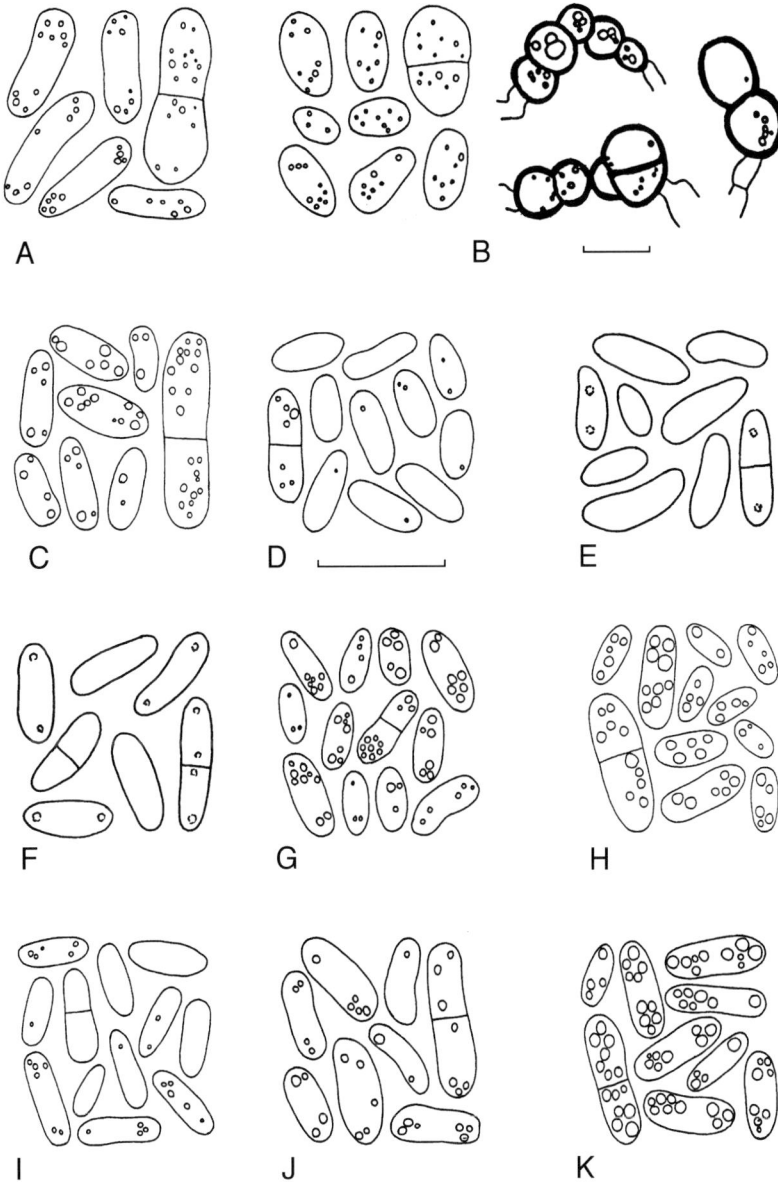

Fig. 30. Shape of conidia and some hyphal chlamydospores. **A** *Phoma acetosellae*. **B** *Phoma arachidicola*. **C** *Phoma argillacea*. **D** *Phoma artemisiicola*. **E** *Phoma cannabis*. **F** *Phoma caricae-papayae*. **G** *Phoma cucurbitacearum*. **H** *Phoma destructiva* var. *diversispora*. **I** *Phoma digitalis*. **J** *Phoma foveata*. **K** *Phoma heliopsidis*. Bars 10 μm.

Phoma acetosellae (A.L. Sm. & Ramsb.) Aa & Boerema, Fig. 30A

> *Phoma acetosellae* (A.L. Sm. & Ramsb.) Aa & Boerema apud de Gruyter,
> Boerema & van der Aa *in* Persoonia **18**(1): 16. 2002.
> > = *Phyllosticta acetosellae* A.L. Sm. & Ramsb. *in* Trans. Br.
> > mycol. Soc. **4**: 173. 1912.

Diagn. data: Boerema *et al.* (1980) (characteristics *in vitro*, conidial dimensions), van der Aa & Vanev (2002) (herb specimens, *in vivo* and *in vitro*
descriptions), de Gruyter *et al.* (2002) (fig. 2: shape of conidia, reproduced in
present Fig. 30A; cultural descriptions, partly adopted below).

Description in vitro

Pycnidia globose to subglobose with usually 1(–2) papillate ostiole(s),
90–350 μm diam., glabrous or sparsely covered by mycelial hairs. Conidial
matrix rosy buff to salmon.
    Conidia mainly aseptate, ellipsoidal to allantoid, usually with small guttules,
6.5–11.5 × 2.5–3 μm; some 1-septate conidia, up to 13 × 5 μm, may occur.
    Colonies on OA *c.* 2.5–3 cm diam. after 7 days, irregular, grey olivaceous
to olivaceous, citrine near margin, with finely floccose to woolly, white aerial
mycelium. On MA similar, but growth rate 2–2.5 cm and aerial mycelium more
compact, reverse olivaceous black to leaden black, pale luteous near margin.
Pycnidia scattered or in concentric zones, on the agar or submerged, as well as
in aerial mycelium, solitary or confluent. Representative culture CBS 631.76.

Ecology and distribution

In Europe a common fungus on ageing leaves of *Rumex acetosella*
(Polygonaceae): leaf spot, stem necroses. The fungus is also recorded in North
America (USA) and can probably be found everywhere the host occurs. For
comparison with other *Phoma*-like fungi described from *Rumex* spp., see
Boerema *et al.* (1980).

Phoma arachidicola Marasas *et al.*, Fig. 30B
Teleomorph: *Didymella arachidicola* (Khokhr.) Taber *et al.*

> *Phoma arachidicola* Marasas, Pauer & Boerema *in* Phytophylactica **6**: 200.
> 1974.

Diagn. data: Marasas *et al.* (1974) (figs 1–11; morphological characteristics *in
vivo* and *in vitro*, taxonomic position and relationships; disease symptoms),
Taber *et al.* (1984) (figs 1A–C, 2A–C, 3A–B, 4A–C, 5A, B: morphology of pycnidia and conidia *in vivo* and *in vitro*; characteristics and synonymy of teleomorph), Punithalingam (1982b) (figs A–E; CMI description sub
Didymosphaeria arachidicola (Khokhr.) Alcorn *et al.* with drawings of ascocarp,
asci, ascospores, chlamydospores and two-celled conidia as occurring *in vivo*

and on water agar with autoclaved groundnut leaf; cultural characters on potato dextrose agar, 'conidia continuous'; phytopathological notes), Phipps (1985) (characteristics *in vivo* and *in vitro*; probable long-distance dissemination by a hurricane), Rai & Rajak (1993) (cultural characteristics), Rai (1998) (effect of different physical and nutritional conditions on the morphology and cultural characters), Noordeloos *et al.* (1993) (figs 6, 7: drawing of conidia and chlamydospores; cultural description, partly adopted below; dendritic crystals consisting of 'pinodellalides A and B'), de Gruyter *et al.* (2002) (figs 1, 30: shape of conidia and chlamydospores, reproduced in present Fig. 30B; literature data on characteristics *in vivo* and *in vitro*, partly quoted below).

Description in vitro

Pycnidia (sub-)globose or ellipsoid with usually 1 ostiole, 110–180(–240) μm diam., glabrous. Conidial matrix buff.

Conidia mainly aseptate, subglobose to broadly ellipsoidal with numerous small guttules, 4–14 × 3–5 μm, commonly 5–7 × 3–4 μm; only a few 1-septate conidia, commonly 10–12 × 4–5 μm.

Chlamydospores (sub-)globose or ellipsoidal, 5–15 μm diam., usually only transversal septate, occasionally with a longitudinal septum, brownish with some relatively small guttules, single or in chains, intercalary or terminal.

Colonies on OA *c.* 3.5 cm diam. after 7 days, regular, palid, sometimes almost white but usually rosy buff or light saffron–rosy buff with white, sparse aerial mycelium; reverse similarly coloured, sometimes with cinnamon or olivaceous concentric zones that correspond with developing chlamydospores. On MA growth rate *c.* 3 cm, with compact, adpressed aerial mycelium and dark reverse; in this medium white fan-shaped or plumose, dendritic crystals are formed in 1 week (pinodellalides A and B, see Noordeloos *et al.*, 1993). Pycnidia usually in concentric rings, solitary or confluent. Representative culture CBS 315.90.

In vivo *(Arachis hypogaea)*

Pycnidia (on leaf blotches, scattered, immersed in the necrotic tissue) similar to those *in vitro*, 80–200 μm diam.

Conidia predominantly 1-septate, (7–)12–16(–17.5) × (3–)4–5(–6) μm.

Pseudothecia (not always occurring; mostly on detached leaflets, immersed in the necrotic tissue) subglobose, sometimes short beaked, (60–)70–140(–150) μm diam. Asci cylindrical to cylindrical-clavate, 37–58(–60) × 11–15(–17) μm. Ascospores more or less biseriate in the ascus, ellipsoidal, septate in the middle, upper cell wider, constricted at the septum, (12.5–)13–16 × 5–6.5(–7) μm (for detailed description and illustrations, see Punithalingam, 1982b sub *Didymosphaeria arachidicola*).

Ecology and distribution

Widespread in groundnut-growing regions of the world (*Arachis hypogaea*, Leguminosae), Africa, North and South America, Asia: net blotch, web blotch or leaf blotch of groundnuts. The disease is characterized by diffuse tan-coloured specks or streaks on the leaflets that merge to form circular, tan-coloured to dark brown blotches with greyish margins. A complete disintegration of the leaves is often the result. In Russia the anamorph erroneously has been referred to as *Ascochyta adzamethica* Schosch. and *Ascochyta arachidis* Woron., synonyms respectively of the plurivorous *Phoma exigua* Desm. var. *exigua* (this section) and *Phoma sorghina* (Sacc.) Boerema *et al.* (sect. *Peyronellaea*, **D**).

Phoma argillacea (Bres.) Aa & Boerema, Fig. 30C

Teleomorph: ***Didymella applanata* (Niessl) Sacc.**

> *Phoma argillacea* (Bres.) Aa & Boerema apud de Gruyter, Boerema & van der Aa *in* Persoonia **18**(1): 17. 2002.
> > ≡ *Phyllosticta argillacea* Bres. *in* Hedwigia **1894**: 206. 1894.
> > ≡ *Ascochyta argillacea* (Bres.) Bond.-Mont. *in* Mater. mikol. Obslêd. Ross. **5**(4): 21. 1924 [misapplied, see note].
> > H ≡ *Ascochyta argillacea* (Bres.) Grove, Br. Coelomycetes **1**: 313. 1935 [misapplied, see note].

Diagn. data (Nearly all under *Didymella applanata*): Koch (1931) (figs 1–11, tabs 1–8; history of the disease in America and Europe; disease symptoms on canes, leaves and buds; synonymy and characteristics of teleomorph, proof of connection with a *Phoma* anamorph: 'the actively parasitic stage'; description of anamorph *in vivo* and *in vitro* with table on dimensions of pycnidia and conidia), Corbaz (1957) (teleomorph description and development of *Phoma* anamorph in culture), Corlett (1974) (figs 1–6; Fungi Can. description with drawings of infected cane, ascocarp with wall cells, ascus, ascospores and conidia with conidiogenous cells; notes on the disease), Punithalingam (1982a) (figs A–F; CMI description with drawings of infected cane, ascus, ascospores, conidia, conidiogenous cells and photographs of pseudothecium and ascospores; phytopathological notes), van der Aa & Vanev (2002) (herb. specimens), de Gruyter *et al.* (2002) (fig. 3: conidial shape, reproduced in present Fig. 30C; identity of anamorph; cultural descriptions, partly adopted below).

Description in vitro

Pycnidia globose to subglobose with 1(–3) non-papillate or papillate ostioles, 40–320 μm diam., glabrous or sparsely covered by mycelial hairs around the ostiole. Conidial matrix buff to rosy buff.

   Conidia generally aseptate, ellipsoidal to allantoid, usually with small guttules, 4–7.5(–13) × 2–4 μm; some 1-septate conidia, up to 13 × 4 μm, may occur.

   Colonies on OA 2–3.5 cm diam. after 7 days, regular, olivaceous buff/pale luteous to citrine/olivaceous, with very sparse, velvety, white aerial mycelium. On MA 1.5–2.5 cm after 7 days with compact aerial mycelium, finely floccose to woolly, white to pale grey olivaceous; colony greenish olivaceous to grey olivaceous, ochraceous near margin. Pycnidia scattered on the agar or sub-merged. Representative culture CBS 205.63.

In vivo *(Rubus idaeus)*

Pycnidia (scattered on stem lesions throughout the summer and autumn, immersed in the cortex with erumpent ostioles, also scattered in necrotic lesions on leaves) similar to those *in vitro*, up to 260 μm diam.

Conidia similar to those *in vitro*, mainly aseptate, on infected canes usually 5–11 × 2–4 μm, mostly shorter, 5–8 × 3–4 μm on leaves.

Pseudothecia (on grey patches on stems late in autumn, gregarious, subepi-dermal in the cortex with erumpent ostioles, usually intermingled with pycnidia) subglobose, up to 270 μm diam. Asci cylindrical to subclavate, (50–)60–65(–75) × 10–13(–15) μm. Ascospores almost biseriate in the ascus, obovoid to oblong, septate in the middle, sometimes inequilateral, upper cell wider, constricted at the septum, (12–)13.5–16.5(–18) × (5–)5.5–7 μm (for detailed descriptions and illustrations see Punithalingam, 1982a and Corlett, 1974).

Ecology and distribution

A cosmopolitan pathogen of raspberry (*Rubus idaeus*, Rosaceae), well known under the teleomorphic name, but until recently with an unnamed *Phoma*-anamorph. The disease is called cane blight or spur blight, but leaves may also be affected, showing irregular or 'V'-shaped leaf necroses (the basionym of *Phoma argillacea* was described from such leaf necroses). The fungus is also recorded occasionally from other species of *Rubus*. The misapplications in *Ascochyta* refer to a quite different species, *A. idaei* Oudem.

Phoma artemisiicola Hollós, Fig. 30D

V   *Phoma artemisiicola* Hollós in Mat. Természettud. Közl. **35**: 40. 1926 [as 'artemisiaecola']; not *Phoma artemisiicola* M.T. Lucas & Sousa da Câmara in Agronomia lusit. **16**: 90. 1954.

Diagn. data: de Gruyter *et al.* (2002) (fig. 17: conidial shape, reproduced in present Fig. 30D; cultural descriptions, partly adopted below).

Description in vitro

Pycnidia globose to subglobose with 1(–3) non-papillate or slightly papillate ostiole(s), 80–280 μm diam., glabrous. Conidial matrix buff to rosy buff.

Conidia generally aseptate, ellipsoidal to allantoid, usually non-guttulate or with small indistinct guttules, (3–)5–6.5(–8.5) × 1.5–3 μm; some 1-septate conidia of similar size may occur.

Colonies on OA 5–7 cm diam. after 7 days, regular to irregular, fulvous/umber to dull green with a pale luteous to amber tinge, due to diffusible pigment(s); aerial mycelium sparse, felty, white to pale olivaceous grey. On MA slower growing, 3–4 cm after 7 days, with compact, finely woolly, white/buff to pale (grey) olivaceous aerial mycelium; the reverse on this medium shows a crystallization of the pigment(s) into green to yellow-green needle-like crystals. Probably anthraquinones are involved. Application of a drop of NaOH on colonies on OA and MA results in a violet/red discoloration of the pigment(s). Pycnidia scattered, on the agar or submerged. Representative culture CBS 102636.

Ecology and distribution

This fungus has been recorded on dead stems of the wild *Artemisia vulgaris* (Compositae) in Hungary and cultivated plants of *Artemisia dracunculus* in France (kitchen-garden). In the latter case the fungus was thought to be the cause of premature death of the plants. It may be that some records of *Phoma artemisiae* Auct. also refer to this species.

Phoma cannabis (L.A. Kirchn.) McPartl., Fig. 30E
Teleomorph: ***Didymella cannabis* (G. Winter) Arx**

> *Phoma cannabis* (L.A. Kirchn.) McPartl. *in* Mycologia **86**(6): 871. 1994.
> > ≡ *Depazea cannabis* L.A. Kirchn. *in* Lotos **6**: 183. 1856.
> > ≡ *Phyllosticta cannabis* (L.A. Kirchn.) Speg. *in* Atti Soc. crittogam. ital. 'II' **3**(1): 67. 1881 [as '(Lasch)'; corrected by Saccardo (1884: 53)].
> > H ≡ *Ascochyta cannabis* (L.A. Kirchn.) Voglino *in* Annali R. Accad. Agric. Torino **55**: 199. 1913 [as '(Speg.)']; not *Ascochyta cannabis* Lasch *in* Klotzschii Herb. mycol. [Ed. Rabenh.] Cent. 11, No. 1059. 1846 [≡ *Septoria cannabis* (Lasch) Sacc. (1884: 557)].
> > = *Erysiphe communis* var. *urticearum* Westend. *in* Bull. Acad. r. Sci. Lett. Beaux Arts Belg. **21**(2): 231. 1854 [powdery mildew interpretation based on the presence of spider mite webbings, see McPartland, 1994].
> > = *Diplodina parietaria* f. *cannabina* Höhn. *in* Verh. zool.-bot. Ges. Wien **60**: 314. 1910.
> > = *Diplodina cannabicola* Petr. *in* Annls mycol. **19**: 122. 1921.

Diagn. data: Röder (1937) (figs 1–5: perithecia, pycnidia and chlamydospores; single ascospore cultures showed the connection between teleomorph and anamorph; preserved culture CBS 234.37 is now sterile), McPartland (1994) (figs 1–5: asci, ascospores, pycnidia and conidia *in vivo*, the latter reproduced in present Fig. 30E; detailed descriptions, partly adopted below, nomenclature of both morphs, geographic distribution and designation of neotype anamorph), van der Aa & Vanev (2002) (note on culture CBS 234.37).

Description in vivo *(Cannabis sativa; sporulating cultures were not at our disposal)*

Pycnidia (immersed, then erumpent, in dense groups on brown-black spots on leaves and stems) globose to subglobose, (65–)130(–180) μm diam., with usually 1 distinct slightly papillate ostiole; walls made up of 3–4 layers of cells. Conidial matrix whitish to light brown.

Conidia variable, oval to ellipsoidal, sometimes biguttulate, aseptate or 1-septate, mostly 3–8 × 2–3 μm.

Chlamydospores (close to the pycnidia in host tissue, causing the dark brown discoloration) intercalary or terminal, solitary or in chains globose to oval, 8–17 μm diam.

Pseudothecia (immersed, then erumpent on dead stems) subglobose with papillate ostiole, (90–)135(–180) μm diam. Asci clavate to cylindrical, 50–85 × 9–10 μm, 8-spored, more or less biseriate. Ascospores subovoid to oblong, submedially 1-septate with two cells unequal in size, constricted at the septum, (11–)13.5(–17) × (4.5–)5.5(–8) μm.

Ecology and distribution

This pathogen is especially known as causal organism of a brown-black leaf spot of hemp, *Cannabis sativa* (Cannabaceae); but similar spots may occur on the stems. The records are from various countries in Eurasia and also from North America (USA).

Phoma caricae-papayae (Tarr) Punith., Fig. 30F

> *Phoma caricae-papayae* (Tarr) Punith. *in* Trans. Br. Mycol. Soc. **75**: 340. 1980.
>> ≡ *Ascochyta caricae-papayae* Tarr, Fungi Pl. Dis. Sudan 53. 1955.
>> *H*= *Ascochyta caricae* Pat. *in* Bull. Soc. Mycol. Fr. **7**: 178. 1891; not *Ascochyta caricae* Rabenh. *in* Bot. Z. Berl. **9**: 455. 1851.
>> ≡ *Phoma caricae* Punith. *in* CMI Descr. Pathog. Fungi Bact. 634. 1979 [as '(Pat.) Punithalingam comb. nov.'].
>> *H*= *Phoma caricina* J.C.F. Hopkins *in* Trans. Rhodesia Sc. Ass. **35**: 130. 1938; not *Phoma caricina* (Thüm.) Sacc. *in* Sylloge Fung. **3**: 164. 1884.

Diagn. data: Punithalingam (1979b) (figs 30A–D: pycnidia and conidia *in vivo*, the latter reproduced in present Fig. 30F; CMI description as *Phoma caricae* comb. nov., partly adopted below; phytopathological notes), Punithalingam (1980a) (nomenclatural note, introduction of *Phoma caricae-papayae* comb. nov.).

Description in vivo *(Carica papaya; cultures were not at our disposal)*

Pycnidia (immersed on leaves, leaf stalks and fruits) subglobose, 100–120 μm diam., with flat dark-celled ostiole; walls made up of 2–3 layers of cells.

Conidia shortly cylindrical to ellipsoidal or slightly reniform, aseptate or 1-septate, mostly (7–)8–10(–12) × 2.5–3.5 μm.

Ecology and distribution

The diseases caused by this pathogen of papaw, *Carica papaya* (Caricaceae), is known as black leathery fruit rot, trunk rot and brown or white leaf spot. In phytopathological literature the fungus was usually classi-fied in *Ascochyta*. The records are from Africa, Asia, Australasia and Oceania, Central and South America. In India it has been claimed that the pathogen produces also a *Didymella* (*Mycosphaerella*) teleomorph, but this still needs further confirmation.

Phoma cucurbitacearum (Fr. : Fr.) Sacc., Fig. 30G

Teleomorph: *Didymella bryoniae* (Auersw.) Rehm

Phoma cucurbitacearum (Fr. : Fr.) Sacc. *in* Sylloge Fung. **3**: 148. 1884 [according to Seymour, Host Index Fungi N. Am. there also should exist the combination '*Phoma cucurbitacearum* (Fr. : Fr.) Curtis', but this could not be confirmed].

≡ *Sphaeria cucurbitacearum* Fr. : Fr., Syst. mycol. **2** [Sect. 2]: 502. 1823 [type material not known to be in existence; the interpretation as anamorphic is confirmed by a collection in Schweinitz herbarium, PH (Philadelphia)].

≡ *Laestadia cucurbitacearum* (Fr. : Fr.) Sacc. *in* Sylloge Fung. **2**: xxxiii [Add. vol. **1**]. 1883 [description refers to teleomorph, basionym erroneously listed as '*Sphaeria cucurbitacearum* Schweinitz'; the collection of *Sph. cucurbitacearum* Fr. in Schweinitz herbarium, PH, is predominantly anamorphic with only a few unidentified immature ascomata].

≡ *Sphaerella cucurbitacearum* (Fr. : Fr.) Cooke *in* J. Bot., Lond. **21**: 67[–71]. 1883 [with reference to the combina-tion *Laestadia cucurbitacearum* made by Saccardo, 1884].

= *Phyllosticta cucurbitacearum* Sacc. *in* Michelia **1**(2): 145. 1878 [according to Catalogue Strains ATCC ed. 9: 127. 1970 once transferred to '*Phoma cucurbitacearum* (Sacc.) Kurata', but this could not be confirmed].

= *Phyllosticta orbicularis* Ellis & Everh. *in* J. Mycol. **4**: 10. 1888.

= *Ascochyta cucumis* Fautrey & Roum. *in* Revue mycol. **13**: 79. 1891.

≡ *Mycosphaerella cucumis* (Fautrey & Roum.) W.F. Chiu & J.C. Walker *in* J. agric. Res. **78**: 98. 1949 [name of anamorph, in spite of attribution to a teleomorphic genus: Art. 59.3].

= *Phyllosticta citrullina* Chester *in* Bull. Torrey bot. Club **18**: 374. 1891.

≡ *Ascochyta citrullina* (Chester) C.O. Sm. *in* Delaware Coll. agric. Exp. Stn Bull. **70**: 7. 1905.

≡ *Diplodina citrullina* (Chester) Grossenb. *in* Tech. Bull. N.Y. State agric. Exp. Stn **9**: 226. 1909 [as '(C.O. Smith)'].

= *Ascochyta bryoniae* Kabát & Bubák apud Bubák & Kabát *in* Sber. K. böhm. Ges. Wiss. [Math.-naturw. Kl.] **1903**(11): 3. 1904.

= *Ascochyta melonis* Potebnia *in* Annls mycol. **8**: 63. 1910.

Θ= *Ascochyta bryoniae* H. Zimm. *in* Petrak, Fl. Boh. et Morav. No. 954. 1914 [herb. name without description].

= *Diplodina cucurbitae* Nevod. apud Byzova *et al. in* Fl. spor. Rast. Kazakh. **5**(2): 319. 1968.

Diagn. data: Boerema & van Kesteren (1972) (nomenclature anamorph), Punithalingam & Holliday (1972b) (figs A–F; CMI description, anamorph, teleomorph; phytopathological notes), Boerema & van Kesteren (1981a) (confusion with *Phoma exigua* Desm. var. *exigua*), Lee *et al.* (1984) (figs 1–6, anamorph, teleomorph and conidial pleomorphism: large 1–2-septate, aseptate-macrotype, aseptate-microtype; seed infection), Corlett *et al.* (1986) (figs 1–6; Fungi Can. description, anamorph, teleomorph; notes on disease), Keinath *et al.* (1995) (morphological, pathological and genetic differentiation between *Didymella bryoniae*/anam. *Phoma cucurbitacearum* and other *Phoma* spp.), de Gruyter *et al.* (2002) (fig. 26: conidial shape, reproduced in present Fig. 30G; cultural descriptions, partly adopted below), van der Aa & Vanev (2002) (*Phyllosticta* synonyms).

Description in vitro

Pycnidia globose to irregular, with 1(–2), sometimes papillate ostiole(s), later developing into an elongated neck, 80–380 μm diam., glabrous or with mycelial outgrowths. Conidial matrix white to buff [Pseudothecia also may develop, hardly distinguishable from the pycnidia; see the description *in vivo*].

Conidia variable in shape, subglobose to ellipsoidal or allantoid, with several small guttules, usually aseptate, 4–8 × 2–3 μm, but some larger 1-septate conidia may be present, up to 10 × 4.5 μm.

Colonies on OA and MA fast growing, after 7 days the plates may be nearly full-grown; growth rate on OA already 5–7 cm diam. after 5 days, regular, colourless/dull green to olivaceous/olivaceous grey; with woolly to floccose, white to olivaceous grey aerial mycelium; reverse buff to dull green/olivaceous, to leaden grey/leaden black. Pycnidia [and pseudothecia] developing on the agar, solitary to confluent; small pycnidia also in the aerial mycelium. Representative cultures CBS 133.96, CBS 109171.

In vivo *(especially* Cucumis sativus*)*

Pycnidia (in yellow/brown lesions on stems and leaves, subepidermal, usually followed by pseudothecia; also on infected seedlings and in dark cracked sunken lesions on fruits) subglobose to flattened ellipsoidal with a distinct ostiole, 120–190 μm diam.

Conidia extremely variable in size and septation. Sometimes they are mostly aseptate with some 1-septate and a few 2-septate, but usually they are mostly 1(–2)-septate, with a small percentage unicellular. The dimensions are commonly (6–)8–10(–13) × (2.5–)3–4(–5) μm, but the septate ones can be larger, up to 20–24 μm (ascochytoid). Pycnidia on seed coats usually contain only small aseptate conidia, (3.5–)4–8(–8.5) × 2–3 μm, thus resembling those *in vitro*.

Pseudothecia (in stems, leaves and fruits, subepidermal, together with pycnidia) globose to subglobose with somewhat conical neck, (125–)140–200(–215) μm diam. Asci cylindrical to subclavate, (50–)60–70(–90) × (9–)10–15(–13) μm, 8-spored, biseriate. Ascospores (13–)14–18 × 4–6(–7) μm, ellipsoidal to nearly obovoid with rounded ends, 1-septate, faintly guttulate (for more detailed descriptions and illustrations see Punithalingam & Holliday, 1972b and Corlett *et al.*, 1986).

Ecology and distribution

A cosmopolitan seed-borne pathogen of Cucurbitaceae, especially cucumber, *Cucumis sativus*, melon and muskmelon, vars of *Cucumis melo*, pumpkin, *Cucurbita pepo*, and watermelon, *Citrullus vulgaris*. The disease, known as gummy stem blight, includes a variety of symptoms which are referred to as leaf spot, stem canker, vine wilt and black fruit rot. The name of the disease refers to the gummy exudate on stems and fruit lesions. The fungus occasionally also has been isolated from plants of other families (e.g. Solanaceae, Caricaceae and Primulaceae). The cosmopolitan distribution of the fungus may explain the recorded variation in pathogenicity and the extreme conidial variability of the anamorph *in vivo*. This extreme variability could also explain why *Phoma cucurbitacearum* has often been confused with other species of *Phoma*, such as *Phoma exigua* Desm. var. *exigua* (this section) and *Phoma herbarum* Westend. (sect. *Phoma*, **A**).

Phoma destructiva var. *diversispora* Gruyter & Boerema, Fig. 30H

> *Phoma destructiva* var. *diversispora* Gruyter & Boerema apud de Gruyter, Boerema & van der Aa *in* Persoonia **18**(1): 28. 2002.
>> [‡ = ? *Phoma exigua* var. *capsici* L.Z. Liang *in* Acta Microbiol. sin. **31**(2) : 161. 1991 (not validly published, no Latin description and type indication, Arts 36, 37)].

Diagn. data: Dorenbosch & van Kesteren (1978) (fig. 1; first description and illustration of the disease symptoms caused by this fungus on tomato leaves), de Gruyter *et al.* (2002) (fig. 9: conidial shape, reproduced in present Fig. 30H;

description as separate variety of *Phoma destructiva* Plowr. on account of conidial and genetical characteristics; cultural descriptions, partly adopted below).

Description in vitro

Pycnidia globose to irregular with 1–3 papillate ostiole(s), 90–260 μm diam., glabrous. Conidial matrix rosy buff to rosy vinaceous.

Conidia mainly aseptate, subglobose to ellipsoidal or allantoid, with several distinct guttules, 4–7(–9) × 1.5–2.5(–3) μm; a number of larger 1-septate conidia are always produced, 8.5–11.5 × 2–3.5 μm.

Colonies on OA 4.5–5 cm after 7 days, regular, dull green at centre, with (finely) floccose, olivaceous grey aerial mycelium. On MA somewhat faster growing, 5–5.5 cm after 7 days, with compact woolly, pale olivaceous grey aerial mycelium; reverse dull green to olivaceous buff near margin, leaden grey to olivaceous black at centre. Pycnidia abundant, scattered or obviously concentrically zoned, both on and in the agar and in aerial mycelium, solitary or confluent. Representative culture CBS 162.78.

Ecology and distribution

Since 1977 this fungus has been frequently recorded on tomato crops, *Lycopersicon esculentum* (Solanaceae), in glasshouses in The Netherlands (compare Boerema & van Kesteren, 1980). It causes light brown necroses on leaves, leaf stalks and stems, with dark pycnidia often in concentric rings: necrotic spot. It may also cause fruit rot. This pathogen demonstrates the close relationship between sections *Phoma* and *Phyllostictoides*; see also the Note under *Phoma medicaginis* var. *medicaginis* (this section). Typical isolates of *Ph. destructiva* Plowr. var. *destructiva* [fruit rot and foliar lesions of tomato and paprika, *Capsicum annuum*; apparently common in (sub-)tropical regions and probably of American origin] produce pycnidia with only aseptate conidia and therefore must be classified in sect. *Phoma* (**A**). Although similar in cultural characters, isolates of *Ph. destructiva* var. *diversispora* always produce in addition to aseptate conidia, a number of somewhat larger 1-septate conidia. The subspecific classification is supported by AFLP studies (Abeln *et al.* 2002), the typestrain of *Ph. destructiva* var. *diversispora* was genetically different from two typical strains of var. *destructiva*, but they all clearly belonged to one cluster. It is plausible that var. *diversispora*, just like the type variety also may attack *Capsicum annuum*. The cultural description of *Ph. exigua* var. *capsici* (Liang, 1991), frequently isolated from seed of *C. annuum* in Beijing, China, shows much resemblance to the characteristics of *Ph. destructiva* var. *diversispora* (van der Aa *et al.*, 2000: 453).

Phoma digitalis Boerema, Fig. 30I

Phoma digitalis Boerema apud Boerema & Dorenbosch *in* Versl. Meded. plziektenk. Dienst Wageningen **153** (Jaarb. 1978): 19[–20]. 1979.

≡ *Ascochyta molleriana* G. Winter *in* Bolm Soc. broteriana
1883 [= Contr. Fl. mycol. Lusit. V]: 26. 1884; not *Phoma
molleriana* (Thüm.) Sacc. *in* Sylloge Fung. **3**: 110. 1884 [≡
Ceuthospora molleriana (Thüm.) Petr.].

Diagn. data: Boerema & Dorenbosch (1979) (taxonomic position, confusing
with *Ascochyta digitalis* (Fuckel) Fuckel), Allescher (1898–1901) (p. 641, characteristics *in vivo*), Ondřej (1970) (tab. 5, 2, conidial dimensions *in vivo* and *in
vitro*), de Gruyter *et al.* (2002) (fig. 10: conidial shape, reproduced in present
Fig. 30I; cultural descriptions, partly adopted below).

Description *in vitro*

Pycnidia globose/subglobose to irregular, with usually 1 indistinct non-papillate
or papillate ostiole, relatively small, 40–120 μm diam. Conidial matrix buff.
    Conidia generally aseptate, ellipsoidal to allantoid, with or without small
guttules, (4–)5–7(–8.5) × (1.5–)2–3(–3.5) μm; some 1-septate conidia, up to
10 μm, may occur.
    Colonies on OA 4.5–5 cm after 7 days, regular, colourless to (rosy) buff or
dull green to olivaceous, with finely floccose white aerial mycelium. On MA
slower growing, 3.5–4 cm after 7 days; reverse apricot to saffron, dull green to
hazel/olivaceous at centre, salmon near margin. Pycnidia scattered, on the agar
or submerged, as well as in aerial mycelium, solitary or confluent.
Representative cultures CBS 229.79, CBS 109180.

Ecology and distribution

Widespread on *Digitalis* spp. (Scrophulariaceae), especially *D. purpurea* in
Europe: leaf spot, mainly seed-borne. The fungus is also found in New Zealand
and probably occurs everywhere on the host. Often erroneously identified in
old literature as *Ascochyta digitalis* (Fuckel) Fuckel (= *Ramularia* sp. cf. type).
The conidia of *Ph. digitalis in vivo* are mostly 9.5–12 × 3.5 μm and initially
always aseptate.

Phoma exigua Desm. var. *exigua*, Fig. 29, Colour Plate I-A, B

Phoma exigua Desm. *in* Annls Sci. nat. (Bot.) III, **11**: 282. 1849, var.
exigua [as 'Var. a'].
            ≡ *Phoma exigua* Desm. f. sp. *exigua* Malcolmson & E.G.
            Gray *in* Trans. Br. mycol. Soc. **51**: 619. 1968.
            = *Phyllosticta cytisi* Desm. *in* Annls Sci. nat. (Bot.) III, **8**: 34.
            1847; not *Phoma cytisi* Sacc. *in* Sylloge Fung. **2**: 258. 1883.
            = *Sphaeria (Depazea) vagans* subsp. *malvaecola* Fr., Syst.
            mycol. **2** [Sect. 2]: 532. 1823 [as 'd. *Malvaecola*'].
        V = *Phyllosticta destructiva* Desm. *in* Annls Sci. nat. (Bot.) III,
            8: 29. 1847, var. *destructiva* [as 'var. *malvarum*']; not

Phoma destructiva Plowr. *in* Gdnrs' Chron. II [New Series],
16: 620. 1881 (sect. *Phoma*, **A**).

H ≡ *Ascochyta destructiva* (Desm.) Höhn. *in* Hedwigia **60**:
165. 1919; not *Ascochyta destructiva* Kabát & Bubák *in*
Sber. K. böhm. Ges. Wiss. [Math.-naturw. Kl.]
1903(11): 4. 1904 [synonym listed below].

‡ ≡ *Peyronellaea destructiva* (Desm.) Goid. *in* Atti Accad.
naz. Lincei Rc. VIII, **1**: 455. 1946 [genus then not valid,
Art. 43].

Θ = *Phyllosticta destructiva* var. *lycii* Desm. *in* Annls Sci. nat.
(Bot.) III, **8**: 29. 1847 [without separate description].

H ≡ *Ascochyta lycii* (Desm.) ex Höhn. *in* Hedwigia **60**:
165–166. 1919; not *Ascochyta lycii* Sacc., E. Bommer &
M. Rousseau *in* Bull. Soc. r. Bot. Belg. **26**: 220. 1887
[synonym listed below].

= *Phyllosticta sambuci* Desm. *in* Annls Sci. nat. (Bot.) III, **8**:
34. 1847; not *Phoma sambuci* Pass. *in* J. Hist. nat.
Bordeaux **1885**: 135. 1885.

= *Phoma westendorpii* Tosquinet apud Westendorp *in* Bull.
Acad. r. Sci. Lett. Beaux-Arts Belg. II, **2**: 564. 1857.

≡ *Phyllosticta nupharis* Allesch. *in* Rabenh. Krypt.-Flora [ed.
2], Pilze **6** [Lief. 61]: 133. 1898 [vol. dated '1901']; not
Phyllosticta westendorpii Thüm. *in* Byull. mosk. Obshch.
Ispȳt. Prir. **55**: 229 [= Pilzflora Sibiriens : 304]. 1880.

= *Phyllosticta cynarae* Westend. *in* Bull. Acad. r. Sci. Lett.
Beaux-Arts Belg. II, **2**: 568. 1857.

H ≡ *Ascochyta cynarae* (Westend.) H. Zimm. *in* Verh. naturf.
Ver. Brünn **52**: 100. 1913; not *Ascochyta cynarae* Died.
in Krypt.-Flora Mark Brandenb. 9, Pilze **7** [Heft 2]: 381.
1912 [vol. dated '1915'] [synonym listed below].

Θ = *Phyllosticta daturae* Westend., Herbier L. Pire, 1860 [herb.
name without description].

= *Phoma dilleniana* Rabenh. *in* Fungi europ. exs./Klotzschii
Herb. mycol. Cont. [Ed. Rabenh.] Cent. 10, No. 960.
1866; *in* Hedwigia **5**: 192. 1866.

Θ = *Phyllosticta destructiva* f. *althaeae-rosae* Thüm. *in* Herb.
mycol. oecon. Fasc. 8, No. 369. 1875; *in* Mycoth. univ.
Cent. 13, No. 1299. 1879 [herb. name without description].

V Θ = *Phoma herbarum* f. *stramonii* Thüm. *in* Mycoth. univ. Cent.
7, No. 677. 1877 [as 'strammonii'; without description].

= *Phyllosticta cytisella* Sacc. *in* Michelia **1**(2): 17. 1878.

H = *Phyllosticta coryli* Sacc. & Speg. apud Saccardo *in* Michelia
1(2): 138. 1878; not *Phyllosticta coryli* Westend. *in* Bull.
Acad. r. Sci. Lett. Beaux-Arts Belg. **19**(9): 121. 1851 [=
Asteroma coryli (Fuckel) B. Sutton, Coelomycetes : 496.
1980].

= *Phyllosticta calycanthi* Sacc. & Speg. *in* Michelia **1**(2): 139. 1878.

= *Phyllosticta capparidis* Sacc. & Speg. *in* Michelia **1**(2): 139. 1878.

= *Phyllosticta sonchi* Sacc. *in* Michelia **1**(2): 141. 1878.

  ≡ *Ascochyta sonchi* (Sacc.) Grove *in* J. Bot., Lond. **60**: 48. 1922.

= *Phyllosticta althaeina* Sacc. *in* Michelia **1**(2): 143. 1878.

= *Phyllosticta alismatis* Sacc. & Speg. *in* Michelia **1**(2): 144. 1878.

  ≡ *Ascochyta boydii* Grove *in* J. Bot., Lond. **56**: 315. 1918; not *Ascochyta alismatis* Trail *in* Scott. Nat. **3**: 188. 1887 and not *Ascochyta alismatis* Ellis & Everh. *in* J. Mycol. **1889**: 148. 1889.

= *Phyllosticta phaseolina* Sacc. *in* Michelia **1**(2): 149. 1878.

H = *Phyllosticta filipendulae* Sacc. & Speg. *in* Michelia **1**(2): 150. 1878; not *Phyllosticta filipendulae* Sacc. *in* Michelia **1**(2): 145. 1878.

  ≡ *Phyllosticta filipendulina* Sacc. & Speg. *in* Michelia **1**(2): corrigenda after p. 275. 1878 [replaced name].

= *Phyllosticta chenopodii* f. *chenopodii-polyspermi* Sacc. *in* Michelia **1**(2): 150. 1878.

H ≡ *Phyllosticta chenopodii* Sacc. *in* Sylloge Fung. **3**: 55. 1884 [name change]; not *Phyllosticta chenopodii* Westend. *in* Bull. Acad. r. Sci. Lett. Beaux-Arts Belg. II, **2**: 567. 1857 [= *Phoma heteromorphospora* Aa & Kesteren, sect. *Heterospora*, **B**].

= *Phyllosticta glechomae* Sacc. *in* Michelia **1**(2): 151. 1878.

  ≡ *Ascochyta glechomae* (Sacc.) Baudyš & Picbauer *in* Acta Soc. Sci. nat. moravo-siles. **3**(2): 30. 1926.

= *Phyllosticta gomphrenae* Sacc. & Speg. *in* Michelia **1**(2): 151. 1878.

= *Phyllosticta lappae* Sacc. *in* Michelia **1**(2): 151. 1878.

H ≡ *Ascochyta lappae* (Sacc.) Jaap *in* Annls mycol. **12**: 26. 1914; not *Ascochyta lappae* Kabát & Bubák *in* Hedwigia **47**: 357. 1908 [synonym listed below].

= *Phyllosticta erythraeae* Sacc. & Speg. *in* Michelia **1**(2): 152. 1878.

= *Phyllosticta capsulicola* Sacc. & Speg. *in* Michelia **1**(2): 152. 1878.

= *Phyllosticta celosiae* Thüm. *in* Jorn. Sci. math. phys. nat. I, **6**(24): 230. 1878.

= *Phyllosticta tropaeoli* Sacc. & Speg. *in* Michelia **1**(2): 152. 1878.

  ≡ *Ascochyta tropaeoli* (Sacc. & Speg.) Bond.-Mont. *in* Notul. syst. Sect. cryptog. Inst. bot. [V.L. Komarov.] Acad. Sci. USSR **4**: 42. 1938.

= *Ascochyta malvicola* Sacc. *in* Michelia **1**(2): 161. 1878.

= *Ascochyta daturae* Sacc. *in* Michelia **1**(2): 163. 1878.

= *Ascochyta phaseolorum* Sacc. *in* Michelia **1**(2): 164. 1878.

= *Ascochyta potentillarum* Sacc. *in* Michelia **1**(2): 170. 1878.

= *Phoma herbarum* f. *absinthii* Sacc. *in* Michelia **1**(5): 523. 1879.

= *Phoma herbarum* f. *marrubii* Sacc. *in* Michelia **1**(5): 523. 1879.

= *Phoma herbarum* f. *urticae* Sacc. *in* Michelia **1**(5): 523. 1879.

= *Phyllosticta verbasci* Sacc. *in* Michelia **1**(5): 531. 1879.

= *Phoma herbarum* f. *calystegiae* Sacc. *in* Michelia **2**(1): 93. 1880.

= *Phoma herbarum* f. *dahliae* Sacc. *in* Michelia **2**(1): 93. 1880.

= *Phoma herbarum* f. *foeniculi* Sacc. *in* Michelia **2**(1): 93. 1880.

= *Phoma herbarum* f. *humuli* Sacc. *in* Michelia **2**(1): 92. 1880; not *Phoma herbarum* f. *humuli* Gonz. Frag. *in* Trab. Mus. nac. Cienc. nat., Madr., Ser. bot. **12**: 30. 1917 [= *Phoma herbarum* Westend., sect. *Phoma*, **A**].

= *Phoma herbarum* f. *medicaginis* Sacc. *in* Michelia **2**(1): 93. 1880; not *Phoma herbarum* f. *medicaginum* Westend. ex Fuckel *in* Jb. nassau. Ver. Naturk. **23–24** [= Symb. mycol.]: 134. 1870 ['1869 und 1870'] [= *Phoma medicaginis* var. *macrospora* Boerema *et al.*, this section].

≡ *Phoma herbarum* f. *medicaginea* Sacc. *in* Sylloge Fung. **3**: 133. 1884 [name change].

= *Phoma herbarum* f. *phytolaccae* Sacc. *in* Michelia **2**(1): 93. 1880.

= *Phoma herbarum* f. *schoberiae* Sacc. *in* Michelia **2**(1): 93. 1880.

= *Phyllosticta eupatorina* Thüm. *in* Hedwigia **19**: 179. 1880.

= *Phyllosticta juliae* Speg. *in* An. Soc. cient. argent. **10**: 28. 1880.

= *Ascochyta nicotianae* Pass. *in* Atti Soc. crittogam. ital. **3**(1): 14. 1881.

= *Phoma herbarum* f. *antirrhini* Sacc. *in* Michelia **2**(2): 337. 1881.

= *Phoma herbarum* f. *aristolochiae-siphonis* Sacc. *in* Michelia **2**(2): 337. 1881.

= *Phoma herbarum* f. *dipsaci* Sacc. *in* Michelia **2**(2): 337. 1881.

= *Phoma herbarum* f. *mercurialis* Sacc. *in* Michelia **2**(2): 337. 1881.

= *Phoma herbarum* f. *solidaginis* Sacc. *in* Michelia **2**(2): 337. 1881.

= *Phyllosticta hortorum* Speg. *in* Atti Soc. crittogam. ital. **3**: 67. 1881.

≡ *Ascochyta hortorum* (Speg.) C.O. Sm. *in* Bull. Del. Univ. agric. Exp. Stn **63**: 19–23. 1904.

= *Phoma niesslii* Sacc. *in* Michelia **2**(3): 618. 1882.

V = *Phyllosticta hualtata* Speg. *in* An. Soc. cient. argent. **13**: 11. 1882 [as '*hualtatae*'].

= *Phyllosticta gillesii* Speg. *in* An. Soc. cient. argent. **13**: 11. 1882.

= *Phyllosticta orontii* Ellis & G.W. Martin *in* Am. Nat. **16**: 1002. 1882.

V = *Phyllosticta solani* Ellis & G.W. Martin *in* Am. Nat. **16**: 1002. 1882 [as '*iolani*'; spelling corrected *in* Am. Nat. **17**: 1166. 1883].

V = *Phyllosticta stigmatophylli* Speg. *in* An. Soc. cient. argent. **13**: 12. 1882 [as '*stigmaphylli*'].

= *Phyllosticta decidua* Ellis & Kellerman *in* Am. Nat. **17**: 1165. 1883.

H = *Phyllosticta chenopodii* Sacc. *in* Sylloge Fung. **3**: 55. 1884; not *Phyllosticta chenopodii* Westend. *in* Michelia **1**(2): 150. 1878 [= *Phoma heteromorphospora* Aa & Kesteren, sect. *Heterospora*, **B**].

= *Phyllosticta filipendulina* var. *ulmariae* Sacc. *in* Sylloge Fung. **3**: 41. 1884.

= *Phoma herbarum* f. *brassicae* Sacc. *in* Sylloge Fung. **3**: 133. 1884.

= *Phoma herbarum* f. *euphrasiae* Sacc. *in* Sylloge Fung. **3**: 133. 1884.

= *Phoma herbarum* f. *helichrysi* Sacc. *in* Sylloge Fung. **3**: 133. 1884.

= *Phoma herbarum* f. *hyoscyami* Sacc. *in* Sylloge Fung. **3**: 133. 1884.

= *Phoma herbarum* f. *lactucae* Sacc. *in* Michelia **1**(5): 523. 1879 [without description] ex Sacc. *in* Sylloge Fung. **3**: 133. 1884.

= *Phoma herbarum* var. *lappae* P. Karst. *in* Meddn Soc. Fauna Flora fenn. **11**: 141. 1884.

= *Phoma herbarum* var. *thulensis* P. Karst. *in* Hedwigia **23**: 39. 1884.

= *Ascochyta althaeina* Sacc. & Bizz. apud Saccardo *in* Atti Ist. veneto Sci. VI, **2**: 244. 1884; *in* Sylloge Fung. **3**: 399. 1884.

= *Phoma alcearum* Cooke *in* Grevillea **13**: 94. 1885.

= *Phyllosticta potamia* Cooke *in* Grevillea **14**: 39. 1885.

V = *Phyllosticta penstemonis* Cooke *in* Grevillea **14**: 90. 1885 [as '*pentastemonis*'].

= *Phyllosticta altheicola* Pass. *in* J. Hist. nat. Bordeaux **1885**: 54. 1885; *in* Act. Soc. linn. Bordeaux **40**: 66. 1886.

= *Phyllosticta cytisorum* Pass. apud Brunaud *in* Revue mycol. **8**: 139. 1886.

= *Ascochyta althaeina* var. *brunneo-cincta* Pass. apud Brunaud *in* Revue mycol. **8**: 141. 1886.

= *Phyllosticta mentzeliae* Ellis & Kellerm. *in* J. Mycol. **2**: 4. 1886.

= *Phyllosticta ivaecola* Ellis & Everh. *in* J. Mycol. **2**: 37. 1886.

Θ = *Phoma herbarum* f. *datiscae-cannabinae* Berl. & Roum. *in* Revue mycol. **9**: 178 ['162']. 1887 [without description].

Θ = *Phoma herbarum* f. *eupatorii-sessilifolii* Berl. & Roum. *in* Revue mycol. **9**: 178 ['162']. 1887 [without description].

Θ = *Phoma herbarum* f. *sempervivi-tectorum* Berl. & Roum. *in* Revue mycol. **9**: 178 ['162']. 1887 [without description].

Θ = *Phoma herbarum* f. *solani-nigricantis* Berl. & Roum. *in* Revue mycol. **9**: 178 ['162']. 1887 [without description].

Θ = *Phoma herbarum* f. *verbenae-paniculatae* Berl. & Roum. *in* Revue mycol. **9**: 178 ['162']. 1887 [without description].

= *Ascochyta lycii* Sacc., E. Bommer & M. Rousseau *in* Bull. Soc. r. Bot. Belg. **26**: 220. 1887.

V = *Phyllosticta dahliicola* Brunaud *in* Bull. Soc. bot. Fr. **34**: 429. 1887 [as '*dahliaecola*'].

 ≡ *Ascochyta dahliicola* (Brunaud) Petr. *in* Annls mycol. **25**: 202. 1927.

= *Phyllosticta fatiscens* Peck *in* Rep. N.Y. St. Mus. nat. Hist. **40**: 58. 1887.

V = *Phyllosticta sagittifoliae* Brunaud *in* Revue mycol. **9**: 13. 1887 [as '*sagittaefoliae*'].

= *Phyllosticta zahlbruckneri* Bäumler, Beitr. KryptogFlora Pressburger Comitates **1**: 7. 1887.

= *Phyllosticta antennariae* Ellis & Everh. *in* J. Mycol. **5**: 9. 1888.

= *Phyllosticta coniothyrioides* Sacc. *in* Malpighia **2**: 19. 1888.

= *Phyllosticta coronaria* Pass. *in* Atti R. Accad. naz. Lincei Rc. IV, **4**: 65. 1888.

= *Phoma herbarum* f. *parietariae* Brunaud *in* Bull. Soc. bot. Fr. **36** [II, **11**]: 338. 1889.

= *Phyllosticta calaminthae* Ellis & Everh. *in* J. Mycol. **5**: 145. 1889.

= *Phyllosticta hydrangeae* Ellis & Everh. *in* J. Mycol. **5**: 145. 1889.

H ≡ *Ascochyta hydrangeae* (Ellis & Everh.) Aksel *in* Trudÿ bot. Inst. Akad. Nauk SSSR II, **11**: 83. 1956; not *Ascochyta hydrangeae* M.A. Arnaud & G. Arnaud, listed below [1924].

= *Phyllosticta orontii* var. *advena* Ellis & Everh. *in* J. Mycol. **5**: 146. 1889.

= *Phoma solanicola* Prill. & Delacr. *in* Bull. Soc. mycol. Fr. **6**: 179. 1890.

≡ *Phoma exigua* var. *solanicola* (Prill. & Delacr.) Popkova,
Malikova & Kovaleva, Abstr. Pap. int. Congr. Pl. Path. 2
(Minneapolis, MN) 0572. 1973; *in* Arch. Phytopath.
PflSchutz **10**: 94. 1974 [as '*Phoma exigua* var. *solani-
cola* Prill. & Delacr.'].

= *Phyllosticta molluginis* Ellis & Halsted *in* J. Mycol. **6**: 33.
1890.

= *Ascochyta althaeina* var. *major* Brunaud *in* Annls Soc. Sci.
nat. Charente-Infér. **26**: 62. 1890.

V = *Phyllosticta dioscoreicola* Brunaud *in* Annls Soc. Sci. nat.
Charente-Infér. **26**: 62. 1890 [as '*dioscoreaecola*'].

= *Phyllosticta otites* Brunaud *in* Act. Soc. linn. Bordeaux **44**
[V, **4**]: 242. 1890.

= *Phoma lyndonvillensis* Fairman *in* Proc. Rochester Acad.
Sci. **1**: 51. 1891.

= *Ascochyta parasita* Fautrey *in* Revue mycol. **13**: 79. 1891.

= *Phyllosticta petasitidis* Ellis & Everh. *in* Proc. Acad. nat. Sci.
Philad. **1891**: 76. 1891.

= *Phyllosticta senecionis-cordati* Allesch. *in* Ber. bot. Ver.
Landshut **12**: 13. 1892.

= *Phyllosticta alismatis* f. *santonensis* Brunaud *in* Annls Soc.
Sci. nat. Charente-Infér. **29**: 99. 1893.

= *Phyllosticta dircae* Ellis & Dearn. *in* Can. Rec. Sci. **5**: 267.
1893.

= *Phyllosticta melampyri* Allesch. *in* Hedwigia **33**: 70. 1894.

= *Phoma periplocae* Brunaud *in* Bull. Soc. Sci. nat. Ouest Fr.
4: 8. 1894.

≡ *Phyllosticta periplocae* (Brunaud) Allesch. *in* Rabenh.
Krypt.-Flora [ed. 2], Pilze **6** [Lief. 59]: 63. 1898 [vol.
dated '1901'].

= *Phyllosticta orbicula* Ellis & Everh. *in* Proc. Acad. nat. Sci.
Philad. **1894** [vol. **46**]: 455. 1895.

≡ *Coniothyrium orbicula* (Ellis & Everh.) Keissl. *in* Annln
naturk. Mus. Wien **35**: 21. 1922 [misinterpretation].

= *Ascochyta pyrina* Peglion *in* Malpighia **8**: 444. 1895 [not
'(Fr.) Pegl.' as listed by Wollenweber *in* Z. ParasitKde **8**:
601. 1936].

= *Phyllosticta alpina* Allesch. *in* Hedwigia **34**: 257. 1895.

= *Phyllosticta eupatorii* Allesch. *in* Hedwigia **34**: 264. 1895.

= *Phyllosticta desertorum* Sacc. *in* Malpighia **10**: 272. 1896.

= *Phyllosticta helianthemicola* Allesch. *in* Ber. bayer. bot.
Ges. **4**: 31. 1896.

= *Phoma strelitziae* var. *major* Tassi *in* Atti r. Accad. Fisiocr.
Siena IV, **8**: 5. 1896.

≡ *Phyllosticta strelitziaecola* Allesch. *in* Rabenh. Krypt.-
Flora [ed. 2], Pilze **7** [Lief. 87]: 780. 1903.

= *Phyllosticta asteris* Bres. *in* Beibl.Hedwigia **36**: 157. 1897.

≡ *Ascochyta asteris* (Bres.) Gloyer *in* Phytopathology **14**: 64. 1924.

‡ ≡ *Peyronellaea asteris* (Bres.) Goid. *in* Atti Accad. naz. Lincei Rc. VIII, **1**: 455. 1946 [genus then not valid, Art. 43].

= *Phyllosticta carpathica* Allesch. apud Sydow *in* Beibl. Hedwigia **36**: 157. 1897 [often cited as 'Allescher & Sydow'].

 ≡ *Ascochyta carpathica* (Allesch.) Keissler *in* Annln naturh. Mus. Wien **35**: 21. 1922 [as '(Allescher & Sydow)'].

= *Phyllosticta hieracii* Allesch. apud Sydow *in* Beibl. Hedwigia **36**: 159. 1897.

= *Phyllosticta inulae* Allesch. apud Sydow *in* Beibl. Hedwigia **36**: 159. 1897.

= *Phyllosticta lampsanae* P. Syd. *in* Beibl. Hedwigia **36**: 159. 1897.

H = *Phoma herbarum* f. *euphrasiae* Bres. apud Bresadola & Saccardo *in* Malpighia **11**: 305. 1897; not *Phoma herbarum* f. *euphrasiae* Sacc. *in* Sylloge Fung. **3**: 133. 1884.

= *Phyllosticta pygmaea* Allesch. apud Allescher & Hennings *in* Biblthca bot. **42**: 10. 1897 [Pilze aus dem Umanakdistrict; Grönlandsexpedition].

= *Phoma herbarum* f. *nicotianae* Roum. *in* Revue mycol. **19**: 152. 1897.

= *Phyllosticta adenostylis* Allesch. *in* Rabenh. Krypt-Flora [ed. 2], Pilze **6** [Lief. 60]: 99. 1898 [vol dated '1901'].

= *Phyllosticta mimuli* Ellis & Fautrey *in* Revue mycol. **20**: 59. 1898.

= *Phyllosticta monardae* Ellis & Barthol. *in* Trans. Kans. Acad. Sci. **16**: 165. 1898.

V = *Phyllosticta penstemonis* f. *penstemonis-azurei* Allesch. *in* Rabenh. Krypt.-Flora [ed. 2], Pilze **6** [Lief. 61]: 135. 1898 [vol. dated '1901'] [as '*pentastemonis*' and '(Allescher)'].

 V ≡ *Phyllosticta penstemonis* var. *major* Allesch. apud Sydow *in* Beibl. Hedwigia **36**: 159. 1897 [as '*pentastemonis*'].

V ‡ ≡ *Phyllosticta penstemonis* var. *penstemonis-azurei* (Allesch.) Cejp *in* Nova Hedwigia **13**: 192. 1967 [as '*pentastemonis*'; invalid cf. Art. 33.2].

V = *Phyllosticta penstemonis* f. *penstemonis-hybridi* Allesch. *in* Rabenh. Krypt.-Flora [ed. 2], Pilze **6** [Lief. 61]: 135. 1898 [vol. dated '1901'] [as '*pentastemonis*'].

= *Phyllosticta stachidis* var. *arvensis* Allesch. *in* Rabenh. Krypt.-Flora [ed. 2], Pilze **6** [Lief. 61]: 151. 1898 [vol. dated '1901'].

= *Ascochyta alceina* Lambotte & Fautrey apud Fautrey *in* Bull. Soc. mycol. Fr. **15**: 153. 1899.

≡ *Diplodina alceina* (Lambotte & Fautrey) Allesch. *in* Rabenh. Krypt.-Flora [ed. 2], Pilze **7** [Lief. 88]: 881. 1903.

= *Phyllosticta datiscae* P. Syd. *in* Beibl. Hedwigia **38**: 135. 1899.

= *Phyllosticta halophila* Speg. *in* An. Mus. nac. Hist. nat. B. Aires **6**: 313. 1899.

= *Phoma herbarum* f. *cannabis* Allesch. *in* Rabenh. Krypt.-Flora [ed. 2], Pilze **6** [Lief. 64]: 330. 1899 [vol dated '1901'].

  ≡ *Plenodomus cannabis* (Allesch.) Moesz & Smarods apud Moesz *in* Bot. Közl. **38**: 70. 1941 [in Petrak, Mycoth. gen. No. 1870 erroneously distributed as '*Plenodomus cannabis* Moesz & Smarods n. spec.'].

H = *Phyllosticta periplocae* Tassi *in* Boll. R. Orto bot. Siena **2**: 144. 1899; not *Phyllosticta periplocae* (Brunaud) Allesch., see above sub *Phoma* [1894].

  ≡ *Phyllosticta tassi* Hollós *in* Magy. bot. Lap. **28**: 52. 1929.

= *Phoma solaniphila* Oudem. *in* Versl. gewone Vergad. wisen natuurk. Afd. K. Akad. Wet. Amst. **9**: 297. 1900.

H = *Phyllosticta gei* Bres. *in* Hedwigia **39**: 325. 1900; not *Phyllosticta gei* Thüm. *in* Byull. mosk. Obshch. Isp. Prir. **56**: 130. 1881.

V = *Phyllosticta mucunae* Ellis & Everh., N. Am. Phyllost.: 48. 1900 [as '*mucuneae*'].

Θ = *Phyllosticta vignae* Ellis & Everh., N. Am. Phyllost.: 60. 1900 [without description].

= *Phyllosticta nymphaeacea* Ellis & Everh., N. Am. Phyllost.: 73. 1900.

= *Phyllosticta pucciniospila* C. Massal. *in* Atti Ist. veneto Sci. **59**: 687. 1900 [also erroneously as '*Phyllosticta puccinio-phila*'].

= *Phoma crassipes* f. *foliicola* Tassi *in* Boll. Lab. Orto bot. Siena **1900**: 17. 1900.

  ≡ *Phyllosticta tassiana* Allesch. *in* Rabenh. Krypt.-Flora [ed. 2], Pilze **7** [Lief. 86]: 757. 1903.

= *Phyllosticta canescens* Ellis & Everh. *in* Bull. Torrey bot. Club **27**: 54. 1900.

= *Phyllosticta ariaefoliae* f. *ulmifolia* Bres. apud Krieger *in* Fungi saxon. No. 1632. 1901.

= *Phyllosticta aloides* Oudem. *in* Beih. bot. Zentbl. **1902**: 12. 1902; *in* Ned. kruidk. Archf III, **2**(3): 742. 1902.

‡ ≡ *Ascochyta aloides* (Oudem.) Aksel *in* Trudÿ bot. Inst. Akad. Nauk SSSR Pl. Crypt. II, **11**: 83. 1956 [without reference to basionym, not validly published, Art. 33.2].

  ≡ *Ascochyta akselae* Melnik *in* Nov. Sist. niz. Rast. **7**: 247. 1970.

V = *Phyllosticta berlesiana* var. *socialis* Ferraris *in* Malpighia **16**: 17. 1902 [as 'berleseana'].

H = *Phyllosticta stratiotis* Oudem. *in* Ned. kruidk. Archf III, **2**(3): 747. 1902; not *Phyllosticta stratiotis* Tassi *in* Boll. R. Orto bot. Siena **2**: 114. 1899.

= *Phoma catalpicola* Oudem. *in* Ned. kruidk. Archf III, **2**(3): 737. 1902.

= *Ascochyta destructiva* Kabát & Bubák apud Bubák & Kabát *in* Sber. K. böhm. Ges. Wiss. **1903**(11): 4. 1904 [not '(Desm.) Kabát & Bubák' as sometimes quoted].

= *Ascochyta montenegrina* Bubák *in* Sber. K. böhm. Ges. Wiss. **1903**(12): 13. 1904.

= *Phyllosticta montellica* Sacc. *in* Annls mycol. **3**: 512. 1905.

H = *Ascochyta lycii* Rostr. *in* Bot. Tidsskr. **26**: 311–312. 1905; not *Ascochyta lycii* Sacc., E. Bommer & M. Rousseau *in* Bull. Soc. r. Bot. Belg. **26**: 220. 1887 [synonym listed above].

= *Diplodina althaeae* Hollós *in* Annls hist.-nat. Mus. natn. hung. **4**: 344. 1906.

= *Diplodina hibisci* Hollós *in* Annls hist.-nat. Mus. natn. hung. **4**: 344. 1906.

= *Phyllosticta aricola* Bubák *in* Bull. Herb. Boissier II, **6**: 403. 1906.

= *Phyllosticta malisorica* Bubák *in* Bull. Herb. Boissier II, **6**: 404. 1906.

H = *Phyllosticta berlesiana* Sacc. *in* Annls mycol. **4**: 491. 1906; not *Phyllosticta berlesiana* Allesch. *in* Rabenh. Krypt.-Flora [ed. 2], Pilze **6** [Lief. 59]: 59. 1898 [vol. dated '1901'] [as 'berleseana'].

= *Phyllosticta scrophulariae-bosniacae* Bubák *in* Bull. Herb. Boissier II, **6**: 406. 1906.

= *Phoma herbarum* var. *daturae* Potebnia *in* Annls mycol. **5**: 14. 1907.

H = *Phyllosticta cinchonae* Koord. *in* Verh. K. Akad. Wet. Amst. [Afd. Natuurk.] Sect. 2, **13**: 203. 1907; not *Phyllosticta cinchonae* Pat. *in* Bull. trimest. Soc. mycol. Fr. **8**: 135. 1892.

= *Phyllosticta taraxaci* Hollós *in* Annls Mus. Hist. nat. Hung. **5**: 456. 1907.

≡ *Ascochyta taraxaci* (Hollós) Grove *in* J. Bot., Lond. **60**: 48. 1922.

= *Phyllosticta balsaminae* Voglino *in* Atti Accad. Sci., Torino **43** ['1907–1908']: 93. 1908.

= *Phyllosticta heterospora* Speg. *in* Revta Mus. La Plata **15**: 33. 1908.

= *Phyllosticta malkoffii* Bubák *in* Annls mycol. **6**: 24. 1908.

= *Ascochyta lappae* Kabát & Bubák *in* Hedwigia **47**: 357. 1908.

= *Phyllosticta abutilonis* Henn. *in* Hedwigia **48**: 13. 1908.
= *Phyllosticta bletiae* H. Zimm. *in* Verh. naturf. Ver. Brünn **47** ['1908']: 84. 1909.
= *Phyllosticta stachidis* var. *annua* H. Zimm. *in* Verh. naturf. Ver. Brünn **47** ['1908']: 88. 1909.
= *Ascochyta malvae* H. Zimm. *in* Verh. naturf. Ver. Brünn **47** ['1908']: 94–95. 1909.
= *Phyllosticta mulgedii* Davis *in* Trans. Wis. Acad. Sci. Arts Lett. **16**: 761. 1909.
= *Ascochyta abutilonis* Hollós *in* Annls hist.-nat. Mus. natn. hung. **7**: 53. 1909.
= *Phoma herbarum* f. *antherici* Hollós *in* Annls hist.-nat. Mus. natn. hung. **8**: 3. 1910.
= *Phyllosticta erodii* Speg. *in* An. Mus. nac. Hist. nat. B. Aires III, **20**: 334. 1910.
= *Phyllosticta sordida* Speg. *in* An. Mus. nac. Hist. nat. B. Aires III, **23**: 113. 1912.
H = *Phyllosticta lychnidis* Bondartsev *in* Izv. imp. S.-Petersb. bot. Sada **12**: 102. 1912; not *Phyllosticta lychnidis* (Kunze & J.C. Schmidt : Fr.) Ellis & Everh., N. Am. Phyllost.: 59. 1900 [probably also *Ph. exigua*].
= *Phyllosticta dzumajensis* Bubák *in* Centbl. Bakt. ParasitKde Abt. 2, **31**: 498. 1912.
= *Ascochyta cynarae* Died. *in* Krypt.-Flora Mark Brandenb. 9, Pilze **7** [Heft 2]: 381. 1912 [vol. dated '1915'].
H = *Ascochyta lycii* Died. *in* Krypt.-Flora Mark Brandenb. 9, Pilze **7** [Heft 2]: 391. 1912 [vol. dated '1915']; not *Ascochyta lycii* Sacc., E. Bommer & M. Rousseau *in* Bull. Soc. r. Bot. Belg. **26**: 220. 1887, nor *Ascochyta lycii* Rostr. *in* Bot. Tidsskr. **26**: 311–312. 1905 [both synonyms listed above].
H = *Ascochyta malvae* Died. *in* Krypt.-Flora Mark Brandenb. 9, Pilze **7** [Heft 2]: 391. 1912 [vol. dated '1915']; not *Ascochyta malvae* H. Zimm. *in* Verh. naturf. Ver. Brünn **47**: 94–95. 1908 [synonym listed above].
V ≡ *Ascochyta malvarum* Mig. *in* Thomé KryptogFlora Pilze **4**(1): 281. 1921 [as '*malvacum*'].
= *Phyllosticta bonanseana* Sacc. *in* Annls mycol. **11**: 547. 1913.
= *Phyllosticta phlomidis* Bondartsev & Lebedeva *in* Mater Mikol. Obslêd. Ross. **1**: 53. 1914.
= *Phoma herbarum* f. *verbasci* Gonz. Frag. *in* Trab. Mus. nac. Cienc. nat., Madr., Ser. bot. **7**: 38. 1914.
= *Phoma linicola* Bubák *in* Annls naturh. Hofmus., Wien **28**: 203. 1914; not *Phoma linicola* É.J. Marchal & Verpl. *in* Bull. Soc. r. Bot. Belg. **59**: 22. 1926 [= *Phoma exigua* var. *linicola* (Naumov & Vassiljevsky) P.W.T. Maas, this section]; not *Phoma linicola* Naumov *in* Mater. Mikol. Fitopat. Ross. **5**: 3. 1926 [= *Macrophomina* sp.?].

= *Phoma herbarum* var. *dulcamaricola* Bubák *in* Bot. Közl. **1915**: 63. 1915.

= *Phyllosticta menthae* Bres. *in* Annls mycol. **13**: 104. 1915.

= *Phyllosticta valerianae* A.L. Sm. & Ramsb. *in* Trans. Br. mycol. Soc. **5**: 158. 1915.

= *Phyllosticta polemonii* A.L. Sm. & Ramsb. *in* Trans. Br. mycol. Soc. **5**: 244. 1916.

= *Phoma tuberosa* Melhus, Rosenbaum & E.S. Schultz *in* J. agric. Res. **7**: 251. 1916.

= *Phyllosticta mercurialicola* C. Massal. in Saccardo, Fungi veronensis, Madonna Verona 10. 1918; also *in* Sylloge Fung. **25**: 36. 1931.

= *Phyllosticta hydrocotyles* A.L. Sm. *in* Trans. Br. mycol. Soc. **6**: 153. 1919.

= *Phyllosticta disciformis* var. *brasiliensis* Speg. *in* Bolm Acad. nac. Cienc. Córdoba **32**: 517. 1919.

= *Phyllosticta aconitina* Petr. *in* Annls mycol. **19**: 87. 1921.

= *Ascochyta carpathica* f. *caulivora* Grove *in* J. Bot., Lond. **60**: 46. 1922.

= *Ascochyta cyphomandrae* Petch *in* Ann. R. bot. Gdns Peradeniya **7**: 313. 1922.

= *Ascochyta hydrangeae* M.A. Arnaud & G. Arnaud *in* Revue Path. vég. Ent. agric. Fr. **11**: 56. 1924.

= *Ascochyta capsici* Bond.-Mont. *in* Bolez. Rast. **12**: 7. 1923.

V = *Phyllosticta sinapidis* Bond.-Mont. *in* Bolez. Rast. **12**: 70. 1923 [as 'sinapi'].

H = *Phyllosticta crotalariae* Speg. *in* An. Mus. nac. Hist. nat. B. Aires **31**: 425. 1922; not *Phyllosticta crotalariae* Sacc. & Trotter *in* Sylloge Fung. **22**: 129. 1903.

= *Phyllosticta crinodendri* Speg. *in* Revta chil. Hist. nat. **27**: 58. 1924.

V = *Phyllosticta gueldenstaedtiae* Murashk. *in* Trudÿ sib. sel'.-khoz. Akad. **5**: 2. 1925 [as 'güldenstädtiae'].

= *Ascochyta solani-tuberosi* Naumov *in* Bolez. Rast. **14**: 142. 1925.

= *Phoma hibisci* Hollós *in* Mat. természettud. Közl. [Magy. tudom. Akad.] **35**(1): 41. 1926.

= *Phyllosticta anagallidis* Hollós *in* Mat. természettud. Közl. [Magy. tudom. Akad.] **35**(1): 45. 1926.

= *Phyllosticta polygoni* Hollós *in* Mat. természettud. Közl. [Magy tudom. Akad.] **35**(1): 47. 1926.

H = *Phyllosticta bellidis* Hollós *in* Mat. természettud. Közl. [Magy. tudom. Akad.] **35**(1): 45. 1926; not *Phyllosticta bellidis* Bond.-Mont. *in* Bolez. Rast. **12**: 70. 1923 [= *Phoma bellidis* Neergaard, sect. *Phoma*, **A**].

= *Phyllosticta vernonicida* Speg. *in* Boln Acad. nac. Cienc. Córdoba **29**: 168. 1926.

= *Phyllosticta circuligerens* Tehon & E.Y. Daniels *in* Mycologia **19**: 120. 1927.

= *Phyllosticta cardaminis-amarae* Petr. *in* Annls mycol. **25**: 229. 1927.

= *Diplodina cynarae* Kill. & Maire *in* Bull. Soc. Hist. nat. Afr. N. **19**: 22. 1928.

= *Gloeosporium melleum* Dearn. & Overh. *in* Mycologia **20**: 241. 1928.

= *Phyllosticta staticis-gmelini* Lobik, Mat. mikol. flore Kumy obsled. 1925. Pyatigorsh: 38. 1928.

= *Phyllosticta suaedae* Lobik *in* Bolez. Rast. **17**: 164. 1928.

= *Phyllosticta alliicola* Lobik *in* Morbi Plantarum, Leningrad **17**: 165. 1928.

H = *Phyllosticta salviae* Lobik *in* Bolez. Rast. **17**: 169. 1928; not *Phyllosticta salviae* Hollós *in* Mat. természettud. Közl. [Magy. tudom. Akad.] **35**(1): 48. 1926.

= *Phyllosticta calycanthicola* Hollós *in* Magy. bot. Lap. **28**: 51. 1929.

H = *Phyllosticta xanthosomatis* Petr. & Cif. *in* Annls mycol. **28**: 28. 1930; not *Phyllosticta xanthosomatis* Sacc. *in* Annls mycol. **11**: 548. 1913.

= *Phyllosticta kentiae* Unamuno *in* Bolm R. Soc. esp. Hist. nat. **30**: 293. 1930.

= *Ascochyta hibisci-cannabini* Khokhr. apud Tranzschel, Gutner & Khokhrjakov *in* Trudÿ Inst. nov. Bast Raw material Vashnil, Leningrad **1**: 131. 1933.

= *Diplodina malvae* f. *lavaterae* Grove, Br. Coelomycetes **1**: 335. 1935.

V = *Phyllosticta falcatae* Ziling *in* Trudÿ bot. Inst. Akad. Nauk SSSR II, **3**: 689. 1936 [as '*falcata*'].

= *Phyllosticta balcanica* Bubák & Picbauer *in* Annls mycol. **35**: 139. 1937.

= *Phyllosticta aecidiicola* Hulea *in* Buletin şti Acad. Republ. pop. rom. [Sect. Scient.] **22**: 210. 1939.

= *Ascochyta adzamethica* Schosch. *in* Izv. gruzinsk. Opyt. Sta. Zashch. Rast., Ser. A, Phytopath. **2** [Bull. georg. Exp. Stn Pl. Prot., Ser. A Phytopath. **2**]: 272. 1940.

V = *Phyllosticta scrophulariaecola* Petr. *in* Annls mycol. **39**: 259. 1941 [as '*scrophulariicola*'].

= *Phyllosticta senecionicola* Petr. *in* Annls mycol. **39**: 257. 1941.

= *Ascochyta oro* Viégas *in* Bragantia **5**: 725. 1945.

= *Phyllosticta pogostemonis* Khokhr. *in* Bot. Mater. Otd. spor. Rast. bot. Inst. Akad. Nauk SSSR **7**: 145. 1951.

H = *Phyllosticta alternantherae* Negru, Contrtiuni bot. Cluj 63. 1958; not *Phyllosticta alternantherae* Bat. *in* Bolm Secr. Agric. Ind. Com. Est. Pernambuco **19**: 5. 1952.

Θ = *Phyllosticta calthae* Ellis & Everh. in Cash, Record Fungi Ellis **2**: 300. 1953 [herbarium name].

Θ = *Phyllosticta cynoglossi* Ellis & Everh. in Cash, Record Fungi Ellis **2**: 302. 1953 [herbarium name].

= *Phyllosticta elettariae* S. Chowdhury *in* Lloydia **21**: 152. 1958.

= *Ascochyta urenae* Sawada *in* Spec. Publs Coll. Agric. natn Taiwan Univ. **8**: 152. 1959.

= *Ascochyta sidae* Sawada *in* Spec. Publs Coll. Agric. natn Taiwan Univ. **8**: 152. 1959.

= *Phyllosticta actinidiae* Ablak. & Koval *in* Bot. Mater. Gerb. bot. Inst. V.A. Komarova **13**: 243. 1960.

= *Phyllosticta lactucae* I.E. Brezhnev *in* Bot. Mater. Otd. spor. Rast. bot. Inst. Nauk SSSR **14**: 211. 1961.

= *Phyllosticta hoveniae* Gucevič *in* Izv. Akad. Nauk armyan. SSR, Biol. Nauki **15**: 67. 1962.

V = *Phyllosticta penstemonitis* Sandu & Serea apud Sandu-Ville, Lazar, Hatmanu & Serea *in* Lucr. stiint. Inst. agron. Isasi **1962**: 92. 1962 [as '*pentastemonitis*'].

= *Phyllosticta hydrangeae-quercifoliae* M. Bechet & Crişan *in* Studii Cerc. Biol. Cluj **14**: 170. 1963.

= *Phyllosticta circaeae* Melnik *in* Nov. Sist. niz. Rast. **1965**: 149. 1965.

= *Phyllosticta gerberae* Dzhalag. *in* Nov. Sist. niz. Rast. **1965**: 156. 1965.

= *Phyllosticta myosotidicola* Nelen *in* Nov. Sist. niz. Rast. **1966**: 225. 1966.

= *Phyllosticta monardicola* Cejp *in* Ceská Mykol. **20**: 210. 1966.

= *Phyllosticta erigerontis* Tasl. *in* Mikol. Fitopatol. **1**: 109. 1967.

= *Phyllosticta armeniaca* Tasl. *in* Mikol. Fitopatol. **1**: 111. 1967.

= *Phyllosticta caucasica* Tasl. *in* Mikol. Fitopatol. **1**: 111. 1967 [March].

= *Phyllosticta arborea* Cejp *in* Nova Hedwigia **13**: 184. 1967.

H = *Phyllosticta caucasica* Cejp *in* Nova Hedwigia **13**: 187. 1967 [May]; not *Phyllosticta caucasica* Tasl., see above.

≡ *Phyllosticta cejpii* M. Morelet *in* Bull. Soc. Sci. nat. Archéol. Toulon Var **176**: 7. 1968.

= *Phyllosticta daturicola* Cejp *in* Nova Hedwigia **13**: 188. 1967.

= *Diplodina abutilonis* S. Ahmad *in* Biologia, Lahore **13**: 34. 1967.

= *Phyllosticta sisymbrii* Byzova apud Byzova *et al. in* Fl. spor. Rast. Kazakh. **5**(1): 80. 1967.

= *Phyllosticta balloteacola* Cejp apud Cejp, Dolejš & Zavřel *in* Zpr. Vlastiv. Úst. Olomouci, Cislo **143**: 2. 1969.

= *Phyllosticta torilidis* Cejp apud Cejp, Dolejš & Zavřel *in*
Zpr. Vlastiv. Úst. Olomouci, Cislo **143**: 12. 1969.

H = *Phyllosticta celosiae* Cejp *in* Mycol. Pap. **117**: 3. 1969; not
Phyllosticta celosiae Thüm. *in* Jorn. Sci. math. phys. nat. I,
6(24): 230. 1878 [synonym listed above].

= *Phyllosticta doellingeriae* Cejp *in* Zpr. Vlastiv. Úst.
Olomouci, Cislo **143**: 6. 1969.

= *Phyllosticta hypericicola* Cejp *in* Nova Hedwigia **18**: 563.
1969.

= *Phyllosticta lythri* Cejp *in* Nova Hedwigia **18**: 564. 1969.

H = *Phyllosticta macrospora* Cejp *in* Zpr. Vlastiv. Úst. Olomouci,
Cislo **143**: 6. 1969; not *Phyllosticta macrospora* Ellis &
Everh. *in* Proc. Acad. nat. Sci. Philad. **1894** [vol. **46**]: 355.
1894 [= *Fusicoccum* sp., see van der Aa & Vanev, 2002].

= *Phyllosticta rosae-sinensis* Cejp *in* Mycol. Pap. **117**: 4.
1969.

= *Phyllosticta zambiensis* Cejp *in* Mycol. Pap. **117**: 5. 1969.

= *Phyllosticta gaillardiae* Movss. *in* Nov. Sist. niz. Rast. **6**
['1969']: 197. 1970.

= *Phyllosticta pimpinellae* Cejp, Fassat. & Zavřel *in* Zpr.
Vlastiv. Úst. Olomouci, Cislo **153**: 4. 1971.

= *Phyllosticta poterii* Cejp, Fassat. & Zavřel *in* Zpr. Vlastiv.
Úst. Olomouci, Cislo **153**: 5. 1971.

= *Phyllosticta telekiae* Cejp, Fassat. & Zavřel *in* Zpr. Vlastiv.
Úst. Olomouci, Cislo **153**: 8. 1971.

H = *Phyllosticta magnoliae* Shreem. *in* Indian J. Mycol. Pl. Path.
3 ['1973']: 114. 1974; not *Phyllosticta magnoliae* Sacc. *in*
Michelia **1**(2): 139. 1878.

H = *Phyllosticta carthami* Cejp & Dolejš *in* Fol. Mus. Rer. natur.
Bohemiae occident. Plzen, Bot. **7**: 4. 1976; not *Phyllosticta
carthami* Tropova apud Khokhrjakov *in* Bolez. vred.
maslich. kult. **1**, 2 : 34. 1934.

H = *Phyllosticta coriandri* Cejp & Dolejš *in* Fol. Mus. Rer. natur.
Bohemiae occident. Plzen, Bot. **7**: 5. 1976 ; not
Phyllosticta coriandri Khokhr. *in* Bot. Mater Otd. spor. Rast.
bot. Inst. Akad. Nauk SSSR **7**: 144. 1951.

= *Phyllosticta galinsogae* Cejp & Dolejš *in* Fol. Mus. Rer.
natur. Bohemiae occident. Plzen, Bot. **7**: 6. 1976.

Θ = '*Ascochyta malvacearum* Kabát & Bubák' [in herb. Bubák,
BPI; never published herbarium name].

= '*Phoma* Stamm *S 298*' Rothweiler & Tamm *in* Experientia
22: 750–752. 1966 [prod. Phomine or Cytochalasine-B;
comp. Binder & Tamm *in* Angew. Chem. int. Edn **12**:
370–380. 1973].

Diagn. data: Boerema & Höweler (1967) (fig. 1: pycnidial shape *in vitro*; figs
2, 3: ostiole – see present Fig. 29B – and conidiogenous cells with conidia,

drawn after electron micrographs; fig. 4: variability of conidia in shape, size, septation and presence of guttules; pl. 1 upper part: type specimen in PC; pl. 2 fig. 1: electron micrograph of conidiogenesis; pl. 3 figs 1–4: agar plate cultures; pl. 4 figs 1–7: colour plate of the characteristic oxidation reaction, E→α→β, in a plate culture and in purified extract; tab. 1: résumé of the species and variety concept; tab. 2: sources of isolates; synonymy), Boerema (1969) (differentiation between variety and forma specialis), Dorenbosch (1970) (pl. 1 fig. 2: colour plate of the oxidation reaction with addition of a drop of NaOH: bluish green spot of α pigment, gradually enlarging and turning red, β pigment; pl. 4 figs 3, 4, tab. 1: characteristics *in vitro* with photographs of mycelial mat on OA and MA; isolates from soil), Boerema & Dorenbosch (1973) (pl. 1a–e: cultures on OA; tab. 1: synoptic table on characters *in vitro* with drawing of conidia; tab. 3: sources of isolates), Boerema & Bollen (1975) (pl. 19D, E, pl. 20A, pl. 21D, pl. 22A, pl. 23B, C; electron micrographs of conidiogenesis, secondary septation and germination of conidia), Boerema (1976) (tab. 3; cultural characteristics and differential criteria), Neergaard (1977) (occurrence on seed), Domsch *et al.* (1980) (fig. 281; occurrence in soil; literature references), Sutton (1980) (fig. 227C: drawing of conidia; 'known in herb. CMI from almost 200 collections'), Morgan-Jones & Burch (1988d) (fig. 1A–b: drawings of pycnidia and conidiogenous cells with conidia, see present Fig. 29A, C, portion of pycnidial wall and inflated hyphae; pls 1A–F, 2A–F: photographs of agar plate cultures, hyphae, pycnidial cluster, wall cells and conidia; history; cultural and morphological descriptions), Boerema & Hamers (1990) (phytopathological and taxonomic notes), Rai & Rajak (1993) (cultural characteristics; NB: the study sub *Ph. exigua* var. *exigua* in Rai (1998) must be questioned as it includes isolates producing brown to red pigment changing greenish blue with NaOH), van der Aa *et al.* (2000) (fig. 1A–C: shape of pycnidia, ostiole, conidiogenous cells and conidia, reproduced in present Fig. 29; synonymy; cultural description, partly adopted below; tab. 1: diagnostic features of the varieties recognized *in vitro*), Abeln *et al.* (2002) (AFLP finger prints), van der Aa & Vanev (2002) (data on hosts and type specimens of the various *Phyllosticta* synonyms listed).

Description in vitro

Pycnidia globose to subglobose or irregular with usually 1(–2) non-papillate ostiole(s), 75–200 µm diam., glabrous, solitary or confluent. Conidial matrix white to yellowish or rosy buff/salmon or rosy vinaceous.

Conidia variable in shape and dimensions, subglobose, ellipsoidal to oblong, or allantoid, usually with small guttules, mainly aseptate, (2.5–)4–7(–12) × 2–3.5 µm, but always also some 1(–2)-septate conidia (5.5–)7–10(–13) × 2.5–3.5(–5) µm.

Colonies on OA (2.5–)5–8.5 cm diam. after 7 days [on MA (2.5–)4–7.5 cm], with irregularly scalloped or lobed margin and floccose, white to pale olivaceous grey aerial mycelium, colourless or with various grey-greenish tinges, olivaceous grey to greenish olivaceous/grey olivaceous or olivaceous to olivaceous black, usually colourless, towards margin; reverse similar. Pycnidia scarce to scattered, partly submerged in the agar. In the mycelium olivaceous swollen cells may be formed, constricted at the septa.

Typical strains are characterized by the production of the colourless antibiotic metabolite E (derived from *exigua*), with the NaOH spot test demonstrable by an oxidation reaction (E$^+$ reaction): a blue green discoloration of the agar, changing to brown red. However, not all isolates show a strong reaction with NaOH[11]). Representative culture CBS 431.74.

In vivo

Pycnidia (as commonly occurring on leaves and stems of living or withering plants, usually irregularly scattered, seldom arranged concentrically, sometimes on typical leaf spots, but more often on old leaf spots caused by other organisms or on necrotic tips and borders of leaves or on irregular lesions on stems) mostly 100–200 μm diam., globose or somewhat depressed, with a distinct ostiole, lined with papillate subhyaline cells.

Conidia usually ellipsoidal with rounded ends, but also irregular in shape, aseptate and 1(–2)-septate in variable percentages, usually somewhat larger than those *in vitro*, aseptate mostly 5.5–11 × 2–4(–6) μm, 1(–2)-septate mostly 7.5–14(–16) × 2.5–4(–6) μm.

Ecology and distribution

A worldwide recorded wound and weak parasitic soil fungus, which in Eurasia has been isolated from more than 200 different host genera. The fungus is an opportunistic plant parasite which may cause necroses on leaves and stems, and may produce a rot on fleshy roots and tubers and at the bases of leaves and stems. Some examples: speckle disease of beans, *Phaseolus vulgaris* (Leguminosae; small specks and spots on older leaves, stems and pods), root rot of carrots, *Daucus carota* (Umbelliferae; see also 'Miscellaneous' under *Phoma rostrupii* Sacc.), black root rot of chicory, *Cichorium intybus* (Compositae), gangrene of potatoes, *Solanum tuberosum* (Solanaceae; see also this section under *Phoma foveata* Foister), black rot of dahlia tubers, *Dahlia* cultivars (Compositae), foot rot of lettuce, *Lactuca sativa* (Compositae).

On dying plant substrates, in the close proximity of soil, it is the most common pycnidial fungus found in Europe. The extensive synonymy (more than 300 names, which are listed above in chronological sequence) can be explained by the unlimited plurivorous character of the fungus and its extreme variability in size and septation of conidia. Notable are the many specific 'host-indicating names', formerly described in the genus *Phyllosticta*. As a producer of the notorious cytochalasins A and B (= phomine) the fungus was initially known as '*Phoma* Stamm S 298' (Boerema & Dorenbosch, 1977).

[11] Curiously, most host-specific varieties of *Ph. exigua* do not produce antibiotic E at all: less struggle for life against other fungi and bacteria? Otherwise some plurivorous wound and weak parasitic strains of *Ph. exigua* also do not produce E ('var. *inoxydabilis*', see van der Aa *et al.*, 2000: 452).

The present concept of the species separates a number of host-specific varieties, which are listed in Table 3. The subspecific taxonomy is supported by amplified fragment length polymorphism (AFLP) studies, *see* Abeln *et al.* (2002).

Phoma exigua var. *diversispora* (Bubák) Boerema

> *Phoma exigua* var. *diversispora* (Bubák) Boerema apud Boerema & van Kesteren *in* Gewasbescherming **11**: 122. 1980.
> ≡ *Phoma diversispora* Bubák *in* Öst. bot. Z. **55**: 78. 1905.

Diagn. data: Boerema (1973) (first annotation of this fungus in The Netherlands, often confused with *Ascochyta phaseolorum* = *Phoma exigua*), Boerema & van Kesteren (1980) (fig. 4: disease symptoms on bean; classification as a variety of *Ph. exigua*; occurrence on *Vigna unguiculata* in Kenya), Boerema *et al.* (1981a) (figs 1–6; *in vivo* and *in vitro* descriptions, comparison with two other bean pathogens), van der Aa *et al.* (2000) (differentiating diagnostic features within the *P. exigua* complex), Abeln *et al.* (2002) (AFLP fingerprint).

Description in vitro *(see also Table 3)*

Pycnidia similar to those of the type var. *exigua*.

Conidia, the aseptate vary more widely in width than those of the type var. *exigua* ('*diversispora*').

Chlamydospores, unicellular in short series may be found, best observed on water agar (WA), 10–25 μm diam.

Colonies on OA fast growing, 6–8.5 cm diam. after 7 days, grey olivaceous to olivaceous. On MA 3.5–6 cm, buff to grey olivaceous/olivaceous, aerial mycelium velvety to floccose/woolly, pale olivaceous grey/grey olivaceous. Reverse usually dark olivaceous black. No production of the antibiotic E. Representative culture CBS 102.80.

Ecology and distribution

The primary host of this pathogenic variety is probably the cowpea, *Vigna unguiculata* (Leguminosae), originally native of Central and West Africa. However, in Western Europe and East Africa it is particularly known as seed-borne pathogen of dwarf beans or snap beans, *Phaseolus vulgaris* (Leguminosae). The African genus *Vigna* and the American genus *Phaseolus* are closely related and generally susceptible to their mutual pathogens. Affected hosts show a black discoloration of stem nodes and petioles: black node disease.

Note

In South America similar disease symptoms on beans are caused by the related *Phoma exigua* var. *noackiana* (Allesch.) Aa *et al.*, treated below.

Table 3. Diagnostic features *in vitro* of the varieties of *Phoma exigua*.

Variety	Margin colony OA, MA	Growth rate (cm) OA	MA	E	Aerial mycelium OA	Colony OA	Reverse OA	Colony MA	Reverse MA
exigua	Irregular	(2.5–)5–8.5	(2.5–)4–7.5	+	Variable	Variable	Variable	Variable	Variable
diversispora	Regular to slightly irregular	6.5–8	3.5–6	–	Velvety to floccose/woolly, pale olivaceous grey/grey olivaceous	Grey olivaceous to olivaceous	Olivaceous to leaden grey	Buff to grey olivaceous/olivaceous black	Leaden grey to olivaceous black, or umber/honey to olivaceous
forsythiae	Regular	6.5–8.5	6.5–8.5	–	Velvety/finely floccose/woolly, white to (pale) olivaceous grey, partly tufted	Colourless to grey olivaceous	Olivaceous grey	Olivaceous grey	Leaden grey, grey olivaceous at margin
heteromorpha	Irregular	4–5	2–3	–/+	Floccose, white, or pale olivaceous grey/glaucous grey	Grey olivaceous to olivaceous grey/olivaceous black	Grey olivaceous to olivaceous grey/olivaceous black	Grey olivaceous to olivaceous grey/olivaceous black	Leaden grey/leaden black, partly grey olivaceous, sometimes salmon near margin
lilacis	Regular, on MA irregular	6–8	6–8	–	Typical compact tufts, white	Greenish olivaceous	Greenish olivaceous	Greenish olivaceous/grey olivaceous, olivaceous near margin	Olivaceous, leaden grey
linicola	Irregular	2–4.5	2–4.5	+	Velvety to floccose, white to olivaceous grey	Olivaceous grey/olivaceous black	Olivaceous grey/olivaceous black	Olivaceous grey/olivaceous black	Olivaceous grey/olivaceous black

noackiana	Regular to slightly irregular	7.5–8.5	6.5–7.5	–	Velvety to floccose, olivaceous grey, with compact tufts of white aerial mycelium	Olivaceous to iron grey, or grey olivaceous to olivaceous	Olivaceous to leaden grey/ leaden black	Greenish olivaceous to olivaceous	Leaden grey to olivaceous black, olivaceous at margin
populi	Irregular	4–6.5	2–4	–	Floccose, pale olivaceous grey to glaucous grey	Colourless to grey olivaceous at centre	Colourless to pale olivaceous grey, grey olivaceous/ olivaceous at centre	Grey olivaceous, citrine to greenish olivaceous at margin	Leaden grey to olivaceous black, greenish olivaceous/grey olivaceous near margin
viburni	Regular to slightly irregular	6.5–8.5	6.5–8.5	–	Tufted velvety to finely floccose, white to (pale) olivaceous grey	Colourless to grey olivaceous/ olivaceous grey	Olivaceous grey	Olivaceous grey, sometimes with buff to saffron	Leaden grey/grey olivaceous at margin

Phoma exigua var. *forsythiae* (Sacc.) Aa *et al.*

> *Phoma exigua* var. *forsythiae* (Sacc.) Aa, Boerema & Gruyter *in* Persoonia
> **17**(3): 452. 2000.
>> ≡ *Phyllosticta forsythiae* Sacc. *in* Michelia **1**(1): 93. 1877.
>> ≡ *Ascochyta forsythiae* (Sacc.) Höhn. apud Zimmermann *in*
>> Verh. naturf. Ver. Brünn **47**: 36. 1909.
>> H = *Ascochyta forsythiae* Died. *in* Krypt.-Flora Mark Brandenb.
>> 9, Pilze **7** [Heft 2]: 383. 1912 [vol. dated '1915'].

Diagn. data: van der Aa *et al.* (2000) (typification and differentiating diagnostic features within the *Ph. exigua* complex), Abeln *et al. (2002)* (AFLP fingerprint), van der Aa & Vanev (2002) (herb. specimens).

Description in vitro *(see also Table 3)*

Pycnidia and conidia similar to those of the type var. *exigua*.

Colonies on OA fast growing, 6.5–8.5 cm diam. after 7 days, velvety to finely floccose/woolly, partly tufted, mainly (pale) olivaceous grey aerial mycelium. No production of antibiotic E. Representative culture CBS 100354.

The cultural characteristics of this variety resemble very much those of var. *viburni* (treated below), but they are genetically different (Abeln *et al.*, 2002).

Ecology and distribution

This variety is frequently isolated from weakened and badly growing shrubs of *Forsythia* hybrids (Oleaceae) in Europe. The fungus has been found on dead leaves and may occur in association with circular leaf spots, but most characteristic are dead flower buds encircled by brown bark lesions and with discoloration of the wood.

Phoma exigua var. *heteromorpha* (Schulzer & Sacc.) Noordel. & Boerema

> *Phoma exigua* var. *heteromorpha* (Schulzer & Sacc.) Noordel. & Boerema
> *in* Versl. Meded. plziektenk. Dienst Wageningen **166** (Jaarb. 1987): 109.
> 1989 ['1988'] [erroneously without page indication of the basionym; reference added in printed Errata slip].
>> ≡ *Phoma heteromorpha* Schulzer & Sacc. *in* Hedwigia **23**:
>> 107[–108]. 1884.
>> ≡ *Ascochyta heteromorpha* (Schulzer & Sacc.) Curzi *in* Boll.
>> R. Staz. Patol. veg. **13**: [380–]399[–426]. 1933.
>> = *Phoma oleandrina* Delacr. *in* Bull. Soc. mycol. Fr. **21**:
>> [186–]190. 1905.
>> [=? *Phoma herbarum* f. *vincae* Brunaud *in* Actes Soc. Linn.
>> Bordeaux **40** [= Ser. IV, **10**]: 75. 1886.]

[=? *Phyllosticta vincae-majoris* Allesch. *in* Rabenh. Krypt.-
Flora [ed. 2], Pilze **6** [Lief. 61]: 155. 1898 [vol. dated
'1901'].]

[=? *Phyllosticta vincae-minoris* Bres. & K. Krieg *in* Hedwigia
39: 325. 1901.]

[=p.p. *Phoma exigua* var. *inoxydabilis* Boerema & Vegh apud
Vegh, Bourgois, Bousquet & Velastegui *in* Bull. trimest.
Soc. Mycol. Fr. **90**: 130. 1974.]

Diagn. data: Curzi (1933) (figs 5–7, 13 pls 12, 13 figs 1–3, 14–16, 17 figs 1–2;
morphology, taxonomy and disease symptoms on oleander, pathogenicity to
other plants), Mercier & Metay (1977) (photographs 1–8; disease symptoms,
biology and distribution, susceptibility of cultivars), Keim (1979) (fig. 1; pathogenicity on oleander and characteristics *in vitro*), Noordeloos & Boerema
(1989a) (classification as a separate pathogenic variety of *Ph. exigua*), van der
Aa *et al.* (2000) (differentiating diagnostic features within the *Ph. exigua* complex), Abeln *et al.* (2002) (AFLP fingerprint).

Description in vitro *(see also Table 3)*

Pycnidia and conidia differ from those of the type var. *exigua* by their extreme
morphological variability ('*heteromorpha*'), but always within the scope of the
Ph. exigua-concept.

Colonies on OA moderately fast growing, 4–5 cm diam. after 7 days, on
MA slower growing, 2–2.5(–3) cm diam.; always relatively dark, grey olivaceous to olivaceous black, with often white aerial mycelium. Usually no
demonstrable production of antibiotic E, but some strains showed on MA a
slight positive reaction with NaOH. Representative culture CBS 443.94.

Ecology and distribution

A noxious pathogen of oleander, *Nerium oleander* (Apocynaceae), recorded in
France, Italy and the USA, but which probably occurs everywhere the host is
commercially grown: dieback (canker), leaf necrosis. AFLP fingerprints (Abeln
et al., 2002) showed that var. *heteromorpha* belongs genetically to the same
group as var. *linicola*; the hosts of both varieties are of Mediterranean origin.

This pathogenic variety of *Ph. exigua* proved to be also involved with stem
blight and leaf spot of *Vinca* spp. (also Apocynaceae), especially very common
on *V. minor* (Vegh *et al.*, 1974). However, the *Vinca* disease may be also
induced by other E⁻ strains of *Ph. exigua* ('var. *inoxydabilis*' l.c.). See the discussion by van der Aa *et al.* (2000: 452–453) and the study of the relevant
Vinca disease by Jansen (1965).

Phoma exigua var. *lilacis* (Sacc.) Boerema

> *Phoma exigua* var. *lilacis* (Sacc.) Boerema *in* Phytopath. mediterr. **18**: 105[–106]. 1980 ['1979'].
>
> ≡ *Phoma herbarum* f. *lilacis* Sacc. *in* Michelia **2**(1): 93. 1880.

Diagn. data: Boerema (1980) (figs 1A, B, 2A, B; disease symptoms on lilac, classification of the pathogen as a variety of *Ph. exigua*), Boerema *et al.* (1984) (history of the disease in The Netherlands), van der Aa *et al.* (2000) (differentiating diagnostic features within the *Ph. exigua* complex), Abeln *et al.* (2002) (AFLP fingerprint).

Description in vitro *(see also Table 3)*

Pycnidia and conidia similar to those of the type var. *exigua*.

Colonies of this variety on OA and MA can be differentiated from var. *exigua* by a rather fast growth rate, 6–8 cm diam. after 7 days; on OA they show abundant compact tufted, white aerial mycelium covering the entire greenish olivaceous colony. No production of antibiotic E. Representative culture CBS 569.79.

Ecology and distribution

This variety is known as a pathogen of lilac, *Syringa vulgaris* (Oleaceae): Damping-off of seedlings, leaf necroses, dieback of shoots. The fungus is also found on seed capsules, which indicate the possibility of transmission by seed. It is probably not always a primary pathogen and may follow bacterial blight caused by *Pseudomonas syringae* v. Hall. Occasionally this var. *lilacis* has been isolated from necrotic tissue of a *Forsythia* hybrid (also Oleaceae). Confirmed records are from France, Germany, Italy, The Netherlands, USA and New Zealand. The fungus probably occurs wherever the host, originally native of Central and south-eastern Europe, is cultivated.

Phoma exigua var. *linicola* (Naumov & Vassiljevsky) P.W.T. Maas

> *Phoma exigua* var. *linicola* (Naumov & Vassiljevsky) P.W.T. Maas *in* Neth. J. Pl. Path. **71**: 118. 1965.
>
> ≡ *Ascochyta linicola* Naumov & Vassiljevsky *in* Mater. Mikol. Fitopat. Ross. **5**: 3. 1926.
>
> ≡ *Phoma exigua* f. sp. *linicola* (Naumov & Vassiljevsky) Malc. & E.G. Gray *in* Trans. Br. mycol. Soc. **51**: 619. 1968.
>
> H= *Phoma linicola* É.J. Marchal & Verpl. *in* Bull. Soc. r. Bot. Belg. **59**: 22. 1926; not *Phoma linicola* Bubák *in* Annln naturh. Mus. Wien **28**: 203. 1914 [= *Ph. exigua* Desm. var. *exigua*]; not *Phoma linicola* Naumov *in* Mater. Mikol. Fitopat. Ross. **5**: 3. 1926.
>
> ≡ *Phoma belgica* Cash apud Trotter *in* Sylloge Fung. **26**: 934. 1972.

= *Diplodina lini* Moesz & Smarods *in* Magy. bot. Lap. **29**:
35–38. 1930.

Diagn. data: Maas (1965) (fig. 1A–C; classification of this footrot fungus of flax
as pathogenic variety of *Ph. exigua*, documentation of synonymy and literature), Neergaard (1977) (occurrence on seed), Malone (1982) (figs 1–3; test
methods for seed infection; description and illustrations of colonies, pycnidia
and conidia), van der Aa *et al.* (2000) (differentiating diagnostic features within
the *Ph. exigua* complex), Abeln *et al.* (2002) (AFLP fingerprint).

Description in vitro *(see also Table 3)*

Pycnidia and conidia similar to those of the type var. *exigua*.

Colonies on OA relatively slow growing, 2–4.5 cm diam. after 7 days,
compact, distinct pigmented olivaceous grey to olivaceous black. NaOH
reaction is positive, production of the antibiotic E. Representative culture
CBS 116.76.

Ecology and distribution

A variety parasitizing on cultivated flax, *Linum usitatissimum* (Linaceae):
damping-off, foot rot, dead stalks. The fungus is seed-borne and probably
occurs everywhere the host is cultivated. Checked records are from East and
West Europe and New Zealand. It is occasionally also isolated from other
plants. AFLP studies (Abeln *et al.*, 2002) indicate a close relation to var. *heteromorpha*, pathogen of the Mediterranean oleander. This matches with the
fact that cultivated flax has been derived from the Mediterranean *Linum
angustifolium*.

Phoma exigua var. noackiana (Allesch.) Aa *et al.*

Phoma exigua var. *noackiana* (Allesch.) Aa, Boerema & Gruyter *in*
Persoonia **17**(3): 450. 2000.
≡ *Phyllosticta noackiana* Allesch. *in* Bolm Inst. agron.
Campinas **9**: 85. 1898.

Diagn. data: Obando-Rojas (1989) ('resembling var. *diversispora* but differing
distinctly by its enzyme composition'), van der Aa *et al.* (2000) (typification and
differentiating diagnostic features within the *Ph. exigua* complex), Abeln *et al.*
(2002) (AFLP fingerprint).

Description in vitro *(see also Table 3)*

Pycnidia and conidia similar to those of the type var. *exigua*.

Chlamydospores, unicellular, separate in series may be found, best
observed on water agar (WA), up to 20 μm diam.

Colonies on OA 7.5–8.5 cm diam. after 7 days, olivaceous to iron grey or grey olivaceous to olivaceous with compact tufts of white aerial mycelium, as well as velvety to floccose olivaceous grey aerial mycelium. On MA somewhat slower growing, 6.5–7.5 cm, colonies greenish olivaceous to olivaceous. No production of antibiotic E. Representative culture CBS 100353.

Ecology and distribution

This variety of *Ph. exigua* is repeatedly isolated from beans, *Phaseolus vulgaris* (Leguminosae) in South and Central America. The plants showed a black discoloration of stem nodes and petioles, i.e. symptoms resembling the black node disease of beans, in (eastern) Africa and (western) Europe caused by *Ph. exigua* var. *diversispora* (apparently indigenous to Africa). Both varieties are genetically different, but belong to the same group (Abeln *et al.*, 2002). It is quite possible that the var. *noackiana* has a much wider host range in America (compare the various American synonyms listed under *Ph. exigua* var. *exigua*, but this can only be established by isolates).

Note

This variety may be called an American cousin of the African *Ph. exigua* var. *diversispora*. It differs only slightly in cultural characters. Obando-Rojas (1989) proved that var. *noackiana* and var. *diversispora* are distinctly different in enzyme composition. Both varieties are genetically different but belong to the same group (Abeln *et al.*, 2002).

Phoma exigua var. *populi* Gruyter & P. Scheer

Phoma exigua var. *populi* Gruyter & P. Scheer *in* J. Phytopathology **146**: 413. 1998.

Diagn. data: Richter (1933) (first description of the disease in Germany, erroneously ascribed to *Phoma urens*), Magnani (1966, 1969) (study of the disease in Italy, also ascribed to *Phoma urens*), Man in't Veld & de Gruyter (1995) (isozyme study showed difference with *Ph. exigua* var. *exigua*), de Gruyter & Scheer (1998) (description of the pathogen as a variety of *Ph. exigua*, differences in pathogenicity and differences in susceptibility of poplar clones), van der Aa *et al.* (2000) (differentiating diagnostic features within the *Ph. exigua* complex), Abeln *et al.* (2002) (AFLP fingerprint).

Description in vitro *(see also Table 3)*

Pycnidia and conidia similar to those of the type var. *exigua*.

Colonies on OA moderately fast growing, 4–6.5 cm diam. after 7 days, on MA at room temperature slower growing, 2–4 cm, floccose, pale olivaceous grey to glaucous grey aerial mycelium; reverse colourless or with pale olivaceous grey tinges. No production of antibiotic E. Representative culture CBS 100167.

Ecology and distribution

The disease symptoms caused by this opportunistic pathogen of poplars
(Salicaceae) are called necrotic bark lesion; especially cultivars of *Populus nigra*
and *Populus* × *euramericana* proved to be susceptible. In European literature
before 1998 the disease has been ascribed to *Phoma urens* Ellis & Everh., an
American species of *Sclerophoma*. The disease symptoms resemble canker of
poplar caused by *Cryptodiaporthe populea* (Sacc.) Butin ex Butin [anam.
Chondroplea populea (Sacc.) Kleb.]. The variety *populi* has also been isolated
from a species of *Salix*. On poplars it is so far recorded in Germany, The
Netherlands and Italy.

Phoma exigua var. *viburni* (Roum. ex Sacc.) Boerema

> *Phoma exigua* var. *viburni* (Roum. ex Sacc.) Boerema apud de Gruyter &
> P. Scheer *in* J. Phytopathology **146**: 414. 1998.
>> ≡ *Ascochyta viburni* Roum. ex Sacc. *in* Sylloge Fung. **3**: 387.
>> 1884; not *Ascochyta viburni* Lasch *in* Fungi europ.
>> exs./Klotzschii Herb. mycol. Cont. [Ed. Rabenh.] Cent. 14,
>> No. 1354. 1850 [without description; = *Phoma macros-
>> toma* Mont. var. *macrostoma*, this section].
>> Θ ≡ *Phyllosticta viburni* Roum. *in* Fungi gall. exs. Cent. 21, No.
>> 2036. 1882; *in* Revue mycol. **4**: 99. 1882 [in both cases
>> without description].
>> ≡ *Phoma viburni* (Roum. ex Sacc.) Boerema & M.J. Griffin *in*
>> Trans. Br. mycol. Soc. **63**: 110. 1974.
>> V = *Phyllosticta roumeguerei* Sacc. *in* Michelia **2**(1): 88. 1880
>> [as 'Roumeguerri']; not *Phoma roumeguerei* Sacc. *in*
>> Michelia **2**(1): 89. 1880.
>> = *Phyllosticta lantanoides* Peck *in* Rep. N.Y. St. Mus. nat.
>> Hist. **38**: 94. 1885.
>> = *Phyllosticta viburnicola* Roum. *in* Revue mycol. **7**: 89.
>> 1885; not *Phoma viburnicola* Oudem. [sect. *Phoma*, **A**].
>> = *Phyllosticta punctata* Ellis & Dearn. *in* Can. Rec. Sci. **5**:
>> 268. 1893.

Diagn. data: Boerema & Griffin (1974) (tabs 1, 2; identification of this
pathogen as a species of *Phoma*, synonymy, disease symptoms, cultural characteristics), Rai & Rajak (1993) (cultural characteristics), de Gruyter & Scheer
(1998) (classification as a variety of *Ph. exigua*), van der Aa *et al.* (2000) (differentiating diagnostic features within the *Ph. exigua* complex), Abeln *et al.*
(2002) (AFLP fingerprint).

Description in vitro *(see also Table 3)*

Pycnidia and conidia similar to those of the type var. *exigua*.
   Colonies on OA and MA relatively fast growing, 6.5–8.5 cm diam. after 7
days, slower growing colony sectors may occur; on OA with tufts of velvety to

finely floccose white to pale olivaceous grey aerial mycelium. No production of antibiotic E. Representative culture CBS 100354.

Ecology and distribution

A common pathogen of cultivated *Viburnum* spp. (Caprifoliaceae). Occasionally it has been isolated from *Lonicera* sp. (also Caprifoliaceae) and some other woody plants (mostly in the neighbourhood of *Viburnum* plants): leaf spot, stem lesions, shoot blackening. Most conspicuous are the necrotic leaf spots with a purplish margin. The records of this fungus are from Europe (Germany, the UK, The Netherlands, France) and North America (Canada, USA); but it is probably found everywhere the hosts are cultivated.

Phoma foveata Foister, Fig. 30J, Colour Plate I-D

> *Phoma foveata* Foister *in* Trans. Proc. bot. Soc. Edinb. **33**: 66–67[–68]. 1940 [vol dated '1943'].
> > ≡ *Phoma solanicola* f. *foveata* (Foister) Malc. *in* Ann. appl. Biol. **46**: 639. 1958.
> > ≡ *Phoma exigua* var. *foveata* (Foister) Boerema *in* Neth. J. Pl. Path. **73**: 192. 1967.
> > ≡ *Phoma exigua* Desm. f. sp. *foveata* (Foister) Malc. & E.G. Gray *in* Trans. Br. mycol. Soc. **51**: 619. 1968.

Diagn. data: Boerema & van Kesteren (1962) (figs 1–3, disease symptoms on leaves and tubers of *Solanum tuberosum*, associated with 'forms of *Phoma solanicola*'; review of literature data; notes on other species of *Phoma* found on potatoes), Boerema (1967b) (classification of *Ph. foveata* as variety of *Ph. exigua*), Boerema & Höweler (1967) (tab. 1, pl. 4 figs 8, 9; delimiting and differentiating criteria against *Ph. exigua*, colour plate of cultures on OA and CA showing red discoloration by anthraquinone pigments and E$^+$ reaction with NaOH), Boerema (1976) (history as principal causal organism of potato gangrene; diagnostic tests based on the production of anthraquinone pigments *in vitro* and *in vivo*: ammonia test, chromatography test, stimulation crystallization process of the pigments by the fungicide thiophate-methyl and the violet line test for differentiation against *Ph. exigua*), Boerema (1977) (fig. 1; finding of the fungus in the Andean region of South America on *Chenopodium quinoa*, an important food plant of the Indians, which is often grown together with the wild potato: intercropping), Otazu *et al.* (1979) (data on an isolate of the fungus (E$^+$ strain) from brown rot of the stems of *Chenopodium quinoa*, cultivated in the Altiplana area of Peru. It was as pathogenic to potatoes as the virulent European isolates, but on *C. quinoa* and *C. album* it was more pathogenic than the European isolates), EPPO (1980b, 1982) (data sheet with two photographs of symptoms of gangrene on potato tubers; notes on geographical distribution, biology and identification methods), Boerema *et al.* (1987) (reclassification of the fungus in species rank on account of the finding in the Andean highlands of South America), Monte *et al.* (1990) (support of the species rank of the fungus by physiological and biochemical study of three isolates), de Gruyter *et al.* (2002) (fig. 12: conidial shape, reproduced in present Fig. 30J; cultural descriptions, partly adopted below).

Description in vitro

Pycnidia globose to subglobose with 1(–3) non-papillate or papillate ostiole(s) (may be absent, or visible only as a pale spot), 75–370 µm diam., glabrous or with mycelial outgrowths. Conidial matrix whitish to pale buff.

Conidia mainly aseptate, ellipsoidal to allantoid, with several small, scattered guttules, (3.5–)5–7(–11.5) × (1.5–)2–3(–3.5) µm, the larger ones occasionally 1-septate.

[Chlamydospores and pseudosclerotia, induced by some isolates of the bacterium *Serratia plymuthica*, have been reported recently in isolates of *Ph. foveata* (Camyon & Gerhardson, 1997). The chlamydospores, 1.8–3.7 µm diam., were produced singly, in chains or clustered. Pseudosclerotia were irregular, 60–340 µm diam., resembling those produced by *Phoma chrysanthemicola* Hollós (sect. *Peyronellaea*, **D**).]

Colonies on OA 7–7.5 cm after 7 days, regular, greenish olivaceous/olivaceous to buff/honey/amber due to the release of pigments, with felty to floccose/woolly, white to (pale) olivaceous grey aerial mycelium. On MA growth rate 6.5–8 cm after 7 days, regular, amber to herbage green, occasionally honey at centre, sienna to rust near margin, due to the pigments. In the agar abundant production of citrine green to yellow-green needle-like crystals, usually also small yellowish to brownish crystals are formed both in the hyphae and in the agar. Pycnidia scattered, both on and in the agar, solitary or confluent.

The pigments and crystals represent several anthraquinones, namely pachybasin, chrysophanol, emodin and phomarin (Bick & Rhee, 1966). With application of a drop of NaOH the pigments discolour to violet/red. The production of pigments and crystals is used in diagnostic tests (EPPO, 1980b, 1982). Some strains of the fungus show with additions of a drop of NaOH also the E$^+$ reaction: a blue green spot, changing to brown red. Representative cultures CBS 341.67, CBS 530.66, CBS 557.97, CBS 109176.

Ecology and distribution

This fungus causes lesions on potato tubers, *Solanum tuberosum* (Solanaceae), in Europe known as gangrene. It has been treated as a pigment producing variety of the ubiquitous *Phoma exigua* Desm. var. *exigua* (this section), which may also cause gangrene-like lesions on potatoes. However, the fungus appeared to be originally indigenous to the Andes regions of South America, causing brown stalk rot of *Chenopodium quinoa* (Chenopodiaceae), a grain commonly grown there in association with potatoes (Otazu *et al.*, 1979). Therefore it deserves the species rank, as also confirmed by Monte *et al.* (1990). At present various potato cultivars show tuber-resistance to this fungus.

Phoma heliopsidis (H.C. Greene) Aa & Boerema, Fig. 30K

> *Phoma heliopsidis* (H.C. Greene) Aa & Boerema apud de Gruyter, Boerema & van der Aa *in* Persoonia **18**(1): 40. 2002.
> > ≡ *Phyllosticta heliopsidis* H.C. Greene *in* Trans. Wisc. Acad. Sci. Arts Lett. **50**: 158. 1961.

Diagn. data: de Gruyter *et al.* (2002) (fig. 22: conidial shape, reproduced in present Fig. 30K; cultural descriptions, partly adopted below).

Description in vitro

Pycnidia globose/subglobose with 1–3(–5) papillate ostiole(s), later developing into an elongated neck, 70–300 μm diam., glabrous or with mycelial outgrowths. Conidial matrix salmon/peach or buff to pale vinaceous.

Conidia mainly aseptate, ellipsoidal to allantoid, with several distinct guttules, (5–)6–8(–10.5) × 1.5–3 μm; 1-septate conidia 9–13 × 2–3.5 μm.

Colonies on OA *c.* 8 cm diam. after 7 days, regular to slightly irregular, colourless with an olivaceous/grey olivaceous to dull green stellate pattern, with velvety, olivaceous grey aerial mycelium. On MA similar growth rate; colonies often olivaceous black in centre and citrine/citrine green near margin; on application of a drop of NaOH a weak reddish colour may appear. Pycnidia scattered, both on and in the agar as well as in aerial mycelium, solitary or confluent. Representative culture CBS 109182.

Ecology and distribution

A pathogen of Compositae, possibly widely distributed in North America, mostly affecting leaves, but also the stem and inflorescences. The records refer to collections on *Heliopsis* spp. in the USA and on *Ambrosia artemisiifolia* (common ragweed) in Canada.

Phoma laundoniae Boerema & Gruyter, Fig. 31A

> *Phoma laundoniae* Boerema & Gruyter apud de Gruyter, Boerema & van der Aa *in* Persoonia **18**(1): 30. 2002.

Diagn. data: de Gruyter *et al.* (2002) (fig. 11: conidial shape, reproduced in present Fig. 31A; cultural descriptions, partly adopted below).

Description in vitro

Pycnidia globose to subglobose, with 1(–2) occasionally papillate ostiole(s), 80–280 μm diam., glabrous. Conidial matrix off-white to primrose.

Conidia ellipsoidal to allantoid, with several small, scattered guttules, mainly aseptate, (5–)6–8.5(–11) × 2–3.5 μm; 1-septate conidia up to 13.5 × 4 μm, sparse.

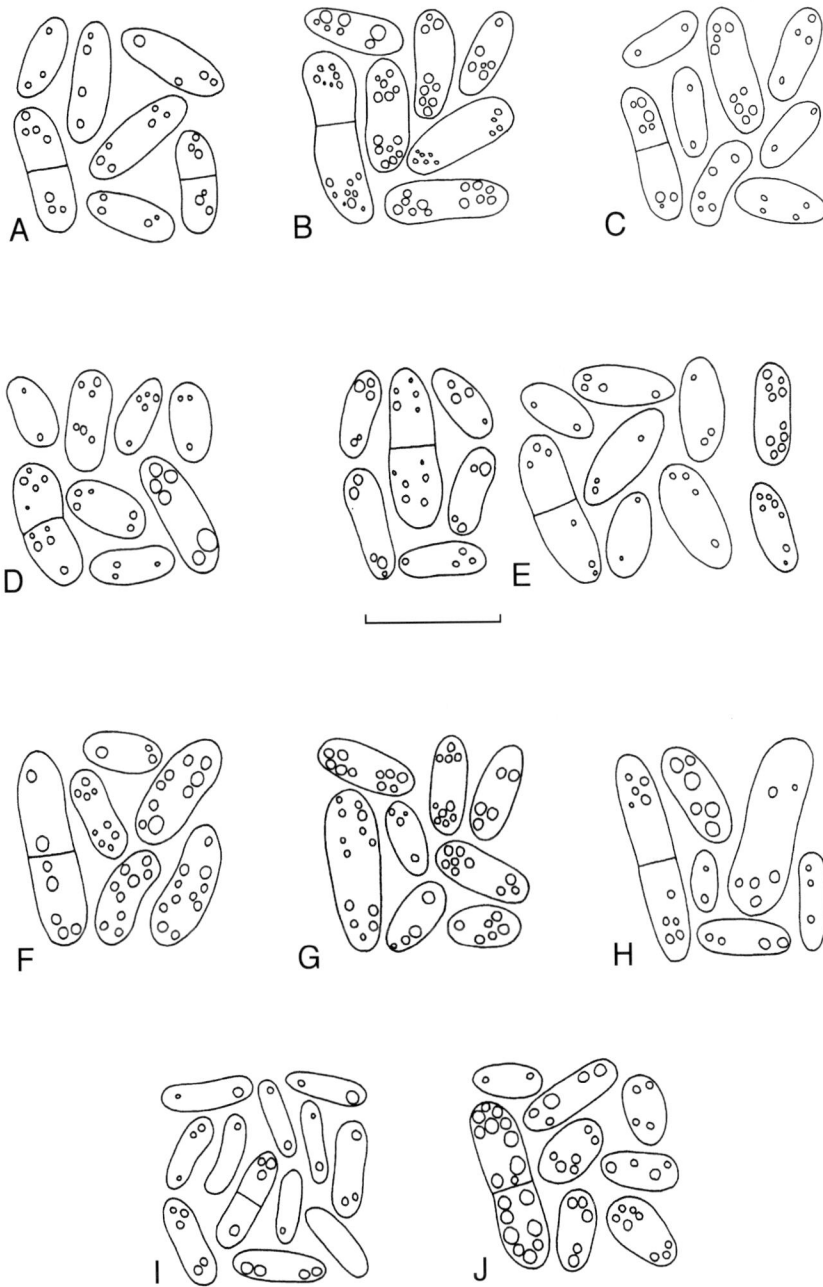

Fig. 31. Shape of conidia and some hyphal chlamydospores. **A** *Phoma laundoniae.* **B** *Phoma ligulicola* var. *ligulicola.* **C** *Phoma ligulicola* var. *inoxydabilis.* **D** *Phoma lycopersici.* **E** *Phoma macrostoma* vars *macrostoma* and *incolorata.* **F** *Phoma matteucciicola.* **G** *Phoma medicaginis* var. *medicaginis.* **H** *Phoma medicaginis* var. *macrospora.* **I** *Phoma nemophilae.* **J** *Phoma nepeticola.* Bar 10 μm.

Colonies on OA *c.* 4.5 cm after 7 days, regular, citrine green to dull green, with felty to finely floccose, grey olivaceous to olivaceous grey aerial mycelium; reverse dull green to olivaceous/olivaceous black. On MA somewhat slower growing, *c.* 4 cm after 7 days, also regular, colony hazel to olivaceous; reverse hazel to grey olivaceous/olivaceous, olivaceous black in centre. Pycnidia in concentric zones, both on and in the agar as well as in aerial mycelium, solitary or confluent. With application of a drop of NaOH on colonies on OA a pale reddish colour may appear. Representative culture CBS 109174.

Ecology and distribution

This fungus has been isolated from lesions on fruits of *Prunus persica* (Rosaceae) in New Zealand. *Phoma* species found in association with peach and other stone fruit trees in New Zealand were formerly often identified as *Phyllosticta circumscissa* Cooke, originally described from apricot in South Australia. Later it was concluded that these records refer to *Phoma pomorum* Thüm. (sect. *Peyronellaea*, **D**), see Pennycook (1989). Quite possibly *Phoma laundoniae* was often also involved in the earlier New Zealand records.

Phoma ligulicola Boerema var. *ligulicola*, Fig. 31B

Teleomorph: *Didymella ligulicola* (K.F. Baker *et al.*) Arx var. *ligulicola*

> *Phoma ligulicola* Boerema apud van der Aa, Noordeloos & de Gruyter *in* Stud. Mycol. **32**: 9. 1990, var. *ligulicola*.
> > ≡ *Ascochyta chrysanthemi* F. Stevens *in* Bot. Gaz. **44**: 246. 1907; not *Phoma chrysanthemi* Voglino *in* Malpighia **15**: 332. 1902 [referring probably to *Phoma ligulicola* var. *inoxydabilis*, see below].

Diagn. data: Baker *et al.* (1949b) (figs 1–3; history of 'Ascochyta chrysanthemi–Ray Blight' in the USA; biology of the fungus and first description of its teleomorph as *Mycosphaerella ligulicola*, with figures of disease symptoms, occurrence and morphology of pycnidia and pseudothecia *in vivo*), Müller & von Arx (1962) (classification of the teleomorph in *Didymella* on account of the revision of the didymospored Pyrenomycetes), Boerema & van Kesteren (1974) (history of the disease in Europe; *Phoma* characteristics of the anamorph; discussion of the newly proposed Italian names of the pathogen, *Didymella chrysanthemi* and *Phoma chrysanthemi*; type material of the latter showed indeed much similarity with *Ascochyta chrysanthemi*), McCoy & Blakeman (1976) (tab. 1, fig. 1; distribution of the fungus in the USA and other areas of the world, and selection for environmental races), Punithalingam (1980b) (fig. A–E; CMI description under *Didymella chrysanthemi* with photographs *in vivo* of pycnidium, conidia, pseudothecium, asci and ascospores; phytopathological notes), EPPO (1980a, 1982) (data sheets under *Didymella chrysanthemi* with photographs of disease symptoms; notes on geographical distribution, biology and identification methods, 'in culture on oatmeal agar the majority of the pycnidiospores remain 1-celled'), Walker & Baker (1983) (figs 1–3; nomenclature of the fungus: rejection of the binomial *Didymella chrysanthemi*, which appeared to be a species of *Mycosphaerella*; doubt about the identity of *Phoma chrysanthemi*), van der Aa *et al.* (1990) (introduction of a new name in *Phoma* for the anamorph of *Didymella liguli-cola*; differentiation of the specific pathogen of florists' chrysanthemum (E[+]) against var. *inoxydabilis* (E[−]) isolated from various other *Compositae*; note on recent misidentifications of both morphs of the fungus), de Gruyter *et al.* (2002) (fig. 5a: shape of conidia, reproduced in present Fig. 31B; cultural descriptions, partly adopted below).

Description in vitro

Pycnidia globose to subglobose, usually with 1, sometimes slightly papillate ostiole, mostly 80–270 μm diam., glabrous or with mycelial outgrowths. Conidial matrix saffron.

Conidia ellipsoidal to oblong, with several small guttules, mainly aseptate, 3.5–7.5(–12) × 2–3(–4) μm, some 1-septate, usually 9–15 × 3–5 μm, but occasionally distinctly large, up to 23 × 8 μm.

Colonies on OA *c.* 7 cm diam. after 7 days, regular to slightly irregular, colourless/greenish olivaceous to dull green olivaceous, often in a zonate pattern; with sparse to abundant, felted to floccose, white to pale olivaceous grey aerial mycelium; reverse grey olivaceous to fawn–hazel or olivaceous grey, more or less discolouring to sienna due to a diffusible pigment. On MA slower growing, 5–6 cm, sometimes with a pale luteous pigment production. Pycnidia developing on and in the agar. All strains showed a positive reaction with the sodium hydroxide test: on application of a drop of NaOH green → red (E[+] reaction). Representative cultures CBS 500.63, CBS 109178.

In vivo *(Dendranthema–Grandiflorum hybrids)*

Pycnidia (aggregated in blackened petals and scattered in brownish black leaf blotches and stem lesions, subepidermal,) of two sizes, small, 72–180 μm, in the petals, and larger, 111–325 μm, in the leaf and stem lesions, depressed globose with 1 inconspicuous ostiole.

Conidia mostly irregular cylindrical-ellipsoidal and extremely variable in dimensions, usually partly aseptate (10–40%), (6–)8.5–13(–22) × 2.5–8 μm, and partly 1(–2)-septate (60–90%), (9–)13–15.5(–23) × (3–)4–5(–6.5) μm. The septation of the conidia should be related to the temperature.

Pseudothecia (occasionally found on old blackened leaf and stem lesions) subglobose and more erumpent than pycnidia, 96–224 μm diam. Asci cylindrical to slightly narrowed near apex, (40–)50–85(–90) × (7–)8–10(–11) μm, 8-spored, irregularly biseriate. Ascospores 12–17 × 4–7 μm, ellipsoid or fusiform, approximately medianly uniseptate, slightly constricted at septum, the upper cell swollen just above the septa, hyaline with guttules [for details and illustrations see Punithalingam, 1980b].

Ecology and distribution

This specific pathogen of florists' chrysanthemum, *Dendranthema–Grandiflorum* hybrids (Compositae, formerly known as, e.g. *Chrysanthemum morifolium* and *C. indicum*), occurs at present nearly everywhere the host is cultivated. The fungus seems to be indigenous to Japan, but was first recorded in the south-eastern USA as the cause of chrysanthemum ray (flower) blight. It may attack all plant parts, roots, stems, leaves and flowers. Cuttings are particularly susceptible; hence the rapid worldwide spread of the fungus since the late 1940s. The suggestion that the disease was present in Europe before the first observations were made in the USA, appeared to be due to confusion with a different teleomorph described in Italy (*Mycosphaerella chrysanthemi* (Tassi) Tomilin, see Walker & Baker, 1983) and the existence of a related fungus occurring on various other wild and cultivated Compositae, distinguished as *Didymella/Phoma ligulicola* var. *inoxydabilis*, see below.

Phoma ligulicola var. *inoxydabilis* Boerema, Fig. 31C

Teleomorph: ***Didymella ligulicola* var. *inoxydabilis* Boerema**

> *Phoma ligulicola* var. *inoxydabilis* Boerema apud van der Aa, Noordeloos & de Gruyter *in* Stud. Mycol. **32**: 10. 1990.
>> = *Phoma chrysanthemi* Voglino *in* Malpighia **15**: 332. 1902 (1901?) [type agrees with *Ph. ligulicola*, see note].
>> ≡ *Phomopsis chrysanthemi* (Voglino) M.E.A. Costa & Sousa da Câmara *in* Port. Acta biol. Ser. B, Sist. (ecol., biogeogr., paleontol.) **3**: 301. 1952 [misapplied].

Diagn. data: Boerema & van Kesteren (1974) (morphological resemblance of *Phoma chrysanthemi* with the anamorph of *Didymella ligulicola*), van der Aa *et al.* (1990) (distinction as E⁻ variety, 'inoxydabilis', which in contrast with the E⁺ type variety, produces *in vitro* both the anamorph and teleomorph; recorded on various wild and cultivated Compositae; diagnostic confusion with the specific pathogen of florists' chrysanthemums), de Gruyter *et al.* (2002) (fig. 5b: conidial shape, reproduced in present Fig. 31C; cultural descriptions, partly adopted below).

Description in vitro

Pycnidia resembling those of the type var. *ligulicola*, but often somewhat larger, 90–400 µm diam., and occasionally with more ostioles (–3). Conidial matrix off-white to buff.

Conidia like those of the type var., but often somewhat smaller, aseptate (3.5–)5–8(–13.5) × 2–3(–4) µm and 1-septate up to 15 × 5 µm; occasionally with only two small, polar guttules.

Pseudothecia in fresh cultures developing intermingled with pycnidia, globose to subglobose with papillate ostiole. Asci and ascospores agree with those of *Didymella ligulicola* var. *ligulicola in vivo*: asci 50–85 × 8–10 µm (mostly

60–65 × 9 µm), ascospores two-celled, 12–17 × 4–7 µm (often 13.5–16.5 × 5.5–7 µm), constricted at septum, but less prominently swollen above the septa than ascospores in the type variety.

Colonies much slower growing than those of the type var., on OA 4.5–5 cm after 7 days, on MA 3.5–4 cm. The grey olivaceous to dull green colonies on OA are covered by floccose white aerial mycelium and do not show a zonate pattern as with the type var. [on MA the aerial mycelium is woolly, white to pale olivaceous grey]. The production of pycnidia is retarded in comparison with the type var., but associated with the development of pseudothecia. No pigment production and no oxidation reaction with application of a drop of NaOH (E$^-$). Representative culture CBS 425.90.

Ecology and distribution

This variety has been found in Europe and Australia, but probably also occurs elsewhere on wild and cultivated Compositae. The various isolates studied were obtained from *Tanacetum* (*Chrysanthemum*/*Pyrethrum*) *cinerariifolium*, feverfew, *Tanacetum* (*Chrysanthemum*/*Pyrethrum*) *parthenium* and zinnia, *Zinnia violacea* (*elegans*). Apart from the frequent production of the teleomorph *in vitro* (fresh cultures), it differs from the type variety *ligulicola* by the absence of antibiotic E (therefore no discoloration with addition of a drop of NaOH, no oxidation-reaction, hence 'inoxydabilis'). It is plausible that *Phoma chrysanthemi* Voglino, described in Italy refers to this variety and not to var. *ligulicola*, which reached Europe only in the late 1940s.

Phoma lycopersici Cooke, Fig. 31D

Teleomorph: ***Didymella lycopersici* Kleb.**

Phoma lycopersici Cooke in Grevillea **13**: 94. 1885.
  H = *Phoma lycopersici* (Plowr.) Jacz. *in* Nouv. Mém. Soc. [imp.] Nat. Mosc. **15**: 350[–351]. 1898.
    ≡ *Sphaeronaema lycopersici* Plowr. *in* Gdnrs' Chron. II [New Series], **16**: 621. 1881.
    = *Ascochyta lycopersici* Brunaud *in* Bull. Soc. bot. Fr. **34** [II, **9**]: 431. 1887.
    = *Ascochyta socia* Pass. *in* Boll. Com. agr. Parmense 1889: 2. 1889; not *Ascochyta socia* (Tassi) Allesch. *in* Rabenh. Krypt.-Flora [ed. 2], Pilze **7** [Lief. 88]: 871. 1903.
    = *Diplodina lycopersici* Hollós *in* Annls hist.-nat. Mus. natn. hung. **5**: 461. 1907.
    = *Phoma ferrarisii* Cif. *in* Staz. Sper. agr. ital. **55**: 149. 1912.
    = *Diplodina lycopersicicola* Bond.-Mont. *in* Mater. mikol. Obslêd. Ross. **5**: 4. 1922.

Diagn. data: Holliday & Punithalingam (1970b) (figs A–D; CMI description with drawings of pycnidium, conidiogenous cells, conidia, ascocarp, asci and ascospores; phytopathological notes), Boerema & Dorenbosch (1973) (pl. 2f, h: cultures on OA; tab. 1: synoptic table on characters *in vitro* with drawings of conidia and stilboid bodies; diagnostic literature), Boerema (1976) (note on confusion with *Phoma exigua* var. *exigua* and *Phoma destructiva*), Sutton (1980) (fig. 228E: drawing of conidia; specimens in CMI collection), Boerema & van Kesteren (1981a) (misidentification as *Phoma exigua* in literature), MorganJones & Burch (1988b) (fig. 1A–E: drawings of pycnidium, wall cells, hyphae with pycnidium initial, conidia; pls 1A–F, 2A–H: photographs of agar plate cultures, pycnidium, wall cells, hyphae and conidia; history; morphological and cultural description; cultural differences with *Ph. exigua* and *Ph. destructiva*; detailed description), de Gruyter *et al.* (2002) (fig. 27: shape of conidia, reproduced in present Fig. 31D; cultural descriptions, partly adopted below).

Description in vitro

Pycnidia globose to subglobose, with 1(–3) non-papillate or slightly papillate ostiole(s), 70–200 μm diam., glabrous or with short mycelial outgrowths. Conidial matrix whitish to buff.

Conidia mainly aseptate, variable in shape, subglobose to ellipsoidal or allantoid, with several small guttules, (3.5–)5–8.5(–10) × 2–3.5(–4.5) μm, 1-septate conidia up to 15.5 × 4.5 μm.

'Stilboid bodies' usually abundantly produced in old cultures of the fungus, more or less clavate, sterile, with the same wall structure as pycnidia and with hyphal outgrowths.

Colonies on OA 6.5–7.5 cm after 7 days, regular, colourless/olivaceous buff to grey olivaceous, with white to olivaceous grey/grey olivaceous aerial mycelium; reverse grey olivaceous/olivaceous grey to olivaceous, olivaceous buff near margin. On MA similar; with application of a drop of NaOH on this medium a yellow/brownish colour may appear. Pycnidia scattered, both on and in the agar, solitary or confluent. Representative culture CBS 378.67.

In vivo (Lycopersicon esculentum)

Pycnidia (in lesions on stems [cankers] and fruits [fruit rot], solitary or gregarious, initially immersed, but becoming erumpent) subglobose, up to 200 μm diam.

Conidia as *in vitro*, aseptate or 1-septate, usually (5–)6–10 × 2–3 μm.

Pseudothecia (only rarely found on dead stems) subglobose, up to 300 μm diam. Asci cylindrical to subclavate, 50–95 × 6–10 μm. Ascospores irregularly biseriate, ellipsoidal, slightly constricted at the septum, 12–18 × 5–6 μm (for illustrations see Holliday & Punithalingam, 1970b).

Ecology and distribution

Widespread on tomato, *Lycopersicon esculentum* (Solanaceae), in Eurasia and Africa: stem and fruit rot (canker). The fungus has often been confused with the plurivorous *Phoma exigua* Desm. var. *exigua* (type of this section) and the 'American' *Phoma destructiva* Plowr. var. *destructiva* (with only aseptate conidia, sect. *Phoma*, **A**) causing fruit rot or leaf and stem blight of tomato. It may also be mistaken for *Phoma destructiva* var. *diversispora* de Gruyter & Boerema (this section), recently described from necrotic spot and fruit rot in tomato crops.

Phoma macrostoma Mont. var. *macrostoma*, Fig. 31E

V   *Phoma macrostoma* Mont. *in* Annls Sci. nat. (Bot.) III, **11**: 52. 1849, var. *macrostoma* [as '*macrostomum*'].

   = *Phyllosticta berberidis* Rabenh. *in* Klotzschii Herb. mycol. [Ed. Rabenh.] Cent. 19, No. 1885. 1854.

   = *Phyllosticta corni* Westend. *in* Bull. Acad. r. Sci. Lett. Beaux-Arts Belg. II, **2**: 567. 1857.

   = *Phoma phyllostictoides* Desm. *in* Pl. cryptog. France II [ed. 3], Fasc. 14, No. 694. 1859.

      ≡ *Ascochyta phyllostictoides* (Desm.) Keissler *in* Annls mycol. **21**: 74. 1923.

   = *Phyllosticta alcides* Sacc. *in* Michelia **1**(2): 135. 1878.

   = *Phyllosticta grossulariae* Sacc. *in* Michelia **1**(2): 136. 1878.

   = *Phyllosticta humuli* Sacc. & Speg. *in* Michelia **1**(2): 144. 1878.

   = *Phyllosticta robiniae* Sacc. *in* Michelia **1**(2): 146. 1878.

   = *Phyllosticta chionanthi* Thüm. *in* Mycoth. univ. Cent. 15, No. 1489. 1879.

   = *Phyllosticta alnigena* Thüm. *in* Hedwigia **19**: 180. 1880.

   = *Phyllosticta pterocaryae* Thüm. *in* Hedwigia **19**: 181. 1880.

   = *Phyllosticta populorum* Sacc. & Roum. apud Roumeguère & Saccardo *in* Revue mycol. **3**: 12. 1881.

      ≡ *Ascochyta populorum* (Sacc. & Roum.) Voglino *in* Annali R. Accad. Agric. Torino **55**: 314[–443]. 1912.

   = *Phoma gentianae* J.G. Kühn *in* Fungi europ. exs./Klotzschii Herb. mycol. Cont. [Ed. Rabenh.] Cent. 29, No. 2893. 1882.

   = *Phoma herbarum* f. *catalpae-capsularum* Sacc. *in* Michelia **2**(1): 93. 1880; *in* Sylloge Fung. **3**: 133. 1884.

   = *Phoma herbarum* f. *rubi* Sacc. *in* Michelia **2**(1): 93. 1880; *in* Sylloge Fung. **3**: 133. 1884.

   = *Phoma pomi* Schulzer & Sacc. *in* Hedwigia **23**: 109. 1884 [as '*Phoma (Aposphaeria?) pomi*']; not *Phoma pomi* Pass. *in* Atti Accad. naz. Lincei Rc. [Cl. Sci. fis. mat. nat.] **4**(2): 96. 1888 [= *Asteromella mali* (Briard) Boerema, see below under *Phyllosticta mali*].

      ≡ *Aposphaeria pomi* (Schulzer & Sacc.) Sacc. *in* Sylloge Fung. **3**: 177. 1884.

   = *Phoma mororum* Sacc. *in* Boll. mens. Bachicolt II, **2**: 53–56. 1884; *in* Sylloge Fung. **3**: 95. 1884.

   = *Phyllosticta amaranthi* Ellis & Kellerman *in* J. Mycol. **1**: 4. 1885.

= *Phoma cicatricum* Pass. *in* Atti Accad. naz. Lincei Rc. [Cl. Sci. fis. mat. nat.] **4**(2): 96. 1888.

V = *Aposphaeria caricae* Pass. *in* Atti Accad. naz. Lincei Rc. [Cl. Sci. fis. mat. nat.] **4**(2): 99. 1888 [as '*Apospheria*'].

= *Phyllosticta saxifragae* Brunaud *in* Annls Soc. Sci. nat. Charente-Infér. **26**: 55. 1890 [description by Saccardo & Sydow listed under *Phyllosticta saxifragicola* Brunaud *in* Sylloge Fung. **14**: 853. 1899].

= *Phoma friesii* Brunaud *in* Bull. Soc. bot. Fr. **36** [= II, **11**]: 337. 1889.

H = *Phyllosticta mali* Prill. & Delacr. *in* Bull. Soc. mycol. Fr. **6**: 181. 1890; not *Phyllosticta mali* Briard, Fl. cryptog. Aube & Suppl. cat. Troyes 79. 1888 [= *Asteromella mali* (Briard) Boerema apud Boerema & Dorenbosch *in* Versl. Meded. plziektenk. Dienst Wageningen **142** (Jaarb. 1964): 149. 1965].

H = *Phyllosticta japonica* Fautrey *in* Revue mycol. **13**: 9. 1891; not *Phyllosticta japonica* Thüm. *in* Instituto Coimbra **28** sub Contr. Fl. mycol. Lusit. III no. 47. 1881 ['1880 e 1881']; quoted *in* Hedwigia **21**: 27–28. 1882; not *Phyllosticta japonica* I. Miyake *in* J. Coll. Agric. Tokyo **2**: 253. 1910, nor *Phyllosticta japonica* Sawada *in* Bull. Govt Forest Exp. Stn Tokyo **45**: 52. 1950 [= *Phomopsis* sp.].

≡ *Phyllosticta humulina* Sacc. & P. Syd. apud Allescher *in* Rabenh. Krypt.-Flora [ed. 2], Pilze **6** [Lief. 64]: 347. 1899 [vol. dated '1901'].

= *Phyllosticta limitata* Peck *in* Rep. N.Y. St. Mus. nat. Hist. **50**: 115. 1897.

≡ *Phoma limitata* (Peck) Boerema apud Boerema & Dorenbosch *in* Versl. Meded. plziektenk. Dienst Wageningen **142** (Jaarb. 1964): 138. 1965.

= *Phyllosticta spaethiana* Allesch. apud Sydow *in* Beibl. Hedwigia **36**: 160. 1897.

= *Phyllosticta mespilina* Montemart. ex Briosi & Cavara, Funghi parass. Fasc. 12, No. 298. 1897.

= *Phyllosticta caraganae* P. Syd. *in* Beibl. Hedwigia **38**: 134. 1899.

= *Phyllosticta cercocarpi* P. Syd. *in* Beibl. Hedwigia **38**: 135. 1899.

H V = *Phyllosticta cydoniicola* Henn. *in* Hedwigia **41**: 158. 1902 [as '*cydoniaecola*'; not *Phyllosticta cydoniicola* Allesch. *in* Beibl. Hedwigia **36**: 158. 1897 [= *Phoma pomorum* Thüm. var. *pomorum*, sect. *Peyronellaea*, **D**].

V = *Phyllosticta bauhiniicola* Henn. *in* Hedwigia **41**: 306. 1902 [as '*bauhinicola*'].

= *Phyllosticta alniperda* Oudem. *in* Ned. Kruidk. Archf III, **2**: 1114. 1904.

Θ = *Phyllosticta grossulariae* var. *ribis-rubri* D. Sacc. *in* Mycoth. Ital., No. 1683. 1905.

= *Phyllosticta mali* var. *comensis* Traverso *in* Malpighia **19**: 141. 1905.

= *Phyllosticta lupulina* Kabát & Bubák apud Bubák & Kabát *in* Öst. Bot. Z. **55**: 77. 1905.

= *Phyllosticta perniciosa* Kabát & Bubák *in* Hedwigia **44**: 350. 1905.

≡ *Phyllosticta apatella* var. *perniciosa* (Kabát & Bubák) Cif. *in* Annls mycol. **20**: 36. 1922.

= *Phyllosticta celtidicola* Bubák & Kabát *in* Annls mycol. **5**: 42. 1907.

= *Phyllosticta adeloica* Speg. *in* Revta Mus. La Plata **25**: 32. 1908.

= *Phyllosticta apicalis* Davis *in* Trans. Wisc. Acad. Sci. **16**: 761. 1909.

V = *Phyllosticta robiniicola* Hollós *in* Annls hist.-nat. Mus. natn. hung. **8**: 2. 1910 [as 'robiniaecola'].

= *Phyllosticta taxi* Hollós *in* Annls hist.-nat. Mus. natn. hung. **8**: 3. 1910.

= *Phyllosticta talae* Speg. *in* An. Mus. Nac. Hist. Nat. B. Aires III, **20**: 340. 1910.

= *Phyllosticta ribiseda* Bubák & Kabát *in* Hedwigia **50**: 39. 1911.

= *Phyllosticta spiraeae-salicifoliae* Kabát & Bubák *in* Hedwigia **50**: 39. 1911.

= *Phyllosticta serebrianikowii* Bubák *in* Hedwigia **52**: 265. 1912.

= *Phyllosticta diedickei* Bubák & Syd. *in* Annls mycol. **13**: 7. 1915.

= *Polyopeus purpureus* var. *verus* A.S. Horne *in* J. Bot., Lond. **58**: 240. 1920.

= *Phyllosticta grossulariae* f. *rubri* Cif. *in* Annls mycol. **20**: 39. 1922 [sometimes cited as 'var. *rubri*'].

= *Phyllosticta angulata* Wenzl *in* Phytopath. Z. **9**: 349. 1936.

= *Phyllosticta physocarpi* H.C. Greene *in* Amer. Midl. Nat. **41**: 737. 1949.

H = *Phyllosticta betulicola* Cejp apud Cejp, Dolejš & Zavřel *in* Zpr. Vlastiv. Úst. Olomouci, Cislo **143**: 3. 1969; not *Phyllosticta betulicola* Vasyagina apud Byzova *et al. in* Fl. spor. Rast. Kazakh. **5**(1): 59. 1967 [= *Asteromella* sp.].

Diagn. data: Boerema & Dorenbosch (1965) (figs 2A–E, 3; discussion of the synonymy, habitat and literature data on occurrence on apple, under *Phoma limitata* (Peck) comb. nov.; with sketches of pycnidia, ostioles, conidia, conidiogenesis and pigmentation of hyphae and photos of cultures on OA), (1970) (figs 1–5, tab. I; identification as *Phoma macrostoma* Mont., a ubiquitous species on woody plants; synonymy, diagnostic characters with figures of pycnidia and conidia *in vivo* and *in vitro*; table on isolate sources), (1973) (pl. 4a, b: cultures on OA; tab. 6a: isolate sources; tab. 1: synoptic table on characters *in vitro* with drawing of conidia), Boerema (1976) (additional synonyms), Sutton, (1980) (fig. 228G: drawing of conidia; 'represented in herb. CMI by almost 100 collections'), Johnston (1981) (tab. 1, fig. 8; occurrence in New Zealand), Stadelmann & Schwinn (1982) (fig. 3b, pycnidium of *Ph. macrostoma* embedded in apple leaf with conidiophores of the scab-fungus, 'when penetrating the cuticula, the scab coniophores form points of entrance'), White & Morgan-Jones (1984) (fig. 1A–F: drawings of pycnidia, conidiogenous cells, conidia; pls 1A–G, 2A–G: photographs of agar plate cultures, pycnidia, wall cells, mycelium and conidia; history; morphological and cultural description), Loerakker (1986) (frequent occurrence of *Ph. macrostoma* in lesions on shoots of lime trees, caused by the hyphomycete *Cercospora microsora*; thereby the pycnidia sometimes look setose, a mixture formerly described as *Pyrenochaeta pubescens*), Noordeloos & Boerema (1989b) (occurrence in association with lenticel rot of apple), Rai & Rajak (1993) (cultural characteristics), Rai (1998) (effect of different physical and nutritional conditions on the morphological and cultural characters), van der Aa & Vanev (2002) (data on host plants and type specimens of the various *Phyllosticta* synonyms listed), de Gruyter *et al.* (2002) (fig. 6a, conidial shape, reproduced in present collective Fig. 31E; cultural descriptions, partly adopted below).

Description in vitro

Pycnidia globose to irregular with 1–2 non-papillate or papillate, relatively wide ostiole(s) (20–45 μm diam.), sometimes with an elongated neck in a later stage, 80–300 μm diam., glabrous. Conidial matrix salmon to flesh.

   Conidia variable in shape, subglobose, ellipsoidal to oblong, or allantoid, usually with small guttules, mainly aseptate (4–)5–8(–11) × 2–3(–4) μm; 1(–3)-septate conidia 8.5–14 × 2.5–4 μm, in old pycnidia often swollen and dark.

   Colonies on OA 4.5–6 cm after 7 days, regular, peach/sienna to red/blood colour or dark vinaceous, due to a red violet pigment in the cytoplasm and guttules of the hyphal cells; aerial mycelium absent or sparse, finely floccose white to pale olivaceous grey. On MA similar. Pycnidia scattered, both on and in the agar, solitary or confluent. With application of a drop of NaOH on colonies on OA a reddish to purplish colour may occur. Representative culture CBS 529.66.

Ecology and distribution

A cosmopolitan plurivorous weak and wound parasite, especially common on woody members of the Rosaceae. Its epithet refers to the relatively large ostioles of the pycnidia. *In vitro* the fungus is characterized by dull red/violet pigment in the cytoplasm and guttules of the hyphae. Often also colourless sectors occur in cultures: var. *incolorata*, listed below. As opportunistic parasite of woody plants the fungus often occurs on lesions caused by other pathogens. Its pycnidia may intermix with conidiophores of hyphomycetes, such as *Spilocaea pomi* Fr. : Fr. (anamorph of apple scab) and *Cercospora microsora* Sacc. (leaf and shoot spot of lime trees, *Tilia* spp., Tiliaceae; the mixed occurrence described as *Pyrenochaeta pubescens* Rostr.), see Stadelmann & Schwinn (1982) and Loerakker (1986).

Phoma macrostoma var. *incolorata* (A.S. Horne) Boerema & Dorenb., Fig. 31E

V   *Phoma macrostoma* var. *incolorata* (A.S. Horne) Boerema & Dorenb. *in* Persoonia **6**(1): 55. 1970 [as '*macrostomum* var. *incolorata*'].
   ≡ *Polyopeus purpureus* var. *incoloratus* A.S. Horne in J. Bot., Lond. **58**: 240. 1920.
   = *Polyopeus purpureus* var. *latirostratus* A.S. Horne *in* J. Bot., Lond. **58**: 240. 1920.
   = *Polyopeus purpureus* var. *nigrirostratus* A.S. Horne *in* J. Bot., Lond. **58**: 240. 1920.

Diagn. data: Boerema & Dorenbosch (1970) (figs 1–5, tab. I; first distinction of this pigmentless variety in a cultural study from spotted apples in 1920; synonymy and isolates, 'same habitat as var. *macrospora*, but less widely distributed'), (1973) (tab. 6b: isolate sources; pl. 4c: cultural characteristics on OA), Boerema (1976) ('often as a colourless sector (saltant) in the dull red violet coloured colonies of the type variety'), Johnston (1981) (records in New Zealand), Rajak & Rai (1983a) (tab. 1: cholesterol content in dried mycelium), Rai & Rajak (1993) (cultural characteristics; NB: the study sub *Ph. macrostoma* var. *incolorata* in Rai (1998) must be questioned as deviating chlamydospores producing isolates are included), de Gruyter *et al.* (2002) (fig. 6b: conidial shape, reproduced in present collective Fig. 31E; cultural characteristics, partly adopted below).

Description in vitro

The differentiation of this cultural variety is based on the absence of the red/violet pigment in the cytoplasm and guttules of the hyphal cells ('*incolorata*').

Pycnidia and conidia similar to those of the type var. *macrostoma*, but conidial matrix is white to buff/rosy buff.

Colonies on OA and MA show the same growth rate as the type var. On OA they are colourless, but weak grey olivaceous/dull green sectors in a stellate pattern may occur. The NaOH spot test is negative. On MA the colonies are primrose/olivaceous buff often with pale honey/olivaceous sectors. Representative cultures CBS 223.69, CBS 109173.

Ecology and distribution

This cultural variety often occurs as a colourless sector (saltant) in the dull red/violet coloured colonies of the type variety. The absence of pigment should be associated with a lower production of cholesterol (Rajak & Rai, 1983a). In nature the var. *incolorata* appears to be less common than var. *macrostoma*, but it is also ubiquitous on woody plants, incidental on herbaceous substrates and cosmopolitan. It is sometimes confused with *Phoma exigua* Desm. var. *exigua* (this section) and *Phoma pomorum* Thüm. var. *pomorum* (section *Peyronellaea*, **D**).

Phoma matteucciicola Aderkas et al., Fig. 31F

V   *Phoma matteucciicola* Aderkas, Gruyter, Noordel. & Strongman *in* Can. J. Pl. Path. **14**: 227. 1992 [as '*matteuccicola*'].

Diagn. data: von Aderkas & Brewer (1983) (study of a disease of ostrich fern, ascribed to *Phoma exigua* var. *foveata*), von Aderkas *et al.* (1992) (description of the pathogen as a separate species; detailed cultural characteristics, partly adopted below), de Gruyter *et al.* (2002) (fig. 15: shape of conidia, reproduced in present Fig. 31F; additional record on ferns indigenous to Europe).

Description in vitro

Pycnidia globose or flask-shaped with a short neck, usually with 1 ostiole, 200–300 μm diam., glabrous. Conidial matrix white.

Conidia ellipsoidal with several guttules, mainly aseptate 5–10 × 2.5–4 μm, the 1(–2)-septate larger, 9–16 × 3–4 μm.

Colonies on OA very fast growing, after 7 days the plates were already full grown, colourless, mycelium submerged in the agar, which stains vivid yellow-green by diffusible pigment; at first without visible aerial mycelium [after 14 days colonies predominantly grey with thin aerial mycelium arising from denser mycelium submerged in the medium]. On MA growth rate also very fast, pale olivaceous grey, with abundant grey aerial mycelium; reverse at first citrine, later with distinct reddish-brown tinges; crystals as yellow spickles represent the crystallized form of the pigments (anthraquinones). Pycnidia develop at the colony margin on aerial mycelium and in the agar, solitary or confluent. Representative culture CBS 259.92 (CMI 286996).

Ecology and distribution

This fungus was first reported in 1983 (von Aderkas & Brewer, 1983) as causal agent of a midrib rot of fronds of the ostrich fern, *Matteuccia struthiopteris* (Polypodiaceae) in Canada: gangrene disease. The host, well known as a garden fern, is used as a spring vegetable in Canada and USA. Inoculation experiments showed a lethal effect on the gametophyte stage of the fern. The pathogen, initially confused with *Phoma foveata* Foister (this section), has also been recorded from fern nurseries in Switzerland (Grimm & Vögeli, 2000). There it was also shown to be a virulent pathogen of the ferns *Dryopteris filix-mas* and *Blechnum spicant*.

Phoma medicaginis Malbr. & Roum. var. *medicaginis*, Fig. 31G

> *Phoma medicaginis* Malbr. & Roum. apud Roumeguère *in* Fungi gall. exs. Cent. 37, No. 3675. 1886 and *in* Revue mycol. **8**: 91. 1886, var. *medicaginis*.
> ‡ = *Phoma medicaginis* var. *medicaginis* f. *microspora* Rössner *in* Phytopath. Z. **63**: 119. 1968 [without Latin diagnosis and type indication, Arts 36, 37].
> = *Phoma cuscutae* Negru & Verona *in* Mycopath. Mycol. appl. **30**: 308. 1966.
> = *Phoma jatropae* Shreem. *in* Indian. Mycol. Pl. Path. **8**: 220–221. 1978.

Diagn. data: Morgan-Jones & Burch (1987a) (detailed study based on the representative culture CBS 533.66; fig. 1A–F: drawings of pycnidium, wall cells, chlamydospores, conidiogenous cells and conidia; pls 1A–F, 2A–F: photographs of agar plate cultures, pycnidium, wall cells and conidia; history; morphological and cultural descriptions), Boerema *et al.* (1993b) (delimitation as separate small-spored variety, producing only in old pycnidia some 1-septate conidia; synonymy; confirmation study Rössner, see below; evidence of coevolution fungus-host), Rössner (1968) (figs 1–4, tabs 3–5; conidial shape and dimensions in relation with the temperature; differentiation in a '*microspora*'-form and a '*macrospora*'-form), Neergaard (1977) (occurrence on seed), Noordeloos *et al.* (1993) (fig. 5: conidial shape; cultural characteristics partly adopted below, production of dendritic crystals of 'brefeldin A'), Rai & Rajak (1993) (cultural characteristics), de Gruyter *et al.* (2002) (fig. 7a: shape of conidia, reproduced in present Fig. 31G; difficulty of differentiation from var. *macrospora* at room temperature).

Description in vitro

Pycnidia globose/subglobose, initially without a distinct ostiole, later on clearly ostiolate, *c.* 250 μm diam., glabrous. Conidial matrix whitish to pale pink.

Conidia subcylindrical, generally aseptate (see Note), rarely 1-septate, 5–7(–12.5) × 1.5–4 μm.

Chlamydospores are occasionally produced in old culture.

Colonies on OA 3.5–4 cm diam. after 7 days, irregular, sinuate outline, buff at centre, at margin olivaceous or grey olivaceous, distinctly concentrically zonate, sometimes with green olivaceous sectors; aerial mycelium poorly developed, white, floccose. On MA growth rate 4–4.5 cm, with well developed greyish, woolly-floccose aerial mycelium; reverse on this medium shows after 2 or 3 weeks white, bryoid, dendritic crystals (brefeldin A, see Noordeloos *et al.*, 1993). Pycnidia abundant in concentric zones and scattered, also in aerial mycelium and agar, solitary or confluent. Representative culture CBS 533.66.

Ecology and distribution

The type variety *Ph. medicaginis* is a cosmopolitan seed-borne pathogen of lucerne, *Medicago sativa* (Leguminosae): black stem disease. However, this disease is also caused by the more pathogenic *Ph. medicaginis* var. *macrospora* Boerema *et al.* (listed below), which cannot be distinguished from var. *medicaginis* on agar media at room temperature. *Ph. medicaginis.* var. *medicaginis* may also attack other Leguminosae such as yellow trefoil, *Medicago lupulina*, and sweet clovers, *Melilotus* spp. The fungus has also been repeatedly isolated from non-leguminous plants (e.g. under the synonyms *Ph. cuscutae* and *Ph. jatropae*).

Note

The type variety of *Ph. medicaginis* does not produce any septate conidia *in vivo*. At low temperatures this absence of septate conidia is the most conspicuous character distinguishing it from var. *macrospora*, which may produce 10–63% relatively large septate conidia in winter (Rössner, 1968). Both varieties also differ in pathogenicity. On account of the presence of pycnidia with only aseptate conidia this fungus is also incorporated in the key to species of *Phoma* sect. *Phoma* (**A**).

Phoma medicaginis var. *macrospora* Boerema *et al.*, Fig. 31H

> *Phoma medicaginis* var. *macrospora* Boerema, R. Pieters & Hamers *in* Neth. J. Pl. Path. **99**, Suppl. **1**: 19–20. 1993.
>> ≡ *Phoma herbarum* f. *medicaginum* Westend. ex Fuckel *in* Jb. nassau. Ver. Naturk. **23–24** [= Symb. mycol.]: 134. 1870 ['1869 und 1870'] [listed by Saccardo *in* Sylloge Fung. **3**: 133. 1884 as 'f. *medicaginis* Fuckel'].
>> Θ ≡ *Phoma herbarum* f. *medicaginum* Westend. *in* Fungi europ. exs./Klotzschii Herb. mycol. Cont. [Ed. Rabenh.], Cent. 5, No. 455b. 1862 [in phytopathological literature often cited as '*P. herbarum* var. *medicaginis*'] [without description].
>> ‡ = *Phoma medicaginis* var. *medicaginis* f. *macrospora* Rössner *in* Phytopath. Z. **63**: 119. 1968 [without Latin diagnosis and type indication, Arts 36, 37].
>> = *Ascochyta imperfecta* Peck *in* N.Y. St. Mus. Bull. [Bull. N.Y. St. Mus.] **157**: 21. 1912.

Diagn. data: Boerema *et al.* (1993b) (significance of a large-spored variety with a variable proportion of septate conidia, especially when grown at low temperatures: study Rössner, see below; synonymy; pathogenicity in relation to probable origin), Rössner (1968) (figs 1–4, tabs 3–5; conidial shape and dimensions in relation with the temperature; differentiation in a '*microspora*'-form and a '*macrospora*'-form), Neergaard (1977) (occurrence on seed), Sutton (1980) (fig. 228F: drawing of conidia; specimens in CMI collection, apparently inclusive var. *medicaginis*), de Gruyter *et al.* (2002) (fig. 7b: shape of conidia, reproduced in present Fig. 31H; difficulty of differentiation from the type variety *medicaginis* at room temperature).

Description in vitro

Pycnidia similar to those of the type var. *medicaginis*.

Conidia at room temperature not essentially larger than those of the type var.; the only difference is the common presence of a few 1-septate conidia, whereas in the type var. 1-septate conidia are only occasionally found in old cultures.

Chlamydospores, if produced, similar to those of the type var.

Colonies characteristics, e.g. pigment and crystal production, agree also with those of the type var. Representative cultures CBS 404.65, CBS 112.53.

Note

The varietal epithet *macrospora* refers to the relatively large 1–3-septate conidia, up to 28×6 μm ('ascochytoid'), which may be produced in large quantities (up to 63%) at low temperatures, i.e. under winter conditions (first found out by Rössner, 1968); at low temperatures the type variety usually only produces the smaller aseptate conidia. The differences in conidial dimensions and septation at low temperature are also associated with differences in pathogenicity (see below). At common summer temperatures both varieties show also conidial differences *in vivo*, but less significant: in var. *medicaginis* conidia then usually aseptate, mostly $5.5–7 \times 2–2.5$ μm, in var. *macrospora* mostly $6.5–11 \times 2–3$ μm, and with a variable proportion of 1-septate conidia (Boerema *et al.*, 1993b).

Ecology and distribution

Ph. medicaginis var. *macrospora* appears to be relatively strongly pathogenic to lucerne, *Medicago sativa* (Leguminosae), its principal host. It commonly occurs in Eurasia, but is particularly widely distributed in North America (USA and Canada): spring black stem of lucerne. The variety probably originates from the cold mountainous regions in south-western Asia. The fact that only cold-resistant varieties of lucerne (blue alfalfa) are generally grown in North America, may explain why var. *macrospora* appears to be so widely distributed in North America.

Phoma nemophilae Neerg., Fig. 31I

> *Phoma nemophilae* Neerg. *in* Bot. Tidsskr. **44**: 361. 1938.

Diagn. data: Neergaard (1938) (fig. 1: pathogenicity), Neergaard (1977) (occurrence on seed), de Gruyter *et al.* (2002) (fig. 18: shape of conidia, reproduced in present Fig. 31I; cultural descriptions, partly adopted below).

Description in vitro

Pycnidia globose/subglobose to irregular, with 1–5 usually papillate ostioles, later developing to an elongated neck, 60–260 µm diam., glabrous or with some mycelial outgrowths. Conidial matrix off-white.

Conidia cylindrical-allantoid, with several small, scattered guttules, mainly aseptate, 4–9.5 × 1.5–2.5 µm; 1-septate conidia up to 12 × 3.5 µm, sparse.

Colonies on OA and MA 6.5–7.5 cm diam. after 7 days, regular to slightly irregular, grey olivaceous to olivaceous grey with finely floccose, white to olivaceous grey aerial mycelium. Pycnidia scattered, both on and in the agar, solitary or confluent. Application of a drop of NaOH on the colonies on OA and MA results in greenish spots later becoming red (E$^+$ reaction). Representative culture CBS 715.85.

Ecology and distribution

Common in seeds of *Nemophila* spp. (Hydrophyllaceae), especially *N. insignis* and *N. atomaria*, in Europe. Also recorded in North America (USA). May cause damping-off of seedlings and decay of stems and leaves of older plants.

Phoma nepeticola (Melnik) Dorenb. & Gruyter, Fig. 31J
Teleomorph: *Didymella catariae* (Cooke & Ellis) Sacc.

> *Phoma nepeticola* (Melnik) Dorenb. & Gruyter apud de Gruyter, Boerema & van der Aa *in* Persoonia **18**(1): 18. 2002.
>> ≡ *Ascochyta nepeticola* Melnik *in* Novoste Sist. Nizsh. Rast. **1968**: 178. 1968.
>> H ≡ *Ascochyta nepetae* É.J. Marchal & Verpl. *in* Bull. Soc. R. Bot. Belg. **59**: 23. 1927; not *Ascochyta nepetae* Davis, listed below.
>> = *Ascochyta nepetae* Davis *in* Trans. Wisc. Acad. Sci. **19**(2): 711. 1919; not *Phoma nepetae* Sousa da Câmara *in* Bolm Agric. Lisb. II, Ser. 1: 32. 1936 [≡ *Phomopsis nepetae* (Sousa da Câmara) Sousa da Câmara *in* Agron. lusit. **11**: 59. 1949], nor *Phoma nepetae* I.E. Brezhnew *in*

Uchen. Zap. Leningr. Gos. Univ. Ser. Biol. **7**: 181. 1939 [*H*; agrees with *Phoma leonuri* Letendre, sect. *Plenodomus*, **G**].

Diagn. data: de Gruyter *et al.* (2002) (fig. 4: conidial shape, reproduced in present Fig. 31J; anamorph-teleomorph connection; cultural descriptions, partly adopted below).

Description in vitro

Pycnidia globose to subglobose, with usually 1 non-papillate or slightly papillate ostiole, 70–240 μm diam., glabrous or sparsely covered by mycelial hairs. Conidial matrix buff to rosy buff.

Conidia subglobose to ellipsoidal, usually with small guttules, mainly aseptate (4–)5–7(–11.5) × 2.5–5 μm; 1-septate conidia larger, 9.5–14.5 × 2.5–5 μm, ellipsoidal to allantoid, also with small guttules.

Pseudothecia. Fresh isolates, started from single and multi ascospores of *Didymella catariae*, also produced some pseudothecia intermingled with pycnidia in cultures on OA. They were similar in appearance to those on the host (see description below).

Colonies on OA 4–4.5 cm diam. after 7 days [6.5–7.5 cm after 14 days], irregular due to recolonizing sectors, greenish olivaceous/citrine to grey olivaceous, olivaceous buff near margin, with sparse, finely floccose, white to pale grey, olivaceous aerial mycelium. On MA similar, but with compact aerial mycelium. Pycnidia scattered or in concentric zones, on the agar or submerged, solitary or confluent. On application of a drop of NaOH on colonies on OA a weak reddish non-specific colour may develop. Representative culture CBS 102635.

In vivo *(Nepeta cataria)*

Pycnidia (on leaf necroses and dry stems, subepidermal/semi-immersed, scattered) variable in dimensions, 80–200(–300) μm diam., depressed globose with more or less papillate ostiole. Pycnidial wall thin on leaves, thicker on stems. Conidia subcylindrical or sometimes slightly flexuous, mainly 1-septate, 8–15(–17) × (2.5–)3–4.5(–5) μm.

Pseudothecia (on dead stems, subepidermal, scattered or crowded) globose to subglobose, relatively small, 120–200 μm diam. with papillate pore. Asci subclavate, (52–)76–96 × (12.5–)13.5–17.5(–20) μm. Ascospores biseriate, ellipsoidal, septate in the middle and with rounded to acute ends, constricted at the septum, (13.5–)16–18.5 × 5–7(–8) μm (information additional to original description).

Ecology and distribution

A common pathogen of *Nepeta cataria* (Labiatae), a medicinal herb (cat mint) indigenous to the eastern Mediterranean, but becoming naturalized throughout Europe, and also known in North America. The fungus is also recorded on other species of *Nepeta*, and apparently widely distributed in Eurasia and

North America (Canada, USA). According to Mel'nik (2000 translated from
Mel'nik, 1977) the fungus should also affect other Labiatae, such as *Leonurus
cardiaca* and *Mentha* spp. This is quite plausible, but still needs to be confirmed
by pathogenicity tests. Mel'nik listed the anamorph under *Ascochyta leonuri*
Ellis & Dearn., as distinct from *Phoma leonuri* Letendre (sect. *Plenodomus*, **G**;
teleomorph *Leptosphaeria slovacica* Picb.).

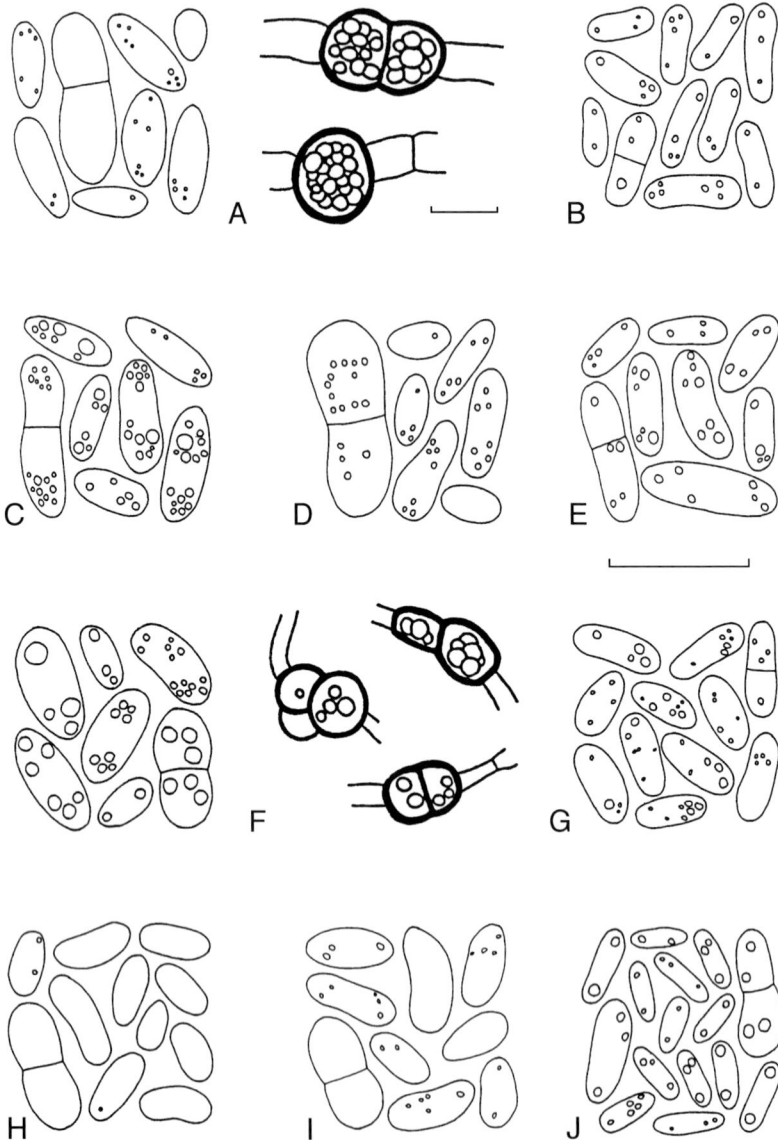

Fig. 32. Shape of conidia and some hyphal chlamydospores. **A** *Phoma pinodella*. **B** *Phoma
rhei*. **C** *Phoma rudbeckiae*. **D** *Phoma rumicicola*. **E** *Phoma sambuci-nigrae*. **F** *Phoma sojicola*.
G *Phoma strasseri*. **H** *Phoma tarda*. **I** *Phoma telephii*. **J** *Phoma valerianellae*. Bars 10 μm.

Phoma pinodella (L.K. Jones) Morgan-Jones & K.B. Burch, Fig. 32A

> *Phoma pinodella* (L.K. Jones) Morgan-Jones & K.B. Burch *in* Mycotaxon
> **29**: 485. 1987.
>> ≡ *Ascochyta pinodella* L.K. Jones *in* Bull. N.Y. St. agric. Exp.
>> Stn **547**: 10. 1927.
>> ≡ *Phoma medicaginis* var. *pinodella* (L.K. Jones) Boerema
>> apud Boerema, Dorenbosch & Leffring *in* Neth. J. Pl. Path.
>> **71**: 88. 1965.
>> = *Phoma trifolii* E.M. Johnson & Valleau *in* Bull. Ky agric.
>> Exp. Stn **339**: 73–74. 1933.

Diagn. data: Jones (1927) (pl. 1, tab. 2; description as *Ascochyta pinodella*; colour plate of culture on OA; table with range of conidial dimensions), Boerema *et al.* (1965b) (fig. 2A–E; identity of this foot rot fungus of pea with the common black stem fungus of red clover; similarity and confusion with the black stem fungus of lucerne; introduction of the infraspecific combination *Phoma medicaginis* var. *pinodella*), Dorenbosch (1970) (pl. 4 figs 5, 6: cultural characteristics on OA and MA; pl. 5 figs 1, 2: crystal production on MA; pl. 6 fig. 4: shape of chlamydospores; synoptic table 1: differential criteria against other soil-borne *Phoma* spp.), Punithalingam & Gibson (1976) (figs A–D; CMI description with sketches of pycnidium, conidia and chlamydospores; phytopathological notes), Boerema & Verhoeven (1979) (p. 167, notes on diagnostic literature), Domsch *et al.* (1980) (fig. 284: chlamydospore chains, conidia; occurrence in soil, literature references), Sutton (1980) (fig. 228H: drawing of conidia; specimens in CMI collection), White & Morgan-Jones (1987a) (fig. 1A–F: drawings of pycnidium, pycnidial wall and micropycnidial wall with conidiogenous cells, chlamydospores and conidia; pl. 1A–H: photographs of agar plate cultures, pycnidium and micropycnidia, wall cells, chlamydospores and conidia; history; morphological and cultural description), Morgan-Jones & Burch (1987a) (recombination of the fungus in species rank as *Phoma pinodella*), Boerema *et al.* (1993b) (arguments for the classification in species rank; phylogenetic relation and probable American origin), Rai & Rajak (1993) (cultural characteristics; NB: the studies sub *Ph. medicaginis* var. *pinodella* in Rai (1998) must be questioned as they include various isolates from living leaves of non-leguminous plants which are not checked on the production of dendritic crystals), Noordeloos *et al.* (1993) (figs 1, 8, 9: dendritic crystals, conidia and chlamydospores; cultural characteristics partly adopted below, production of 'pinodellalide A and B'), Bowen *et al.* (1997) (figs 1–9: micrographs of pseudothecia, asci and ascospores obtained in a single Australian isolate of *Phoma pinodella*; the asci and ascospores were similar to those of *Mycosphaerella pinodes*, but considerably larger), de Gruyter *et al.* (2002) (figs 13, 31: shape of conidia and chlamydospores, reproduced in present Fig. 32A).

Description in vitro

Pycnidia globose or irregular, with usually 1 non-papillate or slightly papillate ostiole, 96–320 μm diam., glabrous. Near the apex of a pycnidium sometimes micropycnidia occur, 30–50 μm diam. Conidial matrix whitish to pale buff.

Conidia oval to ellipsoidal with several polar guttules, mainly aseptate, occasionally 1-septate, rather variable in size, 4–7(–7.5) × 2–3.5 μm.

Chlamydospores abundant, intercalary or terminal, solitary or in chains, (sub-)globose to subcylindrical, 8–20 μm diam.

Colonies on OA 5–6.5 cm diam. after 7 days, regular, greenish olivaceous, yellowish olivaceous or olivaceous, margin pale; distinctly radially filamentous; aerial mycelium practically absent. On MA similar, but slower growing, 5–5.5 cm diam. after 7 days; the reverse on this medium shows conspicuous white, bryoid to dendritic crystals (pinodellalide A and B, see Noordeloos et al., 1993). Pycnidia abundant in concentric zones, but also scattered in the aerial mycelium and agar, solitary or confluent. Representative culture CBS 531.66.

Ecology and distribution

A well known pathogen of Leguminosae, esp. *Pisum sativum*: black stem, foot rot, leaf spot. It is in fact plurivorous and isolated from a wide range of plants. Being seed-borne, it is now distributed worldwide in arable soils. The fungus is often confused with *Mycosphaerella pinodes* (Berk. & A. Bloxam.) Vestergr., anam. *Ascochyta pinodes* L.K. Jones, which agrees in host range, disease symptoms, production of chlamydospores and dendritic crystals of pinodellalide A and B (Noordeloos et al., 1993). However, the cultural characteristics of *M. pinodes* are different, mature conidia are always septate (ascochytoid) and ascomata of the teleomorph usually also develop in fresh cultures. Both fungi are probably related (Boerema et al., 1993b). In this context it is very notable that the asci and ascospores reported by Bowen et al. (1997) in cultures of an Australian isolate of *Ph. pinodella* were similar morphologically to those of *M. pinodes in vitro*, albeit considerably larger.

Phoma rhei (Ellis & Everh.) Aa & Boerema, Fig. 32B

> Phoma rhei (Ellis & Everh.) Aa & Boerema apud de Gruyter, Boerema & van der Aa in Persoonia **18**(1): 42. 2002.
>> ≡ *Ascochyta rhei* Ellis & Everh. in Proc. Acad. nat. Sci. Philad. **1893**: 160. 1893 [as '(Ellis & Everh.) comb. nov.', but based on a illegitimate name (see after this) which means (Art. 58.1) that it has to be treated as a new name dating from 1893].
>> H ≡ *Phyllosticta rhei* Ellis & Everh. in J. Mycol. **5**: 145. 1889; in Proc. Acad. nat. Sci. Philad. **1891**: 77. 1891 [complementary description]; not *Phyllosticta rhei* Roum. in Revue mycol. **9**: 152. 1887.
>> = *Phyllosticta halstediana* Allesch. in Rabenh. Krypt.-Flora [ed. 2], Pilze **6** [Lief. 61]: 144. 1898 [vol. dated '1901'] [as 'new name']; with reference to the collector of the specimen noted in the complementary description].

Diagn. data: de Gruyter *et al.* (2002) (fig. 25: shape of conidia, reproduced in present Fig. 32B; note on nomenclature; cultural descriptions, partly adopted below).

Description in vitro

Pycnidia globose to subglobose, with 1(–2) non-papillate or papillate ostiole(s), 70–280 μm diam., glabrous or with some hyphal outgrowths. Conidial matrix white.

Conidia cylindrical to ellipsoidal/allantoid, with several small guttules, mainly aseptate, (3.5–)5–8(–10.5) × 1.5–3 μm, 1-septate conidia up to 18 × 3 μm, sparse.

Colonies on OA *c.* 7 cm diam. after 7 days, regular, olivaceous buff to greenish olivaceous/grey olivaceous with dull green/olivaceous stellate pattern or zones; aerial mycelium sparse, felty to finely floccose, white to pale olivaceous grey. On MA slower growing, 6–6.5 cm diam. after 7 days, with coarsely floccose to woolly aerial mycelium; reverse leaden grey to leaden black/olivaceous black, honey/citrine near margin. Pycnidia scattered, both on and in the agar as well as in aerial mycelium. Representative culture CBS 109177.

Ecology and distribution

A cosmopolitan pathogen of cultivated rhubarb plants, *Rheum* spp. (Polygonaceae): leaf spot. The second complementary description by Ellis & Everhart was listed separately in Saccardo's Sylloge Fungorum, which explains the 'new name' introduced by Allescher. The occasionally septate conidia are the reason for its subsequent classification into *Ascochyta*.

Phoma rudbeckiae Fairman, Fig. 32C

> *Phoma rudbeckiae* Fairman *in* Proc. Rochester Acad. Sci. **1**: 51. 1890.
> > = *Phyllosticta rudbeckiae* Ellis & Everh. *in* Proc. Acad. nat. Sci. Philad. **1895**: 430. 1895.
> > > ≡ *Ascochyta rudbeckiae* (Ellis & Everh.) H.C. Greene *in* Am. Midl. Nat. **41**: 753. 1949.

Diagn. data: Seaver (1922) (disease symptoms; suggestion of *Phoma-Phyllosticta* synonymy), van der Aa & Vanev (2002) (description *in vivo*), de Gruyter *et al.* (2002) (fig. 16: conidial shape, reproduced in present Fig. 32C; cultural descriptions, partly adopted below).

Description in vitro

Pycnidia globose to subglobose, with 1(–2) non-papillate ostiole(s), 60–360 μm diam., glabrous or sparsely covered by short mycelial hairs. Conidial matrix salmon to saffron.

Conidia ellipsoidal to allantoid, usually with several distinct guttules, aseptate, (5.5–)6.5–11 × 2–4 μm, or 1-septate, 9–14.5 × 3.5 μm.

Colonies on OA *c.* 6.5 cm diam. after 7 days, slightly irregular, honey to pale luteous due to a diffusible pigment and with ochraceous tinges due to abundant pycnidia; aerial mycelium woolly to floccose, whitish. On MA slower growing, *c.* 5 cm after 7 days, with compact woolly to floccose white/salmon to pale olivaceous grey aerial mycelium; reverse pigmented amber to ochraceous/fulvous and partly umber, with citrine green to yellow-green needle-like crystals. Pycnidia mainly on the agar, solitary or confluent. With application of a drop of NaOH on the colonies violet/red discoloration of the pigments occur. Representative culture CBS 109180.

Ecology and distribution

A specific pathogen of *Rudbeckia* spp. (Compositae), esp. *R. lacinata*, indigenous to North America, like the hosts: leaf spot. Now also found in Europe. The leaf spots are rather large, opaque-blackish and with a clearly defined outline. Mature pycnidia on the spots may contain a high percentage of 1-septate conidia, 8–12 × 2–3 μm. On dead tissue the pycnidia usually contain relatively small aseptate conidia, 4–6 × 2–3 μm.

Phoma rumicicola Boerema & Loer., Fig. 32D

> *Phoma rumicicola* Boerema & Loer. apud Boerema, Loerakker & Laundon *in* N.Z. Jl Bot. **18**: 473. 1980.

Diagn. data: Boerema *et al.* (1980) (figs 1–4: disease symptoms, cultures on OA, shape of pycnidia and conidia; description *in vitro*), de Gruyter *et al.* (2002) (fig. 21: conidial shape, reproduced in present Fig. 32D; cultural descriptions, partly adopted below).

Description in vitro

Pycnidia globose to irregular, with 1–4 non-papillate or papillate ostioles, relatively large, 130–650 μm diam., glabrous. Conidial matrix off-white to pale vinaceous.

Conidia ellipsoidal, with several small, scattered guttules, aseptate or occasionally 1-septate, 5–9(–15) × 2–5 μm.

Colonies on OA 5.5–7 cm after 7 days, regular, colourless to buff to pale grey olivaceous/olivaceous grey, partly citrine or cinnamon near margin; aerial mycelium felty to floccose, white to pale olivaceous grey to grey olivaceous. The reverse is colourless to grey olivaceous/olivaceous grey or dull green with partly vinaceous buff, olivaceous black at centre. On MA slower growing, 4–5 cm after 7 days, dull green to citrine. Pycnidia only abundantly on OA, scattered, mainly on the agar, solitary or confluent. Representative culture CBS 683.79.

Ecology and distribution

This fungus is probably a common pathogen of *Rumex obtusifolius* (Polygonaceae) wherever grown: leaf spot. It remained unrecognized because it produces inconspicuous pycnidia *in vivo* and the disease symptoms are very similar to those of *Ramularia rubella* (Bonord.) Nannf.: small, purplish red spots which expand to 1 cm in diameter and have rusty brown, slightly sunken centres and dark purplish margins. The presence of *Ph. rumicicola* can be easily confirmed by isolation. *In vitro* it shows some resemblance with *Phoma acetosellae* (A.L. Sm. & Ramsb.) Aa & Boerema (this section), common on *Rumex acetosella*. However, it can be differentiated by its faster growth rate, and larger conidia, see also Boerema *et al.* (1980).

Phoma sambuci-nigrae (Sacc.) E. Monte *et al.,* Fig. 32E

> *Phoma sambuci-nigrae* (Sacc.) E. Monte, Bridge & B. Sutton *in* Mycopathologia **115**: 102. 1991.
>> ≡ *Phoma herbarum* f. *sambuci-nigrae* Sacc. *in* Sylloge Fung. **3**: 133. 1884.
>> ≡ *Phoma exigua* var. *sambuci-nigrae* (Sacc.) Boerema & Höweler *in* Persoonia **5**(1): 26. 1967.
>> = *Phyllosticta sambucina* Allesch. ex Mig. *in* Thomé KryptogFlora Pilze **4**(1): 33. 1921; not *Phoma sambucina* Sacc. *in* Michelia **2**(1): 97. 1880 [≡ *Phomopsis sambucina* (Sacc.) Traverso *in* Fl. ital. Cryptog. **2**(1): 269. 1906].

Diagn. data: Boerema & Höweler, 1967 (pl. 2 fig. 2, pl. 3, fig. 6; disease symptoms and growth habit on CA; synonymy, classification as a specialized variety of *Phoma exigua*), Monte *et al.* (1991) (variety raised to species rank on account of comparative study with several other *Phoma* taxa *in vitro*), de Gruyter *et al.* (2002) (fig. 19: conidial shape, reproduced in present Fig. 32E; cultural descriptions, partly adopted below).

Description in vitro

Pycnidia globose to subglobose, with 1–3 non-papillate ostioles, 80–240 μm diam., glabrous. Conidial matrix off-white to buff.

Conidia variable in shape, subglobose, ellipsoidal to oblong or allantoid, usually with small guttules, mainly aseptate, (3.5–)5–8(–10.5) × 2–3.5 μm.

Colonies on OA 7–8 cm diam. after 7 days, regular to slightly irregular, greenish olivaceous to grey olivaceous/olivaceous grey and greenish olivaceous/citrine near margin. On MA similar. Pycnidia scattered, both on and in the agar, solitary or confluent. Application of a drop of NaOH on the colonies on OA and MA results in an E$^+$ reaction: greenish, then red. Representative culture CBS 629.68.

Ecology and distribution

Widespread on elder, *Sambucus nigra* (Caprifoliaceae), in Eurasia: leaf spot, shoot dieback. Recent comparative studies have shown that this pathogen of elder is most uniform and stable in its cultural characteristics. Therefore it deserves the species rank in spite of its morphological similarity with the ubiquitous *Phoma exigua* Desm. var. *exigua*, the type species of this section.

Phoma sojicola (Abramov) Kövics *et al.*, Fig. 32F

> *Phoma sojicola* (Abramov) Kövics, Gruyter & Aa *in* Mycol. Res. **103**: 1066. 1999.
>> *V* ≡ *Ascochyta sojicola* Abramov, Bolezni i Vrediteli Soievykh Bobov no Dal'nem Vostoke: 62. 1931 [as '*sojaecola*'].
>> *H V* ≡ *Ascochyta sojicola* Nelen *in* Nov. Sist. Niz. Rast. **14**: 105. 1977 [as '*sojaecola*'; erroneously also listed with the author citation 'Abramov ex Nelen'].

Diagn. data: Kövics *et al.* (1999) (figs 1, 2: conidia and chlamydospores; 'most common coelomycetous pathogen of soybean'; discussion of history, pathogenicity and nomenclature; detailed cultural description, partly adopted below), de Gruyter *et al.* (2002) (figs 14, 32: shape of conidia and chlamydospores, reproduced in present Fig. 32F; possible confusion with other *Phoma* spp., causing similar disease symptoms).

Description in vitro

Pycnidia (sub-)globose or irregular, with 1 or more ostiole(s), 86–345 μm diam., glabrous. Conidial matrix rosy buff to vinaceous buff.

Conidia oblong to ellipsoidal, mainly aseptate, 5–8 × 2–3.5 μm, with several small and/or large guttules, occasionally 1-septate, 8–12.5 × 3.5–5 μm.

Chlamydospores intercalary or terminal, solitary or in chains, sometimes aggregated, (sub-)globose to cylindrical, 8–16 × 7–8 μm.

Colonies on OA 5.5–6.5 cm diam. after 7 days [on MA similar], regular, colourless to pale olivaceous grey or greenish olivaceous to grey olivaceous; aerial mycelium poorly developed, white to pale olivaceous grey. Pycnidia scarce on OA, on the surface and in the agar at the margin of the colony, sometimes aggregated in distinct sectors, solitary or confluent. Distinctive are white dendritic crystals, but only produced in fresh isolates and especially on MA. With application of a drop of NaOH on colonies on MA a pale, non-specific, brownish red colour may appear. Representative culture CBS 100580.

Ecology and distribution

This seed-borne fungus appeared to be the most common *Phoma* species involved with the leaf and pod spot disease of soybean, *Glycine max* (Leguminosae), in Eurasia. Pathogenicity tests, however, have shown that various plurivorous species of *Phoma*, e.g. *Ph. exigua* var. *exigua* and *Ph. pinodella* (both this section) may cause similar symptoms on soybean, see Kövics *et al.* (1999).

Phoma strasseri Moesz, Fig. 32G

> *Phoma strasseri* Moesz in Bot. Közl. **22**: 45. 1924.
> > *H* ≡ *Phoma menthae* Strasser in Verh. zool.-bot. Ges. Wien **60**: 317. 1910; not *Phoma menthae* Roum. *in* Revue mycol. **9**: 26. 1887.

Diagn. data: Horner (1971) (figs 1, 2; disease symptoms; identification and pathogenicity), Melouk & Horner (1972) (growth in culture and pathogenicity to peppermint), de Gruyter *et al.* (2002) (fig. 20: conidial shape, reproduced in present Fig. 32G; cultural descriptions, partly adopted below).

Description in vitro

Pycnidia globose with 1–3 usually non-papillate ostioles, mostly 100–230 μm diam., glabrous or with mycelial outgrowths. Conidial matrix buff to rosy/salmon.

Conidia ellipsoidal with several small, scattered guttules, mainly aseptate, 4.5–6.5 × 2–3 μm; occasionally 1-septate of similar size.

Colonies on OA and MA fast growing, regular, completely filling the plates in 7 days; growth rate after 5 days already 6–7 cm, colourless to olivaceous/grey olivaceous with poorly developed whitish or olivaceous grey aerial mycelium. Some strains showed a positive reaction with the sodium hydroxide test: on application of a drop of NaOH green → red (E$^+$ reaction). Representative culture CBS 261.92.

Ecology and distribution

A serious pathogen of mint, *Mentha* spp. (Labiatae), found in Europe, Japan, New Zealand and North America: rhizome and stem rot. Occasionally the fungus has also been isolated from other Labiatae, namely *Monarda didyma* (North America) and *Stachys officinalis* (Bulgaria). There is also a report from *Valeriana* sp., but that appeared to be based on a coincidental isolation.

Phoma tarda (R.B. Stewart) H. Verm., Fig. 32H

> *Phoma tarda* (R.B. Stewart) H. Verm., Coffee Berry Dis. Kenya: 14. 1979 [Thesis Agric. Univ. Wageningen].
> > ≡ *Ascochyta tarda* R.B. Stewart *in* Mycologia **49**: 430. 1957.
> > = *Ascochyta coffeae* Henn. *in* Hedwigia **41**: 307. 1902; not *Phoma coffeae* Delacr. *in* Bull. Soc. Mycol. Fr. **13**: 122. 1897.

Diagn. data: Stewart (1957) (fig. 1: disease symptoms; description as a species of *Ascochyta*, 'which immature spores may be predominantly or entirely aseptate'), de Gruyter *et al.* (2002) (fig. 23: conidial shape, reproduced in present Fig. 32H; cultural descriptions, partly adopted below).

Description in vitro

Pycnidia globose to subglobose with non-papillate or papillate ostioles, 120–255 µm diam., glabrous. Conidial matrix white.

   Conidia subglobose to ellipsoidal/allantoid, eguttulate or with some small guttules, mainly aseptate, (3–)4–7(–9) × 2–3 µm, 1-septate conidia of similar size.

   Colonies on OA and MA fast growing, the plates in 7 days full-grown, growth rate 5.5–7.5 cm after 5 days, slightly irregular, olivaceous grey to grey olivaceous/dull green; aerial mycelium floccose to woolly, olivaceous grey to smoke grey. Reverse olivaceous grey to olivaceous/olivaceous black to leaden black. In the mycelium somewhat dark, olivaceous swollen cells occur. Pycnidia scattered, mostly on the agar, solitary or confluent. With the NaOH spot test on OA a pale purplish grey non-specific colour may appear. Representative culture CBS 109183.

Ecology and distribution

Recorded as a noxious pathogen of Arabian or arabica coffee, *Coffea arabica* (Rubiaceae), in Africa (Ethiopia, Kenya, Cameroon): leaf blight and stem dieback. The epithet *tarda* refers to the 'late appearance' of septa in the conidia. The fungus has also been recently isolated from coffee shrubs in Brazil, and appears to have been first described in that country. In the description by Stewart (1957) it is noted that pseudothecia of an unnamed species of *Didymella* ('*Mycosphaerella*') in natural infections occur frequently. There should be marked differences in susceptibility of *C. arabica* selections.

Phoma telephii (Vestergr.) Kesteren, Fig. 32I

> *Phoma telephii* (Vestergr.) Kesteren *in* Neth. J. Pl. Path. **78**: 117. 1972.
>> ≡ *Ascochyta telephii* Vestergr. *in* Öfvers. K. VetensAkad. Förh. **54 (1897)**: 41. 1897.
>> = *Ascochyta sedi-purpurei* Rothers *in* Zashch. Rast. Vredit. **6**: 263. 1929.
>> = *Phoma tabifica* Kesteren *in* Gewasbescherming **2**: 74. 1971.

Diagn. data: Vestergren (1897) (original description on *Sedum telephium*), van Kesteren (1972) (fig. 1: disease symptoms; nomenclature of the pathogen; description *in vivo* and *in vitro*), de Gruyter *et al.* (2002) (fig. 8: shape of conidia, reproduced in present Fig. 32I; cultural descriptions, partly adopted below).

Description in vitro

Pycnidia globose/subglobose to irregular, with 1(–2) sometimes papillate ostiole(s), 50–350 μm diam., glabrous or with short mycelial outgrowths. Conidial matrix white to salmon.

Conidia ellipsoidal to allantoid, with several small, scattered guttules, mainly aseptate, (3.5–)5–7(–9.5) × 2.5–3.5 μm; 1-septate conidia 6.5–13.5 × 3–4.5 μm.

Colonies on OA 4–5.5 cm after 7 days, regular, greenish olivaceous/grey olivaceous to olivaceous and olivaceous buff to citrine near margin; aerial mycelium sparse felty, (pale) olivaceous grey. On MA somewhat slower growing, 3.5–4.5 cm, with woolly to floccose, white to olivaceous grey aerial mycelium. Pycnidia scattered, both on and in the agar as well as in aerial mycelium. Representative cultures CBS 760.73 and CBS 109175.

Ecology and distribution

A common pathogen on *Sedum telephium* and other *Sedum* spp. (Crassulaceae) indigenous to Europe. The fungus causes sunken purple spots on stems and leaves: purple blotch disease. The perennial plants may suffer seriously from this disease.

Phoma valerianellae Gindrat *et al.*, Fig. 32J

> *Phoma valerianellae* Gindrat, Semecnik & Bolay *in* Revue hort. suisse **40**: 350–351. 1967 ['rejected' by Gindrat *in* Revue hort. suisse **41**: 181. 1968, but validly published, see Boerema *in* Persoonia **6**(1): 43–44. 1970].
>
> > H = *Phoma valerianellae* Boerema & C.B. de Jong *in* Phytopath. Z. **61**: 368–369. 1968.
> >
> > = *Phoma herbarum* f. *valerianae* Sacc. *in* Michelia **2**(2): 337. 1881.

Diagn. data: Gindrat *et al.* (1967) (figs 1–4; pathogenicity on seedlings of corn salad, description and drawings of pycnidia and conidia), Boerema & de Jong (1968) (figs 1–4: disease symptoms on seedlings; fig. 5: cultures on OA; figs 6, 9: drawings of mycelia, pycnidia, conidia and conidiogenous cells; table on susceptible Valerianaceae; detailed description *in vivo* and *in vitro*), Boerema (1970) (synonymy), Neergaard (1977) (occurrence on seed), de Gruyter *et al.* (2002) (fig. 24: shape of conidia, reproduced in present Fig. 32J; cultural descriptions, partly adopted below).

Description in vitro

Pycnidia globose to subglobose with usually 1–2(–5) papillate ostiole(s), 60–285 μm diam., glabrous or sparsely covered by short mycelial hairs. Conidial matrix white to pale luteous/ochraceous.

Conidia ellipsoidal to cylindrical, usually with small guttules, generally aseptate, 3.5–5.5(–7) × 1–2 μm, occasionally some 1-septate conidia, up to 9 × 3 μm, occur.

Colonies on OA 7–7.5 cm after 7 days, regular, dull green with (finely) floccose, white to olivaceous grey aerial mycelium; reverse of colony grey olivaceous to buff, leaden grey at centre. On MA somewhat faster growing 7.5–8 cm, with compact floccose aerial mycelium. Pycnidia scattered, on the agar or submerged, as well as in aerial mycelium. Representative culture CBS 329.67.

Ecology and distribution

A seed-borne pathogen of Valerianaceae in Europe. It is particularly common and widespread on corn salad, *Valerianella locusta* var. *oleracea*, and other species of *Valerianella*. The fungus may attack roots, stems and leaves in the seedling stage: damping-off. On *Valeriana* spp. it may be confused with a seed-borne saprophytic species, *Phoma valerianae* Henn. (sect. *Phoma*, **A**).

13 F *Phoma* sect. *Sclerophomella*

Phoma sect. *Sclerophomella* (Höhn.) Boerema *et al.*

Phoma sect. *Sclerophomella* (Höhn.) Boerema, Gruyter & Noordel. apud Boerema *in* Mycotaxon **64**: 331. 1997.
> ≡ *Sclerophomella* Höhn. *in* Hedwigia **59**: 237. 1917 [conidiogenesis misinterpreted, *see e.g.* Boerema, 1976: 298].
> = *Sclerochaetella* Höhn. *in* Hedwigia **59**: 251. 1918 ['*chaetella*' refers to a discordant setae-bearing element, *see* Boerema, Loerakker & Hamers, 1996: 177–178].

Type: *Sclerophomella complanata* (Tode : Fr.) Höhn. ≡ *Phoma complanata* (Tode : Fr.) Desm.

The species of section *Sclerophomella* have thick-walled pseudoparenchymatous pycnidia (Fig. 33A, B) *in vivo* superficially resembling the thick-walled pycnidial phenotype in species of *Phoma* sect. *Plenodomus* (**G**). Just as in sect. *Plenodomus* the pycnidium is initially closed, the opening occurs only late in the growing process, i.e. the pycnidium has a pore instead of a predetermined ostiole. In some species there is also a retarded development of the pycnidial cavity, forming of a 'pycnosclerotium'. This contains a compact mass of cells that later disintegrate (histolysis) (Fig. 33C). However, scleroplectenchyma, characteristic of sect. *Plenodomus* is always lacking. With Lugol's iodine (JKJ) the contents of the cells usually become red, but not the cell walls as in scleroplectenchyma.

Members of sect. *Sclerophomella* belong to the ascomycetous genus *Didymella* Sacc. ex Sacc. and not to *Leptosphaeria* as in sect. *Plenodomus*.

The hyaline conidia of species in sect. *Sclerophomella* are mainly aseptate *in vitro*. Secondary septation of conidia may occur, being a typical phenomenon in the type species of the section, *Phoma complanata*. In host tissue, the pycnidia of this species usually contain only aseptate conidia of 'common *Phoma*-size' (see the generic characters in the Introduction). But sometimes a high percentage of the conidia becomes larger and 1-septate. They may be distinctly larger, ascochytoid, resembling the large conidial dimorph in sect.

Heterospora (**B**). Old pycnidia often contain many swollen, dark, 1-septate conidia, as in species of sect. *Peyronellaea* (**D**). *In vitro* the conidia of *Ph. complanata* are highly variable, but usually aseptate in fresh cultures; in old cultures some ascochytoid conidia may occur.

The section includes species with and without chlamydosporal structures. If present, the chlamydospores may be one-celled, solitary or in series, but also multicellular and micro-pseudosclerotioid. Swollen cells occasionally occur.

Diagn. lit.: Boerema (1976) (tab. 9, 1, conidial variability *in vitro*), Cerkauskas (1985) (detailed description of the type species), Boerema *et al.* (1986) (provisional introduction of the present section; pycnidial characteristics), Boerema *et al.* (1996) (fig. 13, conidial variability *in vivo*), Boerema *et al.* (1998) (precursory contribution to this chapter), de Gruyter *et al.* (2002) (additional species).

Synopsis of the Section Characteristics

- Pycnidia relatively thick-walled, pseudoparenchymatous, both *in vitro* and *in vivo* (no scleroplectenchyma), poroid, glabrous; sometimes initially containing a compact mass of cells (pycnosclerotia) which later disintegrates.
- Conidia mostly aseptate *in vitro*, but sometimes mixed with 1- or more septate conidia, ranging from normal size for *Phoma* to extremely large ascochytoid; *in vivo* a high percentage may be septate, normal size or ascochytoid.
- Chlamydospores, if present, simple or multicellular (alternarioid, pseudosclerotioid); occasional occurrence of swollen cells.
- Micro-pseudosclerotia present in one species.
- Teleomorph, if known, *Didymella*.

Key to the Species Treated of Section *Sclerophomella*

Differentiation based on the characteristics *in vitro*

Two insufficiently known species of this section so far can be differentiated only by their specific host relations and characteristics *in vivo*, see 'Possibilities of identification *in vivo*', p. 301.

1. a. Conidia small, not exceeding 5.5 μm, aseptate...**2**
    b. Conidia larger, aseptate or septate **3**

2. a. Conidia cylindrical-oblong, 3–4 × 0.5–1 μm; growth rate slow on OA and MA, *c.* 2 cm after 7 days, a diffusible pigment produced on OA and MA, staining the agar ochraceous to ochre..*Ph. incompta*
        [Specific pathogen of the olive, *Olea europaea*: shoot wilt. Known from southern Europe (Crete, Greece and Italy).]
    b. Conidia ellipsoidal, sometimes reniform, 4–5.5 × 1.5–2.5 μm; growth rate moderate to fast on OA and MA, 4.5–7 cm after 7 days; diffusible pigment absent ....*Ph. dictamnicola*

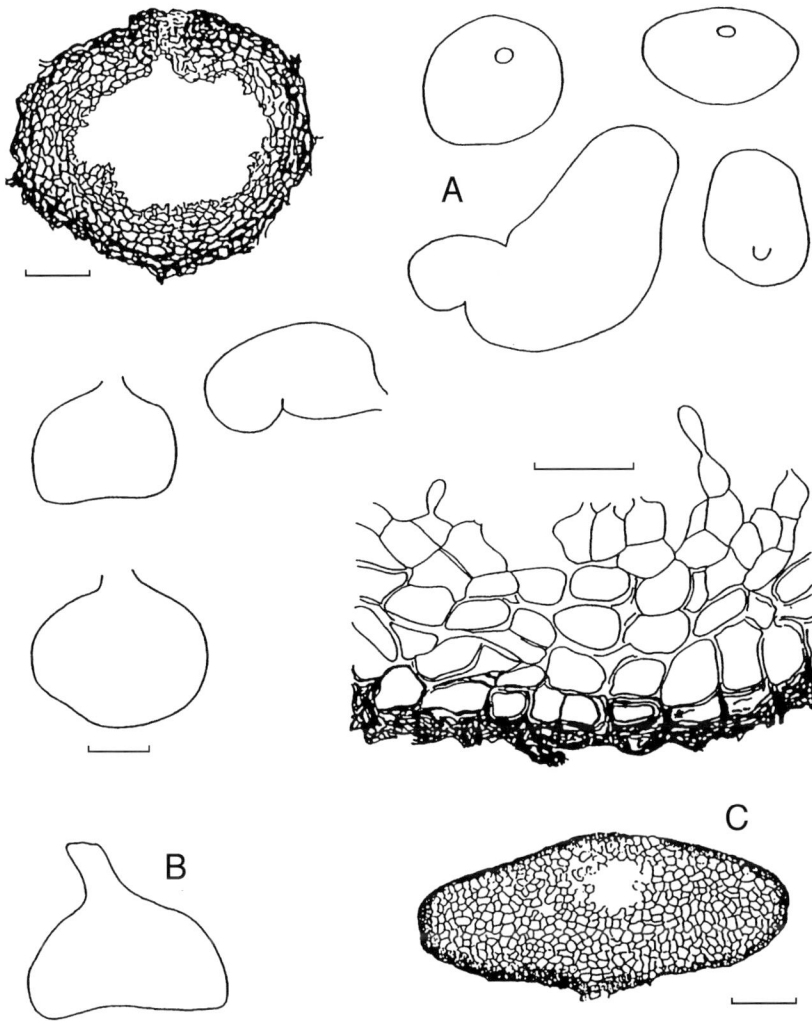

Fig. 33. A *Phoma complanata*, type species of *Phoma* sect. *Sclerophomella*. Vertical section of thick-walled pycnidium *in vivo*, shape of pycnidia *in vitro* and detail of pycnidial wall with conidiogenous cells. **B** *Phoma nigrificans*. Pycnidium with elongated neck. **C** *Phoma versabilis*. Vertical section of pycnosclerotium, the central cells gradually disintegrate, which process finally results in a pycnidial cavity. Bar vertical section 50 μm, bar pycnidial shapes 100 μm and bar detail pycnidial wall 10 μm.

[Specific pathogen of the firework-plant, *Dictamnus albus*: leaf spot. Recorded from Europe and North America.]

3. a. Growth rate slow, *c.* 3 cm on OA after 7 days and *c.* 2 cm on MA, colonies producing micro-pseudosclerotia, in the agar; conidia aseptate, 4.5–11 × 2.5–4 μm or 1(–2)-septate up to 16 × 3.5 μm..............*Ph. gentianae-sino-ornatae*

[Isolated from rotting roots of *Gentiana sino-ornata*. Thus far only known from Scotland.]

   b.   Growth rate faster, 3.5–8 cm on OA after 7 days; micro-pseudosclerotia absent; conidia aseptate, sometimes with some larger septate conidia.........................**4**

4. a.   NaOH reaction positive, yellow-green, later red (E$^+$ reaction), specific on semi-parasitic Scrophulariaceae **5**
   b.   NaOH reaction negative or not specific; not on semi-parasitic Scrophulariaceae..**6**

5. a.   Dendritic crystals present; conidia aseptate, mostly 4–7.5 × 2–3.5 μm, occasionally septate, ascochytoid, up to 18 × 8 μm..
         ....*Ph. alectorolophi*, teleom. *Did. alectorolophi*
         [Necrophyte of semi-parasitic Scrophulariaceae on Gramineae: *Melampyrum, Rhinanthus* and *Pedicularis* spp. Probably widespread in Europe.]
   b.   Crystals absent; conidia aseptate, 3.5–6 × 1–2 μm ....*Ph. sylvatica*
         [Apparently a specific necrophyte of *Melampyrum* spp. (Scrophulariaceae, semi-parasitic on Gramineae). Probably widespread in Europe.]

6. a.   Pycnidia often show a retarded development of the pycnidial cavity, i.e. formation of 'pycnosclerotia', containing a compact mass of cells which afterwards disintegrates[12]; growth rate *c.* 4 cm on OA and *c.* 3 cm on MA after 7 days; conidia 4.5–8 × 2–3 μm ....*Ph. versabilis*
   b.   Typical 'pycnosclerotia' absent..**8**

7. a.   Conidia relatively uniform oblong-ellipsoidal to allantoid, usually with small polar guttules and always aseptate, 4.5–10.5 × 1.5–4 μm. .... *Ph. nigrificans*
         [Recorded from wild and cultivated crucifers in northern Eurasia. Apparently a cold-tolerant opportunistic parasite. Repeatedly found on horse radish, *Armoracia rusticana*, and winter oilseed rape, *Brassica napus* var. *oleifera*: black leg symptoms.]
   b.   Conidia more variable, mainly aseptate but (esp. in fresh cultures) also some 1-septate (often relatively large), without guttules or with several scattered guttules **8**

8. a.   Conidia usually without guttules, variable ellipsoidal to subcylindrical, aseptate 4–10.5 × 2–5 μm, septate usually similar but also significantly larger, 12–20.5 × 3.5–5 μm ....*Ph. protuberans*
         [Specific pathogen of *Lycium halimifolium*: leaf spot. Recorded in Europe and North America.]
   b.   Conidia usually with several, scattered or polar guttules, highly variable in shape and size **9**

---

[12] Pycnosclerotia *in vitro* are also found in a species treated under sect. *Pilosa* (**I**) on account of somewhat 'hairy' pycnidia and relation with a teleomorph in *Pleospora*. Growth rate 7–8 cm on OA after 7 days. Conidia 4.5–7 × 1.5–2.5 μm *Ph. typhina*, teleom. *Pleospora typhicola*. [Commonly associated with leaf spots and decayed leaves, leaf sheaths and stems of *Typha* spp. in Europe.]

9. a.   Growth rate 6–8 cm on OA after 7 days; conidia mainly aseptate, usually 4–10
         × 2–3 μm, occasionally also distinctly larger, 1-septate, ascochytoid, up to 34
         × 10 μm ....*Ph. complanata*
         [Common on last year's dead stems of wild umbelliferous plants in Eurasia
         and North America. Known as seed-borne pathogen of parsnip, *Pastinaca
         sativa*, parsley, *Petroselinum crispum*, and carrot, *Daucus* carota: canker, leaf
         spot.]

   b.   Growth rate 3.5–4.5 cm on OA after 7 days; conidia mainly aseptate,
         usually 7–9 × 2–3 μm, the 1-septate conidia measure 9.5–13 ×
         2.5–3.5 μm ....*Ph. polemonii*
         [Specific pathogen of *Polemonium* spp.: leaf spot. Recorded in Eurasia and
         North America.]

Possibilities of identification *in vivo*

The species of sect. *Sclerophomella* are characterized by the ability to produce
thick-walled pseudoparenchymatous poroid pycnidia and therefore are also
usually well recognizable *in vivo*. The various species so far distinguished
through comparative study *in vitro* show restricted and well-defined host
ranges, which facilitates their identification *in vivo*.

Two insufficiently known species, treated in this chapter, can be differentiated
only by their specific host relations, typical thick-walled 'Sclerophomella'-pycni-
dia and their conidial dimensions:

1. Pycnidia globose (*in vitro* with elongated neck) 130–180 μm diam.
Conidia cylindrical, aseptate, 7.5–13.5 × 2.5–4.5 μm...............*Ph. boerhaviae*
[Specific pathogen of *Boerhavia diffusa* (Nyctaginaceae): twig blight. Recorded
from south-west Asia (India, Pakistan), but probably also elsewhere on the
host.]

2. Pycnidia subglobose (with at the sides dark twisting hyphae) 200–350 μm
diam.
Conidia notable broad, ovate-ellipsoidal, mostly 6–8 × 3.5 μm ....
....*Ph. syriaca* [? teleom. *Did. syriaca*]
[Thus far only known from *Phlomis brevilabris* (Labiatae) in the subalpine
region of Mt Sania, Lebanon.]

Distribution and Facultative Host Relations in Section *Sclerophomella*

Plurivorous
                    Ph. versabilis
                    [Recorded on dead stems of diverse herbaceous plants in
                    Europe.]

With special host relation:

Cruciferae,
e.g. *Thlaspi arvense* and   *Ph. nigrificans*
the cultivated *Armoracia*   (teleom. *Did. macropodii*)
rusticana and *Brassica*
napus var. *oleifera*   [Recorded from wild and cultivated crucifers in northern
                                              Eurasia: black leg symptoms. Apparently a cold-tolerant
                                              opportunistic parasite.]

Gentianaceae
Gentiana sino-ornata   *Ph. gentianae-sino-ornatae*
                                              [So far only recorded in Scotland; isolated from roots of living
                                              plants showing a severe root rot and blackening of tissue.]

Labiatae
Phlomis brevilabris   *Ph. syriaca*
                                              (teleom. *Did. syriaca*)
                                              [So far only recorded in a subalpine region of Lebanon.]

Nyctaginaceae
Boerhavia diffusa   *Ph. boerhaviae*
                                              [In south-west Asia (India, Pakistan) found in association
                                              with dark lesions girdling the twigs: twig blight.]

Oleaceae
Olea europaea   *Ph. incompta*
                                              [Known as a specific pathogen of the olive in southern
                                              Europe (Crete–Greece and Italy), causing a progressive
                                              withering of young shoots, which later die without defolia-
                                              tion; shoot wilt.]

Polemoniaceae
Polemonium spp., esp.   *Ph. polemonii*
P. caeruleum   [Specific pathogen found in Europe and North America:
                                              leaf spot.]

Rutaceae
Dictamnus albus   *Ph. dictamnicola*
                                              [Frequently recorded as a specific pathogen in Europe and
                                              North America, causing irregular spots on the tips or mar-
                                              gins of leaves; leaf spot.]

Scrophulariaceae
Pedicularis, Rhinanthus   *Ph. alectorolophi*
and *Melampyrum* spp.   (teleom. *Did. alectorolophi*)
                                              Ph. sylvatica
                                              (? teleom. *Did. winteriana*)
                                              [Frequently recorded on dead tissue of species of these
                                              semi-parasitic genera on Gramineae in Europe; *Ph. sylvat-
                                              ica* is apparently restricted to *Melampyrum* spp.]

Solanaceae
Lycium halimifolium   *Ph. protuberans*
                             [Specific pathogen found in Europe and North America:
                             leaf spot.]

Typhaceae
Typha angustifolia and   *Ph. typhina* (on account of somewhat hairy pycnidia
Typha latifolia   and relation with the teleomorph *Pleospora typhicola*
                             treated under sect. *Pilosa*, **I**)
                             [Commonly found in Europe in association with leaf spots
                             and on dead leaves, leaf sheaths and stems.]

Umbelliferae,
e.g. the cultivated *Pastinaca*   *Ph. complanata*
sativa, Petroselinum crispum
and *Daucus carota*   [Very common on last year's dead stems of wild umbellifer-
                             ous plants in temperate Eurasia and North America.
                             Known as seed-borne pathogen of the above listed culti-
                             vated plants (lesions on petioles and roots); canker.]

Descriptions of the Taxa

Mainly based on characteristics in culture; with additional morphological data
in vivo (incl. teleomorphs). Two species are only known *in vivo*.

Phoma alectorolophi Boerema et al., Fig. 34A

Teleomorph: ***Didymella alectorolophi* Rehm sensu lato.**

Phoma alectorolophi Boerema, Gruyter & Noordel. *in* Persoonia **16**(3):
366. 1997.

Diagn. data: Corbaz (1957) (char. teleomorph; variability conidia *in vivo* and *in
vitro* sub *Didymella alectorolophi* and *D. pedicularidis* Arx), Corlett (1981)
(variability *in vivo*), Boerema *et al.* (1997) (original diagnosis of anamorph *in
vitro*), Boerema & de Gruyter (1998) (figs 3A, B: shape of conidia *in vitro* and
in vivo, reproduced in present Fig. 34A; *in vitro* and *in vivo* descriptions, partly
adopted below).

Description in vitro

Pycnidia globose to irregular, with 1–2 often indistinct, sometimes papillate
pores, 90–310 μm diam., pale yellow brown, glabrous, solitary or confluent,
abundant on and in the agar. Walls made up of 2–5 layers of cells, the outer
layer pigmented. Micropycnidia present, 60–90 μm diam., often remaining
sterile. Conidial matrix white to salmon.

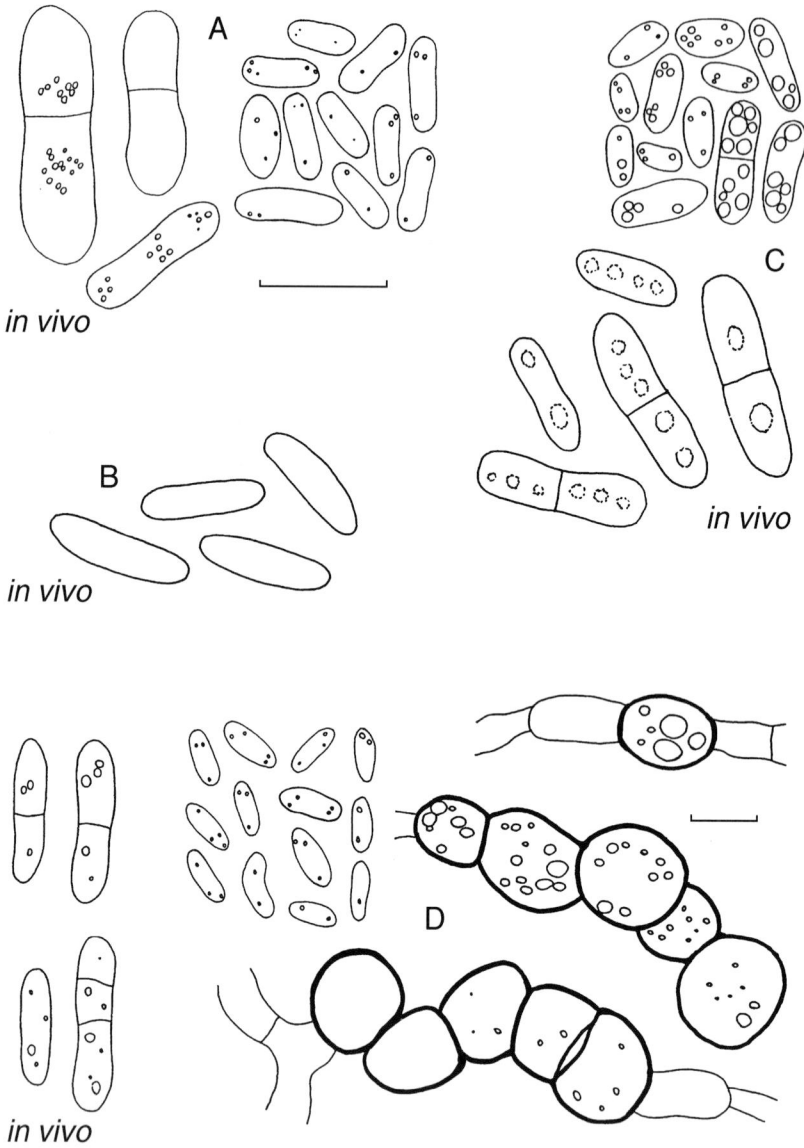

Fig. 34. A *Phoma alectorolophi*. Conidia usually aseptate, but occasionally, especially *in vivo*, some larger 1-septate conidia occur. **B** *Phoma boerhaviae*. Conidia *in vivo*. **C** *Phoma complanata*. *In vitro* the conidia are highly variable in shape and size, but mostly aseptate; *in vivo* they are usually also aseptate, but sometimes a high percentage becomes large and 1-septate. **D** *Phoma dictamnicola*. Conidia *in vitro* relatively small, always aseptate; pycnidia *in vivo* always contain relatively large conidia, mostly 1(–2)-septate. Usually intercalary chains of chlamydospores develop *in vitro*. Bar 10 μm.

Conidia variable in shape and size, eguttulate or with a few small guttules, cylindrical to oblong ellipsoidal, usually aseptate (4–)6–9(–14) × 2–4(–6) µm, but mostly 4–7.5 × 2–3.5 µm. Occasionally also some large 1-septate conidia (ascochytoid), (10–)14–18 × (4–)5–6(–8) µm.

Colonies on OA *c.* 6.5–7 cm diam. after 7 days, regular, colourless to pale greenish olivaceous at margin, aerial mycelium floccose, white; reverse also colourless to pale greenish olivaceous at margin. Representative culture CBS 132.96.

In vivo *(Rhinanthus sp.)*

Pycnidia (on dry calyces, capsules, peduncles and stems) subglobose to flattened, up to 300 µm diam., usually followed by pseudothecia (single identity proved by Corbaz, 1957). Pycnidial primordia stromatic (pycnosclerotia, often indistinguishable from immature ascocarps). The pycnidia usually produce only aseptate conidia *in vivo.* They are mostly oval to cylindrical and less variable than those *in vitro,* (4–)5–7(–9) × 2–2.5(–4) µm. So far larger ascochytoid conidia have only occasionally been found in old pycnidia.

Pseudothecia (on dead stems) subglobose to flattened, mostly 250–300 × 140–160 µm, laterally and basally thick-walled–pseudoparenchymatous, unstable (cells easily come off). Pseudoparaphyses filiform, 2–3 µm wide, septate at intervals of about 10 µm, persisting. Ascospores (16–)18–21(–24) × (4.5–)5–7.5(–8) µm, obovoid to oval, 1-septate, constricted at the septum, upper cell usually larger and wider than the lower cell. In addition to immature ascomata with only pseudoparaphyses, similar structures containing numerous microconidia (or spermatia) were occasionally observed, 2–3.5 × 1–1.5 µm.

Ecology and distribution

Reported and isolated from dead tissue of plants of three genera of semi-parasitic Scrophulariaceae: *Melampyrum, Pedicularis* and *Rhinanthus* (all semi-parasites on roots of Gramineae) in Europe (Switzerland and The Netherlands). The anamorph has repeatedly been confused with two other species of sect. *Sclerophomella,* namely *Phoma complanata* (Tode : Fr.) Desm. and *Phoma sylvatica* Sacc.

Immature ascomata with pseudoparaphyses formerly repeatedly have been interpreted as pycnidia with straight cylindrical conidia: '*Phoma deusta* Fuckel' in Jb. nassau. Ver. Naturk. **23–24**: 377. 1870 ['1869'; = Symb. mycol.], '*Sphaeronaema rhinanthi* Lib.' in Pl. cryptog. Ard. Fasc. 1, No. 63. 1830 ≡ *Zythia rhinanthi* (Lib.) Fr. in Summa veg. Scand. **2**: 408. 1849, and '*Phoma melampyri* P. Karst.' in Acta Soc. Fauna Fl. fenn. **27**(4): 14. 1905 (compare Jaczewski, 1898 and Grove, 1935).

Phoma boerhaviae Shreem., Fig. 34B

> *Phoma boerhaviae* Shreem. *in* Indian J. Mycol. Pl. Path. **2**: 84. 1972.
> = *Phoma nyctaginea* var. *boerhaviae* S. Ahmad *in* Sydowia
> **2**: 78. 1948.

Diagn. data: Shreemali (1972) (fig. 1: shape of pycnidium and conidia, some reproduced in present Fig. 34B; original diagnosis), Boerema (1986b) (p. 31: synonymy and misidentification), Boerema & de Gruyter (1998) (classification in sect. *Sclerophomella*).

Description in vitro *(adopted from Shreemali (1972) and cf. dried culture)*

Pycnidia globose with elongated neck, 168×130.5 μm (av. 162.6 μm) diam., glabrous, dark brown to black, solitary, on and in the agar, thick-walled pseudoparenchymatous, persistent.

Conidia cylindrical, eguttulate, aseptate, $7.5–13.5 \times 2.5–4.5$ μm (av. 10.6 $\times$ 3.8 μm).

Colonies on 'Asthana and Hawker's medium' with light brown to dark brown hyphae, richly branched, poorly septate, 3.8–2.7 diam. wide. Representative dried culture CMI 130821.

In vivo *(Boerhavia diffusa) (adopted from Shreemali, 1972 and Ahmad, 1948)*

Pycnidia (in grey-dark lesions all around the twigs and on dead branches) 'pin head size', containing elongated ellipsoidal aseptate conidia, often $7.5–10.5 \times 3.5–4.5$ μm.

Ecology and distribution

Known from *Boerhavia diffusa* (Nyctaginaceae) in south-west Asia (India, Pakistan), but probably also elsewhere occurring on the host. It produced leaf spots and dark grey lesions girdling the twigs: twig blight. The fungus has been confused with the plurivorous, soil-borne opportunistic plant pathogen *Phoma multirostrata* (Mathur *et al.*) Dorenb. & Boerema (sect. *Phoma*, **A**).

Phoma complanata (Tode : Fr.) Desm., Figs 33A, 34C

> *Phoma complanata* (Tode : Fr.) Desm. *in* Annls Sci. nat. (Bot.) III, **16**: 299–300. 1851.
>> ≡ *Sphaeria complanata* Tode, Fungi mecklenb. Sel. **2**: 22. 1791.
>> : Fr., Syst. mycol. **2** [Sect. 2]: 508. 1823.
>> ≡ *Sclerophomella complanata* (Tode : Fr.) Höhn. *in* Hedwigia **59**: 238. 1918 (1917?).

[≡ '*Plenodomus complanatus* (Tode : Fr.) H. Ruppr. comb.
nov.': manuscript name in herb. B; not published.]
= *Pyrenochaeta rivini* Allesch. apud Sydow *in* Beibl.
Hedwigia **36**: 161. 1897 [cf. isotype, S; with discordant
setae-bearing element].
    ≡ *Sclerochaetella rivini* (Allesch.) Höhn. *in* Hedwigia **59**:
    251. 1918.
    ≡ *Diploplenodomus rivini* (Allesch.) Petr. *in* Annls mycol.
    42: 62. 1944.
= *Phoma anethicola* Allesch. *in* Rabenh. Krypt.-Flora [ed. 2],
Pilze **6** [Lief. 63]: 298. 1898 [vol. dated '1901'].
Θ ≡ *Phoma herbarum* var. *anethi* Westend. apud von
Thümen *in* Fungi austr. Cent. 10, No. 982. 1874 [with-
out description].
= *Phoma punctoidea* P. Karst. *in* Acta Soc. Fauna Flora fenn.
27(4): 7. 1905 [cf. type; H].

Diagn. data: Dennis (1946) (cultural characteristics), Boerema (1976) (fig. 7:
cultures on OA; tab. 9: synoptic table on cultural characters with drawing of
conidia), Sutton (1980) (fig. 226E: drawing of conidia; specimens in CMI collection), Rajak & Rai (1983b, 1984) (influence of pH and nutrient composition
on development of porus), Cerkauskas (1985) (figs 1–5; cause of canker of
parsnip, disease symptoms, detailed description), Cerkauskas (1987) (tabs 1–3;
pathogenicity and survival), Rai & Rajak (1993) (cultural characteristics; NB:
the study sub *Ph. complanata* in Rai (1998) must be questioned as it includes
one isolate from a non-umbelliferous plant), Boerema *et al.* (1996) (fig. 13:
drawing of large 1-septate conidia *in vivo*, some reproduced in present Fig.
34C), Boerema *et al.* (1997) (with reference to the occurrence of large 1-sep-
tate conidia as in sect. *Heterospora*, **B**), Boerema & de Gruyter (1998) (figs
1A, 6: vertical section and shape of pycnidia, pycnidial wall with conidiogenous
cells and conidia, reproduced in present Figs 33A and 34C; *in vitro* and *in vivo*
descriptions, partly adopted below).

Description in vitro

Pycnidia globose to irregular, mostly 80–240 μm diam., with usually 1 non-
papillate pore [on MA sometimes with pronounced neck], glabrous, finally olivaceous black, solitary or confluent, on and in the agar. Walls made up of 2–6
layers of cells, outer layers pigmented. Conidial matrix buff to flesh coloured.
    Conidia highly variable in shape and size, usually with several small gut-
tules, they may be subglobose ellipsoidal but also cylindrical, to fusiform,
mostly aseptate 3–11 × 1.5–4 μm, in some strains usually 5–10 × 2–3 μm, in
others 3–8 × 3.5–4 μm. Occasionally in fresh cultures some larger 1-septate
conidia occur, up to 16 × 4 μm; in older cultures particularly large conidia,
22–34 × 6–10 μm, may be present (ascochytoid).
    Colonies on OA 6–8 cm diam. after 7 days, regular, colourless to buff to
greenish olivaceous, with finely floccose to woolly, sometimes compact white to

pale olivaceous grey aerial mycelium; reverse primrose to salmon or citrine green to olivaceous in centre. Representative culture CBS 633.68.

In vivo *(Umbelliferae)*

Pycnidia (on dead stems and in lesions on leaves, petioles and roots, scattered or aggregated, immersed or partly immersed) subglobose with a dark brown to black outer wall and a central pore, up to 300–400 μm diam.

Conidia usually aseptate, 5–9 × 2–3.5 μm, but sometimes a high percentage of the conidia becomes large and 1-septate, often (10–)12–15(–16) × 2.5–3.5(–4) μm. Pycnidia on old stem lesions may contain distinctly large 1-septate ascochytoid conidia, resembling those found in old cultures. The conidial mass usually darkens with age to brown or black; the conidia then mostly appear 1-septate, swollen and dark.

Ecology and distribution

A very common fungus in temperate Eurasia and North America on last year's dead stems of wild Umbelliferae. A seed-borne pathogen of parsnip (*Pastinaca sativa*), parsley (*Petroselinum crispum*) and carrot (*Daucus carota*): canker (lesions on petioles and roots), leaf spot. Inoculation experiments have demonstrated the occurrence of some host-adapted forms (varieties), which may have been promoted by seed transmission. Records of *Ph. complanata* on non-umbellifers refer to related, but different species.

Phoma dictamnicola Boerema *et al.*, Fig. 34D

> *Phoma dictamnicola* Boerema, Gruyter & Noordel. apud de Gruyter & Noordeloos *in* Persoonia **15**(1): 90. 1992.
>> ≡ *Ascochyta nobilis* Kabát & Bubák *in* Öst. bot. Z. **54**: 3. 1904; not *Phoma nobilis* Sacc. *in* Michelia **2**(3): 616. 1882 [= *Phomopsis* sp.]; not *Phoma nobilis* Thüm. *in* Instituto, Coimbra **28** sub Contr. Fl. myc. Lusit. III n. 562. 1881 ['1880–81'], quoted *in* Hedwigia **2**: 24. 1882 [in both cases as '*nobile*'].
>> = *Phyllosticta dictamni* Fairman *in* Annls mycol. **8**: 324. 1910; not *Phoma dictamni* Fuckel *in* Jb. nassau. Ver. Naturk. **23–24** [= Symb. mycol.] 125. 1870 ['1869 und 1870'].
>> = *Diplodina dictamni* Kabát & Bubák *in* Hedwigia **52**: 349. 1912; not *Phoma dictamni* Fuckel, see above.

Diagn. data: de Gruyter & Noordeloos (1992) (figs 20, 23, 24: shape of conidia and chlamydospores *in vitro* and shape of conidia *in vivo*, all reproduced in present Fig. 34D; cultural descriptions, partly adopted below), Boerema *et al.* (1997) (reference to the occurrence of large 1-septate conidia as in sect.

Heterospora, **B**), Boerema & de Gruyter (1998) (differentiation against other species of sect. *Sclerophomella*).

Description in vitro

Pycnidia subglobose, 250–450 μm diam., prolonged closed, glabrous, greenish olivaceous, solitary or in clusters of 2–3, mainly in the agar, thick-walled. Exudate not observed.

Conidia variable, ellipsoidal to reniform, eguttulate, always aseptate and relatively small, 4–5.5 × 1.5–2.5 μm.

Chlamydospores greenish olivaceous, unicellular, usually intercalary in short chains, globose 8–12 μm diam.

Colonies on OA 4.5–5 cm diam. after 7 days, regular, with a conspicuous greyish-olivaceous woolly felted mycelial mat, total colour greyish olivaceous to greenish olivaceous at margin; reverse beige, towards centre greenish olivaceous [on MA faster growing, 5.5–7 cm diam., with irregular margin]. Pycnidia abundant, especially towards the margin of the colonies, usually in the agar. Representative culture CBS 507.91.

In vivo (Dictamnus albus)

Pycnidia (in irregular leaf spots and on dead stems) subglobose-conical with a central pore, dark brown to black in colour. Conidia extremely variable. Those from pycnidia on dead stems resemble the conidia *in vitro*, usually aseptate, 4–5 × 2 μm, but sometimes also larger, 6–8 × 3–4 μm and then often 1-septate. The pycnidia in the leaf spots always contain relatively large conidia, usually partly aseptate, (8–)11–14(–14.5) × 3–3.5 μm, but mainly 1(–2)-septate, 13.5–15.5(–16) × 3.5–4(–4.5) μm (ascochytoid).

Ecology and distribution

A specific pathogen of *Dictamnus* spp., esp. of the 'firework-plant', *Dictamnus albus* (Rutaceae), frequently recorded in Eurasia and North America: leaf spot. The specimens on which the above synonyms are based illustrate the extreme variability of the conidia of this fungus *in vivo*: large or relatively small, aseptate and/or one-septate. *In vitro* the conidia are always small and aseptate. This may lead to confusion with a spermatial state, *Asteromella dictamni* Petr., occurring on old leaf spots incited by a different pathogen, namely *Septoria dictamni* Fuckel, teleomorph *Mycosphaerella dictamni* Petr. (see van der Aa & Vanev, 2002, sub *Phyllosticta dictamnicola* Lobik and *Phyllosticta guceviczii* Zhilina).

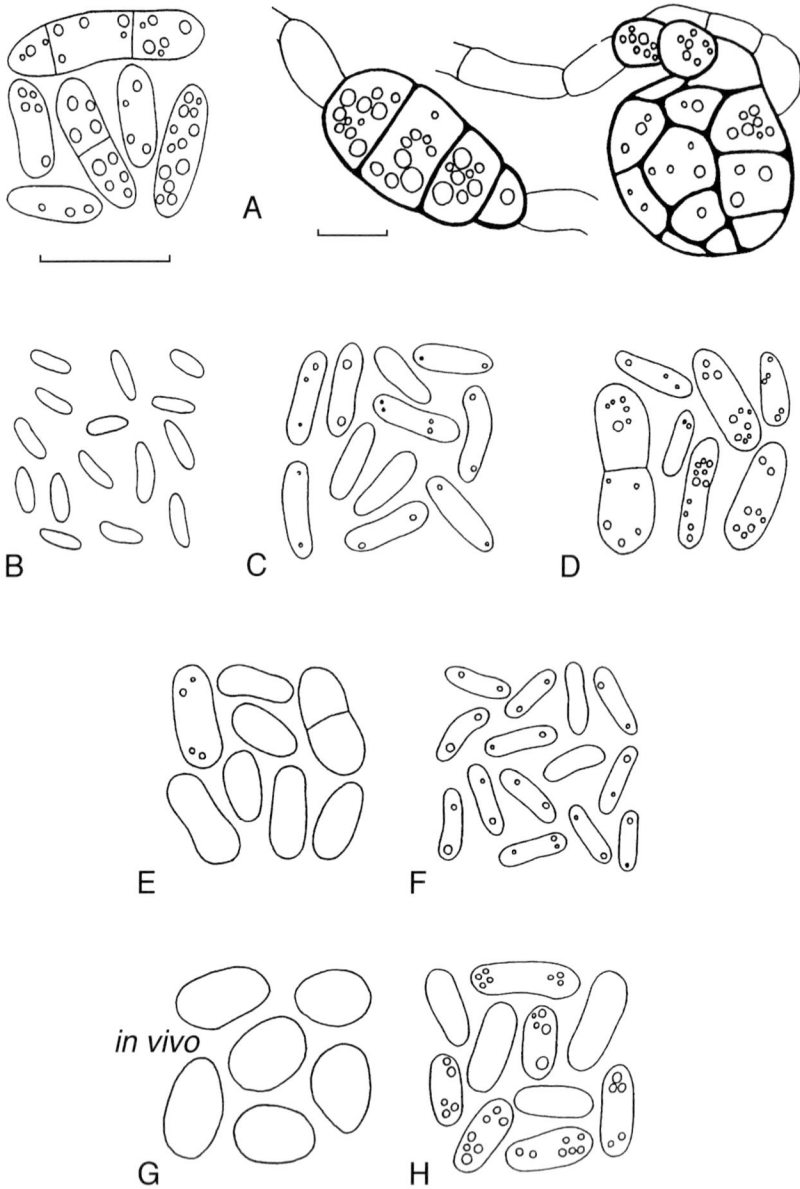

Fig. 35. **A** *Phoma gentianae-sino-ornatae.* Conidia *in vitro* variable in size, occasionally 1(–2)-septate. Micro-pseudosclerotia and chlamydospores are formed in the agar. **B** *Phoma incompta.* Conidia *in vitro* aseptate and very small (similar *in vivo*). **C** *Phoma nigrificans.* Conidia *in vitro* always aseptate (similar *in vivo*). **D** *Phoma polemonii.* Conidia *in vitro* variable (similar *in vivo,* but often only aseptate or only 1-septate). **E** *Phoma protuberans.* Conidia *in vitro* variable (similar *in vivo*). **F** *Phoma sylvatica.* Conidia *in vitro* always aseptate and relatively small (similar *in vivo*). **G** *Phoma syriaca.* Conidia *in vivo.* **H** *Phoma versabilis.* Conidia *in vitro* variable in size, always aseptate (similar *in vivo*). Bars 10 μm.

Phoma gentianae-sino-ornatae Punith. & R. Harling, Fig. 35A

Phoma gentianae-sino-ornatae Punith. & R. Harling *in* Mycol. Res. **97**: 1299. 1993.

Diagn. data: Punithalingam & Harling (1993) (figs 1–6, tabs 1, 2; original description, illustrations of conidia, conidiogenous cells, micro-pseudosclerotia and chlamydospores; comparison with other *Phoma* spp.), Boerema & de Gruyter (1998) (fig. 2A, B: shape of conidia, multicellular chlamydospore and micro-pseudosclerotium, all reproduced in present Fig. 35A; cultural descriptions, partly adopted below).

Description in vitro

Pycnidia subglobose, 180–250 μm diam., with 1 papillate pore, glabrous, finally olivaceous black, usually solitary, on and in the agar. Wall made up of 2–7 layers of cells, outer layers pigmented. Conidial matrix whitish.

Conidia ellipsoidal to ovoid, sometimes curved, guttulate, usually aseptate, (4.5–)6–9.5(–11) × 2.5–4 μm, occasionally 1(–2)-septate, up to 16 × 3.5 μm.

Chlamydospores and micro-pseudosclerotia are formed in the agar, dark olivaceous to olivaceous black; chlamydospores unicellular to multicellular, 8–20 μm diam., pseudosclerotia 35–110 μm diam.

Colonies slow growing, on OA *c.* 3 cm diam. after 7 days and *c.* 2 cm on MA, regular, colourless to grey olivaceous or vinaceous buff to tawny, with woolly to floccose, white to smoke grey aerial mycelium; reverse colourless to olivaceous black. Representative culture IMI 341116 (CBS 878.97).

In vivo *(Gentiana sino-ornata)*

The fungus produces in rotting roots of living hosts numerous micro-pseudosclerotia (on and within the infected tissue), similar to those *in vitro*. Pycnidia have not been recorded *in vivo*.

Ecology and distribution

So far this fungus is only recorded in Scotland, where it has been isolated from rotting and blackening roots of *Gentiana sino-ornata* (Gentianaceae) in a nursery. The disease, first noticed in 1989, may be confused with an attack by *Macrophomina phaseolina* (Tassi) Goid. (charcoal rot). The latter fungus produces similar minute pseudosclerotia, see Punithalingam & Harling (1993), Holliday & Punithalingam (1970c) and Punithalingam (1982d: 267).

Phoma incompta Sacc. & Martelli, Fig. 35B

Phoma incompta Sacc. & Martelli *in* Sylloge Fung. **10**: 146. 1892.

Diagn. data: Malathrakis (1979) (figs 9–16, 25–38, tabs II–VII, XVII–XXVIII, pycnidia *in vivo*, mycelial growth and pycnidial development on different agar media; effect of temp., pH, C and N), de Gruyter & Noordeloos (1992) (fig. 19: shape of conidia *in vitro*, reproduced in present Fig. 35B; cultural descriptions, partly adopted below), Boerema & de Gruyter (1998) (differentiation against other species of sect. *Sclerophomella*).

Description in vitro

Pycnidia subglobose, 50–300 × 50–250 μm, prolonged closed, covered by mycelial hairs, pore visible as a pallid spot, dark olivaceous to rusty blackish, solitary but usually confluent in concentric rings, mainly on, but also in the agar. Thick-walled. Conidial matrix whitish to pale violaceous.

Conidia slenderly cylindrical, always aseptate, small, eguttulate, 3–3.5 × 1 μm.

Colonies slow growing, on OA *c.* 2 cm diam. after 7 days (14 days *c.* 4 cm diam.), regular, chestnut-brown, towards margin ochraceous, due to diffusible pigment(s), with or without sparse woolly grey olivaceous aerial mycelium; reverse chestnut-ochraceous or with weak greenish tinge, distinctly zonated from pycnidial concentric rings. On MA growth rate also *c.* 2 cm; abundant whitish olivaceous aerial mycelium, reverse between sepia and olivaceous, but outer margin more like citrine. In the mycelium swollen cells may occur. Representative cultures CBS 467.76, CBS 526.82.

In vivo *(Olea europaea)*

The pycnidia produced on the dead shoots of heavily infected olive trees are often aggregated, strikingly black, globose and poroid. The conidia are usually somewhat larger than those *in vitro*, mostly 3–5 × 1.5–2 μm.

Ecology and distribution

A specific pathogen of the olive (*Olea europaea*, Oleaceae) causing shoot wilt and recorded in southern Europe (Crete–Greece, Italy). The main symptoms are a progressive withering of young shoots, which later die without defoliation, and dark discoloration of the xylem which extends all along the infected branches (Malathrakis, 1979). In the rainy season infection occurs through wounds, especially leaf scars. The disease may be confused with verticillium wilt of olive caused by *Verticillium dahliae* Kleb. (Tosi & Zazzerini, 1994).

Phoma nigrificans (P. Karst.) Boerema *et al.,*[13] Figs 33B, 35C
Teleomorph: *Didymella macropodii* Petr.

> *Phoma nigrificans* (P. Karst.) Boerema, Loer. & Wittern *in* Jl Phytopath. [=
> Phytopath. Z.] **115**: 270. 1986.
>> V ≡ *Sphaeronaema nigrificans* P. Karst. *in* Meddn Soc. Fauna
>> Flora fenn. **16**: 17. 1888 [as '*Sphaeronema*'].
>> [Θ ≡ '*Rhynchophomella nigrificans* P. Karst.', herbarium name
>> (H, Helsinki), not published.]
>> = *Plenodomus macropodii* Petr. *in* Hedwigia **68**: 237. 1929.

Diagn. data: Boerema *et al.* (1986) (figs 1–3; conidial dimensions, pycnidial shape, see present Fig. 33B, and pycnidial wall with conidiogenous cells and conidia; growth rate at different temperatures; hosts; teleomorph and synonymy of anamorph), Boerema *et al.* (1996: 168) (misidentification anamorph as *Plenodomus*), Marcinkowska & de Gruyter (1996) (occurrence on *Thlaspi arvense* in Poland; detailed description *in vitro*, partly adopted below; comparative pathogenicity tests with *Phoma lingam* isolates), Boerema & de Gruyter (1998) (fig. 7: conidial shape, reproduced in present Fig. 35C; differentiation against other species of sect. *Sclerophomella*).

Description in vitro

Pycnidia globose to subglobose or irregular shaped, 50–330 μm diam., with 1, initially indistinct opening (pore), sometimes papillate, glabrous, greenish olivaceous to olivaceous black, solitary or sometimes aggregated, on and in the agar. Wall up to 8 cells thick, outer layers pigmented. Conidial matrix whitish.

Conidia oblong-ellipsoidal to allantoid, with usually small polar guttules, always aseptate, relatively large, 5–10 × 1.5–4 μm (av. 6.4–6.6 × 1.9–2.7 μm).

Colonies on OA 5–5.5 cm diam. after 7 days, regular, colourless to greenish olivaceous or grey olivaceous, with floccose to woolly, pale olivaceous grey aerial mycelium; reverse colourless to greenish olivaceous or grey olivaceous. Representative cultures CBS 100190 and CBS 100191.

In vivo *(Cruciferae)*

Pycnidia (on the base of dead stems and in black discoloured stem lesions; scattered or in groups, first immersed, later superficial) depressed globose, usually with a conspicuous neck ('*Plenodomus*'-like), black, thick-walled (mostly 50–70 μm, up to 120 μm at the base), pseudoparenchymatous. Conidia

---

[13]We have not adopted van der Aa & Vanev's interpretation (2002) of *Phyllosticta armoraciae* (Cooke) Sacc., based on *Ascochyta armoraciae* Cooke, as synonyms of *Ph. nigrificans*. Cooke's fungus (or fungi) occurred on white spots, whereas lesions caused by *Ph. nigrificans* are black discoloured. Besides van der Aa & Vanev also note that Cooke's fungus 'will need to be recollected and cultured before it can be identified'.

oblong-ellipsoidal to subcylindrical, mostly 6–8.5(–10) × 1.5–2.5(–3) μm, eguttulate or with a small guttule at each end.

Pseudothecia (on basal parts of dead stems, usually together with pycnidia) relatively large, often 300–450 μm diam., depressed globose with a short papillate and poroid neck, thick-walled (40–70 μm thick) pseudoparenchymatous. Asci clavate to short cylindrical, mostly 65–80 × 10–12 μm, relatively thick-walled, 4–8-spored, irregularly biseriate. Ascospores cylindrical to ellipsoidal, straight or slightly curved, broadly rounded at both ends, mostly 14.5–19 × 4–5.5 μm, unequally 2-celled, upper cell sometimes wider, only slightly constricted at the septum. Pseudoparaphyses scarce, atypical, firm filiform, septate and branched. (For detailed description see the original diagnosis in Petrak, 1929: 219.)

Ecology and distribution

Recorded from wild and cultivated crucifers (Cruciferae) in northern Eurasia. Apparently a cold-tolerant opportunistic parasite. In northern Germany it has often been isolated from winter oilseed rape (*Brassica napus* var. *oleifera*) with black leg symptoms resembling those caused by *Phoma lingam* (Tode : Fr.) Desm. (sect. *Plenodomus*, **G**; teleom. *Leptosphaeria maculans* (Desm.) Ces. & De Not.), see Jedrycza *et al.* (1995). The primary host may be horse radish, *Armoracia rusticana*.

Phoma polemonii Cooke, Fig. 35D, Colour Plate I-E

> *Phoma polemonii* Cooke *in* Grevillea **13**: 94. 1885 (cf. original specimen, see Grove, 1935: 98).
>> H = *Phoma polemonii* Oudem. *in* Versl. Meded. K. Akad. Wet. [Afd. Natuurk.] reeks 3, **2**: 161. 1885.
>>> ≡ *Phoma oudemansii* Berl. & Voglino *in* Sylloge Fung. **10**: 174. 1892.
>>> = *Ascochyta polemonii* Cavara *in* Revue mycol. **21**: 104. 1899.
>> H = *Ascochyta polemonii* Rostr. *in* Bot. Tidsskr. **26**: 311. 1905.

Diagn. data: de Gruyter *et al.* (2002) (fig. 28: conidial shape, reproduced in present Fig. 35D; cultural descriptions, partly adopted below).

Description in vitro

Pycnidia globose/subglobose to irregular, 95–220 μm diam., at maturity with 1(–3) papillate pore(s), also often developing into elongated necks, citrine/honey, later olivaceous to olivaceous black, solitary or confluent, scattered both on and in the agar. Wall composed of 2–7 layers of relatively thick-walled pseudoparenchyma, outer layer(s) pigmented. Conidial matrix whitish/smoke grey.

Conidia ellipsoidal to allantoid with several small, scattered guttules, mainly aseptate, (5.5–)7–9(–12) × 2–3(–4) μm; the 1-septate conidia measure 9.5–13 × 2.5–3.5 μm.

Colonies on OA 3.5–4.5 cm diam. after 7 days, regular to slightly irregular, citrine green/greenish olivaceous to herbage green, with coarsely floccose, white/pale olivaceous grey to citrine/grey olivaceous aerial mycelium [on MA slower growing, 2.5–3.5 cm after 7 days, often in a zonate pattern]. With the NaOH spot test a non-specific pale sienna to rust colour may appear. Representative culture CBS 109181.

In vivo *(Polemonium caeruleum)*

Pycnidia on leafspots thin-walled with a high percentage of 1(–2)-septate conidia, (10–)12–13(–14) × 2.5–3(–4) μm; as necrophyte on old stems and leaves pycnidia relatively thick-walled and containing only smaller aseptate conidia 5–8(–10) × 2–2.5(–3) μm, often biguttulate.

Ecology and distribution

A specific pathogen of *Polemonium* spp. (Polemoniaceae), widespread in Europe and also found in North America (USA). The fungus causes brown-yellow leaf spots, and colonizes fading leaves and stems. Most records and synonyms refer to the perennial *P. caeruleum* (Jacob's ladder).

Phoma protuberans Lév., Fig. 35E

Phoma protuberans Lév. *in* Annls Sci. nat. (Bot.) III, **5**: 281. 1846.
            = *Phyllosticta lycii* Ellis & Kellerm. *in* Am. Nat. **17**: 1166. 1883.

Diagn. data: Boerema *et al.* (1997) (synonymy, characters *in vitro* and *in vivo*), de Gruyter *et al.* (2002) (fig. 29: conidial shape, reproduced in present Fig. 35E; cultural descriptions, partly adopted below).

Description in vitro

Pycnidia irregularly globose, 90–210 μm diam., glabrous, closed or with 1 non-papillated pore, greenish olivaceous to olivaceous, later olivaceous black, solitary to confluent, in the agar and in aerial mycelium. Walls made up of 2–12 layers of cells, outer layers pigmented. Conidial matrix buff to salmon.

Conidia variable ellipsoidal to subcylindrical usually without guttules, aseptate or septate; aseptate 4–10.5 × 2–5 μm, septate usually similar, but also significantly larger 12–20.5 × 3.5–5 μm (ascochytoid).

Colonies on OA 4–7 cm diam. after 7 days, regular, colourless-pale primrose to grey olivaceous, with floccose, white to pale olivaceous grey aerial mycelium [on MA about similar, but darker without primrose tinge; growth rate 4.5–6.5 cm]. Representative culture CBS 381.96.

In vivo *(Lycium halimifolium)*

Pycnidia (scattered in circular lesions, initially brown but turning pale-yellow or whitish) with a distinct conical pore, and containing mainly variable sized aseptate conidia with only a few 1-septate ones, 6–10 × 2–3 μm. Large ascochytoid conidia are so far not recorded in the fields.

Ecology and distribution

A specific pathogen of *Lycium halimifolium* (Solanaceae) occasionally found in Europe and North America: leaf spot. The shrubby solanaceous host is indigenous to southern Eurasia; the fungus probably occurs wherever the host is planted or naturalized.

Phoma sylvatica Sacc., Fig. 35F

Probable teleomorph: **Didymella winteriana (Sacc.) Petr.**

V *Phoma sylvatica* Sacc. *in* Michelia **2**(2): 337. 1881; *in* Sylloge Fung. **3**: 128. 1884 [as '*silvatica*'].

> V ≡ *Plenodomus sylvaticus* (Sacc.) H. Ruppr. *in* Sydowia **13**: 21. 1959 [as '*silvatica*'; misapplied].
>
> Θ = *Phoma herbarum* f. *melampyri* Westend. *in* Herb. crypt. [Ed. Beyaert], Fasc. 23, No. 1133. 1857 [as 'α *Melampyri*'; without description].

Diagn. data: Boerema (1970) (p. 36: synonymy), Boerema & van Kesteren (1981b) (confusion with two species of sect. *Plenodomus*), Boerema *et al.* (1996) (p. 179: documentation of herbarium specimens), Boerema & de Gruyter (1998) (fig. 4: shape of conidia, reproduced in present Fig. 35F; *in vitro* and *in vivo* descriptions, partly adopted below).

Description in vitro

Pycnidia globose to subglobose, 110–330 μm diam., with some mycelial outgrowths and 1(–2) often indistinct, non-papillate or slightly papillate pore(s), greenish olivaceous to olivaceous black, solitary or confluent, in more or less concentric zones, on and in the agar. Walls usually made up of 5–9 layers of cells but occasionally up to 20 layers of cells were observed, outer layers pigmented. Exudate not observed.

Conidia cylindrical, sometimes slightly allantoid with usually 2 small polar guttules, always aseptate, 3.5–6 × 1–2 μm.

Colonies on OA 7–7.5 cm diam. after 7 days, regular to slightly irregular, colourless to rosy buff, but grey olivaceous to olivaceous grey at centre, with floccose, (pale) olivaceous grey aerial mycelium; reverse colourless–rose buff with grey olivaceous to olivaceous grey centre. All strains tested showed a E$^+$ reaction on addition of a drop of NaOH (greenish → red). Representative cultures CBS 874.97, CBS 135.93.

In vivo *(Melampyrum sylvaticum)*

Pycnidia (on dead stems, scattered or in groups) mostly 150 μm diam., with relatively thick walls made up of polygonal cells. Conidia similar to those *in vitro*, mostly (3.5–)4–5 × 1–1.5(–2) μm.

[Pseudothecia of *Didymella winteriana* also occur on dead stems, often in close association with the pycnidia of *Ph. sylvatica* and are comparable in size and anatomical appearance with those of *D. alectorolophi* (see *Phoma alectorolophi*, this section) but differ by a thinner peridium, *c.* 20 μm thick, shorter asci, usually 45–60 μm long, and smaller ascospores, often 15–18 × *c.* 4 μm. For a fuller description, see Munk, 1957: 337.]

Ecology and distribution

Widespread recorded on different species of *Melampyrum* (Scrophulariaceae, semi-parasites on roots of Gramineae) in Europe. *Ph. sylvatica* is often confused with *Phoma petrakii* Boerema & Kesteren (1981b), also commonly occurring on *Melampyrum* spp. and producing similar small aseptate conidia. The latter, however, is easily to differentiate by its stable, relatively large scleroplectenchymatous pycnidia (*Phoma* sect. *Plenodomus*; see Boerema *et al.* 1994). The combination *Plenodomus sylvaticus* was based on a misidentified collection of *Ph. petrakii*. A single identity of *Ph. sylvatica* with *Didymella winteriana* is plausible, but has not yet been proved with isolates in pure culture.

Phoma syriaca (Petr.) Boerema *et al.*, Fig. 35G

Probable teleomorph: **Didymella syriaca Petr.**

Phoma syriaca (Petr.) Boerema, Loer. & Hamers *in* Persoonia **16**(2): 180. 1996.

≡ *Plenodomus syriacus* Petr. in Sydowia **1**: 42. 1947.

Diagn. data: Petrak (1947) (original diagnoses of *Ph. syriaca* and the supposed teleomorph *Did. syriaca*), Boerema *et al.* (1996) (misidentification anamorph as *Plenodomus* cf. original collection with both morphs in Herb. Petrak; conidial shape drawn in present Fig. 35G), Boerema & de Gruyter (1998) (classification of anamorph in sect. *Sclerophomella*; plausibility of teleomorph).

Description in vivo *(Phlomis brevilabris) (adopted from Petrak, 1947)*

Pycnidia (subepidermal, scattered or arranged in small groups on dead stems) subglobose with at the sides dark twisting short-celled hyphae, 200–350 μm diam., with 1 pore initially closed.

Conidia notably broad, ovate-ellipsoidal; occasionally with somewhat truncate ends, mostly 6–8 × 3.5–6 μm.

[Pseudothecia (also subepidermal on dead stems in association with above
pycnidia) mostly 200–300 μm diam., depressed globose, thick-walled-pseudo-
parenchymatous. Asci initially clavate, later cylindrical, mostly 80–110 ×
23–28 μm, thick-walled, 8-spored, more or less biseriate. Ascospores straight or
slightly curved, 18–23(–27) × 9–11 μm, 1-septate at about the middle, upper
cell wider than lower cell, slightly constricted to scarely constricted at the septum. Pseudoparaphyses scarce, filiform, septate, but soon dissolving.]
Representative specimens (type specimens) of both morphs: two packets of
Flora syriaca No. 1340 in Herb. Petrak (W).

Ecology and distribution

Thus far only known from *Phlomis brevilabris* (Labiatae) in the subalpine
region of Mt Sania (1700–1900 m) in Lebanon, but probably also elsewhere
with the host.

Phoma versabilis Boerema *et al.*, Fig. 33C, 35H

> *Phoma versabilis* Boerema, Loer. & Hamers *in* Persoonia **16**(2): 154. 1996.
> [Θ= '*Plenodomus cardaminus* H. Ruppr.': manuscript name in
> herb. B, not published.]

Diagn. data: Boerema *et al.* (1996) (original description; fig. 3: pycnoscle-
rotium, reproduced in present Fig. 33C; conidia and conidiogenous cells),
Boerema & de Gruyter (1998) (fig. 5: conidial shape, reproduced in present
Fig. 35H; *in vitro* and *in vivo* descriptions, partly adopted below).

Description in vitro

Pycnidia globose to globose-depressed, 100–260 μm diam., prolonged closed,
honey/citrine to olivaceous/olivaceous black, solitary to confluent, abundant on
and in the agar. Initially containing a compact mass of cells (pycnosclerotia).
Walls of mature pycnidia with 3–8 layers of cells, outer layers pigmented.
Conidial matrix buff to rosy buff.

Conidia ellipsoidal to ovoid, eguttulate or with several guttules, always
aseptate, 4.5–8 × 2–3 μm.

Colonies on OA 4–4.5 cm diam. after 7 days, regular, colourless to dark
herbage green/dull green at margin, with finely floccose to woolly, white to pale
olivaceous grey aerial mycelium; reverse also colourless to herbage and dull
green. With the NaOH spot test a non-specific reddish discoloration may occur.
Representative culture CBS 876.97.

In vivo

Pycnidia (on last year's stems) globose-depressed, immersed, dark brown; just like *in vitro* initially containing a compact mass of cells (pycnosclerotia) which afterwards disintegrates.

Conidia similar to those *in vitro*, mostly 5–7 × 2–2.5 μm.

Ecology and distribution

Recorded on dead stems of various herbaceous plants in Europe, e.g. *Cardamine* (Cruciferae) and *Silene* (Caryophyllaceae) species. The fungus has much in common with two other species of sect. *Sclerophomella*, namely *Phoma alectorolophi* Boerema *et al.* and *Phoma sylvatica* Sacc., both found on semi-parasitic Scrophulariaceae.

14 G *Phoma* sect. *Plenodomus*

Phoma sect. *Plenodomus* (Preuss) Boerema *et al.*

> *Phoma* sect. *Plenodomus* (Preuss) Boerema, Kesteren & Loer. *in* Trans. Br. mycol. Soc. **77**: 61. 1981.
> ≡ *Plenodomus* Preuss *in* Linnaea **24**: 145. 1851.
> = *Diploplenodomus* Died. *in* Annls mycol. **10**: 140. 1912.
> = *Leptophoma* Höhn. *in* Sber. Akad. Wiss. Wien (Math.-naturw. Kl. Abt. I) **124**: 73. 1915.
> = *Deuterophoma* Petri *in* Boll. R. Staz. Patol. veg. Roma II, **9**(4): 396. 1929.
> Type: *Plenodomus rabenhorstii* Preuss = *Phoma lingam* (Tode : Fr.) Desm., anamorph of *Leptosphaeria maculans* (Desm.) Ces. & De Not.

The section *Plenodomus* is characterized by the ability to produce scleroplectenchyma in the pycnidial wall (Fig. 36), i.e. a tissue of cells with uniformly thickened walls, similar to the sclerenchyma in higher plants (evolutionary convergence). The thickening of the cell walls may be so extensive that only a very small lumen remains, as in stone cells of fruit and seed. At the base of the pycnidia the thick-walled cells are often elongated and arranged in more or less diverging rows. The pycnidia always remain long closed; the final formation of a pore may be accompanied by the development of a long neck. Sometimes thick-walled pycnidia open by rupture. The cavity in mature pycnidia is usually irregular due to the occurrence of thin-walled seriate-cellular protrusions. The pycnidia are mostly glabrous, but sometimes have hyphal outgrowths; one species produces setose pycnidia.

In some species the production of scleroplectenchyma occurs as a secondary process. Their pycnidia are initially thin-walled with 'common' pseudoparenchymatous wall structure (type I). Thick-walled scleroplectenchymatous pycnidia (type II) may gradually arise from them (I→II), but these may not develop until the next year *in vivo* (characteristic for some

Phoma *Identification Manual* (G.H. Boerema *et al.*)

pathogens: pycnidia-I then occur in association with disease symptoms and pycnidia-II on old dead host tissue). In several other species of the section the production of scleroplectenchyma already starts in the pycnidial primordia. Occasionally only sterile 'pycnosclerotia' (type III) develop in species of both categories.

Species producing scleroplectenchyma only after a long period in old pycnidia *in vivo*, usually develop only pycnidial type I under normal laboratory conditions *in vitro*. Species immediately producing scleroplectenchymatous pycnidia *in vivo* often remain sterile in culture.

The conidia are mostly one-celled, but secondary septation may occur, especially *in vivo*. The conidia always remain aseptate *in vitro*. One species also produces conidia directly from the mycelium (Fig. 46B, *Phialophora*-synanamorph). Arthrospores have also been observed in a few species. Chlamydospores are usually absent, but characteristic for one species. Swollen cells occasionally occur.

Most members of the section occur on herbaceous plants (Group A). The scleroplectenchymatous pycnidia then develop mostly on last year's dead stems; one species produces pycnidia on sclerotia. A small number of species are found on deciduous trees and shrubs (Group B), with development of scleroplectenchymatous pycnidia on bark and bare wood. The majority of the species in Group A are metagenetically related to members of the ascomycetous genus *Leptosphaeria* Ces. & De Not. with scleroplectenchymatous ascocarp wall structure, i.e. *Leptosphaeria* sensu stricto (gr. *doliolum* L. Holm). Heterothallism may be responsible for the great variability of the conidia in some of these species and the occasionally observed geographical differences in the occurrence of pycnidia-II and the teleomorph.

Diagn. lit.: Boerema & van Kesteren (1964) (synonymy and description of types I, II and III of '*Plenodomus lingam*', the type species of the present section), Boerema (1976) (discussion of some species studied in culture by Dr R.W.G. Dennis), Boerema *et al.* (1981b) (formal introduction of the present section, characteristics of five species *in vitro*), Boerema & van Kesteren (1981b) (nomenclatural notes on five species), Boerema & Loerakker (1981) (confusion of names with respect to anamorph-teleomorph connections), Boerema & Loerakker (1985) (characteristics of five taxa *in vitro*), Boerema *et al.* (1994) (precursory contribution to this chapter), Boerema *et al.* (1996) (misapplications; excluded species), Boerema & de Gruyter (1999) (additional species).

Synopsis of the Section Characteristics

- Pycnidia simple or complex, sometimes thin-walled (type I), but always also more or less thick-walled by the production of scleroplectenchyma (type II), sometimes then remaining sterile (pycnosclerotia, III), poroid (sometimes developing a long neck) or opening by rupture, mostly glabrous, sometimes with hyphal outgrowths (pilose), occasionally setose; some species with only thin-walled pycnidia *in vitro*.

- Conidia mostly aseptate, but *in vivo* a variable number sometimes becomes 1-septate.
- Production of conidia freely on hyphae (*Phialophora* synanamorph) and fragmentation of hyphae (arthrospores) may occur.

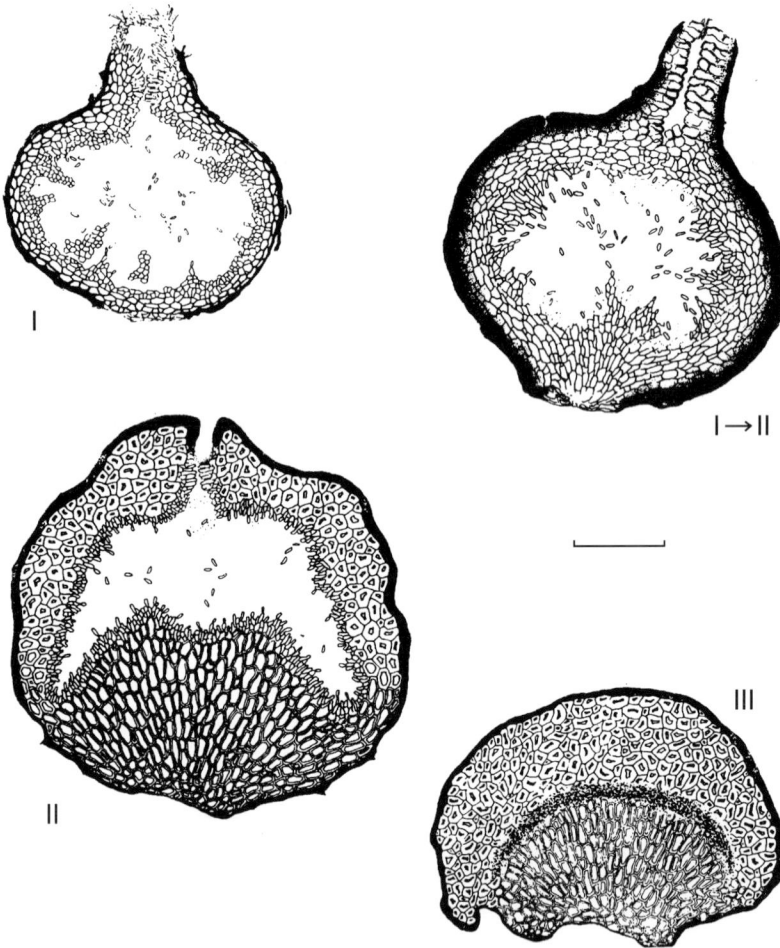

Fig. 36. *Phoma lingam*, type species of *Phoma* sect. *Plenodomus*. Vertical sections of its pycnidial types on cabbage. Thin-walled pycnidia-I occur on seed and in association with disease symptoms. Note the series of cells protruding into the pycnidial cavity. On old stem lesions the pycnidia may become thick-walled by the development of new cell layers, I→II. Note the fan-like arrangement of elongated cells at the base of the pycnidium and the hyphal outgrowths in the papilla. Secondary thickening of the cell walls in the peridium results in scleroplectenchyma as in the relatively large pycnidia-II found on old dead stems or roots at the end of the next season. Note the resemblance with stone cells in higher plants. The process of wall thickening may also occur in pycnidial initials: development of sterile pycnosclerotia-III. Bar 100 μm.

- Chlamydospores usually absent, but conspicuous in one species; occasional occurrence of swollen cells.
- True sclerotia are characteristic for one rare species (*Sclerotium* synanamorph).
- Teleomorphs of species occurring on herbaceous plants belong to *Leptosphaeria* Ces. & De Not. sensu stricto (gr. *doliolum* L. Holm).

Key to the Species and Varieties Treated of Section *Plenodomus*

Differentiation based on characteristics of scleroplectenchymatous pycnidia *in vivo* (pycnidia-II)

Just as in other sections of *Phoma*, species differentiation and delimitation in the section *Plenodomus* is primarily based on comparative study *in vitro*. However, for identification of the species in this section the characteristics of the scleroplectenchymatous pycnidia (II) *in vivo* are essential. Determinations may be checked with the data on growth *in vitro*. Direct identification *in vitro* is difficult and sometimes impossible or not workable; see the discussion under 'Possibilities of identification *in vitro*' on p. 328 As also shown in this key, most species of sect. *Plenodomus* have a restricted host range and/or distribution. Therefore, the extensive and detailed host-fungus index on p. 329 is also very important for identification.

1. a. Scleroplectenchymatous pycnidia (II) on dead stems (occasionally roots) of herbaceous plants **2**
    b. Scleroplectenchymatous pycnidia (II) on dead bark or wood (occasionally petioles) of deciduous trees and shrubs **24**

2. a. Pycnidia II developing directly from the host tissue **3**
    b. Pycnidia II arising from sclerotia on the host, with their outer-surface a clear continuation of the sclerotial rind, globose to subglobose with a papillate pore, variable in size; conidia 2–4 × 1–1.5 μm...
        ....*Ph. korfii*, synanam. *Sclerotium orobanches*
        [On dead stems, flowers and roots of Orobanchaceae in North America (*Epifagus virginianus*) and Europe (*Orobanche major*). *In vitro* also pycnidia-II, usually solitary but under certain conditions also borne on sclerotia; occasionally also pycnosclerotia (III) and thin-walled pycnidia (I).]

3. a. Pycnidia-II variable in size, relatively large, often exceeding 500 μm diam.; sometimes only pycnosclerotia (III) **4**
    b. Pycnidia-II usually not exceeding 500 μm diam. **6**

4. a. All scleroplectenchyma cells in mature pycnidia having about the same wall thickness; pycnidia-II mostly 600–700 μm diam., usually with convex base of elongated cells and a long broad cylindrical neck up to 700 μm, usually with a somewhat swollen top; conidia mostly 4–5 × 1.5–2 μm............*Ph. macrocapsa*
        [On dead stems of *Mercurialis perennis* (Euphorbiaceae) throughout Europe. Apparently a specific necrophyte. *In vitro* also pycnidia-II, but usually with less pronounced neck.]

    b.   Scleroplectenchyma cells partly with an exceptionally thickened wall and a very small lumen (like stone cells of plants); in addition to pycnidia-II, scleroplectenchymatous pycnosclerotia (III) frequently occur....................................**5**

5. a.   Conidia small, usually not exceeding 4.5 μm in length, mostly 3.5–4.5 × 1–1.5 μm; pycnidia-II mostly 300–700 μm diam., not or only slightly papillate, with narrow pore or opened by rupture; sometimes only pycnosclerotia (III) .... *Ph. lingam*, teleom. *Lept. maculans* [Cosmopolitan seed-borne pathogen of cultivated *Brassica* spp. and other Cruciferae: dry rot and canker (Am.: black leg), leaf spots. Pycnidia-II and pycnosclerotia (III) develop on last year's dead stems and occasionally roots. In association with disease symptoms, on seed, and *in vitro*, only pycnidia-I occur.]

    b.   Conidia up to 6 μm in length, mostly 4.5–6 × 1.5–2.5 μm; pycnidia-II mostly 350–800 μm diam., usually with long tubular necks of various lengths, up to 800 μm; sometimes only pycnosclerotia (III) ....*Ph. sclerotioides* [On roots and occasionally stems of various herbaceous plants in arctic regions of Eurasia and North America. Especially known as destructive low temperature pathogen of forage legumes: brown root rot. Produces also pycnidia-II and pycnosclerotia (III) *in vitro*.]

6. a.   Pycnidial base convex with conspicuous diverging rows of elongated cells **16**

    b.   Pycnidial base less thickened, cells at base similar to scleroplectenchyma cells of side walls **7**

7. a.   Pycnidia-II globose-papillate **8**

    b.   Pycnidia-II with pronounced necks **14**

8. a.   Conidia very small, mostly 2–3 × 0.5–1 μm; pycnidia-II variable, relatively large with slightly papillate pore, often aggregated or irregular with the appearance of being multilocular ....*Ph. astragalina* [Plurivorous necrophyte, especially common in the mountainous regions of south-west Asia; especially often on *Astragalus* spp. (Leguminosae). Produces only pycnidia-II, but scleroplectenchyma not so conspicuous as in most other species of the section. Isolates on OA at room temperature remained sterile.]

    b.   Conidia larger, length mostly between 3 and 9 μm **9**

9. a.   Conidia often exceeding 5 μm in length **10**

    b.   Conidia usually not exceeding 5 μm in length **12**

10. a.   Pycnidia-II often coalesced forming large elongated aggregates up to 1000 μm (pycnidial stromata); pycnidia initially relatively small, closed or papillate-poroid with somewhat protruding lip; conidia variable in shape and dimensions, mostly 4–7 × (1.5–)2 μm ....*Ph. agnita*, teleom. *Lept. agnita* [Plurivorous, but especially often on dead stems of Compositae. Necrophytic. In Europe very common. *In vitro* also pycnidia-II, often confluent.]

    b.   Pycnidia usually solitary **11**

11. a.   Conidia uniform subcylindrical, 5–7 × 1.5–2 μm; pycnidia-II with flattened base and papillate or truncate-conoid porus ....*Ph. ruttneri* [? teleom. *Lept. affinis*] [On dead stems of *Rhinanthus* spp. (Scrophulariaceae), Europe. Probably a

specific necrophyte. Only scleroplectenchymatous pycnidia are known. Isolates of the possible teleomorph remained sterile.]

b. Conidia variable, mostly 5–9 × 1.5–3 μm; pycnidia-II usually non-papillate with narrow pore ....*Ph. macdonaldii,* teleom. *Lept. lindquistii* [Cosmopolitan seed-borne pathogen of the commercial sunflower, *Helianthus annuus* (Compositae): black stem, black spot. In association with disease symptoms only pycnidia-I. *In vitro* pycnidia I→II.]

12. a. Pycnidia occasionally setose, i.e. with a number of stiff setae around the porus, but often also without any trace of setae; pycnidia papillate and initially thin-walled: I→II; conidia mostly 3.5–4.5 × 1.5 μm..........................*Ph. drobnjacensis* [Recorded on various wild and cultivated *Gentiana* spp. (Gentianaceae) in Europe. Reported to be pathogenic: leaf spot. *In vitro* pycnidia initially also thin-walled: I→II, mostly with pilose neck.]

b. Pycnidia glabrous..**13**

13. a. Pycnidia-II subglobose-conical with broad base and usually slightly papillate pore, mostly 150–250 μm diam.; conidia mostly 4 × 1.5 μm.............. ....*Ph. conferta,* teleom. *Lept. conferta* [Recorded on dead stems of various wild Cruciferae in Europe; especially common on *Berteroa incana*. Necrophytic. *In vitro* also pycnidia-II, but more irregular in shape.]

b. Pycnidia-II depressed globose with flattened base, explicitly papillate, usually with a dark-lined tube-shaped porus, mostly 200–350 μm diam.; conidia mostly 4–5 × 1–1.5 μm.............................*Ph. petrakii* [? teleom. *Lept. suffulta*] [On dead stems of *Melampyrum* spp. (Scrophulariaceae) throughout Europe. Specific necrophyte. Isolates on OA at room temperature remained sterile.]

14. a. Neck up to 500 μm long, pilose; conidia mostly 4–5.5 × 1.5–2.5 μm ....*Ph. leonuri,* teleom. *Lept. slovacica* [On dead stems of *Leonura cardiaca* and *Ballota nigra* throughout Europe; occasionally also on other Labiatae. Apparently a specific necrophyte. *In vitro* pycnidia-II with very long pilose necks.]

b. Neck shorter and distinct papillate **15**

15. a. Neck usually no longer than 60 μm, i.e. short papillate; conidia 4–4.5 × 1.5 μm, biguttulate..........................*Ph. pimpinellae,* teleom. *Lept. pimpinellae* [On dead stems of *Pimpinella anisum* (Umbelliferae); a necrophyte so far only known from Israel. *In vitro* pycnidia I→II, often globose with a long neck, but also irregular without clear pore.]

b. Neck up to 200 μm long, bare or semi-pilose **16**

16. a. Conidia usually exceeding 5 μm in length, mostly 5–6 × 1.5–2 μm ....*Ph. congesta,* teleom. *Lept. congesta* [On dead stems of *Achillea* spp. (Compositae); specific(?) necrophyte, so far only known from southern Europe. *In vitro* also pycnidia-II, usually with semi-pilose neck.]

b. Conidia usually not exceeding 5 μm in length **17**

17. a. Conidia conspicuous 2–4 guttulate, mostly 4–5 × 1.5–2 μm, but occasionally up to 6 μm long ....*Ph. veronicicola*

[On dead stems of various wild and cultivated *Veronica* spp. (Scrophulariaceae) in Europe. Apparently a specific necrophyte. *In vitro* also pycnidia-II, sometimes with pilose neck.]

    b.   Conidia eguttulate or with 2 inconspicuous small guttules, mostly 3–4 × 1–1.5 μm...*Ph. longirostrata*
[On dead stems of Ranunculaceae, esp. *Aconitum* and *Ranunculus* spp.; specific necrophyte, only known from southern Europe. *In vitro* usually pycnidia-II with several necks, transfers soon become sterile.]

18. a.   Conidia relatively small, not exceeding 5 μm in length and 2 μm in width **19**
    b.   Conidia usually larger **21**

19. a.   Pycnidia-II depressed globose with flattened base and rather sharply delimited papillate neck of variable length; conidia mostly 3.5–5 × 1–1.5 μm....
....................*Ph. acuta* subsp. *errabunda*, teleom. *Lept. doliolum* subsp. *errabunda*
[Plurivorous. Necrophytic. In Europe recorded on dead stems of quite different herbaceous plants. Apparently also widespread in North America (cf. records teleomorph). *In vitro* also pycnidia-II, usually with somewhat pilose necks.]
    b.   Pycnidia-II usually more globose and less depressed, mostly with a conspicuous neck **20**

20. a.   Pycnidia-II more or less subglobose with elongated neck up to 400 μm; conidia mostly 3.5–5 × 1.5 μm, eguttulate..
.........................*Ph. acuta* subsp. *acuta*, teleom. *Lept. doliolum* subsp. *doliolum*
[Common on dead stems of *Urtica* spp. Widespread in Europe and probably also in North America (cf. records teleomorph). *In vitro* also pycnidia-II but usually they remain very small with pilose necks; transfers soon become sterile. A specific pathogenic form of *Ph. acuta* subsp. *acuta* occurs on cultivated phloxes (Polemoniaceae): f. sp. *phlogis* (teleomorph unknown).]
    b.   Pycnidia-II depressed globose to subglobose with flattened base and a pronounced neck up to 200 μm; conidia mostly 4–4.5 × 1–1.5 μm, with 1–2 small guttules.....................................*Ph. sublingam*, teleom. *Lept. submaculans*
[Probably widespread in Europe and North America. Recorded on *Sisymbrium* spp. and other Cruciferae. Reported to be pathogenic: symptoms resembling the dry rot and canker disease of brassicas. Pycnidia-II develop on last year's dead stems. In association with disease symptoms, on seed, and *in vitro* only pycnidia-I occur.]

21. a.   Conidia aseptate **22**
    b.   Conidia occasionally also 1-septate **25**

22. a.   Conidia mostly not exceeding 8 μm in length **23**
    b.   Conidia (6–)8–10(–12) × 2–2.5 μm; occasionally 1-septate and longer, up to 16 μm; pycnidia-II variable in shape with flattened base, often subglobose-papillate, but also with a long neck, usually at one side ....
....*Ph. piskorzii*, teleom. *Lept. acuta*
[On dead stems of *Urtica* spp.; especially inside stem cavity. Although only occasionally recorded, probably widespread in Europe and North America (cf. records of teleomorph). Specific necrophyte. Isolates on OA at room temperature remained sterile.]

23. a.   Conidia often pluriguttulate, variable in shape and dimensions, usually oblong to ellipsoidal and always one-celled, mostly 5–8 × 2–2.5 µm; pycnidia-II with flattened base and pronounced cylindrical neck ....
....*Ph. sydowii* [? teleom. *Lept. senecionis*]
[On dead stems of *Senecio* spp. throughout Europe. Specific necrophyte, but the fungus is occasionally also recorded on other Compositae. *In vitro* also pycnidia-II.]

   b.   Conidia eguttulate or inconspicuously biguttulate **24**

24. a.   Pycnidia-II depressed globose, usually with irregular deformed flattened base and short papillate neck; conidia variable in dimensions, mostly (4–)5–7 × 2–2.5 µm, sometimes much larger, (7–)8–12(–16) × 2–3 µm and then often 1-septate..*Ph. doliolum*, teleom. *Lept. conoidea*
[Plurivorous necrophyte. In Europe recorded on dead stems of various herbaceous plants. Probably also widespread in North America (cf. records of teleomorph). *In vitro* pycnidia-II with conidia of the smaller dimensions; neck usually somewhat pilose.]

   b.   Pycnidia-II conoid with flattened base and conspicuous beak-like elongated neck; initially subglobose-papillate and thin-walled: I→II; conidia variable 4–8 × 2–3 µm and/or 4–6 × 2.5–4 µm ....*Ph. pedicularis*
[Plurivorous necrophyte in arctic-alpine regions of Europe. Recorded on dead stems, leaves and seed capsules of quite different herbaceous plants, but especially often on *Pedicularis* and *Gentiana* spp. Pycnidia *in vitro* also I→II; usually they remain subglobose-papillate and small.]

25. a.   Pycnidia-II with flattened base and long neck, usually at one side **22b**

   b.   Pycnidia-II with irregular base and short papillate neck **24a**

26. a.   Pycnidia-II mostly not exceeding 250 µm in diam. **27**

   b.   Pycnidia-II commonly exceeding 250 µm in diam. **28**

27. a.   Conidia very small, mostly 2–3 × 1–1.5 µm; pycnidia-I→II, subglobose with gradually developing cylindrical neck...........*Ph. tracheiphila*, synanam. *Phialophora* sp.
[Subepidermal on stems of *Citrus* trees in the Mediterranean and Black Sea areas. Noxious pathogen: mal secco disease. *In vitro* pycnidia often remain incomplete, thin-walled and open irregularly at maturity; usually abundant development of conidia from free conidiogenous cells formed on the aerial mycelium: *Phialophora*-synanamorph.]

   b.   Conidia longer, 4.5–6.5 × 1–2 µm; pycnidia-II, subglobose with flattened or somewhat pointed base and broad papillate porus, often with protruding lip .....
....*Ph. coonsii*
[Occasionally found on bark and wood of *Malus pumila* in North America and Japan. Reported to be pathogenic: bark canker. *In vitro* on OA some pycnidia-II may occur after 1–2 months.]

28. a.   Pycnidia-II mostly 250–350 µm diam. **29**

   b.   Pycnidia-II larger, often 400–500 µm diam. **30**

29. a.   Pycnidia globose-papillate with flattened or somewhat pointed base, initially thin-walled and gradually becoming scleroplectenchymatous: I→II; conidia mostly 3–4 × 1–2 µm...........................*Ph. enteroleuca* var. *enteroleuca* and var. *influorescens*

[On bark and wood of various deciduous trees and shrubs in Europe and North America. Opportunistic parasite. *In vitro* also initially thin-walled, I→II. The two distinguished varieties, i.e. strains with and without production of a fluorescing metabolite, are difficult to differentiate *in vivo*.][14]

b.   Pycnidia very thick-walled, mostly irregular-subglobose with flattened, often somewhat elongated base; protopycnidia already thick-walled and sclero-plectenchymatous: II; conidia 2.5–4 × 1–1.5 μm ....*Ph. intricans*
[Common on necrotic bark of *Salix alba* (Salicaceae) throughout Europe, but occasionally also isolated from other trees, e.g. *Acer* and *Malus* spp. Opportunistic pathogen: bark canker. *In vitro* also pycnidia-II, but usually relatively small.]

30. a.   Pycnidia globose-papillate or with a short cylindrical neck, initially thin-walled: I→II; conidia mostly 3.5–5 × 2–2.5 μm ....*Ph. rubefaciens*
[Isolated from bark and wood of various deciduous trees; apparently widespread in Europe. Recorded in association with fruit spot of apple: opportunistic pathogen. *In vitro* pycnidia initially also thin-walled, I→II.]

b.   Pycnidia usually globose-papillate with flattened base, at length often collapsing and becoming discoid or pezoid; protopycnidia already scleroplectenchymatous: II, sometimes remaining sterile: pycnosclerotia (III); conidia very variable, often 4.5–6 × 2–3 and/or 6–10 × 2–3 μm, occasionally 1-septate ....*Ph. pezizoides*
[On wood and occasionally petioles of deciduous trees and shrubs, mostly in the vicinity of rivers, especially near river banks in central and southern Europe and northern USA of North America. Apparently necrophytic. Isolates on OA at room temperature remained sterile.]

Possibilities of identification *in vitro*

Direct and exact determination of isolates is only feasible with a limited number of species in section *Plenodomus*. On the usual agar media fewer than half the taxa described in this paper form the scleroplectenchymatous pycnidia (type II) characteristic of sect. *Plenodomus*. Some species, including the type species of the section, produce 'common' thin-walled pseudoparenchymatous pycnidia (type I) *in vitro*. These may only be recognized as *Plenodomus*-like by the late development of a pore and the occurrence of thin-walled seriate cellular protrusions in the pycnidial cavity. Other species remain sterile on the usual agar media, or develop some pycnidia only after a long time (several months). One species usually produces only a few abnormal pycnidial bodies *in vitro*, but commonly forms a *Phialophora*-synanamorph. Table 4 displays diagnostic features of some species that produce pycnidia readily on OA. More information

---

[14] The pycnidia and conidia of *Ph. enteroleuca* s.l. *in vivo* show much resemblance to those produced in cultures of..*Ph. etheridgei*
[Characteristics *in vivo* unknown; described from isolates obtained from the bark of black galls and cankers of *Populus tremuloides* (Salicaceae) in Canada. Culture morphology, conidial length to width ratio, and colony colour reaction to NaOH showed differences between isolates of *Ph. etheridgei* and *Ph. enteroleuca*.]

on their cultural characteristics can be found in the synoptic tables published by Boerema (1976), Boerema *et al.* (1981b) and Boerema & Loerakker (1985). For the pathogens included in section *Plenodomus,* the disease symptoms may be an important indication of the identity of isolates.

Distribution and Facultative Host Relations in Section *Plenodomus*

(With annotation of the pycnidial types I, II, III and occurrence of teleomorph.)

A. On herbaceous plants

Plurivorous	Especially in mountainous regions of south-western Asia and central Europe:
(But often with	Ph. astragalina
host-preference,	Widespread in temperate regions of Europe and North
see below.)	America:

        Ph. acuta subsp. *errabunda*
        (teleom. *Lept. doliolum* subsp. *errabunda*)
        Ph. doliolum
        (teleom. *Lept. conoidea*)
[All three on dead stems; pycnidia-II, sometimes with teleom.]
Especially in Arctic-alpine regions of Eurasia:
        Ph. pedicularis
[On dead stems, occasionally leaves and seed capsules: pycnidia-I→II.]
Especially in Arctic regions of Eurasia and North America:
        Ph. sclerotioides
[On roots, occasionally stems; pycnidia-II, III.]

With specific or preferred host:

Compositae	
esp. *Achillea* spp.	*Ph. congesta*
	(teleom. *Lept. congesta*)
	[In southern Europe frequently found on dead stems of different species; teleom. also on *Erigeron canadensis.*]
esp. *Eupatorium*	*Ph. agnita*
cannabinum	(teleom. *Lept. agnita*)
	[In Europe frequently found on dead stems; pycnidia-II, often with teleom.; occasionally also recorded on hosts other than crucifers.]
Helianthus annuus	*Ph. macdonaldii*
	(teleom. *Lept. lindquistii*)
	[Worldwide known as causal organism of black stem or black spot: lesions on stems, petioles, leaves and inflorescenses; pycnidia-I in association with disease symptoms, pycnidia-II and teleom. on dead host material.]

Table 4. Diagnostic features *in vitro* of some species and varieties of *Phoma* sect. *Plenodomus*, sporulating well on OA at 20–22°C (the rare species *Ph. korfii*, which also produces pycnidia readily on OA, is not included in this table).

Pycnidia I (→II) (pycnidia initially always type I)		Pycnidia II (scleroplectenchymatous from the start)	
Growth rate 1–1.5 cm (diam. after 7 days in darkness)			
Yellow-red pigment (discoloration at reverse) *Ph. rubefaciens*		Pycnidia small, often only after 1–2 months; with rubigenous grains *Ph. coonsii*	Pycnidia large, developing quickly *Ph. macrocapsa*
Growth rate 1.5–2.5 cm			
Yellow or red pigment purplish-blue with NaOH *Ph. enteroleuca* var. *influorescens*	No pigmentation *Ph. lingam* (teleom. *Lept. maculans*) *Ph. sublingam* (teleom. *Lept. submaculans*) (very similar *in vitro*)	[Cultures more or less yellow coloured]	
Brown pigment orange with NaOH *Ph. etheridgei*		Growth rate on MA reduced (1–2 cm) *Ph. sydowii* [? teleom. *Lept. senecionis*]	Yellow pigment in cell walls *Ph. veronicicola* Yellow pigment fading in daylight *Ph. pedicularis*
		[Pycnidia with long necks]	
		Necks semi-pilose *Ph. congesta* (teleom. *Lept. congesta*)	Necks setose *Ph. leonuri* (teleom. *Lept. slovacica*)
		Necks bare, often *pycnosclerotia* (III) E+ reaction with NaOH *Ph. sclerotioides*	
Growth rate 2.5–4.5 cm			
Yellow and/or red pigment purplish-blue with NaOH	Yellow pigment red with NaOH	Red pigment blue with NaOH *Ph. agnita* (teleom. *Lept. agnita*)	Yellow zones *Ph. conferta* (teleom. *Lept. conferta*)
With white crystals *Ph. macdonaldii* (teleom. *Lept. lindquistii*)	With crystals and chains of chlamydospores *Ph. drobnjacensis*	[No pigmentation]	
		Relatively long necks *Ph. longirostrata*	Papillate conidia relatively large *Ph. doliolum* (teleom. *Lept. conoidea*)
With red crystals *Ph. enteroleuca* var. *enteroleuca*	Without crystals and chlamydospores pycnidia with long necks, semi-pilose		
	Ph. pimpinellae (teleom. *Lept. pimpinellae*)	Small pycnidia *Ph. acuta* subsp. *acuta* (teleom. *Lept. doliolum* subsp. *doliolum*)	Conidia relatively small *Ph. acuta* subsp. *errabunda* (teleom. *Lept. doliolum* subsp. *errabunda*)

Ranunculaceae
esp. *Aconitum* and   *Ph. longirostrata*
Ranunculus spp.   [Only recorded in southern Europe; pycnidia-II.]

Scrophulariaceae
Melampyrum spp.   *Ph. petrakii*
                                       [? teleom. *Lept. suffulta*]
                                       [In Europe widespread found on dead stems; pycnidia-II,
                                       often together with supposed teleom.]
Pedicularis spp.   *Ph. pedicularis*
                                       [In arctic-alpine regions of Europe frequently found on
                                       dead stems, leaves and seed capsules; pycnidia-I→II; also
                                       on various other herbaceous plants in those regions.]
Rhinanthus spp.   *Ph. ruttneri*
                                       [? teleom. *Lept. affinis*]
                                       [In Europe only occasionally found on dead stems; pycni-
                                       dia-II; the supposed teleom. is widely distributed in
                                       Europe.]
Veronica spp.   *Ph. veronicicola*
                                       [In Europe widespread recorded on dead stems; pycnidia-II
                                       only.]

Umbelliferae
esp. *Angelica* spp., also   *Ph. acuta* subsp. *errabunda*
often on *Foeniculum vulgare*   (teleom. *Lept. doliolum* subsp. *errabunda*)
                                       Ph. doliolum
                                       (teleom. *Lept. conoidea*)
                                       [In Europe common on dead stems; pycnidia-II and
                                       teleom.; in temperate regions of Europe and North
                                       America both widely distributed necrophytes on herba-
                                       ceous plants.]
Pimpinella anisum   *Ph. pimpinellae*
                                       (teleom. *Lept. pimpinellae*)
                                       [So far only known from Israel, pycnidia-II together with
                                       teleom. on dead blackened stems.]

Urticaceae
Urtica spp., esp. *U. dioica*   *Ph. acuta* subsp. *acuta*
                                       (teleom. *Lept. doliolum* subsp. *doliolum*)
                                       [In Europe and North America frequently found on dead
                                       stems; pycnidia-II and teleom.]
                                       Ph. piskorzii
                                       (teleom. *Lept. acuta*)
                                       [In Europe and North America common on dead stems;
                                       pycnidia-II often only inside the hollow stems, teleom. at
                                       outer side.]

B. On deciduous trees and shrubs

Plurivorous   Widespread in Europe and apparently also common in
                                     North America:
(But often with some   Ph. enteroleuca var. enteroleuca
host-preference,   and var. influorescens
see below.)   Widespread in Europe:
                                             Ph. rubefaciens
                                     [Both on bark, wood and fruits; pycnidia-I→II.]
                                     Especially near river banks in Europe (central and southern
                                     regions) and North America (north-western USA):
                                             Ph. pezizoides
                                     [On decorticated branches, occasionally twigs and petioles;
                                     pycnidia-II.]

With specific or preferred host:

Bignoniaceae
Catalpa bignonioides   Ph. enteroleuca var. enteroleuca
                                     [In Europe frequently associated with wood discoloration
                                     and general blighting or dying of twigs; pycnidia-II; a
                                     plurivorous opportunistic pathogen.]

Caprifoliaceae
Lonicera caprifolia   Ph. enteroleuca var. enteroleuca
                                     [In Europe frequently associated with discoloured wood
                                     and dying of twigs; pycnidia-II; a plurivorous opportunistic
                                     pathogen.]

Rosaceae
Malus pumila   Ph. coonsii
                                     [In North America (USA) and Japan found on bark and
                                     wood; pycnidia-II; may act as opportunistic pathogen, bark
                                     canker.]
                                             Ph. enteroleuca var. enteroleuca
                                     [In Europe and North America found in association with
                                     discoloured wood, dead branches, bark cankers and fruit
                                     rot; pycnidia-II; a plurivorous opportunistic pathogen.]
                                             Ph. rubefaciens
                                     [In Europe recorded in association with red skin necroses
                                     on apples.]

Rutaceae
Citrus spp., esp. C. limonia   Ph. tracheiphila
                                             (synanam. Phialophora sp.)
                                     [In the Mediterranean and Black Sea areas well known as
                                     causal organism of 'mal secco': wilt and dieback associated
                                     with discoloration in the xylem; pycnidia-I→II.]

Salicaceae
Populus tremuloides   *Ph. etheridgei*
                                      [In North America (Canada) isolated from the bark of
                                      black galls and cankers; strongly antagonistic against
                                      Phellinus tremulae; pycnidia-I→II so far only known from
                                      cultures.]
Salix spp., esp. *S. alba*   *Ph. intricans*
                                      [In Europe repeatedly recorded on necrotic bark and in
                                      association with bark cankers; pycnidia-II; occasionally
                                      also on deciduous trees of other genera.]
                                      Ph. pezizoides
                                      [In Europe and North America, especially in the vicinity
                                      of rivers, found on decorticated branches and dead
                                      twigs; pycnidia-II; also on deciduous trees and shrubs of
                                      other genera.]

Ulmaceae
Ulmus spp.   *Ph. enteroleuca* var. *enteroleuca*
                                      and var. *influorescens*
                                      [In Europe and North America recorded in association
                                      with discoloured wood and superficial bark cankers; pycni-
                                      dia-II; a plurivorous opportunistic pathogen.]

Descriptions of the Taxa

Based on characteristics *in vivo* (incl. teleomorphs). Additional data on growth
in culture (used with the differentiation of the taxa).

Phoma acuta (Hoffm. : Fr.) Fuckel subsp. *acuta*, Fig. 37 Ab

Teleomorph: ***Leptosphaeria doliolum* (Pers. : Fr.) Ces. & De Not.
subsp. *doliolum***

> *Phoma acuta* (Hoffm. : Fr.) Fuckel *in* Jb. nassau. Ver. Naturk. **23–24**
> [= Symb. mycol.]: 125. 1870 ['1869 und 1870'] [as '*acutum* 1'],
> subsp. *acuta*.
>
> > ≡ *Sphaeria acuta* Hoffm., Veg. crypt. **1**: 22. 1787, : Fr., Syst.
> > mycol. **2** [Sect. 2]: 507. 1823 [lectotype Sclerom. Suec.
> > Fasc. 4, No. 118, UPS].
> >
> > ≡ *Leptophoma acuta* (Hoffm. : Fr.) Höhn. *in* Sber. Akad.
> > Wiss. Wien [Math.-naturw. Kl., Abt. I] **124**: 71–75. 1915
> > [as '(Fuckel)'].
> >
> > ≡ *Plenodomus acutus* (Hoffm. : Fr.) Bubák *in* Annls mycol.
> > **13**: 29. 1915 [as '(Fuckel)'].
> >
> > *H* ≡ *Plenodomus acutus* (Hoffm. : Fr.) Petr. *in* Annls mycol. **19**:
> > 192. 1921 [as '(Fuckel)'].

= *Phoma hoehnelii* var. *urticae* Boerema & Kesteren apud
Boerema *in* Trans. Br. mycol. Soc. **67**: 299. 1976.

Θ = *Phoma herbarum* var. *urticae* Roum. *in* Fungi gall. exs.
Cent. 11, No. 1017. 1881; *in* Revue mycol. **3**: 30. 1881
[without description].

= *Strigula urticae* Bonord. *in* Bot. Ztg **11**: 292. 1853.

≡ *Clisosporium urticae* (Bonord.) Bonord. *in* Abh. natur-
forsch. Ges. Halle **8**: 140. 1869.

Diagn. data: Riedl (1959) (fig. 8b; anamorph-teleomorph connection and
characteristics), Lucas & Webster (1967) (fig. 5A–C; anamorph-teleomorph
connection and characteristics), Boerema (1976) (figs 1A: drawings of pycnidium with conidia, pseudothecia with ascospores and photographs of cultures
on OA sub *Ph. hoehnelii* var. *urticae* and *Leptosphaeria doliolum* var.
doliolum; tab. 9: synoptic table on characters *in vitro* with drawing of conidia), Shoemaker (1984) (descr. teleomorph sensu lato), Boerema *et al.* (1994)
(fig. 6Cb: pycnidial and conidial shape, reproduced in present Fig. 37Ab;
nomenclature and synonymy; *in vivo* and *in vitro* descriptions adopted
below), Boerema & Gams (1995) (typification), Boerema *et al.* (1996: 181)
(confusion with *Pyrenochaeta fallax* Bres.).

Description in vivo *(Urtica dioica)*

Pycnidia type II (on dead stems, usually in short rows, subepidermal or superficial), subglobose with flattened base and mostly a pronounced neck, up to
400 μm long. The neck is mostly somewhat 'hairy' (semi-pilose) and thin-
walled. Scleroplectenchyma is especially conspicuous at the 'shoulder' near the
neck. At the base of mature pycnidia a convex thickening consisting of diverging rows of elongated cells. Conidial matrix cream or whitish.

Conidia relatively small (3–)3.5–5(–5.5) × 1.5(–2) μm, subcylindrical, usually eguttulate.

Pseudothecia (also on dead stems) 300–450 μm diam., subglobose with
flattened base and a short conical neck. Wall scleroplectenchymatous. Asci
mostly 120–160 × 6–8 μm, 8-spored, biseriate. Ascospores (20–)24–30(–35)
× 4–5(–5.5) μm [length–width ratio 5–5.5], narrowly fusiform-ellipsoidal, 3-
septate with acute end cells, brownish-guttulate, echinulate.

In vitro

Colonies on OA 2.5–3.5 cm diam. after 7 days, aerial mycelium cottony or
somewhat felted, olivaceous grey; reverse sometimes with yellowish/brown
tinges, in ageing cultures often a pinkish discoloration of the agar. Slow development of pycnidia-II on and in the agar, they are very small in comparison
with those *in vivo*, mostly 80–200 μm diam., subglobose, usually with 1 or
more pilose necks. Conidia as *in vivo*. Transfers of this fungus soon become
sterile. Representative culture CBS 505.75.

Ecology and distribution

In Europe very common (anamorph and teleomorph) on dead stems of nettle, esp. *Urtica dioica* (Urticaceae). Reports from North America (Canada) refer to different *Urtica* spp. (teleomorph). In Europe the fungus is occasionally also recorded on some other herbaceous plants (see also below under f. sp. *phlogis*). Since 1976 the anamorph of *Leptosphaeria doliolum* ss. on *Urtica* spp. was commonly known as *Phoma hoehnelii* var. *urticae*, but the present Code has made it possible to reinstate the old species name *Phoma acuta*, see Boerema & Gams (1995). However, it should be noted that *Ph. acuta* and its synonyms *Leptophoma acuta* and *Plenodomus acutus* formerly have been used for different anamorphs in the section *Plenodomus*, compare Boerema & van Kesteren (1981b, tab. 1). In the literature also often has been stated that *Ph. acuta* as described above represents the anamorph of *Leptosphaeria acuta* (Fuckel) P. Karst., which pseudothecia frequently occur at the base of nettle stems. The pycnidia of the latter, *Phoma piskorzii* (Petr.) Boerema & Loer. (this section), usually develop inside the hollow stems and contain significantly longer conidia.

f. sp. ***phlogis*** [Boerema, Gruyter & Kesteren *in* Persoonia **15**(4): 465. 1994]
      ≡ *Phoma phlogis* Roum. *in* Revue mycol. **6**: 160. 1884.
      = *Phyllosticta decussata* P. Syd. *in* Beibl. Hedwigia **36**: 158 1897 [cf. van der Aa & Vanev, 2002].
      V = *Sphaeronaema pirottae* Ferraris *in* Malpighia **16**: 23. 1902 [as 'Sphaeronema'].

In Europe (France, Germany, the UK, Italy, Russia, The Netherlands) frequently found on phloxes, the annual *Phlox drummondii* and hybrids of the perennial *Phlox paniculata* (Polemoniaceae). Indistinguishable from the anamorph on nettle, but never found in association with the teleomorph. Recorded to be pathogenic: stem lesions. Dieback.

***Phoma acuta* subsp. *errabunda* (Desm.) Boerema *et al.*, Fig. 37Aa**

Teleomorph: ***Leptosphaeria doliolum* subsp. *errabunda* Boerema *et al.***

    Phoma acuta subsp. *errabunda* (Desm.) Boerema, Gruyter & Kesteren *in* Persoonia **15**(4): 465. 1994.
        ≡ *Phoma errabunda* Desm. *in* Annls Sci. nat. (Bot.) III, **11**: 282. 1849, syntypes Pl. cryptog. France [ed. 1] Fasc. 38, No. 1870. 1849, and Pl. cryptog. France [ed. 2] Fasc. 30, No. 147. 1849.
        = *Leptophoma doliolum* Höhn. *in* Sber. Akad. Wiss. Wien [Math.-naturw. Kl., Abt. I] **124**: 75. 1915.
          ≡ *Plenodomus doliolum* (Höhn.) Höhn. *in* Ber. dt. bot. Ges. **36**: 139. 1918.
        H ≡ *Plenodomus doliolum* (Höhn.) Petr. *in* Annls mycol. **21**: 125, 211. 1923.

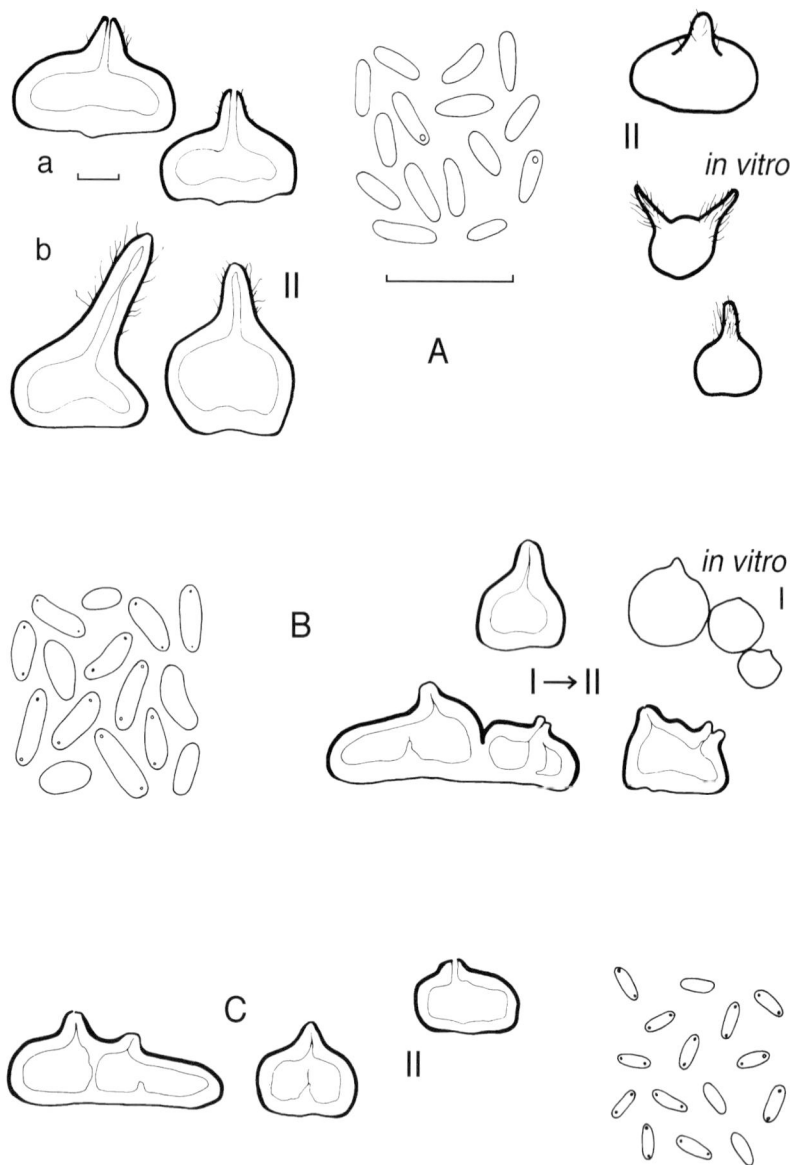

Fig. 37. A Subspecies of *Phoma acuta*. Pycnidia-II with rather sharply delimited poroid papillae or necks of variable length. Above (a) subsp. *errabunda*, pycnidia with flattened base. At bottom (b) subsp. *acuta*, pycnidia usually more globose and less depressed, mostly with a long neck; *in vitro* (right) the pycnidia of this nettle fungus usually remain very small with pilose neck(s). **B** *Phoma agnita*. Pycnidia-I–II initially relatively small, closed or papillate-poroid with somewhat protruding lip, often coalescing to large multilocular pycnidial stromata. Note the variable shape of the conidia. **C** *Phoma astragalina*. Characterized by very small conidia and extremely variable pycnidia-II, often aggregated, or irregular with the appearance of being multilocular. Bar pycnidia 100 μm, bar conidia 10 μm.

  ≡ *Phoma hoehnelii* Kesteren *in* Neth. J. Pl. Path. **78**: 116.
   1972 [as nom. nov. to avoid homonymy with *Phoma*
   doliolum P. Karst., this section].
  = *Phoma oleracea* var. *heraclei-lanati* Sacc. *in* Nuovo G. bot.
   ital. II, **27**: 81.1920.
 [Θ = '*Plenodomus scrophulariae* H. Ruppr.': manuscript name in
   herb. B, not published.]

Diagn. data: Lucas & Webster (1967) (fig. 5D–F; anamorph-teleomorph connection and characteristics), Boerema (1976) (figs 1B, 9: drawings of pycnidium with
conidia and pseudothecium with ascospores and photographs of cultures on OA
sub *Ph. hoehnelii* var. *hoehnelii* and *Leptosphaeria doliolum* var. *conoidea*; tabs
5, 9: sources of isolates; synoptic table on characters *in vitro* with drawing of conidia), Shoemaker (1984) (descr. teleomorph sensu lato), Boerema *et al.* (1994)
(fig. 6Ca: pycnidial and conidial shape, reproduced in present Fig. 37Aa; *in vivo*
and *in vitro* descriptions adopted below; nomenclature anamorph and teleomorph), Boerema *et al.* (1996: 184) (specimen in herb. B).

Description in vivo *(esp.* Angelica sylvestris*)*

Pycnidia always of type II (on dead stems, scattered or in lines, first subepidermal, later superficial), resembling those of *Ph. acuta* subsp. *acuta*, but generally
more depressed globose with a rather sharply delimited poroid papilla or neck
of variable length; the tube-shaped pore is usually dark lined. Conidial matrix
also cream or whitish.
  Conidia similar to those of subsp. *acuta*, (3–)3.5–5(–5.5) × 1–1.5(–2) μm,
subcylindrical and usually eguttulate.
  Pseudothecia (also on dead stems) up to 600 μm diam., generally more
depressed globose and more flattened than those of subsp. *doliolum*. Asci on
average 120–140 × 6–7 μm. Ascospores mostly (18–)20–22(–25) ×
(3–)3.5–4(–5) μm, i.e. usually shorter and thinner than those of subsp.
doliolum [but similar length–width ratio: 5–5.5].

In vitro

Colonies on OA show about the same growth rate as those of *Ph. acuta* /*Lept.
doliolum* sensu stricto. The cultural characters are also nearly similar, but the
pycnidial development on and in the agar is always much more abundant. The
dimensions of pycnidia-II are the same as those *in vivo*. Conidia also as *in vivo*.
Representative culture CBS 617.75.

Ecology and distribution

In Europe and North America a widely distributed plurivorous necrophyte;
especially common on dead stems of Compositae and Umbelliferae. In North
America the teleomorph is most frequently recorded on *Solidago* spp. (Comp.),
in Europe the most common host of the fungus is apparently *Angelica sylvestris*

(Umb.). The teleomorph of the fungus was until recently arranged under *Leptosphaeria doliolum* var. *conoidea* (De Not.) Sacc., at present recognized as a separate species: *Leptosphaeria conoidea* (De Not.) Sacc. The classification of the teleomorph as a separate subspecies of *Leptosphaeria doliolum* (Pers. : Fr.) Ces. & De Not. is concluded from a Canadian study of herbarium material (Shoemaker, 1984) in comparison with cultural studies earlier made in England and at our laboratory (Lucas & Webster, 1967 and Boerema, 1976). *L. doliolum* subsp. *errabunda* is characterized by ascospores resembling those of *Leptosphaeria doliolum* s.s. on *Urtica* spp. in having acute end cells and similar length–width ratio [5–5.5], but differing by relatively small dimensions. The conidia of the anamorphs of both subspecies of *Leptosphaeria doliolum* are similar and significantly smaller than those of the anamorph of *Leptosphaeria conoidea*, *Phoma doliolum* P. Karst. (this section). The ascospores of *L. conoidea* have obtuse ends and are broader than those of both varieties of *L. doliolum* [length–width ratio about 4].

Phoma agnita Gonz. Frag., Fig. 37B

Teleomorph: ***Leptosphaeria agnita* (Desm.) Ces. & De Not.**

Phoma agnita Gonz. Frag. *in* Mems R. Acad. Cienc. Artes Barcelona **15**: 432. 1920.

= *Plenodomus chondrillae* Died. *in* Annls mycol. **9**: 140. 1911; *in* Krypt.-Flora Mark Brandenb. 9, Pilze **7** [Heft 1]: 236. 1912 [vol. dated '1915']; not *Phoma chondrillae* Hollós *in* Annls hist.-nat. Mus. natn. hung. **4**: 337. 1906 [= *Phomopsis chondrillae* (Hollós) Dias & M.T. Lucas *in* Agronomia lusit. **37**: 99. 1975].

Θ = *Phoma acuta* f. *petasites* Roum. *in* Fungi gall. exs. Cent. 11, No. 1007. 1881; *in* Revue mycol. **3**: 30. 1881 [without description].

Diagn. data: Lucas & Webster (1967) (fig. 1A–E: anamorph-teleomorph connection and characteristics; table of some collections grown in culture), Boerema & Loerakker (1985) (synonymy anamorph), Shoemaker (1984) (descr. teleomorph; 'probably exclusively in Europe'), Boerema *et al.* (1994) (fig. 3C: pycnidial and conidial shape, reproduced in present Fig. 37B; *in vivo* and *in vitro* descriptions adopted below).

Description in vivo *(esp.* Eupatorium cannabinum*)*

Pycnidia always of type II (on dead stems, usually densely crowded, first immersed, later superficial), variable in shape and dimensions, initially depressed globose and relatively small, 125–250 μm diam., closed or with a sharply delimited papillate pore with somewhat protruding lip; later often coalescing to large flattened-elongated or irregular, more or less multilocular pycnidial 'stromata', up to 1000 μm diam. Wall made up of polygonal scleroplectenchyma cells of variable dimensions, parallel rows of elongated cells usually occur around the party-walls. Conidial matrix (pale) reddish or amethyst coloured.

Conidia variable, usually oblong-ellipsoidal, 4–7 × (1.5–)2 µm, straight or slightly curved, with 2 inconspicuous guttules.

Pseudothecia (also on dead stems), 350–500 µm diam., depressed globose with a strong ridge, flattened base and a short strongly papillate pore. Wall scleroplectenchymatous. Asci 110–125 × 9–11 µm, 8-spored, ± biseriate. Ascospores 31–35 × 4–5 µm, narrowly subcylindrical, 6-septate, third cell from above slightly swollen, yellowish brown with 2 guttules per cell (for recent description see Shoemaker, 1984).

In vitro

Colonies on OA 3–4 cm diam. after 7 days, greenish-transparent; aerial mycelium sparse, tenuous-felted, whitish to yellowish grey; reverse vinaceous buff, pale salmon, but in the centre yellow or greenish olivaceous. On application of a drop of NaOH the reddish diffusible pigments turn blue. Thick-walled pycnidia, type II, usually occur in abundance on and in the agar, they are more or less pilose but otherwise resemble those *in vivo*, also often confluent, forming irregular or catenate aggregates. Small pycnidia sometimes occur in aerial mycelium. *In vitro* the conidia are even more variable in shape and dimensions than those *in vivo*; being ellipsoidal, oval to ovoid or pyriform and commonly straight, (3.5–)4–6(–7.5) × (1.5–)2(–3) µm. Representative culture CBS 121.89.

Ecology and distribution

This fungus has been found in Europe on dead stems of herbaceous plants of different families, but most records refer to Compositae, especially *Eupatorium cannabinum*. North American records of the fungus are in doubt (cf. Shoemaker, 1984). In nature the development of pseudothecia probably always precedes that of the pycnidia.

Phoma astragalina (Gonz. Frag.) Boerema & Kesteren, Fig. 37C

Phoma astragalina (Gonz. Frag.) Boerema & Kesteren *in* Persoonia **11**(3): 317. 1981.
≡ *Ceuthospora astragalina* Gonz. Frag. *in* Boln. R. Soc. esp. Hist. nat. **18**: 84. 1918.
≡ *Plenodomus astragalinus* (Gonz. Frag.) Petr. apud Rechinger (f.), Baumgartner, Petrak & Szatala *in* Annln. naturh. Mus. Wien **50** ['1939']: 498–499. 1940.
H = *Phoma dianthi* (Bubák) Bubák *in* Annls mycol. **13**: 30. 1915; not *Phoma dianthi* Sacc. & Malbr. *in* Atti R. Ist.

veneto Sci. VI, **1**: 1276. 1883 [= *Phomopsis* sp.]; not
Phoma dianthi Ellis & Everh. *in* Langl. Cat. Pl. Basse-La
32. 1887 [nomen nudum]; not *Phoma dianthi* Lagière *in*
Annls Ec. natn. Agric. Grignon III, **5**: 160. 1946 [=
Phomopsis sp.].

≡ *Plenodomus dianthi* Bubák *in* Annln K.K. naturh.
Hofmus. Wien [Annln Naturh. Mus. Wien] **28**: 204.
1914.

= *Sclerophomella abnormis* Petr. *in* Annls mycol. **21**:
213–214. 1923 [cf. holotype in herb. Petrak, W].

= *Phoma genistaecola* Hollós, *in* Bot. Közl. **25**: 131. 1928.

= *Plenodomus khorasanicus* Petr. apud Rechinger (f.),
Baumgartner, Petrak & Szatala *in* Annln naturh. Mus. Wien
50 ['1939']: 499–500. 1940.

Diagn. data: Boerema & van Kesteren (1981b) (tab. II: host records; pycnidial
and conidial characteristics; nomenclator and synonymy), Boerema (1982b)
(fig. 9c: distribution), Boerema *et al.* (1994) (fig. 3B: pycnidial and conidial
shape, reproduced in present Fig. 37C; *in vivo* description adopted below),
Boerema *et al.* (1996: 175) (additional synonymy; distribution).

Description in vivo *(esp. on* Astragalus *spp.)*

Pycnidia always of type II (mostly on dead stems, scattered or in groups, subepidermal or subcortical), but scleroplectenchyma not as conspicuous as in most
other species of the section, extremely variable, sometimes mainly depressed
globose to ellipsoidal, 150–300 μm diam., but also more irregular in shape and
larger, up to 600 μm diam., usually with a flattened base and slightly papillated
pore; often aggregated or confluent. Wall of mature pycnidia sometimes with
irregular outgrowths bulging into the cavity giving it a multilocular appearance.
Young pycnidia (protopycnidia) are initially completely filled with more or less
parallel rows of somewhat elongated cells. Conidial matrix whitish.

Conidia very small, ellipsoidal, 2–3 × 0.5–1 μm, usually with 2 minute
polar guttules.

In vitro

Isolates on OA made from a fresh collection produced at room temperature
only sterile mycelium. The fungus probably requires a period of low temperature and/or special nutritional conditions for fructification.

Ecology and distribution

This plurivorous necrophyte is frequently found in the mountains of south-
western Asia (Iran, Turkey, Afghanistan), but appears to occur also in the
mountainous area of Central Europe. Most records are from dead stems of
herbaceous plants, especially *Astragalus* spp. (Leguminosae), but the fungus is
also found on dried branches of a shrub.

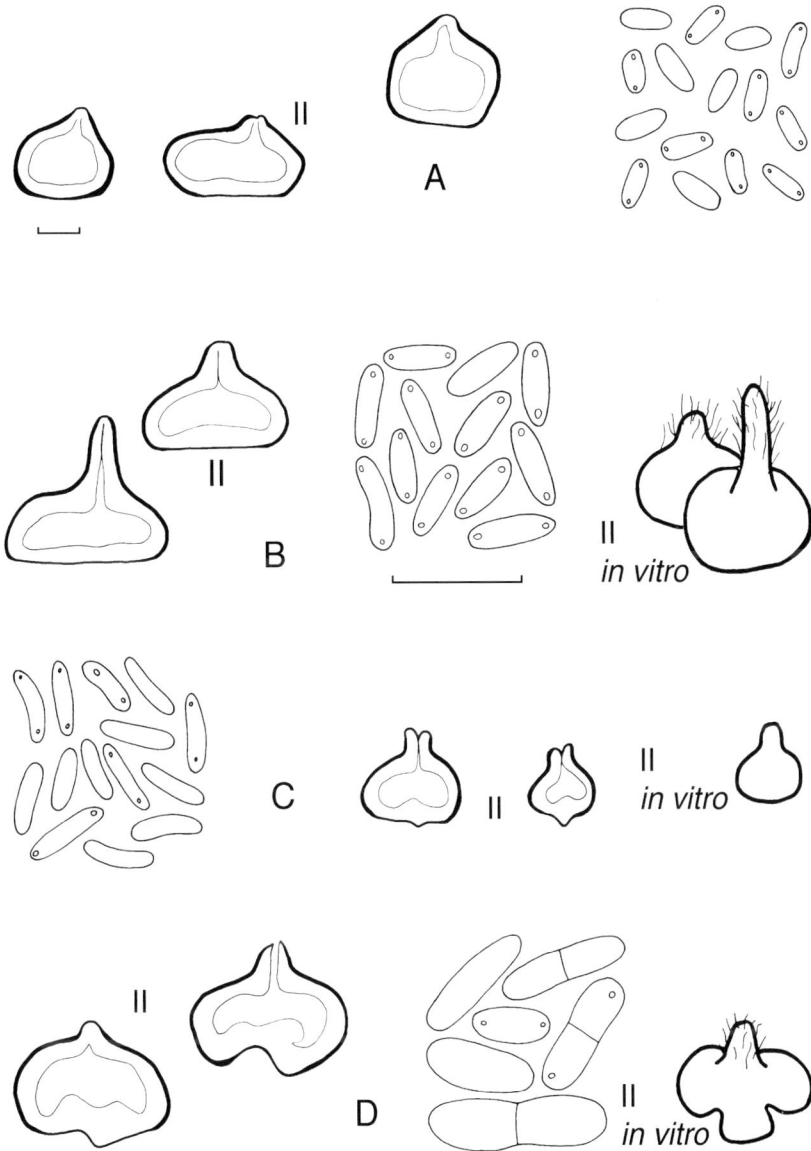

Fig. 38. A *Phoma conferta.* Pycnidia-II subglobose-conical with broad base and usually slightly papillate pore. **B** *Phoma congesta.* Pycnidia-II depressed globose with flattened base and conspicuous beak-like papilla which gradually may develop into a long neck. *In vitro* (right) neck usually pilose. **C** *Phoma coonsii.* Pycnidia-II relatively small, subglobose with a flattened or somewhat pointed base and broad papillate pore, often with a protruding lip. **D** *Phoma doliolum.* Pycnidia-II depressed globose, usually with irregular deformed flattened base and short papillate neck. Conidia variable in dimension, the larger ones often 1-septate. Bar pycnidia 100 µm, bar conidia 10 µm.

Phoma conferta P. Syd. ex Died., Fig. 38A

Teleomorph: *Leptosphaeria conferta* **Niessl ex Sacc.**

Phoma conferta P. Syd. ex Died. *in* Krypt.-Flora Mark Brandenb. 9, Pilze **7** [Heft 1]: 142. 1912 [vol. dated '1915']; not *Phoma conferta* Ellis & Everh. *in* H.L. Jones Bull. Oberlin Coll. Lab. **9**: 7. 1898 [as '*confertum*'] [nomen nudum].

Θ ≡ *Phoma conferta* P. Syd. *in* Mycoth. March. Cent. 43, No. 4291. 1895 [without description].

Diagn. data: Lucas (1963) (fig. 1A–F; anamorph-teleomorph connection and characteristics), Boerema *et al.* (1994) (fig. 4D: pycnidial and conidial shape, reproduced in present Fig. 38A; *in vivo* and *in vitro* descriptions adopted below).

Description in vivo

Pycnidia always of type II (on dead stems, scattered or seriate, at first covered by the epidermis, later superficial), subglobose-conical with broad base and usually slightly papillate pore, 150–250(–400) μm diam. Wall consisting of polygonal scleroplectenchyma cells of variable dimensions; cells at base similar to those at side walls, but with more or less parallel arrangement. Conidial matrix whitish.

Conidia oblong-ellipsoidal (3.5–)4(–5) × 1–1.5(–2) μm, usually with 2 small guttules.

Pseudothecia (also on dead stems) up to 380 μm diam., conical to subglobose. Wall scleroplectenchymatous. Asci 100–140 × 14–18 μm, 8-spored, irregularly quadriseriate. Ascospores 44–52 × 6–7 μm, fusiform, mostly curved, 3-septate, yellowish, guttulate [for recent detailed description see Lucas, 1963].

In vitro

Colonies on OA 3–4 cm after 7 days, cream white; aerial mycelium tenuous, yellow/green; reverse grey with yellow zones. Pycnidia in concentric zones both on and in the agar and in aerial mycelium, resembling those *in vivo*, but more irregular in shape. Conidia as *in vivo*. Representative culture CBS 375.64.

Ecology and distribution

Common in Europe on dead stems of various wild Cruciferae, occasionally also on non-cruciferous plants. The anamorph is often confused with *Phoma sublingam* Boerema (this section), teleom. *Leptosphaeria submaculans* L. Holm. Both fungi commonly occur together on *Berteroa (Farsetia) incana*. They can be easily differentiated by their teleomorphs [*Lept. conferta* is distinctive in having neck-less pseudothecia and 3-septate ascospores]. The conidia of their anamorphs are similar but the pycnidia of *Ph. conferta* are always scleroplectenchymatous, type II, and nearly neckless, whereas the pycnidial types I and II of *Ph. sublingam* usually have a pronounced neck.

Note

The pycnidia-II of *Ph. conferta* may develop simultaneously with the pseudothecia of its teleomorph, but on *Berteroa incana* they are more often found together with the pseudothecia of *Lept. submaculans*. Such a confusing coexistence also occurs with the pycnidia of *Phoma acuta* (Hoffm. : Fr.) Fuckel subsp. *acuta* (this section; teleom. *Leptosphaeria doliolum* (Pers. : Fr.) Ces. & De Not. subsp. *doliolum*) and the pseudothecia of *Leptosphaeria acuta* (Fuckel) P. Karst. (anam. *Ph. piskorzii* (Petr.) Boerema & Loer., this section).

Phoma berteroae Hollós, described *in* Annls hist.-nat. Mus. natn. hung. **6**: 529. 1908, may be conspecific with *Ph. conferta*, but the original material of *Ph. berteroae* appears to have been destroyed during the Second World War (information from Museum of Natural History, Budapest) and therefore could not be checked for the presence of scleroplectenchyma.

Phoma congesta Boerema *et al.*, Fig. 38B

Teleomorph: *Leptosphaeria congesta* M.T. Lucas

Phoma congesta Boerema, Gruyter & Kesteren *in* Persoonia **15**(4): 461. 1994.

Diagn. data: Lucas (1963) (fig. 2; anamorph-teleomorph connection and characteristics), Boerema *et al.* (1994) (fig. 5C: pycnidial and conidial shape, reproduced in present Fig. 38B; *in vivo* and *in vitro* descriptions adopted below).

Description in vivo *(esp. on* Achillea *spp.)*

Pycnidia always of type II (on woody parts of dead stems, gregarious and occasionally confluent), depressed globose with flattened broad base and conspicuous beak-like papilla which gradually may develop into a long neck, mostly 250–500 μm diam. The wall has a somewhat convex thickening at the base, but similar polygonal scleroplectenchyma cells throughout. Conidial matrix white to very pale yellow (ivory).

Conidia oblong-ellipsoidal to subcylindrical, (4–)5–6(–7) × 1.5–2 μm, mostly inconspicuously biguttulate.

Pseudothecia (also on dead stems) 380–400 μm diam., globose with conical neck. Wall scleroplectenchymatous. Asci 92–120 × 12–14 μm, 8-spored, biseriate in the upper part, uniseriate below. Ascospores 24–32 × 7–8 μm, rhomboid-fusiform, 3-septate, yellowish-pale brown (for detailed description see Lucas, 1963).

In vitro

Colonies on OA relatively slow growing, *c.* 2.5 cm after 7 days, regular, cream/yellowish (straw/luteous), translucent; aerial mycelium very tenuous, greenish olivaceous [on MA with short coralloid hyphal branches]; reverse yellowish with broad colourless margin. Abundant production of pycnidia type II, solitary or confluent, mostly (220–)240–480(–550) μm diam., resembling those *in vivo*, but with a semi-pilose beak which may grow out into a long neck (up to 500 μm) [on MA often with several necks]. Conidia similar to those *in vivo*, but occasionally up to 8 μm long. Cultures of this fungus did not survive lyophilization. Dried culture L 993.373.042.

Ecology and distribution

Most specimens of this fungus are found in southern Europe (Italy, Portugal) on dead stems of different *Achillea* spp. (*A. ageratum*, *A. millefolium* and *A. macrophylla*). The teleomorph was first described on stems of *Erigeron canadensis*; therefore the fungus may also occur on other members of the Compositae.

Phoma coonsii Boerema & Loer., Fig. 38C

Phoma coonsii Boerema & Loer. *in* Trans. Br. mycol. Soc. **84**: 289[–290]. 1985.

Diagn. data: Boerema & Loerakker (1985) (figs 1A, 4A, 5A, F: photograph of pycnidial primordium *in vitro*, drawing of pycnidia *in vitro*, photographs of cultures on OA and MA and production of rubiginous verrucose granules on OA; tab. II: synoptic table on characters *in vitro*, with drawing of conidia; nomenclature, distribution), Boerema *et al.* (1994) (fig. 8D: pycnidial and conidial shape, reproduced in present Fig. 38C; *in vivo* and *in vitro* descriptions adopted below).

Description in vivo (Malus pumila)

Pycnidia always of type II (immersed in bark and on wood, usually densely grouped), subglobose with a flattened or somewhat pointed base and a dark papillate pore, often with a protruding lip, 100–150(–250) μm diam. The wall of a mature pycnidium has randomly arranged polygonal scleroplectenchyma cells and is about the same thickness throughout, but may have some invaginations. Pycnidial primordia (protopycnidia) are completely filled with diverging rows of somewhat elongated cells. Conidial matrix dirty white.

Conidia (sub-)cylindrical, often slightly curved, 4.5–6.5 × 1–2 μm, eguttulate and biguttulate.

In vitro

Colonies on OA slow growing, *c.* 1.5 cm diam. after 7 days, pale greenish yellow; aerial mycelium either felted fluffy-cottony or scant, at least in zones; reverse becoming yellow, citrine, amber or dark olivaceous due to diffusible pigment(s), which become reddish (quickly fading) on application of a drop of

NaOH. Rubiginous verrucose granules, up to 15 μm diam., are produced in old colonies. Pycnidia often develop only after 1 or even 2 months. Conidia as *in vivo* [colonies on MA are very slow growing with an odour of lovage or liquorice]. Representative culture CBS 141.84.

Ecology and distribution

In North America (USA) and Japan found on bark and wood of apple, *Malus pumila* (Rosaceae). Reputed to be pathogenic, bark canker, but probably only an opportunistic parasite.

Phoma doliolum P. Karst., Fig. 38D

Teleomorph: **Leptosphaeria conoidea (De Not.) Sacc.**

Phoma doliolum P. Karst. *in* Meddn Soc. Fauna Flora fenn. **16**: 9–10. 1888.

= *Phoma acuta* subsp. *amplior* Sacc. & Roum. *in* Revue mycol. **6**: 30. 1884; *in* Sylloge Fung. **3**: 133–134. 1884 [in both cases as 'Phoma acuta – Ph. amplior'].

≡ *Phoma hoehnelii* subsp. *amplior* (Sacc. & Roum.) Boerema & Kesteren apud Boerema *in* Trans. Br. mycol. Soc. **67**: 299. 1976.

= *Plenodomus microsporus* Berl. *in* Bull. Soc. mycol. Fr. **5**(2): 55–56. 1889.

≡ *Diploplenodomus microsporus* (Berl.) Höhn. *in* Hedwigia **59**: 250. 1918.

= *Phoma la(m)psanae* P. Karst. *in* Acta Soc. Fauna Flora fenn. **27**: 7. 1905.

= *Phoma seseli* Hollós *in* Annls hist.-nat. Mus. natn. hung. **4**: 340. 1906.

= *Diploplenodomus malvae* Died. ex Died. *in* Krypt.-Flora Mark Brandenb. 9, Pilze **7** [Heft 2]: 415. 1912 [vol. dated '1915'].

Θ ≡ *Diploplenodomus malvae* Died. *in* Annls mycol. **10**: 140. 1912 [1 April] [without description].

= *Plenodomus labiatarum* Petr. *in* Annls mycol. **21**: 237[–238]. 1923.

= *Plenodomus aconiti* Petr. *in* Annls mycol. **20**: 151. 1922.

= *Phoma lunariae* Moesz *in* Magy. bot. Lap. **25**: 36. 1926.

= *Phoma origani* Mark.-Let. *in* Bolez. Rast. **16**: 194. 1927.

≡ *Plenodomus origani* (Mark.-Let.) Petr., *in* Ber. bayer. bot. Ges. **24**: 8. 1940 [cf. reprint].

= *Plenodomus vincetoxici* Petr. *in* Kryptog. Forsch., Bayer. bot. Ges. Erforsch. heim. Flora **2**(2): 187–188. 1931.

[Θ = 'Plenodomus galeopsidis H. Ruppr.': manuscript name in herb. B, not published.]

Diagn. data: Boerema (1976) (figs 1C, 10: drawings of pycnidium with conidia, pseudothecium and ascospores and photographs of cultures on OA sub *Phoma hoehnelii* subsp. *amplior* and *Leptosphaeria doliolum* subsp. *pinguicula*; tabs 6, 9: sources of isolates; synoptic table on characters *in vitro* with drawing of conidia), Shoemaker (1984) (descr. and Canadian records of teleomorph), Boerema *et al.* (1994) (fig. 8A: pycnidial and conidial shape, reproduced in present Fig. 38D; synonymy anamorph, nomenclature anamorph and teleomorph; *in vivo* and *in vitro* descriptions adopted below), Boerema *et al.* (1996: 183) (specimen in herb. B).

Description in vivo *(esp.* Foeniculum vulgare*)*

Pycnidia always of type II (on dead stems, scattered or in groups, first subepidermal, later becoming superficial), depressed globose, usually with irregular deformed flattened base and a short papillate neck with tube-shaped pore, relatively large (250–)300–500(–800) × 150–250(–300) μm. Wall explicitly scleroplectenchymatous with a convex or irregular thickening at the base and more or less parallel cell-structure; sometimes also irregular outgrowths from the sidewalls, sometimes making the pycnidial cavity appear multilocular. Conidial matrix cream or off-white.

Conidia ellipsoidal to subcylindrical, extremely variable in size, often (4–)5–7 × 2–2.5(–3) μm, but also much larger, (7–)8–12(–16) × 2–3 μm and then often 1-septate, eguttulate or with 2 small polar guttules.

Pseudothecia (also on dead stems) up to 500 μm diam., depressed globose with flattened base and short conical neck. Wall scleroplectenchymatous. Asci 75–135 × 7–8 μm, i.e. variable in length and relatively broad, 8-spored, uniseriate. Ascospores (18–)20–23(–25) × 5–6.5(–7) μm [generally broader than those of *L. doliolum*], broadly ellipsoidal, 3-septate with obtuse rather than acute end cells, yellowish-brown with 1 guttule per cell, echinulate (for recent detailed description see Shoemaker, 1984).

In vitro

Colonies on OA 3–4 cm diam. after 7 days [on MA slower growing, 1–2 cm], aerial mycelium cottony or somewhat felted, olivaceous/greyish; reverse with buff, salmon or ochraceous discoloration of the agar. Abundant production of relatively large pycnidia on and in the agar, resembling those *in vivo*. Conidia usually 4–6(–7.5) × 1.5–2.5(–3) μm. Representative culture CBS 616.75.

Ecology and distribution

A plurivorous necrophyte, widespread in Europe and North America, and especially common on dead stems of Compositae or Umbelliferae. Many records in Canada (teleomorph) refer to *Solidago* spp. (Comp.) and in Europe the most common hosts (anamorph and teleomorph) are *Angelica* and *Foeniculum* spp. (Umb.). The wide host range is similar to that of the plurivorous *Phoma acuta* subsp. *errabunda* (Desm.) Boerema *et al.* (this section, teleom. *Leptosphaeria doliolum* subsp. *errabunda* Boerema *et al.*). It is possible that both fungi represent originally American and European counterparts. Their pycnidia and pseudothecia are very similar but the conidia of *Ph. doliolum* are comparatively large (and *in vivo* often 1-septate) and the ascospores of *Lept. conoidea* are relatively broad (length–width ratio about 4 instead of 5–5.5 in *Lept. doliolum*, see Shoemaker, 1984). The teleomorphs, however, have often been confused; in an earlier cultural study of these fungi (Boerema, 1976) *Lept. doliolum* subsp. *errabunda* was referred to as *Lept. doliolum* var. *conoidea* (De Not.) Sacc., whereas *Lept. conoidea* was listed under its synonym *Lept. doliolum* subsp. *pinguicula* Sacc.

Phoma drobnjacensis Bubák, Fig. 39A

> *Phoma drobnjacensis* Bubák *in* Bot. Közl. **14**: 63. 1915 [holotype in BPI originally labelled '*Plenodomus drobnjacensis*'].
> > = *Pyrenochaeta gentianae* Chevassut *in* Bull. Soc. mycol. Fr. **81**: 36. 1965.

Diagn. data: Chevassut (1965) (figs 1–3; disease symptoms; description and illustration *in vivo*), Boerema *et al.* (1984) (synonymy), Boerema *et al.* (1994) (fig. 4C: shape of pycnidia, conidia and chlamydospores, reproduced in present Fig. 39A; *in vivo* and *in vitro* descriptions adopted below).

Description in vivo (Gentiana *spp.*)

Pycnidia of type I (leaf spots and basal stem rot, usually aggregated in short rows), subglobose with a short papillate neck, relatively small, (100–)150–200 µm diam.; neck of mature pycnidia black and semi-setose, i.e. often with a number of setae around the porus, setae rigid and septate, usually 45–60 × 3–4 µm. Pycnidia of type II (superficial on dead stems) also subglobose-papillate, but becoming larger, 200–500 µm diam. and usually without any trace of setae. Wall of full-grown pycnidium shows randomly polygonal scleroplectenchyma cells and is about the same thickness throughout. Conidial matrix whitish.
    Conidia oblong-ellipsoidal, sometimes curved, (3–)3.5–4.5(–5) × (1–)1.5(–2) µm, usually biguttulate.

In vitro

Chlamydospores commonly develop in agar cultures; they are unicellular, mostly 10–15 µm diam., usually intercalary in short or long, sometimes branching chains, relatively thick-walled and olivaceous with one or more greenish guttules.
    Colonies on OA 3.5–4.5 cm diam. after 7 days, transparent-yellow; aerial mycelium very tenuous, low, somewhat felted, greyish; reverse conspicuous sulphur yellow to citrine green. Colonies on MA show clearly that the yellowish pigment diffuses from the hyphae into the agar and crystallizes out as complexes of needles; on the surface of the agar they may form greenish yellow scales, 200–600 µm diam. In old cultures on MA the yellow colour changes to

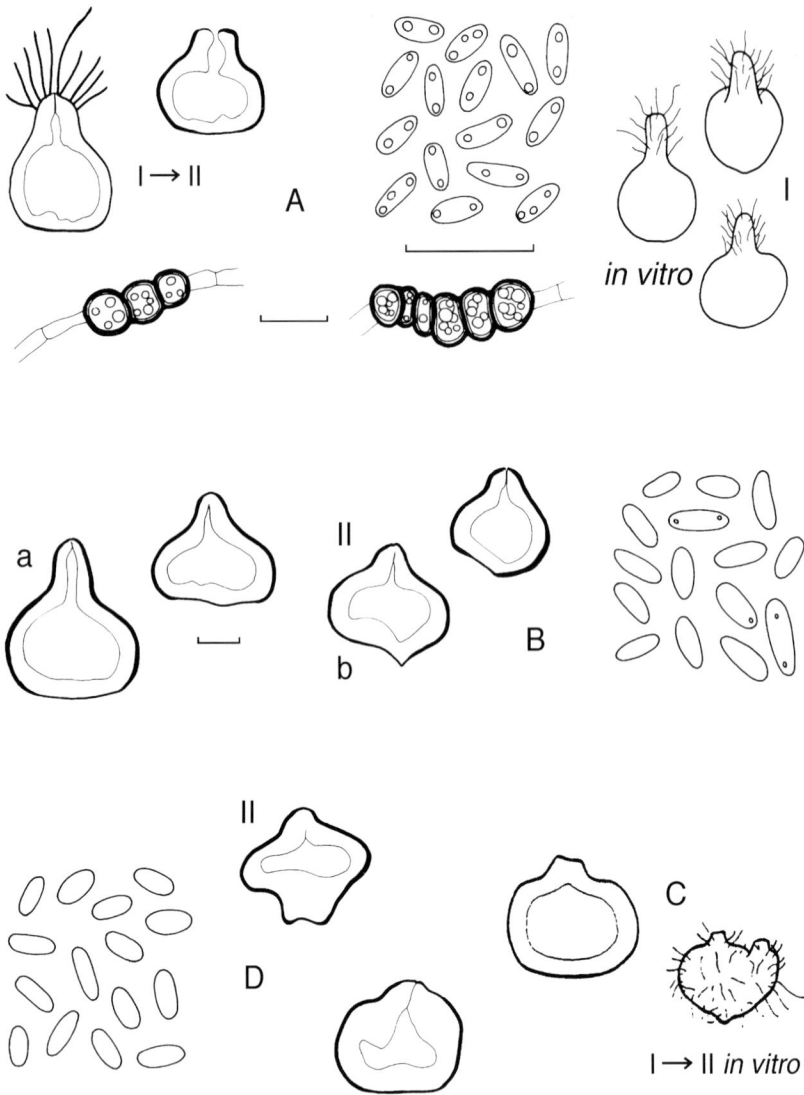

Fig. 39. A *Phoma drobnjacensis.* Pycnidia initially thin-walled I→II; in association with disease symptoms on gentians sometimes with rigid setae on the papilla; *in vitro* (right) usually with a pilose neck. The fungus produces chains of thick-walled chlamydospores. **B** Varieties of *Phoma enteroleuca.* Pycnidia-I→II, globose-papillate with flattened or somewhat pointed base; var. *enteroleuca* (a) produces slightly larger pycnidia than var. *influorescens* (b). **C** *Phoma etheridgei.* Pycnidia-I→II only known from isolates, with hyphal outgrowths in varying degrees and sometimes with a pointed base. Conidia similar to those of *Ph. enteroleuca.* **D** *Phoma intricans.* Characterized by very thick-walled pycnidia-II, subglobose or irregular in shape with narrow pore and flattened, often somewhat elongated base. Bar pycnidia 100 μm, bar conidia and chlamydospores 10 μm.

a reddish hue naturally; this colour change occurs also with addition of a drop of NaOH. Pycnidia may develop abundantly on the agar, I→II, usually with a short cylindrical neck, glabrous or hairy at the apex, i.e. semi-pilose (not setose as *in vivo*), mostly solitary and connected by superficial hyphae. Occasional aberrant small pycnidia occur in aerial mycelium. Conidia as *in vivo*. Representative culture CBS 270.92.

Ecology and distribution

Originally described from *Gentiana asclepidea* (Gentianaceae) in the mountains of Montenegro, former Yugoslavia (semi-setose pycnidia). Recently repeatedly recorded as a pathogen on different cultivated species of *Gentiana* in England, France, Germany and The Netherlands: leaf spot. The fungus has occasionally been isolated from a cultivar of *Lisianthus russellianus* (also Gentianaceae).

Phoma enteroleuca Sacc. var. *enteroleuca*, Fig. 39Ba

Phoma enteroleuca Sacc. *in* Michelia **1**(4): 358. 1878, var. *enteroleuca*.
≡ *Phomopsis enteroleuca* (Sacc.) Wollenw. & Hochapfel *in* Z. ParasitKde **8**: 571. 1936 [misapplied].
= *Phoma armeniacae* Thüm. *in* Mitt. forstl. VersWes. Öst. **11**: 3. 1888.
= *Phoma malvacei* Brunaud *in* Bull. Soc. Sci. nat. Ouest Fr. **3**: 218. 1893.
H = *Phoma berberidicola* Brunaud *in* Act. Soc. Linn. Bordeaux **1898**: 12. 1898 [cf. fresh isol.]; not *Phoma berberidicola* Vestergr. *in* Öfvers. Vetens.-Akad. Förh. **1897**: 38. 1897 [= *Phoma aliena* (Fr. : Fr.) Aa & Boerema, sect. *Phyllostictoides*, **E**].
≡ *Phoma berberidella* Sacc. & P. Syd. *in* Sylloge Fung. **14**: 867. 1899.
= *Phoma macra* P. Syd. *in* Beibl. Hedwigia **38**: 136. 1899.
H = *Phoma cornicola* D. Sacc. apud Saccardo & Sydow *in* Sylloge Fung. **16**: 856. 1902; not *Phoma cornicola* Oudem. *in* Ned. kruidk. Archf III, **2**(1): 234. 1900.
H = *Phyllosticta catalpicola* Oudem. *in* Ned. kruidk. Archf III, **2**(4): 890. 1903; not *Phyllosticta catalpicola* (Schwein.) Ellis & Everh., N. Am. Pyren. 747. 1892 [= *Phomopsis* sp.].
= *Phoma cruris-hominis* Punith. *in* Nova Hedwigia **31**: 135–138. 1979 [cf. holotype].
[= 'Plenodomus chondrillae' sensu Batista & Vital *in* Annais Soc. Biol. Pernamb. **15**: 419. 1957; misapplied; *Plen. chondrillae* Died. = *Phoma agnita* Gonz. Frag., this section.]
[Θ = 'Plenodomus glechomae H. Ruppr.': manuscript name in herb. B., not published.]

Diagn. data: Boerema & Loerakker (1985) (figs 1B–E, 2, 4B, 5B: drawing of conidia and conidiogenous cells of holotype, photographs of pycnidia *in vivo* and *in vitro*, disease symptoms on *Catalpa bignonioides* with pycnidia on leaf scar, drawing of pycnidia *in vitro* and photographs of cultures on OA and MA; tab. 1: host records and isolate sources; tab. 2: synoptic table on characters *in vitro* with drawing of conidia; synonymy; specimens examined, host range; pathogenicity), Boerema (1986a) (fig. 6; role as stress-fungus on trees and shrubs; tab. 1: host list), Boerema *et al.* (1994) (fig. 9Aa: pycnidial and conidial shape, reproduced in present Fig. 39Ba; *in vivo* and *in vitro* descriptions partly adopted below), Boerema *et al.* (1996: 184) (specimen in herb. B).

Description in vivo

Pycnidia initially thin-walled, type I, but gradually becoming scleroplectenchymatous, type II (superficial on dead branchlets, cankers, leaf scars, etc.; usually densely crowded), globose-papillate, mostly 200–400 μm diam. Scleroplectenchymatous wall often with irregular invaginations at the base and at the sides; consisting of polygonal cells of variable size, usually in more or less parallel rows at the base. Conidial matrix pink.

Conidia ellipsoidal or ovoid, relatively small, 3–4(–4.5) × 1–2 μm, eguttulate or with 2 minute polar guttules.

In vitro

Colonies on OA 3.5–4.5 cm diam. after 7 days, pale olivaceous grey, greenish glaucous to greenish yellow; aerial mycelium tenuous, twined or tufted often with red, needle-shaped crystals on the hyphae; reverse buff, straw or amber by a yellow pigment diffusing into the agar, locally red (purplish-blue with addition of a drop of NaOH), in the centre always ochraceous. A special feature is the production of a fluorescing metabolite (reverse under UV-366 nm). Pycnidia-I→II scattered on and in the agar, globose, becoming papillate as *in vivo*. Conidia resembling those *in vivo*. Representative culture CBS 142.84.

Ecology and distribution

Recorded from bark, wood and fruits of various deciduous trees and shrubs in Europe (France, Germany, Italy, The Netherlands) and North America (USA). An opportunistic parasite, e.g. frequently found in association with discoloured wood and dead branches or twigs of *Catalpa bignonioides* (Bignoniaceae), *Lonicera caprifolia* (Caprifoliaceae), *Malus pumila* (Rosaceae) and *Ulmus* spp. (Ulmaceae). The fungus is sometimes confused with *Phoma fimeti* Brunaud (sect. *Phoma*, **A**; easily distinguished by a broader conidium frequently with a large guttule).

The synonym *Phoma cruris-hominis* (type I) refers to an isolate from a lesion on the leg of a woman after a fall she sustained while in Singapore.

Phoma enteroleuca var. *influorescens* Boerema & Loer., Fig. 39Bb

> *Phoma enteroleuca* var. *influorescens* Boerema & Loer. *in* Trans. Br. mycol. Soc. **84**: 290. 1985 [erroneously as '*inflorescens*', corrected in Erratum-slip].
>
> > = *Phyllosticta tweediana* Penz. & Sacc. apud Penzig *in* Atti r. Ist. veneto Sci. VI, **2** ['Funghi Mortola']: 15. 1884.

Diagn. data: Boerema & Loerakker (1985) (figs 1F, 4C, 5C: photograph of pycnidium *in vivo*, drawing of pycnidia *in vitro*, photographs of cultures on OA and MA; tab. 1: isolate sources; tab. 2, synoptic table on characters *in vitro* with drawing of conidia), Boerema *et al.* (1994) (fig. 9Ab: pycnidial and conidial shape, reproduced in present Fig. 39Bb; synonymy; *in vivo* and *in vitro* descriptions adopted below).

Description in vivo

Pycnidia (superficial in groups on dead branchlets, etc.) resembling those of the type variety, initially thin-walled, type I, and later becoming more or less scleroplectenchymatous, type II, globose-papillate, but generally smaller than those of var. *enteroleuca*, mostly 150–350 μm diam., and conidial matrix colourless, not pink.

Conidia similar to those of var. *enteroleuca*, but often somewhat larger, 3–4(–5) × (1–)1.5–2 μm, and acute at one end.

In vitro

Colonies on OA slower growing than var. *enteroleuca*, 2–2.5 cm diam. after 7 days; more greenish/yellow and aerial mycelium not tufted but plumose (red crystals absent); reverse also characterized by a yellow pigment (mostly with a distinct citrine zone), but not fluorescent; red pigment may also be produced (purplish/blue with a drop of NaOH). Pycnidia in zones on and in the agar, globose-papillate as *in vivo*. Conidia as *in vivo*. Representative culture CBS 143.84.

Ecology and distribution

Found on various trees and shrubs in Europe and North America, just like the type variety, but apparently much less common. Opportunistic parasite. Occasionally also on herbaceous plant residues.

Phoma etheridgei L.J. Hutchison & Y. Hirats., Fig. 39C

> *Phoma etheridgei* L.J. Hutchison & Y. Hirats. apud Hutchison, Chakravarty, Kawchuk & Hiratsuka *in* Can. J. Bot. **72**: 1425–1427. 1994.

Diagn. data: Hutchison *et al.* (1994) (figs 2–6; original *in vitro* description; comparison with *Phoma enteroleuca* and other wood-inhabiting *Phoma* spp.

using morphological, physiological and molecular characteristics), Boerema &
de Gruyter (1999) (fig. 1C: pycnidial and conidial shape, reproduced in present
Fig. 39C; cultural descriptions, partly adopted below).

Appearance in vivo *(Populus tremuloides)*

This pycnidial fungus is only known from isolates obtained from the bark of
'black galls' and cankers of American trembling aspen in Canada.

Description in vitro

Pycnidia (immersed to superficial, solitary to gregarious), pseudoparenchymatous [gradually becoming scleroplectenchymatous type II], globose/subglobose
to irregular, sometimes with a pointed base, covered to varying degrees by hyaline to dark-coloured hyphae, 95–270 μm diam. Conidiogenous cells globose
to flask-shaped. Conidial matrix flesh/salmon to pale vinaceous.

   Conidia ellipsoidal to oblong/ovoid or allantoid, 3–4.5(–5) × 1–2 μm,
eguttulate [and with 1–2 minute guttules].

   Colonies on OA *c.* 2 cm after 7 days, regular, with floccose, white aerial
mycelium, colourless to greenish olivaceous, with pale grey olivaceous at centre; reverse similar [on MA greyish yellow to orange-grey, reverse apricot brown
to dark brown with concentric greyish orange to brownish orange zones].
Brownish pigment diffusing into agar, becoming greenish to orange coloured
on application of a drop of NaOH. Pale yellow-coloured crystals may be produced in agar at margin of colony, especially among fresh isolates.
Representative culture DAOM 216539.

Ecology and distribution

This fungus seems to be specific to the bark of black galls ('burls') and related
cankerlike structures on American trembling aspen (*Populus tremuloides*,
Salicaceae). Trees with these stem deformities are occasionally found in western Canada and the Rocky Mountain states of the USA. Such trees showed a
significant decrease or absence of infestation by the aspen decay pathogen
Phellinus tremulae (Bondartsev) Bondartsev & Borisov. This phenomenon may
be due to the presence of *Phoma etheridgei* which proved to be strongly antag-
onistic *in vitro* against *Phellinus tremulae*.

Phoma intricans M.B. Schwarz, Fig. 39D

   Phoma intricans M.B. Schwarz *in* Meded. phytopath. Lab. Willie
   Commelin Scholten **5**: [42–]44. 1922.

Diagn. data: Schwarz (1922) (fig. 10), Boerema *et al.* (1994) (fig. 9B: pycnidial
and conidial shape, reproduced in present Fig. 39D; *in vivo* and *in vitro*
descriptions adopted below).

G Phoma *sect.* Plenodomus   355

Description in vivo *(Salix alba)*

Pycnidia always of type II (immersed in bark and on wood, usually densely crowded), mostly subglobose-papillate with a flattened somewhat elongated base, sometimes without any definite shape, but always with one narrow pore, 250–350 μm diam. The walls of mature pycnidia usually have many invaginations and consist of polygonal scleroplectenchyma cells of quite different sizes. Extensive wall-thickenings at the base of the pycnidial cavities have a diverging parallel structure. Conidial matrix pink-violet.

   Conidia ovoid to ellipsoidal, 2.5–3.5(–4) × 1–1.5(–2) μm, eguttulate.

In vitro

Colonies on OA 5–5.5 cm diam. after 7 days, regular, olivaceous grey with sectoring hyphal and pycnidial zones; reverse olivaceous and greenish black with pale red (coral) margin. With addition of a drop of NaOH the reddish pigment turns blue [on MA orange needle-like crystals were seen on the hyphae]. The pycnidia produced *in vitro* are relatively small, 150–250 μm diam., globose-papillate, scleroplectenchymatous (II); the conidia are similar to those *in vivo*. Representative culture CBS 139.78.

Ecology and distribution

Very common in Europe on varieties and hybrids of *Salix alba* (Salicaceae). Reported to be pathogenic: bark canker, but probably only an opportunistic parasite. The fungus is also incidental occurring on deciduous trees of other genera.

Phoma korfii Boerema & Gruyter, Fig. 40A

Synanamorph: ***Sclerotium orobanches* Schwein. : Fr.**

   Phoma korfii Boerema & Gruyter *in* Persoonia **17**(2): 275. 1999.

Diagn. data: Yáñez-Morales *et al.* (1998) (figs 1–8, 10–13: sclerotia with pycnidia *in vivo* and *in vitro*, sections of pycnidium and pycnidial wall, pycnosclerotium *in vitro*, conidia; fig. 9 and tab. 1: cultural characters under different light/dark and temperature regimes; history of the fungus and detailed descriptions of the sclerotial and pycnidial anamorphs *in vivo* and *in vitro*, partly adopted below), Boerema & de Gruyter (1999) (fig. 1A: pycnidial and conidial shape, reproduced in present Fig. 40A; description of the pycnidial anamorph as a new species of *Phoma* sect. *Plenodomus*, named after Dr R.P. Korf, who first discovered it).

Description in vivo *(cf. Yáñez-Morales* et al. *(1998) on* Epifagus virginianus*)*

Pycnidia II (only on the surface of sclerotia, typically at the base) globose to subglobose with papillate pore, 110–333 μm wide and 130–370 μm high. Wall of the pycnidia clearly merely on extension of the sclerotial rind, with no differ-

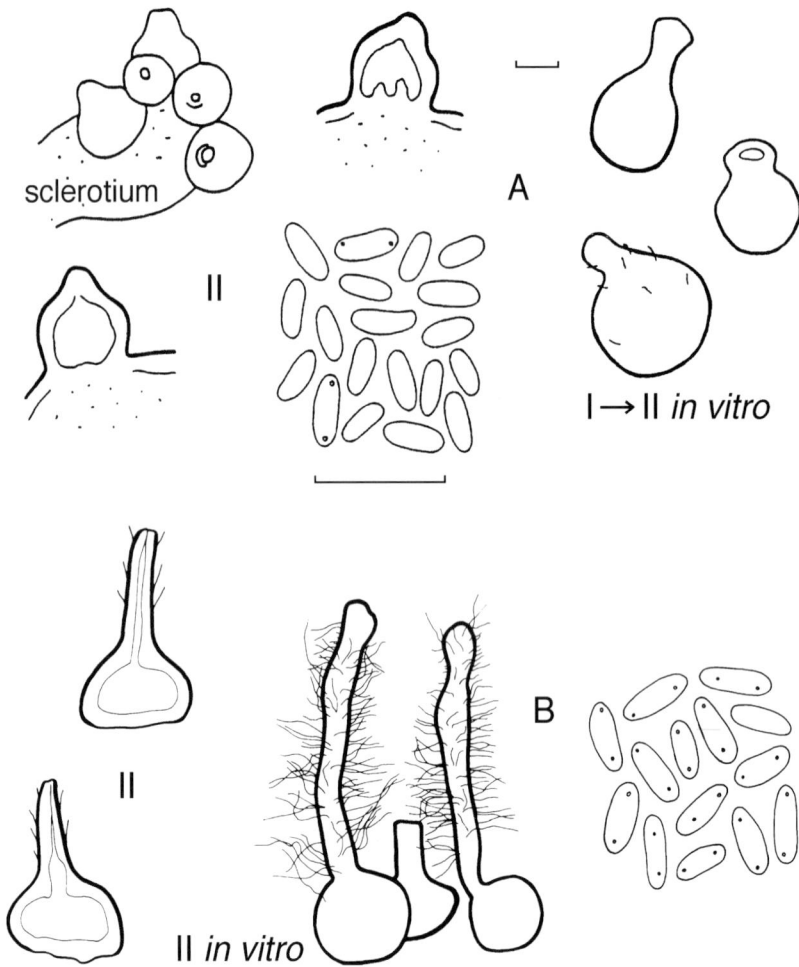

Fig. 40. A *Phoma korfii*. Pycnidia-II *in vivo* always arising from sclerotia (synanam. *Sclerotium orobanches*), globose to subglobose with papillate pore. Pycnidia *in vitro* I→II more irregular, often developing an elongated neck in a later stage. **B** *Phoma leonuri*. Characterized by pycnidia-II with conspicuous long cylindrical necks; *in vivo* semi-pilose but *in vitro* explicitly pilose. Bar pycnidia 100 μm, bar conidia 10 μm.

entiation of lower wall to the pycnidium. Conidial matrix white to pale luteous (*in vitro* also white to buff/rosy buff).

Conidia ellipsoidal to oblong, 2–4 × 1–1.5 μm, sometimes guttulate.

Sclerotia (synanamorph *Sclerotium orobanches*) immersed in or erumpent from or superficial on dead stems, flowers and roots of the host plant, separate or in small clusters, rounded to elongate, often lobulate or twisted to vermiform, smooth or with 1–8 protusions, i.e. pycnidia, (0.5–)1–1.5 × 2–9 mm. In section with a dark-celled cortex and a hyaline-celled medulla.

In vitro

Colonies on OA 4.5 cm diam. after 7 days; aerial mycelium whitish, otherwise colourless; reverse colourless to weak primrose near margin. Under the standard method employed in our culture study (p. 15) the fungus produced only scattered pycnidia I→II, on and in the agar, 110–320 μm diam., irregular to subglobose-papillate, in a later stage with an elongated neck, glabrous or with short mycelial outgrowths. In the aerial mycelium also micropycnidia, up to 50 μm diam. Conidia oblong to ellipsoidal, somewhat larger than those observed *in vivo*: (2–)3–5.5 × (1–)1.5–2 μm, eguttulate. Representative culture CBS 101638.

[Yáñez-Morales *et al.* (1998) obtained cultures on other media and different light/dark and temperature regimes with only sclerotia, or sclerotia and pycnidia, with pycnidia either separate from or developing on sclerotia. Pycnosclerotia (III) were sometimes also present in the cultures.]

Ecology and distribution

The sclerotial anamorph of the fungus was already described by von Schweinitz in 1822 in North America from roots and stems of *Orobanche virginiana* ≡ *Epifagus virginianus*, an achlorophyllous plant (Orobanchaceae) parasitic on the roots of *Fagus grandifolia*. Fries (1828) recorded the sclerotial anamorph from Sweden and Germany on *Orobanche major*, parasitic on the roots of leguminous shrubs. After this the fungus appears to be only very rarely collected in the USA and Canada. The above description is based on a recent collection on *Epifagus virginianus* in the Lloyd-Cornell Preserve, east of Ithaca, New York, USA (Yáñez-Morales *et al.*, 1998).

Phoma leonuri **Letendre, Fig. 40B**

Teleomorph: ***Leptosphaeria slovacica* Picb.**

Phoma leonuri Letendre apud Roum. *in* Fungi gall. exs. No. 3068. 1884; *in* Revue mycol. **6**: 229. 1884.
≡ *Plenodomus leonuri* (Letendre) Moesz & Smarods apud Moesz *in* Magy. bot. Lap. **31**: 38. 1932.
V = *Phoma complanata* var. *acuta* Auersw. *in* Fungi europ. exs./Klotzschii Herb. mycol. Cont. Cent. 4, No. 343. 1861 [as 'complanatum' and var. 'acatum'; obviously introduced as a new variety ('A typica specie, non nisi ostiolo elongato diversa') representing 'Sphaeria acutum Pers. p.p.' on *Ballota nigra* subp. *foetida*].
V = *Phoma acuta* f. *ballotae* Thüm. *in* Verh. K.K. zool.-bot. Ges. Wien **25**: 550. 1875 [as 'acutum Awd', see above].
Θ V = *Phoma acuta* f. *ballotae* P. Syd. *in* Mycoth. March. No. 2571. 1889 [as 'acutum'] [without description].
Θ = *Phoma acuta* f. *ballotae* Allesch. *in* Rabenh. Krypt.-Flora [ed. 2], Pilze **6** [Lief. 63]: 271. 1898 [vol. dated '1901'] [without description].

Diagn. data: Boerema *et al.* (1981b) (figs 1A, B, 2D, 3A: photographs of pycnidia *in vivo* and *in vitro*, drawing of pycnidia *in vitro*, photographs of cultures on
OA and MA; tab. 2: synoptic table on characters *in vitro* with drawing of conidia; anamorph-teleomorph connection; synonymy anamorph and teleomorph;
specimens studied, host range), Boerema *et al.* (1994) (fig. 5B: pycnidial and
conidial shape, reproduced in present Fig. 40B; *in vivo* and *in vitro* descriptions
adopted below).

Description in vivo *(esp.* Leonurus cardiaca*)*

Pycnidia always of type II (on dead stems, usually aggregated in short rows,
subepidermal, later superficial), subglobose with flattened base, mostly
200–300 μm diam., with cylindrical semi-pilose necks, often up to 500 μm
long. Wall at the flattened base and in the 'shoulder'-region strongly thickened,
but otherwise uniformly made up of polygonal scleroplectenchyma cells of differing dimensions. Conidial matrix white/yellowish.
    Conidia oblong to ellipsoidal, (3.5–)4–5.5 × 1.5–2.5 μm, usually with 2
minute polar guttules.
    Pseudothecia (also on dead stems) mostly 275–375 μm diam., subglobose,
non-sulcate, with distinct, occasionally papillate pore. Wall ± scleroplectenchymatous. Asci (60–)75–100 × 5.5–7.5 μm, 8-spored, biseriate in the upper
part, uniseriate below. Ascospores 18–22(–28) × 4.5–5.5 μm, broadly
fusiform, 3-septate with acute end cells, olivaceous yellow (for detailed description see the original diagnosis in Trotter, 1972: 405).

In vitro

Colonies on OA 1–2.5 cm diam. after 7 days; aerial mycelium absent or scarce,
lanose in the centre, whitish or (pale) olivaceous grey, sometimes with dark olivaceous felted sectors; reverse mostly ochraceous or yellow/red due to a somewhat diffusible pigment [no reaction with addition of a drop of NaOH].
Variable production of pycnidia II with conspicuous long pilose necks, resembling those *in vivo*, but less flattened and thickened at the base, at first pale
brown, ultimately dark with greenish tinge. Conidia as *in vivo*. Representative
culture CBS 389.80.

Ecology and distribution

Widespread and common in continental Europe on dead stems of the perennial labiate herbs *Leonurus cardiaca* and *Ballota nigra* subsp. *nigra* and subsp.
foetida. The fungus is occasionally recorded on dead stems of other Labiatae.
Specimens of the anamorph on *Ballota nigra* are formerly usually regarded as
identical with *Phoma acuta* (Hoffm. : Fr.) Fuckel subsp. *acuta* on *Urtica dioica*
(this section; teleom. *Leptosphaeria doliolum* (Pers. : Fr.) Ces. & De Not. subsp.
doliolum).

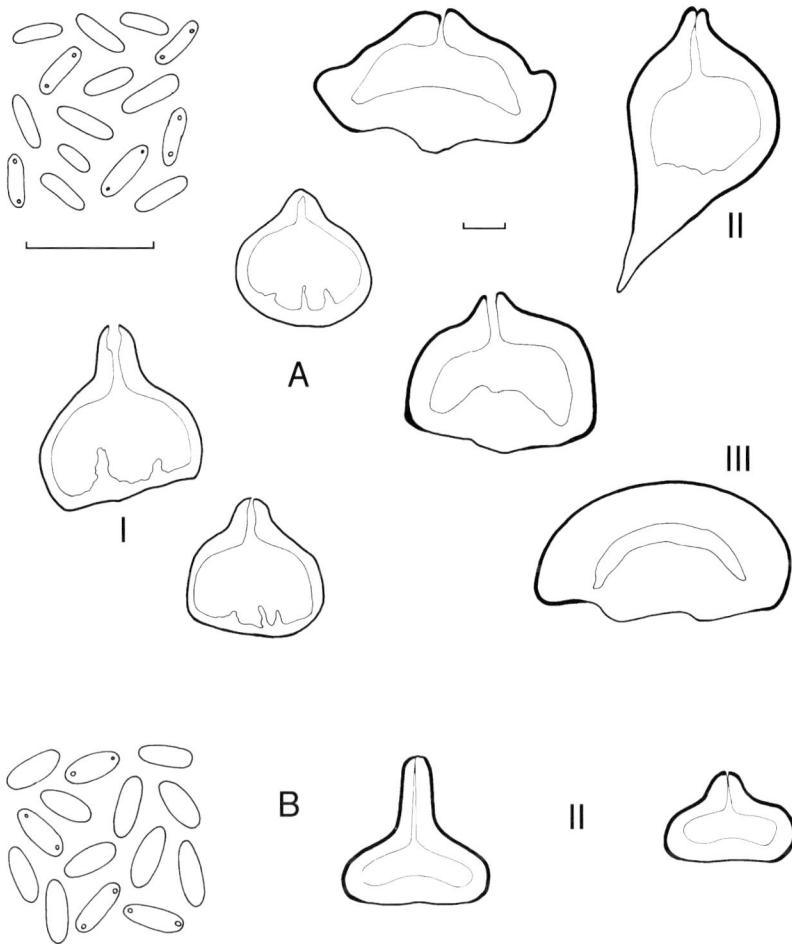

Fig. 41. A *Phoma lingam*. The cabbage pathogen displays a wide diversity in pycnidial types. Pycnidia-I are mostly globose-papillate, but some strains produce pycnidia with elongated swollen necks. The shape of the relatively large scleroplectenchymatous pycnidia-II is variable, sometimes beetroot-like or turban-like, but always neck-less. **B** *Phoma longirostrata*. Pycnidia-II with conspicuous cylindrical poroid neck of various length. Conidia eguttulate or with 2 inconspicuous small guttules. Bar pycnidia 100 μm, bar conidia 10 μm.

Phoma lingam (Tode : Fr.) Desm., Figs 36, 41A

Teleomorph: **Leptosphaeria maculans (Desm.) Ces. & De Not.**

Phoma lingam (Tode : Fr.) Desm. *in* Annls Sci. nat. (Bot.) III, **11**: 281. 1849.

≡ *Sphaeria lingam* Tode, Fungi mecklenb. **2**: 51. 1791.
: Fr., Syst. mycol. **2** [Sect. 2]: 507[–508]. 1823.

≡ *Plenodomus lingam* (Tode : Fr.) Höhn. *in* Sber. Akad. Wiss. Wien [Math.-naturw. Kl., Abt. 1) **120**: 463. 1911.

V = *Depazea brassicicola* Desm. *in* Pl. cryptog. N. France [ed. 1] Fasc. 3, No. 185. 1826 [as '*brassicaecola*'].

   ≡ *Depazea vagans* var. *brassicae* J.J. Kickx, Fl. crypt. Env. Louv. 125. 1835 [as 'γ *Brassicae*'; name change].

   ≡ *Septoria brassicae* Westend. & Wallays *in* Herb. crypt. Belge Fasc. 6, No. 294. 1847 [name change].

   ≡ *Phyllosticta brassicae* Westend. *in* Bull. Acad. r. Sci. Lett. Beaux-Arts Belg. [Bull. Acad. r. Belg. Cl. Sci.] **18**: 397. 1851 [name change] [often erroneously listed as a transfer of *Sphaeria brassicae* Curr. *in* Trans. Linn. Soc. Lond. **22**(4): 334. 1859 = *Asteromella brassicae* (Chevall.) Boerema & Kesteren, spermogonial state of *Mycosphaerella brassicicola* (Duby) Lind, see Boerema & van Kesteren, 1964].

 = *Sclerotium sphaeriaeforme* Lib. *in* Pl. cryptog. Ard., Fasc. 3, No. 237. 1834.

 = *Plenodomus rabenhorstii* Preuss *in* Linnaea, Halle **24**: 145. 1851.

 = *Aposphaeria brassicae* Thüm. *in* Hedwigia **12**: 189–190. 1880.

   ≡ *Phoma brassicae* (Thüm.) Sacc. *in* Sylloge Fung. **3**: 119. 1884.

 = *Phoma densiuscula* Sacc. & Roum. *in* Revue mycol. **6**: 30. 1884 [as '*Phoma (Aposph.)*'] [= Reliq. Libert. Ser. IV No. 86].

 = *Phoma napobrassicae* Rostr. *in* Tidsskr. Landøkon. **11**: 330. 1892.

   ≡ *Phoma lingam* var. *napobrassicae* (Rostr.) Grove, Br. Coelomycetes **1**: 70. 1935.

 = *Phyllosticta turritis* Lobik *in* Bolez. Rast. **17**(3–4): 170. 1928.

Diagn. data: Pound (1947) (figs 1–5, tabs 1–9; variability in isolates of *Phoma lingam* with photographs of cultures on potato-dextrose agar and disease symptoms induced on leaves, stems and pods of brassicas; differentiation of a 'variant' causing mild symptoms and characterized by brownish-yellow pigment produced in culture), Boerema & van Kesteren (1964) (figs 1–3: drawings of pseudoparenchymatous pycnidial type 'I', scleroplectenchymatous pycnidial type 'II' and pycnosclerotia 'III'; discussion of nomenclature and synonymy of anamorph and the connected teleomorph *Leptosphaeria maculans*), Smith & Sutton (1964) (fig. 1: drawings of pycnidium with conidia *in vivo* and pseudothecium with ascospores; tabs 1, 2: measurements of conidia and ascospores; confusion with a saprophytic anamorph), Bontea (1964) (figs 1–5, tabs I–II; cultures on potato agar: pycnidial and mycelial types; pycnidium development *in vitro*: symphogenous), Petrie & Vanterpool (1968) (occurrence of a separate strain of *Lept. maculans*/*Ph. lingam* on stinkweed, which may also infect brassicas), Punithalingam & Holliday (1972a) (figs A–C: drawings of scleroplectenchymatous pycnidium, conidia, pseudothecium, asci and ascospores; CMI description), Boerema (1976) (fig. 14: cultures on OA; tab. 9: synoptic table on characters *in vitro* with drawing of conidia; variability *in vivo* and *in vitro*, confusion of the anamorph with *Ph. herbarum* [sect. *Phoma*, **A**] and *Ph. exigua* var. *exigua* [sect. *Phyllostictoides*, **E**], confusion of the teleomorph with other species of *Leptosphaeria*; importance of scleroplectenchymatous pycnidia and pseudothecia in the epidemiology of the disease), Ndimande (1976) (figs 1–32, tabs 1–27; life cycle; occurrence of microconidia, mycelial fragmentation and swollen cells; 'biotypes'; disease symptoms), Neergaard (1977) (fig. 5.31, occurrence of anamorph on seed of cabbage and other crucifers), McGee & Petrie (1978) (fig. 1, tab. V: cultural differences between the pigment producing avirulent variant and the virulent strains of *Lept. maculans/Ph. lingam*), Sutton (1980) (fig. 228D: drawing of conidia; specimens in CMI collection, records on non-cruciferous hosts probably misapplications), Boerema (1981) (figs 6: differentiation against other species of *Phoma* which occasionally occur on seed of brassicas; fig. 7: differences in pycnidial shape and conidial matrix of deviating variant on brassica seed), Hall (1992) (fig. 1, tab 1; review of data on epidemiology and disease cycle), Boerema *et al.* (1994) (figs 1, 2B: pycnidial and conidial shape, partly reproduced in present Figs 36, 41A; description including the characteristics of the avirulent variant; description of the virulent *Lept. maculans/Ph. lingam* adopted below), Boerema *et al.* (1996) (misapplication of anamorph and teleomorph in literature and exsiccata works), Williams & Fitt (1999) (review of all the methods used to differentiate the virulent *Lept. maculans/Ph. lingam* from the less damaging variant), Shoemaker & Brun (2001) (description and illustration of the avirulent variant as a new species, *Leptosphaeria biglobosa*, with unnamed *Phoma*-anamorph, see discussion under Miscellaneous; figs 8–13: photographs of pseudothecia and ascospores of *Lept. maculans*; documentation of synonyms and herbarium material).

Description in vivo *(esp.* Brassica *spp.)*

Pycnidia of type I (leaf spots, stem and pod lesions, seed, usually solitary and arranged in rows), variable in shape but generally subglobose or flask-shaped with broad base, variable in dimensions, mostly 150–350(–400) µm diam., at maturity usually with 1 distinct poroid papilla (occasionally more), which may grow out into a long neck, sometimes confluent becoming irregular, up to 600 µm diam. with several papillae (or necks), black. Pycnidia of type II (woody parts of last year's dead stems, occasionally roots) highly variable, mostly subglobose with an irregular flattened base, but sometimes beetroot-like, relatively large, (200–)300–700(–1000) µm diam., not or only slightly papillate; with narrow pore or opened by rupture. At length the pycnidia often collapse and become discoid or turban-like. Sometimes they remain closed and sterile: pycnosclerotia, type III. Wall explicitly scleroplectenchymatous: above and around polygonal thick-walled cells with a very small lumen (like stone cells), at the base with a convex thickening consisting of diverging rows of less-thickened elongated cells. Conidial matrix usually red violet (amethyst coloured) but also whitish or pinkish.

Conidia ellipsoidal to subcylindrical, occasionally with 2 small polar gut-tules, (2.5–)3.5–4.5(–5) × 1–1.5(–2) μm.

Pseudothecia (subepidermal on stems during overwintering) up to 600 μm diam., depressed globose with flattened base and inconspicuous conical neck. Wall more or less scleroplectenchymatous. Asci (100–)120–135(–150) × 12–16 μm, 8-spored, quadriseriate above, biseriate below. Ascospores (35–)45–55(–70) × (4.5–)6–7(–8) μm, narrowly fusiform, 5-septate, yellowish-brown with guttules (for detailed descriptions see Holm, 1957; Shoemaker, 1984; Shoemaker & Brun, 2001).

Occasionally multiloculate pycnidia, containing microconidia, *c.* 1.5–3 × 1–1.5 μm, occur side by side with young pseudothecia [the former look like spermogonia, but the microconidia are able to germinate and give rise to nor-mal pycnidial cultures; compare Smith & Sutton, 1964 and Ndimande, 1976]. More often in association with the pseudothecia, old non-scleroplectenchyma-tous pycnidia occur [typical pycnidia of type II usually develop only on stem debris the next year].

In vitro

Colonies on OA slow growing, (1.5–)2.5 cm diam. after 7 days, often irregular with dendritic pattern but also regular; usually with copious aerial mycelium varying in colour, white, grey, greenish (dull green–dark herbage green), yel-lowish (straw, amber, luteous) or brown (fulvous, umber). Production of pycni-dia type I usually abundant on and in the agar, mostly solitary, globose-papillate, black and relatively small, 150–250 μm diam., sometimes larger and/or confluent.

In old cultures the pycnidia occasionally show a thickening of the cell walls in the peridium (→II: ± scleroplectenchymatous). Pound (1947) records the occur-rence of pycnosclerotia in a monoconidial isolate (S 39) on potato-dextrose agar.

Swollen cells, which are terminal or intercalary, solitary or in clusters, may occur in the aerial mycelium.

Arthrospores (fragmentation of hyphae) are also observed, they may be single or several cellular, see Ndimande (1976). Representative cultures CBS 532.66, CBS 475.81.

Ecology and distribution

A cosmopolitan major pathogen of cultivated *Brassica* spp. (Cruciferae): dry rot and canker (Am.: black leg), leaf spots. The fungus is also recorded on various other cultivated and wild Cruciferae. However, on wild crucifers, closely related but different fungi occur, for example *Phoma sublingam* Boerema (this section), teleom. *Leptosphaeria submaculans* L. Holm. The above listed synonyms of the pycnidial anamorph are nearly all described from brassicas, most of them refer to the scleroplectenchymatous type (II) found on last year's stem debris. The pseudoparenchymatous type (I) occurs *in vitro* and in association with the dis-ease symptoms. The fungus is heterothallic and displays some variability in cul-tural characteristics and pathogenicity. It produces a range of phytotoxins, known as sirodesmins. A separate pathogenic strain (with whitish conidial matrix) is known from the weed *Thlaspi arvense*, but occasionally also isolated from *Brassica* spp. Primary infections are usually initiated by ascospores from pseudothecia on residues of cruciferous crops, but the primary inoculum may also be mycelium from seed and conidia from pycnidia on crop residues (II) and seed (I). For other aspects of the epidemiology see Ndimande (1976) and Hall (1992).

Until recently, the concept of *Lept. maculans* included also a 'variant' that was weakly pathogenic on brassicas (non-aggressive or avirulent), faster growing and producing a yellow brown discoloration of the agar media. This variant is now recognized as a separate species, *Leptosphaeria biglobosa* Shoemaker & Brun (2001). Its scleroplectenchymatous pseudothecia differ from those of *Lept. maculans* in having a long neck that is usually swollen at the upper part. The unnamed *Phoma* anamorph of this fungus often also has pycnidia with a somewhat swollen or club-like neck. The conidia are similar to those of *Ph. lingam*, but the conidial matrix is usually reddish brown and not amethyst coloured. *Lept. biglobosa* is listed under 'Miscellaneous', p. 421. For a review article on the various methods used to differentiate *Lept. maculans/Ph. lingam* from the less damaging *Lept. biglobosa* Shoem. & Brun, see Williams & Fitt (1999).

Finally it should be noted that *Ph. lingam* on seed may be confused with saprophytic species as *Phoma herbarum* Westend. (sect. *Phoma*, **A**) and *Phoma glomerata* (Corda) Wollenw. & Hochapfel (sect. *Peyronellaea*, **D**). The statement in old literature that the scleroplectenchymatous pycnidia of *Ph. lingam* also occur on old wet wood refers to their superficial resemblance to pycnidia of *Phoma pezizoides* (Ellis & Everh.) Boerema & Kesteren (this section). For other misapplications see Boerema *et al.* (1996).

Phoma longirostrata Bubák, Fig. 41B

Phoma longirostrata Bubák *in* Bull. Herb. Boissier II, **6**: 476. 1906.

Diagn. data: Boerema *et al.* (1994) (fig. 6B: pycnidial and conidial shape, reproduced in present Fig. 41B; *in vivo* and *in vitro* descriptions adopted below).

Description in vivo *(esp.* Aconitum *sp.)*

Pycnidia always of type II (dead stems, scattered or in groups, subepidermal or superficial), depressed subglobose with flattened base, mostly 250–450 μm diam., with a distinct cylindrical neck of variable length (55–200 μm), occasionally confluent and with 2–several necks. The wall has a somewhat convex thickening at the base, but about the same polygonal scleroplectenchyma cells throughout. Conidial matrix whitish or yellow.

Conidia oblong ovoid or ellipsoidal, eguttulate or with 2 small inconspicuous polar guttules, mostly 3–4(–4.5) × 1–1.5(–2) μm.

In vitro

Colonies on OA 3–3.5 cm diam. after 7 days, regular, translucent, somewhat zonate, advanced zone faintly raised; aerial mycelium cottony but tenuous, olivaceous grey; reverse hardly coloured, zonate. Pycnidia sparse, covered by dense aerial mycelium, solitary or confluent, up to 800 μm diam., pale coloured with 1 to several dark short or long cylindrical necks. The cultures soon became stale and did not survive the lyophilization. Dried culture preserved as L 993.373.103.

Ecology and distribution

In southern Europe (Montenegro, Italy) occasionally on dead stems of Ranunculaceae (*Aconitum* and *Ranunculus* spp.). It should be noted that the plurivorous *Phoma doliolum* P. Karst. (this section; teleom. *Leptosphaeria conoidea* De Not.) has been recorded more frequently on *Aconitum* spp. in Europe, see also Holm (1957) and Shoemaker (1984). The conidia of *Ph. doliolum* are significantly larger than those of *Ph. longirostrata*.

Phoma macdonaldii Boerema, Fig. 42A

Teleomorph: *Leptosphaeria lindquistii* Frezzi

Phoma macdonaldii Boerema in Persoonia **6**(1): 20–21. 1970.

Diagn. data: Frezzi (1964) (figs 1–3; disease symptoms, characteristics pycnidia and conidia), McDonald (1964) (fig. 1A–F; disease symptoms; characteristics pycnidia and conidia), Frezzi (1968) (figs 1–3; disease symptoms; description and illustrations teleomorph), Boerema (1970) (identity), Marič & Schneider (1979) (figs 1–5: disease symptoms associated with pycnidia; fig. 6: drawing vertical section of pycnidium type I *in vivo*; fig. 7: cultures on OA and MA), Boerema *et al.* (1981b) (figs 2E, 3B, 3F: drawing of pycnidia *in vitro*, photographs of cultures on OA and MA, and production of crystals on OA; tab. 1: synoptic table on characters *in vitro*, with drawing of conidia), Marič *et al.* (1981) (fig. 1A–H, tab. 1; occurrence teleomorph in Yugoslavia; change of pycnidium production into pseudothecia development; illustrations of pseudothecium, asci, ascospores, colony habitus and conidia), Boerema (1982b) (fig. 9D: distribution), Gaudet & Schulz (1984) (figs 1–7; association with a stem weevil, SEM figs of conidia), Donald *et al.* (1986) (report teleomorph in USA), Boerema *et al.* (1994) (fig. 4B: pycnidial and conidial shape, reproduced in present Fig. 42A; *in vivo* and *in vitro* descriptions adopted below).

Description in vivo (Helianthus annuus)

Pycnidia of type I (lesions on stems, leaves, etc., solitary, scattered or in rows) subglobose, often 70–170(–200) μm diam., not or only slightly papillate. Pycnidia of type II (on last year's dead stems) also subglobose, but mostly larger, 100–300 μm diam., usually non-papillate with narrow pores. Conidial matrix off-white or red/violet coloured.

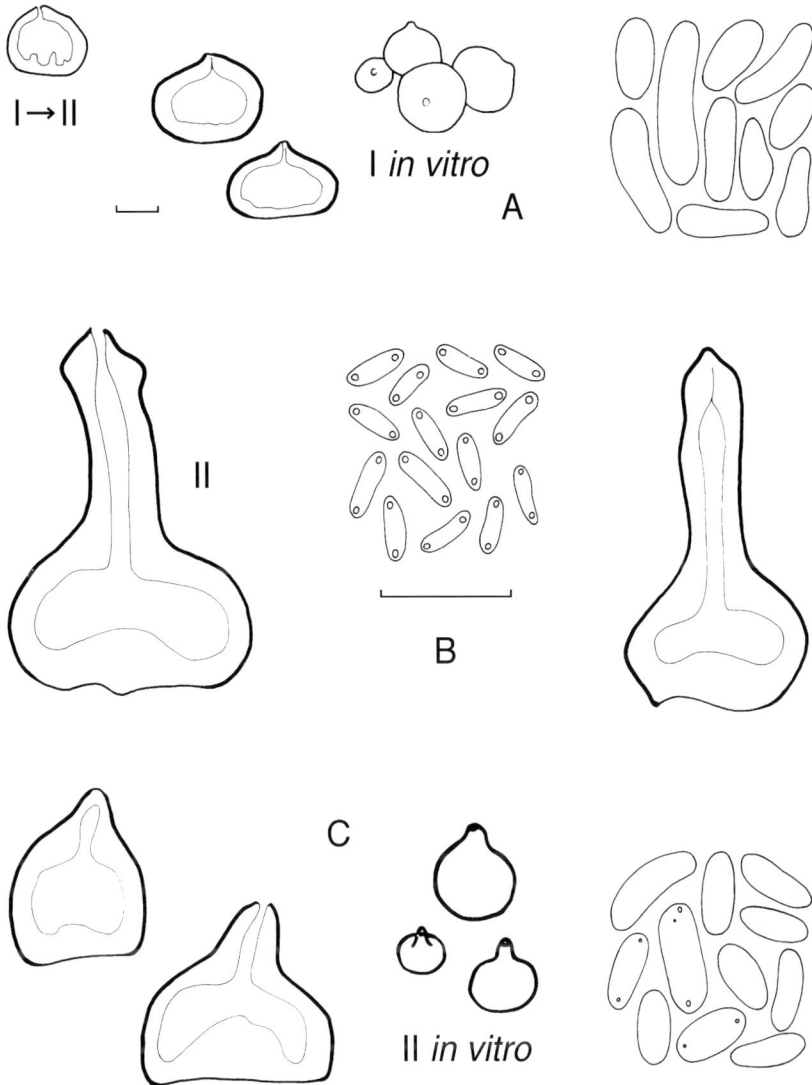

Fig. 42. A *Phoma macdonaldii.* This pathogen of the commercial sunflower produces pycnidia-I in association with disease symptoms; pycnidia-II occur on dead stems. Both pycnidial types are subglobose and not or only slightly papillate. Conidia are highly variable in shape and dimensions. **B** *Phoma macrocapsa.* Remarkable for its large pycnidia, characterized by long broad cylindrical necks with somewhat swollen tops. Conidia usually with two conspicuous polar guttules. **C** *Phoma pedicularis.* Pycnidia-I→II, initially small subglobose-papillate, finally relatively large, conoid with flattened base and conspicuous beak-like elongated neck. Conidia vary markedly in shape and dimensions. *In vitro* the pycnidia remain subglobose-papillate and small. Bar pycnidia 100 μm, bar conidia 10 μm.

Conidia highly variable in shape and dimensions, mostly reniform to ovoid-ellipsoidal, (4.5–)5–9(–10) × 1.5–3(–4) μm, eguttulate.

Pseudothecia (also on dead stems) 130–230 μm diam., depressed globose, not or only slightly papillate. Wall more or less scleroplectenchymatous. Asci 70–145 × 7.5–10.5 μm, 8-spored, irregularly uniseriate. Ascospores 12.5–25 × 3.5–8.5 μm, irregularly fusiform, 1–3-(often 2-)septate, pale yellow with guttules (for detailed description see Frezzi, 1968).

In vitro

Colonies on OA 2.5–4.5 cm diam. after 7 days; aerial mycelium cottony, mostly whitish, greyish or greenish olivaceous; reverse olivaceous, greenish/yellow or glaucous tinged, often with white granular or flaky crystals. In old colonies often pink or red pigments occur, becoming purple or blue with addition of a drop of NaOH (the presence of anthraquinone cynodontin has been demonstrated together with some unknown yellow pigments). Usually abundant production of pycnidia, type I→II, resembling those *in vivo*; conidia highly variable in shape and dimensions as *in vivo*. Representative culture CBS 386.80.

Ecology and distribution

A pathogen of the 'American' sunflower, *Helianthus annuus* (Compositae), causing lesions on stems, petioles, leaves and inflorescences: black stem, black spot. At present the fungus probably occurs wherever the commercial sunflower is cultivated; the records are from Europe (France, Romania, former Yugoslavia), North America (Canada, USA) and South America (Argentina). In association with the disease symptoms only pseudoparenchymatous pycnidia (type I) occur; the scleroplectenchymatous pycnidia (type II) and the perithecia develop on dead host material. [In the past the anamorph has been erroneously referred to as *Phoma oleracea* var. *helianthi-tuberosi* Sacc., a synonym of the ubiquitous saprophyte *Phoma herbarum* Westend., sect. *Phoma*, **A**].

Phoma macrocapsa Trail, Fig. 42B

Phoma macrocapsa Trail *in* Scott. Nat. **8** [II, **2**]: 237. 1886.
≡ *Plenodomus macrocapsa* (Trail) H. Ruppr. *in* Sydowia **13**: 20–21. 1959.

Diagn. data: Grove (1935) (occurrence in England), Boerema *et al.* (1994) (fig. 2A: pycnidial and conidial shape, reproduced in present Fig. 42B; *in vivo* and *in vitro* descriptions adopted below).

Description in vivo *(Mercurialis perennis)*

Pycnidia always of type II (on dead stems, scattered or in groups), depressed globose with flattened base and a conspicuous broad and long cylindrical neck, usually with a somewhat swollen top (phallus-like), mostly

(400–)600–700 µm diam., the neck reaching to a height of 400–700 µm. Wall explicitly scleroplectenchymatous with a convex thickening consisting of diverging rows of somewhat elongated cells at the base. The cell walls of the scleroplectenchyma have the same thickness throughout. Conidial matrix cream or whitish.

Conidia ellipsoidal to subcylindrical, (3–)4–5 × (1–)1.5–2 µm, usually with 2 conspicuous polar guttules.

In vitro

Colonies on OA relatively slow growing, 1–1.5(–2) cm diam. after 7 days, regular, with fine, compact, dark olivaceous grey or greenish grey aerial mycelium; reverse greenish olivaceous with primrose margin. In fresh isolates numerous scattered relatively large scleroplectenchymatous pycnidia develop on the agar, resembling the pycnidia *in vivo*, but usually with a less pronounced neck. Conidia as *in vivo*. Representative culture CBS 640.93.

Ecology and distribution

Very common in Europe on dead stems of *Mercurialis perennis* (Euphorbiaceae). Regarded as a harmless specialized necrophyte, but under some conditions the fungus apparently causes damage to the host. The fungus probably occurs everywhere on the host. 'Remarkable for its large pycnidia' (Grove, 1935).

Phoma pedicularis Fuckel, Fig. 42C

V   *Phoma pedicularis* Fuckel in von Heuglin, Reisen Nordpolarmeer III Beitr. Fauna Fl. Geol. 318–319. 1874 [as 'pedicularidis'].; not *Phoma pedicularis* Wehm. *in* Mycologia **38**: 319. 1946 [= *Phoma herbicola* Wehm., sect. *Phoma*, **A**].

V ≡ *Diplodina pedicularis* (Fuckel) Lind, Rep. scient. Results Norw. Exped. Novaya Zemlya 1, **19**: 21. 1924 [as 'pedicularidis'; misapplied].

V H= *Ascochyta pedicularis* (Fuckel) Arx *in* Proc. K. ned. Akad. Wet. C **66**: 180. 1963 [as 'pedicularidis'; misapplied]; not *Ascochyta pedicularis* (Rostr.) von Arx *in* Verh. K. ned. Akad. Wet. [Afd. Natuurk.] reeks 2, **51**(3) [= Revis. Gloeosporium, ed. 1]: 116. 1957 [comp. Mel'nik, 1977/2000].

= *Phoma barbari* Cooke *in* Grevillea **14**: 3. 1885 [cf. holotype in herb. Cooke, K; comp. van der Aa & van Kesteren, 1971].

= *Phoma lingam* f. *linariae* Sacc. & Paol. apud Saccardo *in* Mém. Soc. r. Bot. Belg. [Supplement to Bull. Soc. r. Bot. Belg.] **28**: 96–97. 1889 ['Myc. Sibir.', reprint: 20–21] [Author citation cf. Saccardo *in* Sylloge Fung. **10**: 175. 1892].

V = *Sphaeronaema gentianae* Moesz *in* Bot. Közl. **14**: 152. 1915 [as '*Sphaeronema*'].

≡ *Plenodomus gentianae* (Moesz) Petr. *in* Annls mycol. **23**: 54. 1925.

= *Phoma prominens* Bres. *in* Studi trent. Cl. II Sci. nat. econ. **7**: 67. 1926.

≡ *Plenodomus prominens* (Bres.) Petr. ex Arx *in* Sydowia **4**: 390. 1950.

= *Plenodomus svalbardensis* Lind *in* Skr. Svalbard Ishavet **13**: 35. 1928 [cf. sample of type collection].

= *Plenodomus karii* Petr. *in* Annls mycol. **34**: 453. 1936.

= *Plenodomus sphaerosporus* Petr. *in* Annln naturh. Mus. Wien **52**: 384–385. 1942.

= *Plenodomus helveticus* Petr. *in* Sydowia **2**: 239–240. 1948.

Θ = *Phoma acuta* f. *gentianae* Roum. *in* Fungi gall. exs. Cent. 11, No. 1009. 1881; *in* Revue mycol. **3**: 30. 1881 [without description].

Θ = *Phoma pedicularis* f. *caulicola* Bres. *in* Malpighia **11**: 305. 1897 [reprint p. 67] [without description].

Diagn. data: Boerema *et al.* (1981b) (figs 1C, 2B, 3C: photographs of pycnidium *in vivo*, drawing of pycnidia *in vitro* and photographs of cultures on OA and MA; tab. I: synoptic table on characters *in vitro* with drawing of conidia; synonymy; specimens studied, host range), Boerema (1982b) (fig. 9: Arctic-alpine distribution), Boerema *et al.* (1994) (fig. 8B: pycnidial and conidial shape, reproduced in present Fig. 42C; synonymy; *in vivo* and *in vitro* descriptions adopted below), Boerema *et al.* (1996: 184) (additional synonym).

Description in vivo

Pycnidia I→II (on dead stems, leaves and seed capsules); initially subglobose-papillate, mostly 200–300 μm diam., relatively thin-walled and often with irregular cellular protrusions into the pycnidial cavity; gradually becoming larger, more conoid with flattened base, 300→600 μm, with conspicuous dark beak-like elongated papillate neck (under snow cover necks can be up to 150 μm long), wall then explicitly scleroplectenchymatous with a more or less convex thickening consisting of diverging rows of elongated cells at the base; occasionally confluent and with 2 or more necks. Conidial matrix whitish to pale ochraceous or primrose.

Conidia vary markedly in shape and size: oblong to ellipsoidal, subcylindrical or allantoid, 4–8(–8.5) × 2–3 μm, and/or oval-ovoid or nearly spherical, 4–6 × (2–)2.5–4(–4.5) μm; usually eguttulate, occasionally with 1–2 minute guttules.

In vitro

Colonies on OA relatively slow growing, 1.5–2 cm diam. after 7 days; aerial mycelium tenuous, translucent, fluffy or downy, whitish grey or olivaceous, often with grey to dark olivaceous felted or compact cottony sectors; reverse pale ochraceous, but below the felted sectors, dark olivaceous, bordered by a narrow yellow red and a narrow white zone [pigment quickly fading in daylight and not reacting to a drop of NaOH]. In fresh isolates usually abundant production of globose-papillate pycnidia I→II; relatively small, 100–200 μm diam., in comparison with the pycnidia *in vivo*. Conidia of two types as *in vivo*, but less variable and always produced in the same pycnidium. Representative culture CBS 390.80.

Ecology and distribution

In Eurasian regions with prolonged snow cover this species is frequently found on dead stems, leaves and seed capsules of various herbaceous plants. Typically it is a plurivorous fungus with an Arctic-alpine distribution. Accidental observations in other regions usually apply to specimens with relatively small pycnidia, like those *in vitro*. Most records refer to last year's stems of Scrophulariaceae (esp. *Pedicularis* and *Veronica* spp.) and Gentianaceae (various *Gentiana* spp.). *Ph. pedicularis* has been confused with *Ascochyta pedicularis* (Rostr.) Arx, which has much larger two-celled conidia (Mel'nik, 2000 translated from Mel'nik, 1977).

Phoma petrakii Boerema & Kesteren, Fig. 43A

Possible teleomorph: ***Leptosphaeria suffulta* (Nees : Fr.) Niessl**

> *Phoma petrakii* Boerema & Kesteren *in* Persoonia **11**(3): 321[–322]. 1981.
>> H ≡ *Plenodomus niesslii* Petr. *in* Annls mycol. **20**: 322–323. 1922; not *Phoma niesslii* Sacc. *in* Michelia **2**(3): 618. 1882 [= *Phoma exigua* Desm. var. *exigua*, sect. *Phyllostictoides*, **E**].
>> [= '*Plenodomus sylvaticus*' sensu von Rupprecht *in* Sydowia **13**: 21. 1959 as '*sylvatica*'; misapplied.]

Diagn. data: Boerema & van Kesteren (1981b) (nomenclature, host-records, confusion with *Phoma sylvatica* Sacc.), Boerema *et al.* (1994) (fig. 5A: pycnidial and conidial shape, reproduced in present Fig. 43A; *in vivo* description adopted below), Boerema *et al.* (1996) (misapplication).

Description in vivo *(Melampyrum spp.)*

Pycnidia always of type II (on dead stems, solitary, scattered or in groups) depressed globose with flattened base and distinct papillate neck usually with dark lined tube-shaped pore, usually 200–350 μm diam. The pycnidia show much resemblance with those of the plurivorous *Phoma acuta* subsp. *errabunda* (this section; teleom. *Leptosphaeria doliolum* subsp. *errabunda*) but differ by the absence of a convex basal wall-thickening; the scleroplectenchymatous wall has uniform thickness. Conidial matrix usually salmon.

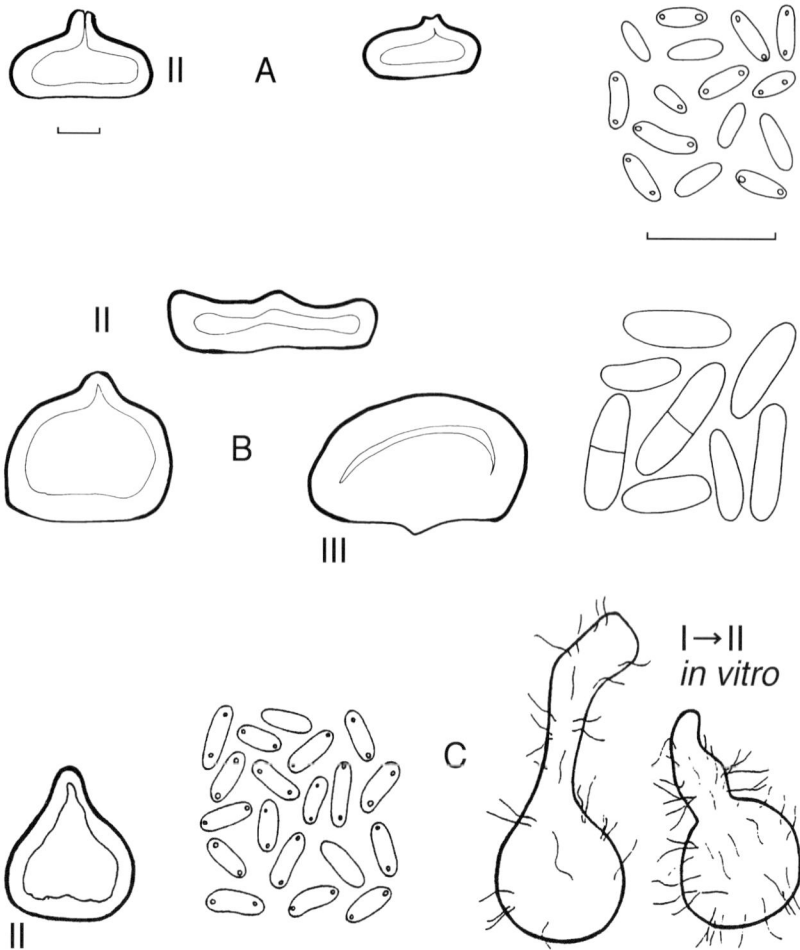

Fig. 43. A *Phoma petrakii.* Pycnidia-II depressed globose with flattened base and distinctly papillate, usually with a dark-lined tube-shaped pore. **B** *Phoma pezizoides.* Produces relatively large pycnidia-II and pycnosclerotia-III. Pycnidia usually globose-papillate with flattened base; at length they often collapse and become discoid or pezizoid. Conidia very variable and occasionally 1-septate. **C** *Phoma pimpinellae.* Pycnidia-II *in vivo* regular subglobose with a distinct papillate neck found together with, and very similar to the ascomata of the teleomorph *Leptosphaeria pimpinellae.* Pycnidia *in vitro* I→II, at first irregular, but later becoming more regular. Bar pycnidia 100 μm, bar conidia 10 μm.

Conidia ellipsoidal to subcylindrical, straight or slightly curved, (3.5–)4–5 × 1–1.5(–2) μm, usually with 2 polar guttules.

[Pseudothecia (also on dead stems) up to 450 μm diam., truncate-conical. Wall scleroplectenchymatous. Asci 80–100 × 6–8 μm, 8-spored, biseriate. Ascospores 25–30 × 4–4.5 μm, narrowly fusiform, 3-septate, central cell nearly as long as end cells, yellowish, without guttules (for recent description of this supposed teleomorph see Shoemaker, 1984).]

In vitro

Repeated attempts to isolate this fungus on OA at room temperature have not been successful. [Also an obvious difference with the plurivorous *Ph. acuta* subsp. *errabunda*, which can always be grown easily in culture.]

Ecology and distribution

Widespread in Europe on *Melampyrum* spp., Scrophulariaceae (*M. nemorosum*, *M. pratense*, *M. sylvaticum* and *M. commutatum*; all semi-parasitic, esp. on roots of Gramineae). Often accompanied with pseudothecia of the above teleomorph, but a metagenetic relationship must still be proved. *Phoma petrakii* has often been confused with *Phoma sylvatica* Sacc. (sect. *Sclerophomella*, **F**; teleom. *Didymella winteriana*), which in Europe also commonly occurs on *Melampyrum* spp. The pycnidia of *Ph. sylvatica* are smaller than those of *Ph. petrakii* and not scleroplectenchymatous, but their conidia have approximately the same dimensions. The combination *Plenodomus sylvaticus* was based on a misidentified collection of *Ph. petrakii*.

Phoma pezizoides (Ellis & Everh.) Boerema & Kesteren, Fig. 43B

> *Phoma pezizoides* (Ellis & Everh.) Boerema & Kesteren *in* Persoonia **11**(3): 322. 1981.
>> ≡ *Aposphaeria pezizoides* Ellis & Everh. *in* Proc. Acad. nat. Sci. Philad. **1894**: 358. 1894.
>> ≡ *Coniothyrium pezizoides* (Ellis & Everh.) Kuntze, Revis. Gen. Pl. **3**(3): 459. 1898.
>> = *Aposphaeria salicum* Sacc. apud Sydow *in* Annls mycol. **1**: 537–538. 1903; *in* Sylloge Fung. **18**: 276. 1906.
>>> ≡ *Plenodomus salicum* (Sacc.) Died. *in* Annls mycol. **9**: 140. 1911.
>> = *Phoma wallneriana* Allesch. *in* Rabenh. Krypt.-Flora [ed. 2], Pilze **6** [Lief. 61]: 175. 1898 [vol. dated '1901'].
>>> ≡ *Plenodomus wallneriana* (Allesch.) Bubák *in* Annls mycol. **13**: 30. 1915.
>> = *Plenodomus helicis* Curzi & Barbaini *in* Atti Ist. bot. Univ. [Lab. crittogam.] Pavia III, **3**: 173. 1927.

Diagn. data: Boerema & van Kesteren (1981b) (tab. III: host records; pycnidial and conidial characteristics; nomenclature and synonymy), Boerema (1982b) (fig. 9D: distribution), Boerema *et al.* (1994) (fig. 9D: pycnidial and conidial shape, reproduced in present Fig. 43B; *in vivo* description adopted below).

Description in vivo *(esp. Salix spp.)*

Pycnidia always of type II (esp. on decorticated branches, often in rows), usually globose-papillate with flattened base, mostly 250–500 μm diam., but also larger, up to 1000 μm diam.; at length the pycnidia often collapse and become discoid or pezizoid. The wall consists of polygonal scleroplectenchyma cells with a more or less parallel arrangement in the thickened, somewhat convex base. Protopycnidia are completely filled with such polygonal cells; the proliferating layer develops initially in a cap-shaped pattern. Sometimes the pycnidial primordia remain closed and sterile, forming pycnosclerotia, type III. Conidial matrix whitish.

Conidia vary markedly in shape and dimensions: ovoid-oval, 4.5–6(–7.5) × 2–3 μm, and ellipsoidal to subcylindrical, 6–10 × 2–3 μm; in both types mostly biguttulate. Occasionally 1-septate conidia also occur, 8–10 × 2–3 μm.

In vitro

Cultures on OA at room temperature produced only sterile mycelium; the fungus apparently needs special conditions for fructification.

Ecology and distribution

In Europe (Austria, Eastern Germany, Italy) and North America (USA, West Virginia) recorded on dead branches, twigs and petioles of various trees and shrubs, especially near river banks. It probably originates from Central Europe. The fungus produces *in vivo* sometimes pycnidia with only the short conidial type and sometimes pycnidia with only the longer conidial type. Various collections, however, bear pycnidia with both types of conidia. The pycnidia of *Ph. pezizoides* superficially resemble the scleroplectenchymatous pycnidia-II of *Phoma lingam* (Tode : Fr.) Desm. (this section) on dead cabbage stems. This explains the statement in old literature that the latter also occurs on wet, old wood.

Phoma pimpinellae Boerema & Gruyter, Fig. 43C
Teleomorph: ***Leptosphaeria pimpinellae*** Lowen & Sivanesan

Phoma pimpinellae Boerema & Gruyter *in* Persoonia **17**(2): 278. 1999; **18**(2): 159. 2003 [info type specimen].

Diagn. data: Lowen & Sivanesan (1989) (fig. 2: pycnidial morphology *in vitro*, conidia and conidiogenous cells; original anamorph-teleomorph description, fig. 1: pseudothecia, ascus and ascospores; cultural study on cornmeal dextrose agar and OA), Boerema & de Gruyter (1999) (fig. 1B: pycnidial and conidial shape, reproduced in present Fig. 43C; cultural descriptions, partly adopted below).

Description in vivo *(cf. Lowen & Sivanesan (1989) on* Pimpinella anisum*)*

Pycnidia of type II (on dead, blackened, still standing stems, immersed becoming superficial), similar to the pseudothecia of teleomorph: globose with flattened base and a distinct papillate neck, wall scleroplectenchymatous to the outside, esp. conspicuous at the 'shoulder' near the neck, up to

300 μm diam. Conidiogenous cells well differentiated, bottle shaped with a long neck, 6–12 × 3(at the base)–1(at the phialidic apex) μm. Conidial matrix pink (rosy vinaceous).

Conidia short cylindrical (oblong), 4–5.5 × 1–2 μm, biguttulate.

In vitro

Colonies on OA regular, 4.5 cm diam. after 7 days; aerial mycelium floccose, white to pale olivaceous grey, but partly citrine green, pale luteous to citrine, due to the release of a diffusable pigment, the colour changes to brick with addition of a drop of NaOH; reverse pale luteous to amber, with olivaceous grey at centre. Pycnidia scattered, on and in the agar and in aerial mycelium as well, irregular to subglobose with long elongated neck and mycelial outgrowths, thick-walled I→II. Conidia oblong to ellipsoidal, somewhat shorter as noted *in vivo*: 3.5–4.5 × 1–1.5(–2) μm, av. 4.0–1.5 μm. Conidiogenous cells globose-papillate to bottle shaped with a long neck 4–6(–8) × 1.5(at the apex)–5 μm. Representative culture CBS 101637.

Ecology and distribution

This fungus is so far only known from Israel, where it was found on dead blackened stems of *Pimpinella anisum* (Umbelliferae). On the holotype substratum the pycnidia occurred together with the pseudothecia of the teleomorph, but it is plausible that both morphs play a different role in the life cycle of the fungus.

Phoma piskorzii (Petr.) Boerema & Loer., Fig. 44A
Teleomorph: *Leptosphaeria acuta* (Fuckel) P. Karst.

> *Phoma piskorzii* (Petr.) Boerema & Loer. in Persoonia **11**(3): 315. 1981.
> ≡ *Diploplenodomus piskorzii* Petr. in Annls mycol. **21**: 123–124[–125]. 1923.

Diagn. data: Lacoste (1965) (pl. II fig. 1; cultural study of teleomorph, influence of temperature and nutrition on fructification; description of pycnidia, conidia and teleomorph), Boerema & Loerakker (1981) (fig. 1a–f: drawings and photographs of pycnidia, pseudothecium coalesced with pycnidium, conidia and asci; typification; confusion with the anamorph of *Leptosphaeria doliolum* (sensu stricto) on nettle), Boerema *et al.* (1994) (fig. 7B: pycnidial and conidial shape, reproduced in present Fig. 44A; *in vivo* description adopted below).

Description in vivo (Urtica dioica)

Pycnidia always of type II (on the base of dead stems, especially inside stem cavity, usually solitary), variable in shape, mostly depressed subglobose with a

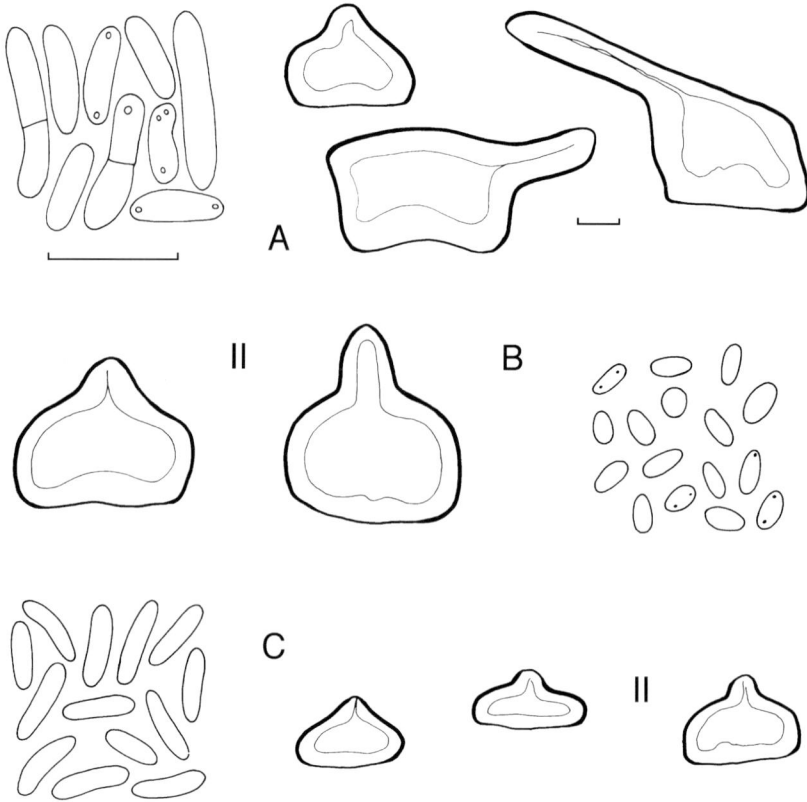

Fig. 44. A *Phoma piskorzii*. Pycnidia-II variable in shape, with flattened base, often subglobose-papillate, but also with a long neck, usually developing at the side. Conidia relatively large, occasionally 1-septate. **B** *Phoma rubefaciens*. Pycnidia initially thin-walled, I→II, relatively large, globose-papillate or with a short cylindrical neck. **C** *Phoma ruttneri*. Pycnidia-II depressed globose with a flattened base and a papillate or truncate-conical pore. Conidia uniform subcylindrical. Bar pycnidia 100 μm, bar conidia 10 μm.

flattened base, 250–500 μm diam., usually papillate, but also with a long poroid neck up to 500 μm, usually developing at one side; occasionally opened by rupture. Wall made up of polygonal scleroplectenchyma cells of variable size; usually a conspicuous convex wall-thickening at the base with diverging rows of more or less elongated cells. Conidial matrix whitish.

Conidia oblong ellipsoidal to subcylindrical, usually (6–)8–10(–12) × 2–2.5(–3) μm, but occasionally also longer up to 16 μm, mostly continuous but occasionally with 1 indistinct septum, often with 1–2 guttules.

Pseudothecia (also at the base of dead stems, on outer surface) up to 500 μm diam. with a prominent truncate-conical neck and almost flattened base. Wall more or less scleroplectenchymatous. Asci 120–150 × 7–12 μm, 8-spored, biseriate above, one and a half seriate below. Ascospores 37–45(–53) × 5–7 μm, narrowly fusiform, (5–)8–13-septate [in North American collections often 7-septate], pale yellow with guttules [for recent detailed description see Shoemaker, 1984].

In vitro

Cultures on OA at room temperature produced only sterile mycelium; the fungus requires low temperature and special nutrition for fructification (study Lacoste, 1965).

Ecology and distribution

In Europe the anamorph of this Urticaceae-fungus is common at the base of dead stems of nettles, particularly *Urtica dioica*, in spring. At first the pycnidia usually develop inside the hollow stems, which explains why they have often been overlooked. The fungus is probably widespread on nettles in the whole of Eurasia. The teleomorph is also recorded in temperate North America; exact records from other continents are wanting. In older literature the anamorph of *Leptosphaeria acuta* is usually referred to as '*Phoma acuta*', '*Leptophoma acuta*', or '*Plenodomus acutus*'. These names, however, refer in fact to the anamorph of a related Urticaceae-fungus *Leptosphaeria doliolum* subsp. *doliolum*. Correct citation *Phoma acuta* (Hoffm. : Fr.) Fuckel subsp. *acuta* (this section). The pycnidia of the latter may be present abundantly on dead stems of nettles, occasionally together with the pseudothecia of *Lept. acuta*. The conidia of *Phoma acuta* are significantly shorter than those of *Ph. piskorzii*.

Phoma rubefaciens Togliani, Fig. 44B

Phoma rubefaciens Togliani *in* Annali Sper. agr. II, **7**: 1626. 1953.

Diagn. data: Boerema *et al.* (1981b) (figs 2A, 3D: drawings of pycnidia *in vitro*, photographs of cultures on OA and MA; tab. 1: synoptic table on characters *in vitro* with drawings of conidia), Boerema *et al.* (1994) (fig. 9C: pycnidial and conidial shape, reproduced in present Fig. 44B; *in vivo* and *in vitro* descriptions, partly adopted below).

Description in vivo *(esp.* Malus pumila*)*

Pycnidia (subepidermal in centre of red fruit spots, immersed in bark and on wood) initially thin-walled, type I, but gradually becoming scleroplectenchymatous, type II; markedly varying in shape and size, developing subepidermal: depressed-globose with inconspicuous papillate pore, mostly 200–250 μm diam., as subemergent bodies, more or less globular with a prominent short neck, up to (350–)400–500 μm diam. Wall of mature pycnidia shows randomly arranged polygonal scleroplectenchyma cells. Conidiogenous cells well-differentiated, cone-shaped. Conidial matrix whitish.
    Conidia ovoid-oval to subcylindrical, 3.5–5 × 2–2.5(–3) μm, eguttulate or with 1–2 minute guttules.

In vitro

Colonies on OA 1–1.5 cm diam. after 7 days; aerial mycelium either very sparse, whitish, greyish, or velvety-cottony, flat, dark olivaceous, often with glaucous grey farineceous sectors; reverse saffron, apricot, yellow/red, below the farineceous sectors dark olivaceous with yellow red margin [chemical analysis revealed the anthraquinone cynodontin and a yellow pigment with a chrysophenol-like u.v.-visible spectrum]. Mostly abundant production of relatively large globose pycnidia in concentric rings on the agar, mostly with a short cylindrical neck, thin-walled or scleroplectenchymatous, I→II, covered by dark hyphae. Representative culture CBS 387.80.

Ecology and distribution

Isolated from bark and wood of various deciduous trees in Europe. The fungus has been recorded in association with red skin necroses on apples, *Malus pumila* (Rosaceae), but is probably only an opportunistic pathogen.

Phoma ruttneri (Petr.) Boerema & Kesteren, Fig. 44C

Probable teleomorph: *Leptosphaeria affinis* P. Karst.

> *Phoma ruttneri* (Petr.) Boerema & Kesteren *in* Persoonia **11**(3): 324. 1981.
> ≡ *Plenodomus ruttneri* Petr. *in* Sydowia **8**: 582–583. 1955.

Diagn. data: Boerema & van Kesteren (1981b), Boerema *et al.* (1994) (fig. 4A: pycnidial and conidial shape, reproduced in present Fig. 44C; *in vivo* description adopted below).

Description in vivo *(Rhinanthus spp.)*

Pycnidia. The two records of this anamorph refer to pycnidial type II (on dead stems, scattered, subepidermal), depressed globose with a distinct papillate or truncate-conical pore, mostly 250–350 μm diam. Wall uniform in thickness and made up of several layers of polygonal scleroplectenchyma cells of variable sizes.

Conidia subcylindrical, usually somewhat curved, 5–7 × 1.5–2 μm, eguttulate.

[Pseudothecia (also on dead stems) up to 400 μm diam., conic with a flattened base and a short truncate-conical neck. Wall ± scleroplectenchymatous. Asci 85–100 × 5–6 μm, 4-spored, biseriate or triseriate above. Ascospores 40–60 × 5–6 μm, narrowly fusiform and somewhat clavate, 3-septate, yellow or nearly colourless, without guttules (for recent description of this supposed teleomorph, see Shoemaker, 1984).]

In vitro

Cultures made from the supposed teleomorph produced only sterile mycelium.

Ecology and distribution

In Europe (Austria, Germany) found in dead stems of *Rhinanthus* spp. (Scrophulariaceae). On the type substratum the pycnidia occur together with pseudothecia of *Leptosphaeria affinis*, which is widespread in Europe on *Rhinanthus minor*. A single identity of both morphs is plausible but has not yet been proved by comparison of cultures.

Phoma sclerotioides Preuss ex Sacc., Fig. 45A

> *Phoma sclerotioides* Preuss ex Sacc., Fungi Herb. Brux. 21. 1892; *in* Sylloge Fung. **11**: 492. 1895.
>
> > Θ ≡ *Plenodomus sclerotioides* Preuss *in* Rabenh., Klotzschii Herb. mycol. No. 1281. 1849 [without description].

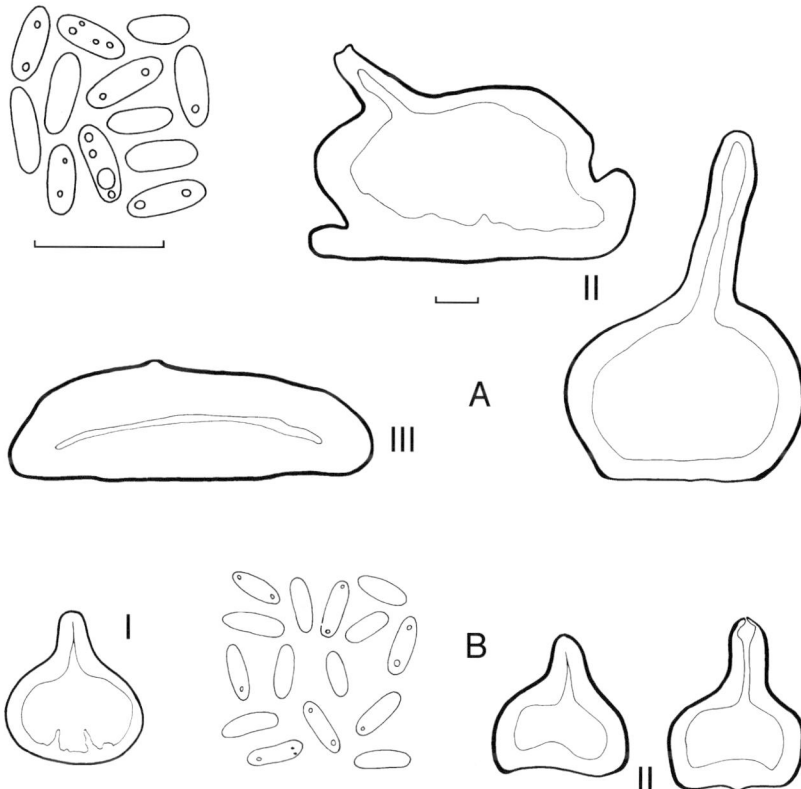

Fig. 45. A *Phoma sclerotioides.* This low temperature pathogen produces large scleroplectenchymatous pycnidia-II and pycnosclerotia-III. The pores of the pycnidia usually develop on long thin tubular necks. **B** *Phoma sublingam.* Pycnidia-I subglobose with broad base and distinct papillate neck, are found in association with disease symptoms. Pycnidia-II, depressed globose to subglobose with flattened base and a pronounced neck, occur on dead stems. Bar pycnidia 100 μm, bar conidia 10 μm.

= *Plenodomus meliloti* Mark.-Let. *in* Bolez. Rast. **16**: 195. 1928.

H = *Plenodomus meliloti* Dearn. & Sanford *in* Annls mycol. **28**: 324–325. 1930.

= *Plenodomus sorghi* Morochk. *in* Trudÿ bot. Inst. Akad. Nauk SSSR **1**: 277–278. 1933.

Diagn. data: Sandford (1933) (figs 1–5; disease symptoms 'brown root rot'; effect temperature and pH on pycnidial development), Salonen (1962) (figs 1–3: agar plate cultures at 7°C and 22°C; shape of pycnidia and conidia), Colotelo & Netolitzky (1964) (figs 1–17: photographs of stages in pycnidial development), Netolitzky & Colotelo (1965) (fig. 1: drawing of conidiogenous cells), Zafar & Colotelo (1978) (processes involved in production of pycnidia in light and darkness), Boerema & van Kesteren (1981b) (figs 1, 2: drawings of pycnidia on dead root and vertical section of pycnidium on OA; tab. IV: documentation of host records; synonymy; occurrence of pycnosclerotia), Boerema (1982b) (fig. 9), Boerema & Loerakker (1985) (figs 3 A–C, 4E, 5D: photographs and drawings of pycnosclerotia, drawing of pycnidium and photographs of cultures on OA and MA; tab. 1: synoptic table on characters *in vitro*, with drawing of conidia), Boerema *et al.* (1994) (fig. 3A: pycnidial and conidial shape, reproduced in present Fig. 45A; *in vivo* and *in vitro* descriptions adopted below).

Description in vivo *(esp. on* Melilotus alba*)*

Pycnidia always of type II (mainly on roots, occasionally on basal stem parts, usually in dense clusters and nearly superficial), subglobose to depressed globose with flattened base (occasionally thickened at basal margin), relatively large, (200–)350–800(–1000) μm diam., initially closed, pores developing as short papillae or, usually, as long thin tubular necks of various lengths (up to 800 μm). Sometimes the pycnidia remain closed and sterile: pycnosclerotia, type III. Walls of mature pycnidia show different scleroplectenchymatous cell structures; on the outside polygonal thick-walled cells with a very small lumen (like stone cells) and on the inside similar polygonal cells with relatively thin walls; at the central base the latter cells may be elongated and form a pallisade. Protopycnidia are at first completely filled with relatively large polygonal thin-walled cells; the proliferate layer, made up of very small cells, arises in the centre and has initially a cap-like shape, the resulting central cavity gradually enlarges, apparently at the cost of the large thin-walled cells. Conidiogenous cells well-differentiated, cone-shaped. Conidial matrix cream or yellowish.

Conidia ellipsoidal to subcylindrical, 4.5–6 × 1.5–2.5(–3) μm, eguttulate or with 1–4 polar guttules.

In vitro

Colonies on OA 1.5–2.5 cm diam. after 7 days; aerial mycelium scarce, cottony or fluffy, green/yellowish; reverse ochraceous, sometimes with a luteous or amber zone. All strains tested produced antibiotic E: on application of a drop of

NaOH green → red (E⁺). Fresh isolates usually produce abundant scleroplectenchymatous pycnidia and pycnosclerotia, II & III, often covered with hyaline, ochraceous or brownish droplets. Conidia as *in vivo*. Representative
culture CBS 144.84.

Ecology and distribution

In the northern parts of Eurasia and North America common on the roots and
occasional on lower stem parts of various herbaceous plants. Well known as
plurivorous low temperature parasite: brown root rot. The fungus is particularly
destructive on herbage legumes (Leguminosae), notably sweet clover, *Melilotus
alba* and lucerne, *Medicago sativa*, following winter dormancy. It may also be
pathogenic on grasses and cereals exposed to low temperature, see Smith
(1987). [A record in India under the synonym *Plenodomus sorghi* (Mathur,
1979) probably refers to *Phoma sorghina* (Sacc.) Boerema *et al.*, sect.
Peyronellaea, **D**.]

Phoma sublingam Boerema, Fig. 45B

Teleomorph: ***Leptosphaeria submaculans* L. Holm**

> *Phoma sublingam* Boerema *in* Versl. Meded. plziektenk. Dienst
> Wageningen **157** (Jaarb. 1980): 24. 1981.
> > ≡ *Plenodomus lunariae* Syd. & P. Syd. *in* Annls mycol. **22**:
> > 264. 1924; not *Phoma lunariae* Moesz *in* Magy. bot. Lap.
> > **25**: 36. 1926 [= *Phoma doliolum* P. Karst., this section].
> > V Θ = *Sphaeronaema senecionis* f. *sisymbri* K. Krieg. *in* Fungi sax.
> > Fasc. 47, No. 2332. 1915 [as 'Sphaeronema'] [without
> > description].

Diagn. data: Boerema (1981) (fig. 6 D, F: drawing of conidia and ascospores;
differentiation against *Phoma lingam*/*Leptosphaeria maculans*), Boerema *et al.*
(1994) (fig. 7A: pycnidial and conidial shape, reproduced in present Fig. 45B;
in vivo and *in vitro* descriptions adopted below).

Description in vivo *(esp. Sisymbrium spp.)*

Pycnidia of type I (lesions on stems and leaves, usually solitary, subepidermal
and arranged in rows) resembling very much those of *Phoma lingam* (p. 341),
the anamorph of *Leptosphaeria maculans*: usually subglobose with a broad
base and a distinct papillate neck, mostly solitary, 200–250 μm diam. Pycnidia
of type II (on last year's dead stems) are smaller than those of *Ph. lingam*,
mostly 200–300 μm diam., depressed globose to subglobose with a pronounced neck; wall explicitly scleroplectenchymatous, the flattened base more
or less thickened by diverging rows of elongated cells. Conidial matrix
whitish/pink.

Conidia ellipsoidal to subcylindrical, (3.5–)4–4.5(–5) × 1–1.5(–2) μm,
usually with 1–2 small guttules.

Pseudothecia (also on dead stems) usually 200–400 μm diam., depressed globose with short broad conical neck [more distinct than in *Lept. maculans*]. Wall scleroplectenchymatous. Asci 110–120 × 16–18 μm, 8-spored, irregularly quadriseriate. Ascospores 54–70 × 6–7 μm, narrowly fusiform, 5-septate with the third cell broader but shorter than all others [in *Lept. maculans* central cells are of equal length], yellowish-brown with guttules (for recent detailed description see Shoemaker, 1984). [Often old pycnidia of type I occur in association with pseudothecia; true scleroplectenchymatous pycnidia usually develop only at the end of the next season.]

In vitro

Colonies on OA *c.* 2.5 cm diam. after 7 days; aerial mycelium greyish with yellowish brown tinges. Variable production of pycnidia-I on and in the agar. Conidia as *in vivo*.

Ecology and distribution

This fungus is in Europe found on various Cruciferae (especially often on *Sisymbrium* spp.) and occasionally also recorded on non-cruciferous plants. It may be associated with disease symptoms resembling the well known dry rot and canker disease of brassicas caused by *Phoma lingam* (Tode : Fr.) Desm. (this section), teleom. *Leptosphaeria maculans* (Desm.) Ces. & De Not. It can likewise be transmitted by seed and may very probably occur in North America [compare Petrie & Vanterpool, 1965 '*Sisymbrium*-strain' of *Ph. lingam*]. Both fungi are without doubt closely related. Their teleomorphs can be easily differentiated [e.g. *Lept. submaculans* has a short broad neck and ascospores with a short swollen third cell], but their type I pycnidia and conidia are similar. The scleroplectenchymatous pycnidia-II of *Ph. sublingam* are distinguished by being smaller with a strongly papillate neck. The teleomorph has often been erroneously identified in the past as *Leptosphaeria conferta* Niessl ex Sacc. (anam. *Phoma conferta* P. Syd. ex Died., this section), which is also common on Cruciferae.

Phoma sydowii Boerema *et al.*, Fig. 46A

Possible teleomorph: ***Leptosphaeria senecionis* (Fuckel) G. Winter**

Phoma sydowii Boerema, Kesteren & Loer. *in* Trans. Br. mycol. Soc. **77**: 71. 1981.

≡ *Sphaeronaema senecionis* Syd. & P. Syd. *in* Annls mycol. **3**: 185. 1905; not *Phoma senecionis* P. Syd. *in* Beibl. Hedwigia **38**: 136. 1899 [sect. *Phoma*, **A**].

≡ *Plenodomus senecionis* (Syd. & P. Syd.) Bubák *in* Annls mycol. **13**: 29. 1915.

H ≡ *Plenodomus senecionis* (Syd. & P. Syd.) Petr. *in* Annls mycol. **19**: 192. 1921.

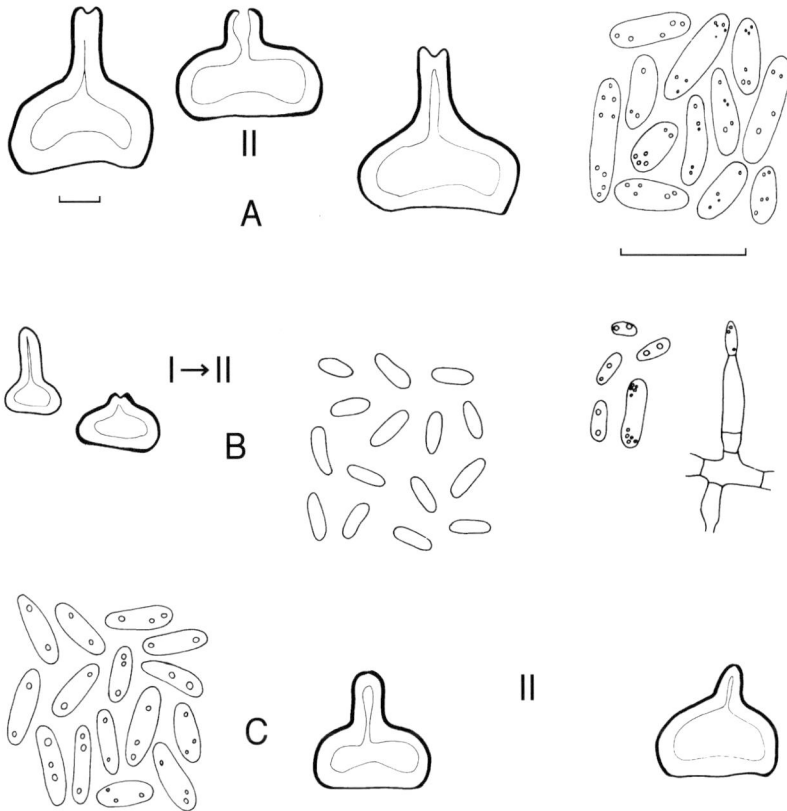

Fig. 46. A *Phoma sydowii.* Pycnidia-II depressed globose with flattened base and pronounced cylindrical neck with tube-shaped pore. Conidia variable and usually pluriguttulate. **B** *Phoma tracheiphila.* This pathogen of citrus trees is characterized by very small pycnidia-I→II, subglobose with a gradually developing cylindrical neck. *In vitro* conidia may arise from free conidiogenous cells on the mycelium (*Phialophora*-synanamorph). **C** *Phoma veronicicola.* Produces only pycnidia-II with a pronounced neck, which usually remains closed for some time; *in vitro,* neck sometimes pilose. Conidia 2–4 guttulate. Bar pycnidia 100 μm, bar conidia 10 μm.

> = *Plenodomus rostratus* Petr. in Annls mycol. **21**: 199–200. 1923; not *Phoma rostrata* O'Gara in Mycologia **7**: 41. 1915.

Diagn. data: Boerema *et al.* (1981b) (figs 1D, 2C, 3E: photograph of pycnidium *in vivo*, drawing of pycnidia *in vitro* and photographs of cultures on OA and MA; tab. 1: synoptic table on characters *in vitro* with drawing of conidia; nomenclature-synonymy; specimens studied, host range), Boerema *et al.* (1994) (fig. 7C: pycnidial and conidial shape, reproduced in present Fig. 46A; *in vivo* and *in vitro* descriptions adopted below).

Description in vivo *(Senecio spp.)*

Pycnidia always of type II (on dead stems generally scattered, subepidermal, later superficial), depressed globose, with flattened base and pronounced cylindrical neck of variable length and with a tube-shaped pore, mostly 300–450 μm diam., neck usually 150–200 μm but also longer up to 600 μm. Wall explicitly scleroplectenchymatous with a flat, or somewhat convex thickening at the base consisting of diverging rows of elongated cells. Conidial matrix pink/whitish.

Conidia irregular, oblong-ellipsoidal to subcylindrical, conspicuously variable in length, (4–)5–8(–8.5) × (1.5–)2–2.5 μm, with 2-several small, more or less polar guttules.

[Pseudothecia (dead stems) 300–350 μm diam., globose with a flattened base and a short truncate-conical neck. Wall scleroplectenchymatous. Asci 80–90 × 11–12 μm, 8-spored, biseriate. Ascospores 24–30 × 6.5–7.5 μm, broadly fusiform, 3-septate, yellowish-brown with guttules (for this supposed teleomorph see Holm, 1957; North American collections differ by broader asci, shorter and broader ascospores without guttules).]

In vitro

Colonies on OA 1–2.5 cm diam. after 7 days, tomentose, translucent, zonated, green, greenish yellow with colourless submerged margin; reverse green, yellowish green or bluish green, sometimes with amber or straw zones. Mostly abundant production of pycnidia on and in the agar; they have about the same dimensions as those *in vivo* and usually also a conspicuous cylindrical neck (often blackened around the pore). Conidia variable as *in vivo*, usually with numerous small guttles. Representative culture CBS 385.80.

Ecology and distribution

Common in Europe on dead stems of wild, perennial species of *Senecio*, but also recorded on other Compositae. Often accompanied with pseudothecia of the above teleomorph, but a metagenetic relation must still be proved by cultural experiments. In the past *Ph. sydowii* was erroneously considered to be conspecific with the nettle fungus *Phoma acuta* (Hoffm. : Fr.) Fuckel subsp. *acuta* (this section; teleom. *Leptosphaeria doliolum* (Pers. : Fr.) Ces. & De Not. subsp. *doliolum*). The latter produces significantly smaller conidia.

Phoma tracheiphila (Petri) L.A. Kantsch. & Gikaschvili, Fig. 46B

Hyphomycetous synanamorph: **Phialophora sp.**

> *Phoma tracheiphila* (Petri) L.A. Kantsch. & Gikaschvili *in* Tr. Inst. Zashch. Rast, Tiflis **5**: 20. 1948.
>
> > ≡ *Deuterophoma tracheiphila* Petri *in* Boll. R. Staz. Patol. veg. Roma II, **9**(4): 396. 1929.

≡ *Bakerophoma tracheiphila* (Petri) Cif. *in* Atti Ist. bot. Univ. Pavia V, **5**: 307. 1946.

Diagn. data: Goidànich & Ruggieri (1947) (figs 1, 2: drawings of series subepidermal pycnidia in *Citrus* branch, mature pycnidium with scleroplectenchymatous wall structure and thin-walled conidiogenous cells, picture of hyphomycetous synanamorph 'del tipo *Phialophora*' developing in culture), Graniti (1955) (pl. I figs 1–3: photographs of pycnidia *in vivo*; pls II figs 1–4, III figs 1–3: sections of pycnidia; pls IV figs 1–3, V figs 1–4: pictures of the hyphomycetous synanamorph *in vitro*), Ciccarone & Russo (1969) (figs 1–5: photographs of pycnidium and details of pycnidial beak and conidiogenous layer; figs 6–14: electron micrographs of conidiogenesis; fig. 15: irregular pycnidia on potato dextrose agar; discussion of systematic history of this pathogen: 'The fungus (type species of *Deuterophoma*) is to be included in the form-genus *Phoma*'), Punithalingam & Holliday (1973b) (figs A–E; CMI description with phytopathological notes), Sutton (1980) (fig. 229D: conidia as developing directly from the hyphae; specimens in CMI collection), Boerema *et al.* (1994) (fig. 8C: pycnidial and conidial shape *in vivo*, hyphal conidia *in vitro*, all reproduced in present Fig. 46B; *in vivo* and *in vitro* descriptions adopted below).

Description in vivo *(esp.* Citrus limonia*)*

Pycnidia I→II (on twigs and branches and around leaf scars, at first covered by the epidermis), mostly 60–135 μm diam., subglobose, at maturity always with a distinct neck, 45–70 μm diam. and up to 250 μm in length (these necks are easily removed with the epidermis, leaving behind widely and irregularly opened pycnidial bodies). Wall of mature pycnidia consists of randomly arranged polygonal scleroplectenchyma cells and is about the same thickness throughout.

Conidia subcylindrical, straight or slightly curved, 2–3(–3.5) × 1–1.5 μm, eguttulate or biguttulate.

In vitro

Colonies of fresh isolates on OA grow well, 2.5–3 cm after 7 days, flat with little aerial mycelium; pigmentation of mycelial mat and medium variable, depending on strain and ranging from pale pink or bright orange to dark olive brown. On application of NaOH the reddish pigments turn blue (the presence of helminthosporin and cynodontin have been demonstrated). Production of pycnidia is scarce, these often remain incomplete, thin-walled (I) and open irregularly at maturity. However, colonies easily produce conidia directly from the mycelium, either from papillate extrusions or from flask-shaped conidiogenous cells (representing the *Phialophora*-synanamorph).

Hyphal conidia variable in shape and size, depending on strain, often 2–2.5 × 1–1.5 μm, but also larger, 3–8 × 1.5–3 μm, usually with 2-several polar guttules. Representative culture CBS 551.93.

Ecology and distribution

This vascular pathogen of lemons and other *Citrus* spp. (Rutaceae) occurs throughout the Mediterranean and Black Sea areas: mal secco disease. The typical symptoms consist of red discoloured strands in the xylem of stems, veinal chlorosis, wilt and shedding of leaves and ultimately dieback of twigs and branches. Infection occurs through stomata and wounds. The disease symptoms are induced by a phytotoxin, 'malseccin' (Nachmias *et al.*, 1997). The production of pycnidia is fluctuating and often rare. An ashy appearance on the stem indicates the presence of pycnidia beneath the epidermis.

Phoma veronicicola Boerema & Loer., Fig. 46C

> *Phoma veronicicola* Boerema & Loer. in Trans. Br. mycol. Soc. **84**: 297. 1985.
>> ≡ *Sphaeronaema veronicae* Hollós in Annls hist.-nat. Mus. natn. hung. **4**: 341. 1906; not *Phoma veronicae* Roum. in Revue mycol. **6**: 160. 1884.

Diagn. data: Boerema & Loerakker (1985) (figs 3D, E, 4D, 5E: photographs of pycnidia *in vivo* and *in vitro*, drawings of pycnidia *in vitro*, photographs of cultures on OA and MA; tab. 1: synoptic table on characters *in vitro*, with drawing of conidia; nomenclature-synonymy; specimens studied, hosts), Boerema *et al.* (1994) (fig. 6A: pycnidial and conidial shape, reproduced in present Fig. 46C; *in vivo* and *in vitro* descriptions adopted below).

Description in vivo *(Veronica spp.)*

Pycnidia always of type II (on dead stems, gregarious, subepidermal or superficial), depressed globose, usually 200–350 μm diam., with a flattened base and cylindrical neck, mostly 150–200 μm long, which remains closed for some time. Wall thickness initially uniform, later more strongly thickened at the flattened base and near the tube-shaped porus; the polygonal scleroplectenchyma cells have different dimensions and may have a somewhat parallel arrangement in the basal thickening. Conidial matrix white/yellowish.

Conidia ellipsoidal to subcylindrical, (3.5–)4–5(–6) × (1–)1.5–2 μm, with 2-several more or less polar guttules.

In vitro

Colonies on OA 1.5–2.5 cm diam. after 7 days; ochraceous, greenish olivaceous or greenish grey; aerial mycelium tenuous or rather compact, cottony or somewhat felted; reverse colourless or yellowish red, but always with a broad amber or green zone. Yellow pigment localized in cell walls, hardly diffusing into medium [no reaction with addition of a drop of NaOH]. Abundant production of pycnidia II on and in the agar, with more or less cylindrical necks as *in vivo*, sometimes pilose; agar close to submerged pycnidia rusty-brown. Conidia similar to those *in vivo*. Representative culture CBS 145.84.

Ecology and distribution

Common in Europe on dead stems of wild and cultivated perennial species of
Veronica (Scrophulariaceae). Regarded as a necrophyte, but possibly weakly
pathogenic. It should be noted that in the mountainous regions of Europe the
plurivorous *Phoma pedicularis* Fuckel (this section) has also been recorded on
Veronica spp.

Species Excluded from *Phoma* sect. *Plenodomus*

(Misapplications in the synonymous genera *Plenodomus*, *Diploplenodomus*,
Leptophoma and *Deuterophoma*.)

In the past various pycnidial anamorphs have been placed erroneously in genera that are now regarded as synonymous with *Phoma* sect. *Plenodomus*.
These misapplications are listed below. For documentation, identification and
discussion see Boerema *et al.* (1996). Twelve misapplications appeared to refer
to *Phoma* species treated in other sections or under Miscellaneous.

Plenodomus-pathogen of apples (Hara), *Plenodomus borgianus* Sacc.,
Plenodomus brachysporus Petr.[15], *Plenodomus cannabis* (Allesch.) Moesz &
Smarods[16], *Plenodomus cenangium* (Corda) Oudem., *Plenodomus chelidonii*
Naumov, *Plenodomus chenopodii* (P. Karst. & Har.) Arx[17], *Plenodomus coco-
genus(a)* Sawada, *Plenodomus corni* Bat. & A.F. Vital, *Plenodomus cruentus*
Syd. & P. Syd.[18], *Plenodomus destruens* Harter, *Plenodomus erythrinae*
Oudem., *Plenodomus eucalypti* J.V. Almeida & Sousa da Câmara,
Plenodomus filarszkyanus (Moesz) Petr.[19], *Plenodomus fusco-maculans* (Sacc.)
Coons, *Plenodomus gallarum* (Lév.) Oudem., *Plenodomus haematites* Petr.[20],
Plenodomus herbarum Allesch., *Plenodomus hoveniae* Gucevič, *Plenodomus
humuli* Kusnezowa, *Plenodomus inaequalis* Sacc. & Trotter, *Plenodomus
macropodii* Petr.[21], *Plenodomus metasequoiae* Gucevič, *Plenodomus molleri-
anus* Bres., *Plenodomus mori* (Mont.) Höhn., *Plenodomus nigricans* Negodi,
Plenodomus oleae Cavara[22], *Plenodomus pyracanthae* Gucevič, *Plenodomus
ramealis* (Desm.) Höhn., *Plenodomus spurius* (Vestergr.) Petr., *Plenodomus
strobilinus* (Desm.) Höhn., *Plenodomus sylvaticus* (Sacc.) H. Ruppr.[23],
Plenodomus syriacus Petr.[24], *Plenodomus valentinus* Caball. and *Plenodomus
verbascicola* (Schwein.) Moesz.

[15] = *Phoma labilis* Sacc. (sect. *Phoma*, **A**)
[16] = *Phoma exigua* Desm. var. *exigua* (sect. *Phyllostictoides*, **E**)
[17] = *Phoma chenopodiicola* Gruyter *et al.* (sect. *Phoma*, **A**)
[18] ≡ *Phoma cruenta* (Syd. & P. Syd.) Boerema *et al.* (Miscellaneous)
[19] ≡ *Phoma filarskyana* (Moesz) Boerema *et al.* (Miscellaneous)
[20] ≡ *Phoma haematites* (Petr.) Boerema *et al.* (Miscellaneous)
[21] = *Phoma nigrificans* (P. Karst.) Boerema *et al.* (sect. *Sclerophomella*, **F**)
[22] = *Phoma glomerata* (Corda) Wollenw. & Hochapfel (sect. *Peyronellaea*, **D**)
[23] ≡ *Phoma sylvatica* Sacc. (sect. *Sclerophomella*, **F**)
[24] ≡ *Phoma syriaca* (Petr.) Boerema *et al.* (sect. *Sclerophomella*, **F**)

Diploplenodomus aggregatus Höhn. and *Diploplenodomus rivini* (Allesch.) Petr.[25]

Leptophoma paeoniae Höhn.[26] and *Leptophoma urticae* (Schulzer & Sacc.) Höhn.

Deuterophoma ulmi (Verrall & C. May) Goid. & Ruggieri and *Deuterophoma*-pathogen of *Chrysanthemum* spp.[27]

---

[25] = *Phoma complanata* (Tode : Fr.) Desm. (sect. *Sclerophomella*, **F**)
[26] = *Phoma nebulosa* (Pers. : Fr.) Berk. (sect. *Phoma*, **A**)
[27] = *Phoma vasinfecta* Boerema *et al.* (sect. *Phoma*, **A**)

15 H *Phoma* sect. *Macrospora*

Phoma sect. *Macrospora* Boerema *et al.*

> *Phoma* sect. *Macrospora* Boerema, Gruyter & Noordel. apud Boerema *in*
> Mycotaxon **64**: 332. 1997.
> Type: *Phyllosticta maydis* Arny & Nelson ≡ *Phoma zeae-maydis* Punith.

Species in this section always produce relatively large conidia, both *in vivo* and
in vitro, (7–)8–19(–25) × (2.5–)3–7(–9) μm. This means that their length and
width exceed the general conidial measurements of the genus (see p. 12). The
conidia are initially aseptate, but they may become 1-septate by secondary
septation. The pycnidia, simple or complex, are thin-walled and pseudoparenchymatous with distinct or inconspicuous ostioles, generally glabrous, but
sometimes with mycelial outgrowths.

So far nine species, studied *in vitro*, have been included in this section.
The type species, *Phoma zeae-maydis* Punith. was originally interpreted as a
large spored species of *Phyllosticta* ('Macrostiteae'). This holds also for some
other species of the section. Due to the large conidia and the occasional
occurrence of a septum, some species were formerly classified in *Ascochyta*
Auct.

A study of ultra structure of conidiogenesis in one of the species presently
arranged under sect. *Macrospora*, namely *Phoma rabiei* (Pass.) Khune,
showed typical characteristics of the genus *Phoma*: a three-layered apical
thickening of the conidiogenous cell prior to the first conidium formation, and
conidial septa attaining from the very start the thickness of a final septum
(Singh *et al.*, 1997).

The section includes species with and without chlamydosporal structures,
the latter may be unicellular in chains or clusters, but also multicellular
(dictyo/phragmosporous). Two species, including the type species, have been
connected with a teleomorph belonging to *Mycosphaerella* Johanson.

Diagn. lit.: Boerema (1997) (introduction of *Phoma* sect. *Macrospora*), Punithalingam (1990) (classification of type species in *Phoma*; teleomorph *Mycosphaerella*), de Gruyter (2002) (precursory contribution to this chapter).

Synopsis of the Section Characteristics

- Pycnidia simple or complex, thin-walled, pseudoparenchymatous, with distinct or inconspicuous ostioles, glabrous, but sometimes with hyphal outgrowths.

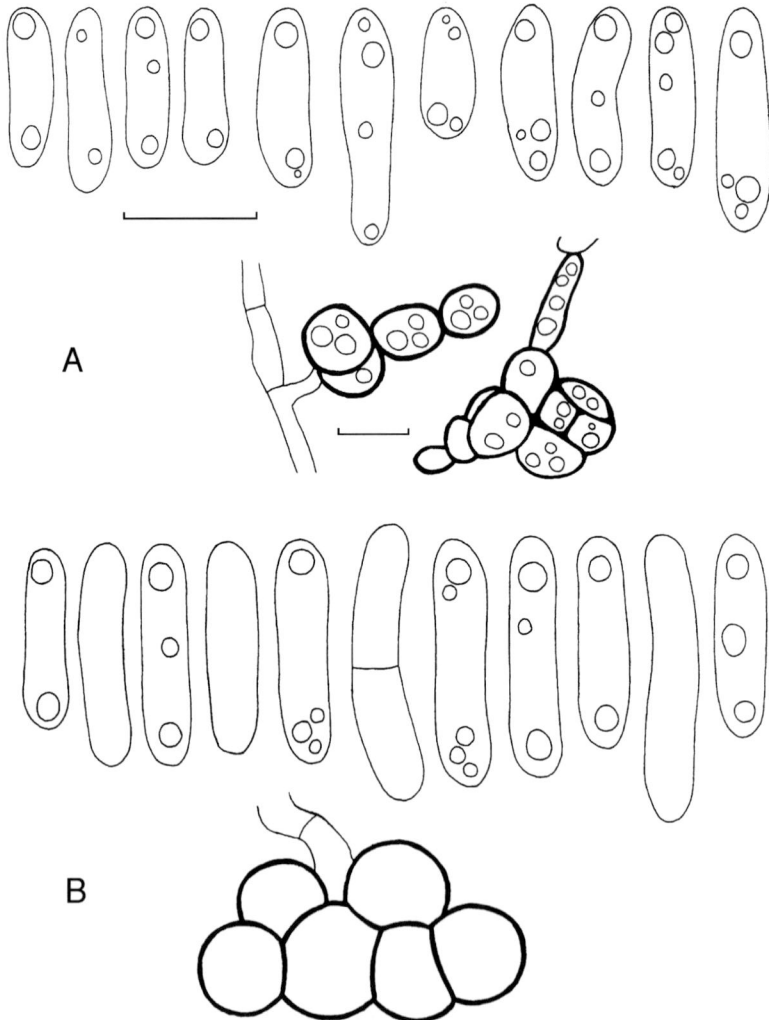

Fig. 47. Variation of conidia and hyphal chlamydospores. **A** *Phoma zeae-maydis*, the type species of *Phoma* sect. *Macrospora*. The relatively large conidia remain aseptate (only when germination is initiated they may become 1-septate). **B** *Phoma andropogonivora*. Conidia rarely two-celled by secondary septation. Bars 10 μm.

- Conidia always relatively large *in vivo* and *in vitro*, mostly 8–19 × 3–7 μm, initially aseptate, but they may become 1-septate.
- Chlamydospores, if present, simple (in chains or clusters) or multicellular (dictyo/phragmosporous).
- Teleomorph *Mycosphaerella*.

Key to the Species Treated of Section *Macrospora*

Differentiation based on the characteristics *in vitro*

1. a. Colonies on OA colourless to greenish or greyish **2**
    b. Colonies on OA red to bluish purple or violet **8**

2. a. Growth rate on OA and MA slow, 1.5–3 cm after 7 days; on OA colonies colourless to pale olivaceous grey, sometimes with a rosy buff tinge, conidia mainly aseptate, 7–15.5 × 3–5.5 μm ....*Ph. rabiei*, teleom. *Myc. rabiei*
    b. Growth rate moderate to fast on OA, > 3 cm after 7 days **3**

3. a. Growth rate fast on OA and MA, 6–8 cm after 7 days....................................**4**
    b. Growth rate moderate, on OA 3.5–6 cm after 7 days, on MA 3–4.5 cm after 7 days..**5**

4. a. Growth rate on OA 6 cm, on MA 6.5–7 cm; on OA colonies dark herbage green to dull green, yellow green near margin, conidia mainly aseptate, 8–15 × 3.5–5.5 μm ....*Ph. boeremae*
        [Recorded on dried stems and seed of cultivated species of *Medicago* in Europe and Australia.]
    b. Growth rate on OA 6–6.5 cm, on MA *c.* 8 cm; on OA colonies colourless/buff to pale olivaceous grey, conidia mainly aseptate, (10.5–)13–17(–21) × (4–)5–6.5 μm..*Ph. commelinicola*
        [Pathogen of Commelinaceae, e.g. *Commelina nudiflora* and *Tradescantia subaspera*, in North and Central America, also recorded in New Zealand.]

5. a. Colonies on OA greenish olivaceous to citrine green, NaOH spot test positive, bluish/green to red (not an E$^+$ reaction), conidia mainly aseptate, hyaline to pale yellowish, (7.5–)8.5–17.5 × 3.5–5.5(–7) μm, ellipsoidal to allantoid, sometimes typical curved, chlamydospores absent.........................*Ph. xanthina*
        [Specific pathogen of *Delphinium* spp. recorded in Europe.]
    b. Colonies on OA colourless/rosy buff to grey olivaceous, with olivaceous grey/olivaceous black sectors, or dark herbage green/dull green to olivaceous, chlamydospores or chlamydospore-like cells present, NaOH spot test negative or pale reddish/brown, not specific...**6**

6. a. Growth rate on OA *c.* 4 cm; colonies colourless to grey olivaceous, with olivaceous grey to olivaceous black sectors, conidia mainly aseptate, (12–)15–17(–25) × 3.5–5(–6.5) μm, chlamydospores may be present, up to 15 μm diam. ....*Ph. zeae-maydis*, teleom. *Myc. zeae-maydis*
        [Pathogen of *Zea mays*, also on *Sorghum* and *Setaria* spp., recorded in North and South America, and in southern Africa.]

b.   Growth rate on OA 4.5–6 cm; colonies dark herbage green/dull green to olivaceous or colourless/rosy buff to olivaceous **7**

7. a.   Growth rate on OA 4.5–5.5 cm; colonies dark herbage green/dull green to olivaceous, conidia aseptate, 10–12.5 × 2.5–3.5 µm, chlamydospores 8–12 µm diam. ....*Ph. gossypiicola*
        [Pathogen of *Gossypium* spp., cosmopolitan in cotton-growing areas.]
    b.   Growth rate on OA 4.5–6 cm (slower growing strains, <4.5 cm occur); colonies colourless to rosy buff, sometimes slightly olivaceous at centre, conidia mainly aseptate, 14–19 × 3.5–4.5 µm, chlamydospore-like cells present, 12–19 µm diam...............................*Ph. andropogonivora* [? teleom. *Myc.* sp.]
        [Pathogen of perennial bunch grasses, *Andropogon gerardii* and *Schizachyrium scoparium* (= *Andropogon scoparius*), in the USA.]

8. a.   Growth rate on OA slow, 1–1.5 cm; colonies olivaceous grey to red/bluish purple due to a pigment production, conidia mainly aseptate, 8.5–12.5(–16) × 3–4.5(–5) µm ....*Ph. chenopodii*
        [Necrophyte on *Chenopodium album* and some other Chenopodiaceae (*Atriplex crassifolia*, *Beta vulgaris*), Eurasia.]
    b.   Growth rate on OA fast; colonies grey olivaceous/olivaceous to red violet due to a pigment production, conidia mainly aseptate, 9.5–13.5 × 5.5–9 µm ....*Ph. necator*
        [Pathogen of *Oryza sativa*, recorded in southern Europe and south-eastern USA.]

Distribution and Host Relations in Section *Macrospora*

Chenopodiaceae
Atriplex crassifolia   *Ph. chenopodii*
Beta vulgaris
Chenopodium album
(main host)   [Necrophyte recorded in Eurasia.]

Commelinaceae
Commelina nudiflora   *Ph. commelinicola*
Tradescantia spp., e.g.
T. subasper   [Pathogen causing leaf necrosis in North and Central America; also recorded in New Zealand.]

Gramineae
Andropogon gerardii   *Ph. andropogonivora*
Schizachyrium scoparium   [? teleom. *Myc.* sp.]
                                           [Causal organism of a leaf spot disease in North America.]
Oryza sativa   *Ph. necator*
                                           [Found in association with wilt symptoms in southern Europe and south-eastern USA.]
Setaria and *Sorghum* spp.   *Ph. zeae-maydis*
Zea maydis (main host)   (teleom. *Myc. zeae-maydis*)
                                           [Causal organism of yellow leaf blight recorded in North and South America and in southern Africa.]

Leguminosae
Cicer arietinum   *Ph. rabiei*
                                 (teleom. *Myc. rabiei*)
                                 [Causal organism of anthracnose or chickpea blight, world-
                                 wide recorded on the host.]
Medicago spp., e.g.   *Ph. boeremae*
M. falcata and *M. littoralis* [Necrophyte, found in Europe and Australia.]

Malvaceae
Gossypium spp.   *Ph. gossypiicola*
                                 [Known as causal organism of wet weather blight, probably
                                 worldwide on the host.]

Ranunculaceae
Delphinium spp.   *Ph. xanthina*
                                 [Recorded in association with leaf and stem necroses in
                                 Europe.]

Descriptions of the Taxa

Phoma andropogonivora (R. Sprague & Rogerson) Gruyter, Fig. 47B

Probable teleomorph: ***Mycosphaerella* sp.**

Phoma andropogonivora (R. Sprague & Rogerson) Gruyter *in* Persoonia **18**(1): 97. 2002.
≡ *Phyllosticta andropogonivora* R. Sprague & Rogerson *in* Mycologia **50**: 6. 1958.
≡ *Ascochyta andropogonivora* (R. Sprague & Rogerson) Morgan-Jones apud Morgan-Jones, Owsley & Krupinsky *in* Mycotaxon **42**: 56. 1991.

Diagn. data: Morgan-Jones *et al.* (1991) (fig. 1A–E: drawings of pycnidia, chlamydospore-like cells and conidia; pl. 1A–H: photographs of agar plate culture, pycnidium, pycnidial wall, chlamydospore-like cells and conidia; history and morphological and cultural descriptions; classification in *Ascochyta* 'mainly on the basis of shape and size of conidia', 'although its conidia are mainly unicellular'. 'In other characteristics, … , it is closely similar to some species of *Phoma*'), de Gruyter (2002) (figs 7, 13: copy of conidia and chlamydospore-like cells, reproduced in present Fig. 47B; some characteristics *in vitro*, adopted below).

Description in vitro *(adopted from Morgan-Jones et al., 1991)*

Pycnidia (sub)globose to irregular, with usually 1–3, non-papillate or very slightly papillate ostiole(s), up to 240 μm diam., glabrous.

Conidia cylindrical, often slightly curved, usually with distinct polar guttules, mainly aseptate, occasionally 1-septate, 14–19 × 3.5–4.5 μm.

Chlamydospores in long intercalary chains or in botryose clusters, subglobose to globose, 12–19 μm diam.

Colonies on PDA (comparable with OA) mostly 4.5–6 cm diam. after 7 days, regular, at first whitish but becoming pale salmony (rosy buff), sometimes with a slightly olivaceous tinge at centre; aerial mycelium densely woolly to felty. Slower growing strains, 4–4.5 cm after 7 days, are less salmon coloured, but with cream patches intermixed with orange-greyish areas and an olive-grey centre. Darker sectors appear, due to the formation of chlamydospore-like cells. Pycnidia abundant, mostly superficial on the agar. Representative dried cultures preserved in AUA (Auburn, USA) (living culture not obtained).

Description in vivo *(esp.* Schizachyrium scoparium*)*

Pycnidia (scattered in necrotic lesions on the leaves, epiphyllous) more or less subglobose, smaller than those in vitro, 70–140 μm diam., with non-papillate ostioles. Conidia similar to those *in vitro*, but generally somewhat shorter and slightly wider than those produced in culture.

[Pseudothecia (mingled with pycnidia in the necrotic lesions) somewhat lenticular, and of similar size and shape as the pycnidia, often about 70 μm diam., with wide ostiole. Asci broadly clavate, often somewhat curved, about 40×12 μm. Ascospores broadly fusoid with a submedian septum and strongly constricted ('isthmus-like'), about $15–16 \times 6–6.5$ μm (cf. Greene, 1960: 85–86).]

Ecology and distribution

This Gramineae-fungus causes a leaf spot on varieties of *Andropogon gerardii* and also on plants of *Schizachyrium scoparium* (≡ *Andropogon scoparius*), both perennial bunch grasses, widely distributed in the Great Plains of the USA (big- or sand blue stem and little blue stem). The collections studied *in vitro* by Morgan-Jones *et al.* (1991) were from Minnesota and North and South Dakota. The *Mycosphaerella* species described above, was found (Greene, 1960) in Wisconsin on dead leaves of *Schizachyrium scoparium* together with 'pycnidia of a *Phyllosticta*', corresponding with those of *Ph. andropogonivora*. In the type collection of the latter on *Andropogon gerardii* a similar, but immature loculoascomycete has been found (Morgan-Jones *et al.*, 1991). A single identity of both morphs is plausible, but must still be proven.

Phoma boeremae Gruyter, Fig. 48A

Phoma boeremae Gruyter *in* Persoonia **18**(1): 91. 2002.
≡ *Macrophoma medicaginis* Hollós *in* Math. Termész Közlém. Magy, Tudom-Akad. **35**(1): 37. 1926; not *Phoma medicaginis* Malbr. & Roum. apud Roumeguère *in* Fungi gall. Exs. Cent. 37, No. 3675. 1886 and *in* Revue mycol. **8**: 91. 1886 [sect. *Phyllostictoides*, **E**].

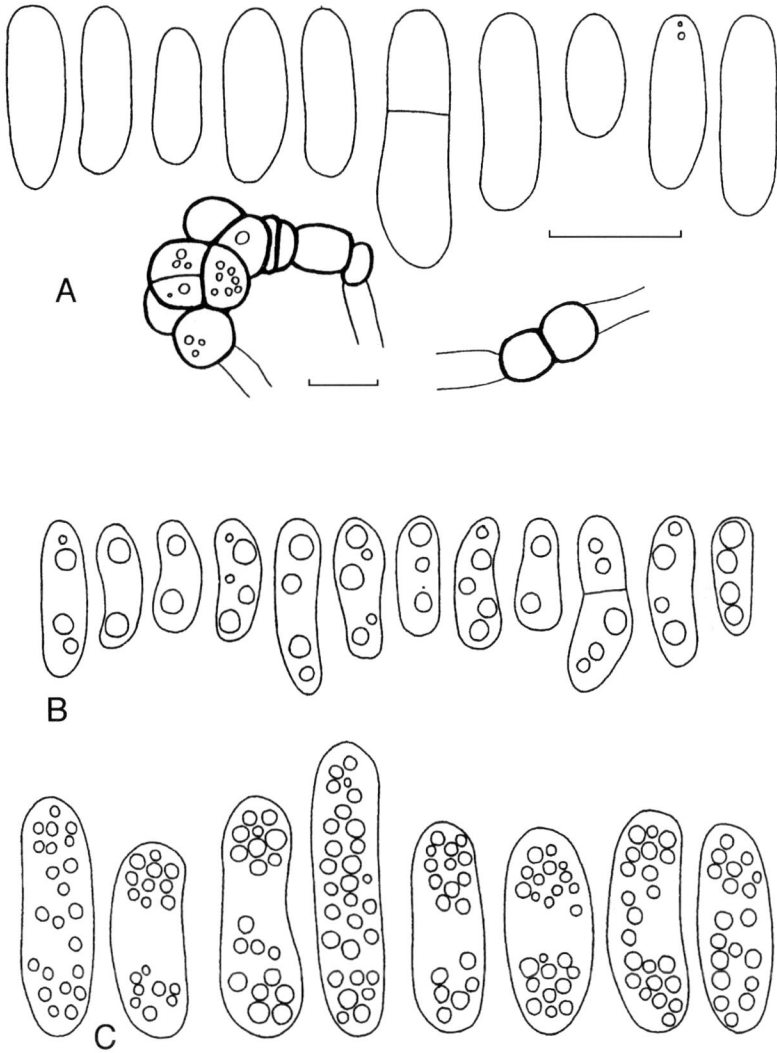

Fig. 48. Variation of conidia and some hyphal chlamydospores. **A** *Phoma boeremae*. **B** *Phoma chenopodii* (conidial septation commonly occurring before germination). **C** *Phoma commelinicola*. Bars 10 μm.

Diagn. data: de Gruyter (2002) (figs 2, 10: shape of conidia and chlamy-dospores, reproduced in present Fig. 48A; cultural descriptions, partly adopted below).

Description in vitro

Pycnidia globose/subglobose to irregular, with 1(–3) non-papillate or papillate ostiole(s), later often developing into an elongated neck, 40–320 μm diam., glabrous or with mycelial outgrowths. Conidial matrix rosy buff to vinaceous.

Conidia oblong to ellipsoidal, eguttulate or with some small guttules, mainly aseptate, 8–15 × 3.5–5.5 μm, occasionally occurring 1-septate conidia somewhat larger.

Chlamydospores intercalary or terminal, unicellular or multicellular-dictyo/phragmosporous, botryoid, pale olivaceous, 6–22 μm diam.

Colonies on OA 6 cm after 7 days, regular, dark herbage green to dull green, yellow green near margin; aerial mycelium floccose, olivaceous grey to dull green. The reverse also dark herbage to dull green and yellow green near margin. On MA faster growing, 6.5–7 cm after 7 days, dull green, citrine near margin and with grey olivaceous to dull green aerial mycelium; reverse dull green to leaden black. Pycnidia scattered, both on and in the agar, solitary or confluent. Representative culture CBS 109942.

Ecology and distribution

The two records so far refer to dried stems and seed of cultivated species of *Medicago* (Leguminosae) in Europe and Australia. Under such dry conditions the conidia may look granulous due to the presence of numerous guttules (compare Boerema *et al.*, 1997: 352–353), but *in vitro* they are usually eguttulate. The fungus is characterized by its dark herbage green to dull green colonies.

Phoma chenopodii S. Ahmad, Fig. 48B

Phoma chenopodii S. Ahmad in Sydowia **2**: 79. 1948.
      H = *Phoma chenopodii* Pavgi & U.P. Singh in Mycopath. Mycol. appl. **30**: 265. 1966.
        = *Phyllosticta bacilliformis* Padwick & Merh in Mycol. Pap. **7**: 4. 1943; not *Phoma bacilliformis* Wehm. in Mycologia **38**: 316–317. 1946 [= *Asteromella* sp.].

Diagn. data: Ahmad (1948) (original description from dead stems of *Chenopodium album* (type) and *Atriplex crassifolia*), Boerema (1984b) (note on p. 33: history, records in Russia and The Netherlands; confusion about its identity), de Gruyter (2002) (fig. 8: shape of conidia, reproduced in present Fig. 48B; description adopted below).

Description in vitro

Pycnidia subglobose, slightly papillate, with inconspicuous ostiole, 100–200 μm diam.

Conidia irregularly subcylindrical to ellipsoidal with several large guttules, mostly aseptate within the pycnidium, but often becoming 1-septate and occasionally 2-septate in the exuding mass (secondary septation preceding germination), 8.5–12.5(–16) × 3–4.5(–5) μm.

Colonies on OA slow growing, 1–1.5 cm after 7 days, regular, olivaceous grey, with a red/bluish purple discoloration due to a pigment in the agar; aerial mycelium sparse, (pale) olivaceous grey; reverse bluish purple. On MA similar, but with irregular outline and less pigmentation. NaOH spot test negative. Representative cultures have been lost.

Ecology and distribution

The main host of this soil- and seed-borne saprophyte is *Chenopodium album*, but it has also been found on some other Chenopodiaceae, e.g. *Atriplex crassifolia* and *Beta vulgaris*. The above synonyms refer to specimens collected in Pakistan and India, but the fungus is also recorded in Russia and The Netherlands. The relatively large conidia (7.5–16 × 3–5 µm *in vivo*) with septation usually occurring immediately before germination, explains why the fungus has sometimes been confused with a true *Ascochyta* occurring on Chenopodiaceae, namely *Ascochyta caulina* (P. Karst.) Aa & Kesteren, the conidial anamorph of *Pleospora calvescens* (Fr.) Tul. & C. Tul. (see Boerema *et al.*, 1987: 199–200).

Phoma commelinicola (E. Young) Gruyter, Fig. 48C

Phoma commelinicola (E. Young) Gruyter *in* Persoonia **18**(1): 93. 2002.
≡ *Phyllosticta commelinicola* E. Young *in* Mycologia **7**: 144. 1915.

Diagn. data: Greene (1960) (p. 88, collection on *Tradescantia subaspera* (cult.) in Madison, Wisconsin, 'the aspect suggests *Ascochyta* … but no septa were noted'), de Gruyter (2002) (fig. 3: shape of conidia, reproduced in present Fig. 48C; cultural descriptions, partly adopted below).

Description *in vitro*

Pycnidia globose to subglobose, with 1(–2) papillate or non-papillate ostiole(s), 70–320 µm diam., glabrous. Conidial matrix rosy buff.
Conidia ellipsoidal, with several small, scattered guttules, mainly aseptate, (10.5–)13–17(–21) × (4–)5–6.5 µm. Rarely occurring 1-septate conidia of similar size.
Colonies on OA 6–6.5 cm after 7 days, regular, colourless/buff to pale olivaceous grey; aerial mycelium woolly, white to pale olivaceous grey; reverse similar. On MA similar but faster growing, c. 8 cm after 7 days; reverse buff to honey, partly olivaceous black. Pycnidia scattered, both on and in the agar as well as in aerial mycelium, solitary confluent; also micropycnidia, 4–70 µm diam. Representative culture CBS 100409.

Ecology and distribution

This pathogenic species, originally described from dead or dying leaves of *Commelina nudiflora* in Puerto Rico, also occurs on other Commelinaceae. The above description is based on an isolate from leaf spots on *Tradescantia* sp. in New Zealand. The genus *Tradescantia* is indigenous to America. A collection on *T. subaspera* was collected in the USA in 1959 (Greene (1960): 'Infection proceeds from the leaf tip inward until the entire leaf becomes dead and brown'). The conidia *in vivo* were aseptate and measured (9.5–)12–15(–20) × (3.5–)5–7 μm.

Phoma gossypiicola Gruyter, Fig. 49A

> *Phoma gossypiicola* Gruyter *in* Persoonia **18**(1): 96. 2002.
>> ≡ *Ascochyta gossypii* Woron. *in* Vêst. tiflis. bot. Sada **35**: 25. 1914; not *Phoma gossypii* Sacc. *in* Michelia **2**(1): 144. 1880 [= *Phomopsis* sp.].
>> H = *Ascochyta gossypii* Syd. *in* Annls mycol. **14**: 194. 1916.
>> [= 'Sphaeronaema allahabadense' sensu Manoharachary & Ramarao *in* Indian Phytopath. **28**: 427–428. 1975; see Boerema & Dorenbosch, 1980.]

Diagn. data: Boerema *et al.* (1973) (p. 133, confusion in literature with *Ph. pomorum* and *Ph. exigua*), Holliday & Punithalingam (1970a) (figs A–C; CMI description sub *Ascochyta gossypii*), Boerema & Dorenbosch (1980) (misidentification of soil isolate), de Gruyter (2002) (figs 6, 12: shape of conidia and chlamydospores, reproduced in present Fig. 49A; cultural descriptions, partly adopted below).

Description in vitro

Pycnidia globose to subglobose, with or without one usually non-papillate ostiole, 100–250 μm diam., glabrous, honey coloured, later olivaceous to olivaceous black. Conidial matrix off-white.

Conidia ellipsoidal, with several small, scattered guttules, aseptate, 10–12.5 × 2.5–3.5 μm.

Chlamydospores olivaceous with greenish guttules, unicellular, usually in chains, globose to elongate, 8–12 μm diam.

Colonies on OA 4.5–5.5 cm after 7 days, regular, dark herbage green/dull green to olivaceous; aerial mycelium sparse velvety, olivaceous grey; reverse grey olivaceous/olivaceous to violaceous grey/leaden grey. On MA slower growing, 3–3.5 cm after 7 days (14 days: 6–7 cm), regular, olivaceous black, grey olivaceous to dull green near margin; aerial mycelium sparse velvety, grey olivaceous; reverse leaden grey to olivaceous/olivaceous black. Pycnidia scattered, both on and in the agar and in aerial mycelium, solitary or confluent. Representative culture CBS 377.67.

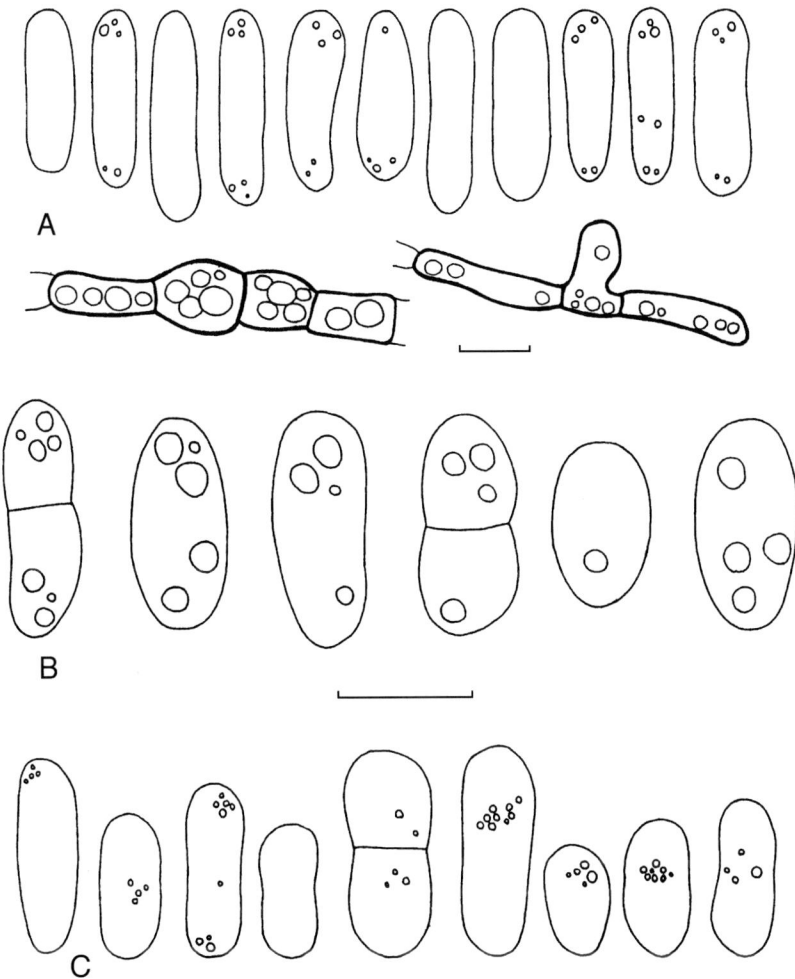

Fig. 49. Variation of conidia and some hyphal chlamydospores. **A** *Phoma gossypiicola* (conidia *in vivo* usually become 1(–2)-septate and larger). **B** *Phoma necator.* **C** *Phoma rabiei.* Bars 10 μm.

Ecology and distribution

This species is a well-known cause of leaf spots and stem cankers on cotton, *Gossypium* spp. (Malvaceae): wet weather blight. As noted above, the relatively large conidia always remain one-celled *in vitro*, but *in vivo* most conidia may become two- or even more-celled and longer, up to 14 μm (Holliday & Punithalingam, 1970a). It appears to be seed- and soil-borne and probably occurs everywhere cotton is grown. Isolates studied were from North America (USA), India and Africa (Sudan). The fungus may also attack some other cultivated crops. However, the literature on its host range must be read with much reserve, because the fungus has been confused with other species such as *Phoma pomorum* Thüm. (sect. *Peyronellaea*, **D**) and *Phoma exigua* Desm. (sect. *Phyllostictoides*, **E**), see documentation in Boerema *et al.* (1973) and Boerema & Dorenbosch (1980).

Phoma necator Thüm., Fig. 49B

> *Phoma necator* Thüm. *in* Lab. Versuchs-Station Wein-Obstbau Klosterneuburg **12** [= Pilze Reispfl.]: 12. 1889 [not '*necatrix*' as erroneously listed in compiling works].
>> *V* ≡ *Phyllosticta necator* (Thüm.) I. Miyake *in* J. Coll. Agric. imp. Univ. Tokyo **2**(4): 268. 1910 [as '*necatrix*'].

Diagn. data: Padwick (1950) (data on occurrence and characters *in vivo*), Bessi & de Carolis (1974) (recorded in Italy; characteristics *in vitro*), de Gruyter (2002) (fig. 9: shape of conidia, reproduced in present Fig. 49B; characteristics *in vitro* (dried culture) adopted below).

Description in vitro

Pycnidia globose to subglobose, with usually one non-papillate or papillate ostiole, 110–325 µm diam., glabrous. Exuding conidial mass not observed.

Conidia subglobose to ellipsoidal with 1–several relatively large guttules, aseptate and 1-septate, 9.5–13.5 × 5.5–9 µm, the 1-septate conidia sometimes larger, up to 16 × 6 µm, constricted at the septum.

Colonies on OA fast growing, irregular, grey olivaceous to olivaceous, with red violet discoloration of the agar due to a pigment; aerial mycelium abundant, grey olivaceous. Pycnidia scattered, often completely covered by aerial mycelium, solitary or confluent. Representative dried culture CBS 3509.

Ecology and distribution

In Austria and Italy found in association with a rapid basal wilt of rice plants, *Oryza sativa* (Gramineae). The fungus is also recorded in south-eastern USA. The conidial dimensions *in vivo* should vary in the range 10–12 × 6–8 µm.

Phoma rabiei (Pass.) Khune ex Gruyter, Fig. 49C
Teleomorph: *Mycosphaerella rabiei* Kovatsch. ex Gruyter

> *Phoma rabiei* (Pass.) Khune ex Gruyter *in* Persoonia **18**(1): 89. 2002.
>> ≡ *Zythia rabiei* Pass. *in* Comment. Soc. crittogam. ital. **2**(3): 437. 1867.
>> ‡ ≡ *Phoma rabiei* (Pass.) Khune in R.S. Mathur, Coelom. India 182–183. 1979 [without reference to basionym: not validly published; Art. 33.2].

‡ *H* ≡ *Phoma rabiei* (Pass.) Khune & J.N. Kapoor *in* Indian
        Phytopath. **33**: [119–]120. 1980 [with citation of
        basionym, but without reference of page of publication:
        also not validly published; Art. 33.2].
    ≡ *Phyllosticta rabiei* (Pass.) Trotter *in* Revue Path. vég. Ent.
        agric. Fr. **9**: 7. 1918.
    ≡ *Ascochyta rabiei* (Pass.) Labr. *in* Revue Path. vég. Ent.
        agric. Fr. **18**: 228. 1931.
    = *Phyllosticta cicerina* Prill. & Delacr. *in* Bull. Soc. mycol. Fr.
        9: 273. 1893.

Diagn. data: Punithalingam & Holliday (1972d) (figs A, B, C, pycnidium and
conidia *in vivo*; CMI description sub *Ascochyta rabiei*; 'some conidia unicellular'), Neergaard (1977) (fig. 5.32: pycnidia and conidia; 'a major pathogen of
the host'), Khune & Kapoor (1980) (fig. 1a, b, c: pycnidium and conidia *in
vivo*; 'conidia occasionally 1-septate'), Singh *et al.* (1997) (figs 1–26: electron
micrographs (transmission and scanning) of conidiogenesis and conidia;
'marked apical thickening of the conidiogenous cells was recorded prior to
conidium formation'; 'conidia septation originated as a result of ingrowth of the
lateral wall and from the very start attained the thickness of a final septum'),
Kaiser (1997) (fig. 1: diagram of life cycle; tabs 1–3: occurrence of the teleomorph and both of its mating types in different countries), de Gruyter (2002)
(fig. 1: shape of conidia, reproduced in present Fig. 49C; cultural descriptions,
partly adopted below).

Description in vitro

Pycnidia globose to subglobose, with usually one non-papillate or papillate
ostiole, 50–160 μm diam., glabrous or with some short mycelial outgrowths,
citrine/honey, later olivaceous to olivaceous black. Conidial matrix buff.
        Conidia ellipsoidal to allantoid, with several small guttules, mainly asep-
tate, 7–15.5 × 3–5.5 μm, 1-septate conidia up to 18 × 5.5 μm.
        Colonies on OA relatively slow growing, 1.5–3 cm after 7 days (14 days:
4.5–5 cm), regular to slightly irregular, colourless to pale olivaceous grey, sometimes with a rosy buff tinge; aerial mycelium sparse to absent; reverse similar.
On MA 1.5–3 cm after 7 days (14 days: 3–4.5 cm), regular to irregular, greenish olivaceous/dull green and somewhat iron grey in centre, buff to rosy buff
near margin; aerial mycelium finely woolly to floccose, white to pale olivaceous
grey; reverse similar. Pycnidia in concentric zones or scattered, both on and in
the agar as well as in aerial mycelium, solitary or confluent. Representative culture CBS 581.83.

Description in vivo (Cicer arietinum)

Pycnidia (in concentric ringes on lesions on stems, leaves and pods, immersed
becoming erumpent) globose, (90–)140–160(–200) μm diam., with non-papil-

late ostioles. Conidia similar to those *in vitro*, usually with some small polar guttules, many aseptate, some 1-septate; usually 6–16 × 3–7 μm.

Pseudothecia (observed on overwintered chickpea debris, especially on pods, and on artificially inoculated stem pieces and leaves) globose or depressed globose, (110–)160–175(–250) μm diam. (height 75–150 μm), with inconspicuous papillate ostioles. Asci cylindrical-clavate, 20–70 × 9–13.5 μm, 8-spored, usually uniseriate, rarely biseriate. Ascospores ovoid, 1-septate, upper cells much larger than the lower cells, strongly constricted at the septum, 12.5–19 × 6.5–7.5 μm (for detailed description and illustrations see Kovatschevski, 1936).

Ecology and distribution

A noxious pathogen of chickpea, *Cicer arietinum* (Leguminosae). The disease, anthracnose, chickpea blight (or 'Ascochyta' blight), is the major disease in most chickpea-growing areas. Being seed-borne, the mycelium may be present in the seed coat and cotyledons, and conidia often contaminate the seed sur- face (Mathur, 1981). Despite conidia usually remaining unicellular *in vivo*, the anamorph has been confused with *Ascochyta pisi* Lib., type species of *Ascochyta* (conidia mainly septate *in vivo* and *in vitro*, 'wall-thickening septa- tion', Boerema, 1984a), see the discussion by Khune & Kapoor (1980). In phy- topathological literature the anamorph is commonly called *Ascochyta rabiei* and 'the Ascochyta pathogen of chickpea'. The fungus shows variation in viru- lence. It appears to be heterothallic, because compatible mating types are required for development of fertile pseudothecia. Both pycnidia and pseudothecia may develop on overwintered chickpea debris, but compared with the pycnidia very few pseudothecia develop. For detailed information on the life cycle of this pathogen see Kaiser (1997).

Phoma xanthina Sacc., Fig. 50

> *Phoma xanthina* Sacc. *in* Michelia **1**(4): 359. 1878.
>> ≡ *Macrophoma xanthina* (Sacc.) Berl. & Voglino *in* Atti Soc. veneto-trent. Sci. nat. **10** ['1886']: 181. 1887.
>> ≡ *Ascochytella xanthina* (Sacc.) Petr. & P. Syd. *in* Annls mycol. **22**: 347. 1924.

Diagn. data: de Gruyter (2002) (fig. 4: shape of conidia, reproduced in present Fig. 50; cultural descriptions, partly adopted below).

Description in vitro

Pycnidia globose to subglobose, with usually one non-papillate, often indistinct ostiole, 100–320 μm diam., glabrous or with mycelial outgrowths, citrine/honey, later olivaceous black. Exuding conidial masses not observed.

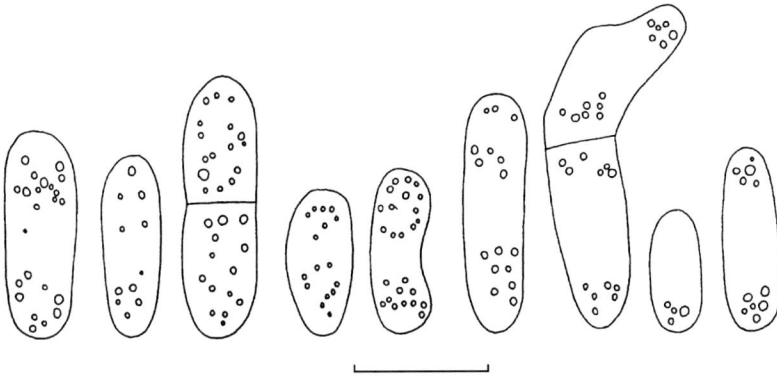

Fig. 50. Variation of conidia in *Phoma xanthina* (conidia *in vivo* usually remain aseptate). Bar 10 μm.

Conidia ellipsoidal to allantoid, sometimes curved, eguttulate or with several scattered guttules, hyaline to pale yellowish, mainly aseptate, (7.5–)8.5–17.5 × 3.5–5.5(–7) μm, occasionally occurring 1-septate conidia of similar size or larger up to 24 μm.

Colonies on OA 3.5 cm after 7 days (14 days: 5.5–6.5 cm), regular, greenish olivaceous/citrine green to grey olivaceous; aerial mycelium velvety to woolly, white to pale olivaceous grey/grey olivaceous; reverse similar. On MA 3.5–4 cm (14 days: 8–8.5 cm), regular, citrine green to herbage green/greenish olivaceous, olivaceous grey/iron grey in centre; aerial mycelium compact woolly, white to grey olivaceous; reverse leaden grey/leaden black, herbage green near margin. Pycnidia both on and in the agar as well as in aerial mycelium, mostly solitary. On application of a drop of NaOH on colonies on OA a bluish/green to red colour appears, a reddish brown colour on MA (not an E⁺ reaction). In the agar media small yellowish to brownish pigmented grains are produced. Representative culture CBS 383.68.

Ecology and distribution

So far only recorded in Europe it is apparently a specific pathogen of *Delphinium* spp. (Ranunculaceae) causing stem and leaf necroses. The latter often start as small leaf spots, which later coalesce. Conidia usually remain unicellular *in vivo*, common dimensions 9–17 × 5–7 μm. It should be noted that another species, *Phoma delphinii* (Rabenh.) Cooke (sect. *Heterospora*, **B**), also frequently occurs on *Delphinium* spp. in Europe.

Phoma zeae-maydis **Punith., Fig. 47A**

Teleomorph: ***Mycosphaerella zeae-maydis* Mukunya & Boothr.**

Phoma zeae-maydis Punith. *in* Mycopathologia **112**: 50. 1990.
≡ *Phyllosticta maydis* Arny & R.R. Nelson *in* Phytopathology
61: 1171. 1971; not *Phoma maydis* Fautrey *in* Revue

mycol. **16**: 161. 1894, not *Phoma maydis* Ellis & Everh. *in*
A. Rep. Del. Agric. Exp. Stn **6** ['1893']: 33. 1895 [nomen
nudum].

Diagn. data: Arny & Nelson (1971) (figs 1–3: pycnidia, conidia; *in vivo* and *in
vitro* description, data on pathogenicity), Mukunya & Boothroyd (1973) (fig.
1A–D: characteristics pseudothecia, asci, ascospores), van der Westhuizen
(1980) (figs 1–7: disease symptoms; pycnidia and conidia *in vivo* and *in vitro*),
Punithalingam (1990) (figs A–D, leaf spots, pycnidium, conidia, ascus and
ascospores; CMI description with notes), de Gruyter (2002) (figs 5, 11: shape
of conidia and chlamydospores, reproduced in present Fig. 47A; cultural
descriptions, partly adopted below).

Description in vitro

Pycnidia globose to subglobose, with usually one papillate ostiole,
120–160(–230) μm diam., glabrous, citrine/honey, later olivaceous to olivaceous black. Exuded conidial masses not observed.
     Conidia ellipsoidal with several small, scattered guttules, aseptate, although
some septa are formed upon germination, (12–)15–17(–25) × 3.5–5(–6.5) μm.
     Chlamydospores may be formed, intercalary or terminal, in short chains or
clustered, globose to subglobose, 8–20 μm diam.
     Colonies on OA 4 cm after 7 days, regular, colourless to grey olivaceous
with olivaceous grey to olivaceous black sectors; aerial mycelium felty to
finely floccose, white to pale olivaceous grey; reverse similar. On MA 3.5 cm
after 7 days, irregular, partly dull green to olivaceous, herbage green near
margin; aerial mycelium compact, finely floccose to woolly, white to olivaceous buff; reverse similar, leaden grey/olivaceous black in centre. Pycnidia
scattered both on and in the agar, solitary or confluent. With a drop of NaOH
a reddish/brown non-specific colour may develop. Representative culture
CBS 588.69.

Description in vivo (Zea mays)

Pycnidia (as tiny pinpoints in necrotic lesions on the leaves, chiefly epiphyllous)
subglobose to globose, 120–160 μm diam., with slightly papillate ostioles.
Conidia similar to those *in vitro*, usually conspicuously biguttulate, aseptate,
but germinating conidia often develop septa.
     Pseudothecia (observed naturally in spring on maize leaf debris, and
obtained artificially in sterilized leaf tissue) subglobose to globose
(86–)90–192(–200) μm diam., initially closed, later with papillate ostioles. Asci
cylindrical to subclavate, 40–65 × 9.5–12 μm, 8-spored, biseriate to irregularly
biseriate. Ascospores ellipsoidal, 1-septate, upper cells usually larger than the
lower cells, constricted at the septum, (14–)16–17(–19) × 4–5(–6) μm (for
detailed descriptions and illustrations see Mukunya & Boothroyd, 1973, and
Punithalingam, 1990).

Ecology and distribution

The main host of this fungus is *Zea mays*: yellow leaf blight of maize, but it is also recorded on *Sorghum* and *Setaria* spp. (all Gramineae). Its distribution includes North America (Canada, USA), South America (Bolivia, Ecuador) and southern Africa. The fungus is probably homothallic; pseudothecia are only known from maize leaves overwintered in the field. Ascospores may be the cause of early infection in the spring whilst conidia cause infections in the growing season. The conidia are usually aseptate *in vivo* and mostly 10–15 × 3–4 μm.

16 | *Phoma* sect. *Pilosa*

Phoma sect. *Pilosa* Boerema *et al.*

> *Phoma* sect. *Pilosa* Boerema, Gruyter & Noordel. apud Boerema *in* Mycotaxon **64**: 332. 1997.
>
> Type: *Phoma betae* A.B. Frank, anamorph of *Pleospora betae* (Berl.) Nevod.

The two species presently included in this section produce regular globoid pseudo-parenchymatous pycnidia with late development of an opening, a pore instead of a predetermined ostiole. Retarded development of the pycnidial cavity may occur ('pycnosclerotia'). *In vivo* the globose or depressed globose pycnidia appear glabrous with a flush central opening, but *in vitro* they are densely surrounded by hyphae emerging from them: 'hairy' or pilose appearance (Fig. 51A); the pore is usually only visible as a light spot in crushed hairy pycnidia.

The conidia remain aseptate *in vitro* and *in vivo*; only when germination is initiated they may become 1-septate.

Both species in this section possess a teleomorph referred to the genus *Pleospora* Rabenh. ex Ces. & De Not. The type species was carefully studied *in vitro* by Monte & Garcia-Acha (1988a, b) and Monte *et al.*(1989). Their electron microscope observations on the conidiogenesis (Monte & Garcia-Acha, 1988b) showed conidial ontogeny in agreement with other species of *Phoma*. Only the intermediate layer in the papilla preceding the initiation of the first conidium, could not be observed.

Diagn. lit.: Boerema & Dorenbosch (1973) (type species, characteristics *in vitro*), Boerema (1984b) (metagenetic relation with species of *Pleospora*), (1997) (formal introduction of this section), (2003) (precursory contribution to this chapter).

Synopsis of the Section Characteristics

- Pycnidia globose to subglobose, *in vitro* covered by hyphae emerging from them (pilose), pseudoparenchymatous, poroid; *in vivo* the pycnidia are glabrous, globose to depressed globose with a central opening.

- Conidia unicellular, both *in vitro* and *in vivo*.
- Chlamydospores, if present, simple; swollen cells commonly occur.
- Teleomorph: *Pleospora*.

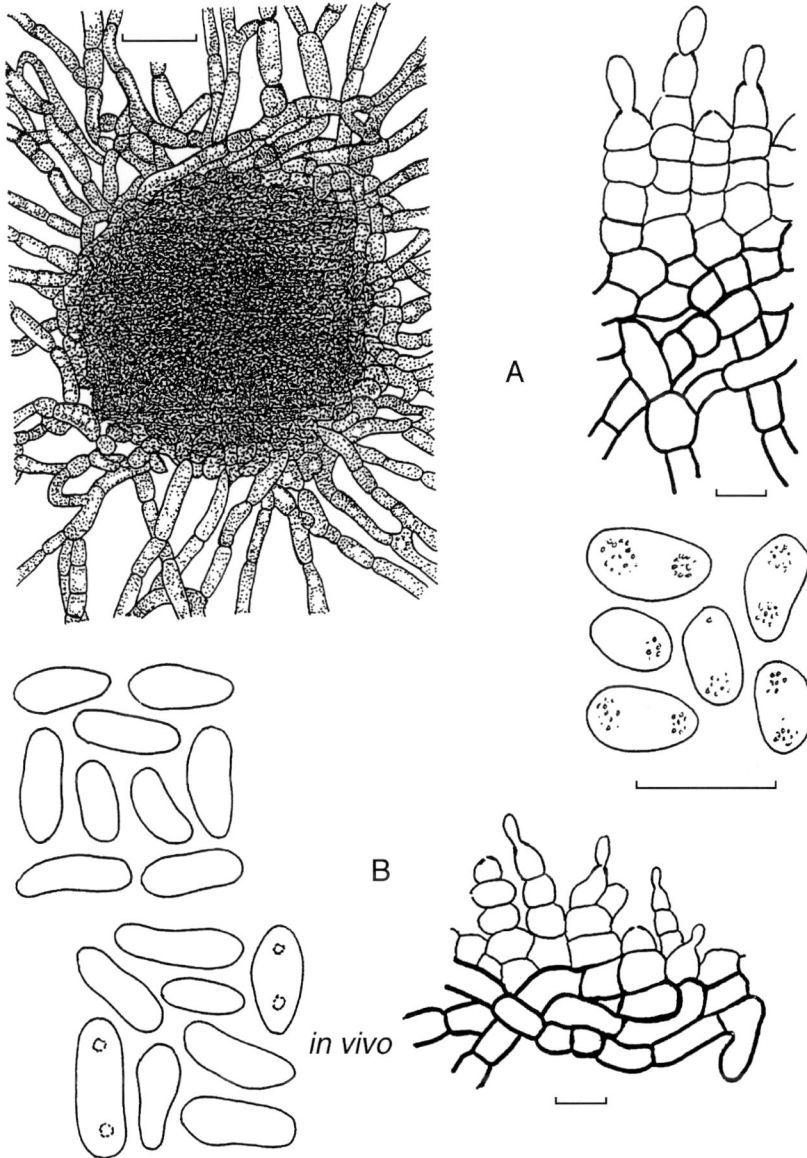

Fig. 51. A *Phoma betae*, type species of *Phoma* sect. *Pilosa*: superficial view of 'hairy' pycnidium *in vitro*, layer of conidiogenous cells and conidia *in vitro*. **B** *Phoma typhina*: conidia *in vitro* and *in vivo*, layer of conidiogenous cells *in vitro* of a thin-walled pycnidium. Bar pycnidium 50 μm, bars conidia and conidiogenous cells 10 μm.

Key to Both Species Treated of Section *Pilosa*

1. a. Conidia relatively broad, globose to subglobose/ellipsoidal, mostly 4–6.5 ×
        2.5–4 μm ....*Ph. betae*, teleom. *Pleosp. betae*
   b. Conidia smaller, ellipsoidal to subcylindrical or allantoid, mostly 4–6 ×
        2–2.5 μm ....*Ph. typhina*, teleom. *Pleosp. typhicola*

Distribution and Facultative Host Relations in Section *Pilosa*

Chenopodiaceae,
e.g. *Beta vulgaris*   *Ph. betae*
     Spinacia oleracea   (teleom. *Pleosp. betae*)
                                      [Worldwide recorded pathogen: especially well known from
                                      beet: black leg, damping-off in seedlings, root rot, storage
                                      rot and leaf spot; mainly seed-borne.]

Typhaceae,
e.g. *Typha angustifolia*   *Ph. typhina*
     Typha latifolia   (teleom. *Pleosp. typhicola*)
                                     [Commonly found in Europe in association with leaf spots
                                     and on dead leaves, leaf-sheaths and stems.]

Descriptions of the Taxa

Phoma betae A.B. Frank, Fig. 51A

Teleomorph: ***Pleospora betae*** **(Berl.) Nevod.**

> *Phoma betae* A.B. Frank in Z. Rübenzucker-Ind. **42**: 904, tab. 20. 1892
> [nom. cons. prop. (Shoemaker & Redhead, 1999); often erroneously listed
> as '(Oud.) Frank' or 'Rostrup'].
>> = *Phyllosticta betae* Oudem. *in* Ned. kruidk. Archf II, **2**(3):
>> 181. 1877.
>> H = *Phoma sphaerosperma* Rostr. *in* Tidsskr. Landfkon. **8**: 746.
>> 1889; not *Phoma sphaerosperma* P. Karst. *in* Hedwigia **24**:
>> 74. 1885; not *Phoma sphaerosperma* Fuckel *in* Fungi rhen.
>> Suppl. Fasc. 5, No. 1945. 1867.
>> = *Phyllosticta tabifica* Prill. *in* Bull. Soc. mycol. Fr. **7**: 19.
>> 1891 [*in* Sylloge Fung. **10**: 180. 1892 listed as '*Phoma tab-
>> ifica* Prill.', interpreted as an unintentional error of citation
>> (Boerema & Dorenbosch, 1973), but see Shoemaker &
>> Redhead, 1999]; not *Phoma tabifica* Kesteren *in*
>> Gewasbescherming **2**: 74. 1971 [= *Phoma telephii*
>> (Vestergr.) Kesteren, sect. *Phyllostictoides*, **E**].
>> = *Phyllosticta spinaciae* H. Zimm. *in* Verh. naturf. Ver. Brünn
>> **47** ['1908']: 87. 1909.

> = *Phoma spinaciae* Bubák & Willi Krieg. apud Bubák *in* Annls mycol. **10**: 47. 1912.
> = *Gloeosporium betae* Dearn. & E.T. Barthol. apud Dearness *in* Mycologia **9**: 356. 1917.

Diagn. data: Booth (1967) (figs A–F: drawings of ascus, ascospores, pycnidial wall with conidiogenous cells and conidia *in vivo* and *in vitro*; CMI description with phytopathological notes), Boerema & Dorenbosch (1973) (pl. 1f, g: cultures on OA; tab. 1: synoptic table on characters *in vitro* with drawing of conidia; nomenclature), Boerema (1984b) (fig. 11: diagrammatic comparison with related pathogens), Boerema *et al.* (1987) (nomenclatural and phytopathological notes), Monte & Garcia-Acha (1988a) (figs 1–18: micrographs of vegetative and reproductive structures *in vitro*; tab. 1: growth rate on different culture media, tabs 3–5: influence of illumination, carbon and nitrogen sources on colonies and/or pycnidial features, tabs 2, 6: pycnidial and conidial sizes in different isolates), Monte & Garcia-Acha (1988b) (figs 1–4: electron micrographs of conidiogenesis), Monte *et al.* (1989) (figs 1–16: microscopic observations on the development of pycnidia and 'holdfasts'), Rai & Rajak (1993) (cultural characteristics; NB: the study sub *Ph. betae* in Rai (1998) must be questioned as it includes one isolate of non-chenopodiaceous plant), Shoemaker & Bissett (1998) (figs 1–5: photographs *in vivo* of ascocarp, asci with ascospores, pycnidium wall with pore and conidiogenous cells; Fungi Canad. with e.g. descriptions of ascocarps and pycnidia *in vivo*, nomenclatural notes and literature references), Boerema (2003) (fig. 1A: drawings of pycnidium, conidiogenous cells and conidia *in vitro*, reproduced in present Fig. 51A; cultural descriptions, partly adopted below).

Description in vitro

Pycnidia globose to subglobose, 100–200(–350) μm diam., densely covered by mycelial 'hairs', with 1(–2) inconspicuous non-papillate pore(s), olivaceous to olivaceous black, solitary or confluent, scattered both on and in the agar. Outer wall formed of 2–3(–4) layers of dark polygonal cells, inner wall made up of 1–3(–4) layers of thin-walled, often radially arranged hyaline cells. With Lugol's iodine the entire outer cells of the pycnidial wall become red (different from the blotting paper effect of cell walls in sect. *Plenodomus*, **G**). Conidial matrix milky white, later rosy buff to ivory.

    Conidia globose to subglobose/ellipsoidal, biguttulate or with two polar concentrations of many small guttules, aseptate, (2.5–)4–6.5(–9.5) × (1.5–)2.5–4(–5.5) μm.

    Colonies on OA 5–7 cm after 7 days, regular to slightly irregular, greenish olivaceous/grey olivaceous to dull green, with finely floccose, pale olivaceous grey/mouse grey to grey olivaceous aerial mycelium. On MA very fine needle-like crystals are produced. Polymorphic swollen cells (pseudo-chlamydospores) commonly occur on both media. Most strains tested on OA and MA show at once an E^+ reaction on addition of a drop of NaOH (immediately greenish blue soon changing to red). Representative cultures CBS 523.66, CBS 109410.

[Monte & Garcia-Acha (1988a) obtained on other media: holdfasts (in water agar after 24 h incubation at room temperature), arthrospores (in static liquid cultures, after initial shaking for 24 h), chlamydospores (in poor culture media with restricted carbon and nitrogen sources) and pycnosclerotia (e.g. in media with methionine).]

Description in vivo *(especially on* Beta vulgaris*)*

Pycnidia (immersed in necrotic tissue of seedlings, leaves, stems and roots; occasionally also in seed clusters) subglobose, up to 250 μm diam., glabrous with a flush circular central pore. Conidia as *in vitro*, subglobose-ellipsoidal, arising from an irregular layer of doliiform conidiogenous cells. These cells sometimes develop alternately or even in a zigzag pattern.

Pseudothecia (occasionally found immersed in necrotic tissue of overwintered seed stalks) subglobose, becoming depressed cupulate, with a short papilla, 200–400 μm diam. Asci cylindrical-clavate, 70–80(–120 when free) × (14–)15–16(–17) μm, 8-spored, overlapping biseriate, separated by numerous paraphyses. Ascospores ellipsoidal to inequilaterally ovoid, 3-septate, with 1 longitudinal septum in the two central cells, constricted at the septa, (18–)20–22(–27) × (7–)8–9(–11) μm (for detailed descriptions and illustrations see Booth, 1967 and Shoemaker & Bissett, 1998).

Ecology and distribution

Worldwide on beet (*Beta vulgaris*) and spinach (*Spinacia oleracea*). Also recorded on *Beta vulgaris* var. *cicla*, *Beta macrorrhiza* and from various wild Chenopodiaceae: black leg, damping-off in seedlings, root rot, storage rot and leaf spot. Mainly seed-borne. The disease symptoms may be confused with those caused by *Ascochyta caulina* (P. Karst.) Aa & Kesteren, teleom. *Pleospora calvescens* (Fr.) Tul. & C. Tul., see Boerema *et al.* (1987). Both fungi may be related (Boerema, 1984b).

Phoma typhina (Sacc. & Malbr.) Aa, Fig. 51B

Teleomorph: *Pleospora typhicola* (Cooke) Sacc.

> *Phoma typhina* (Sacc. & Malbr.) Aa apud van der Aa & Vanev, Revision Phyllosticta CBS: 468. 2002.
>> ≡ *Phyllosticta typhina* Sacc. & Malbr. apud Saccardo *in* Michelia **2**(1): 88. 1880.
>> = *Phoma typharum* Sacc. *in* Sylloge Fung. **3**: 163. 1884.
>>> ≡ *Phyllosticta typharum* (Sacc.) Allesch. *in* Rabenh. Krypt.-Flora [ed. 2], Pilze **6** [Lief. 61]: 166. 1898 [vol. dated '1901'].
>> = *Phyllosticta renouana* Sacc. & Roum. *in* Revue mycol. **6**: 32. 1884.

= *Phoma typhae* Pass. apud Brunaud *in* Act. Soc. Linn. Bordeaux **40**: 20. 1886.
≡ *Phyllosticta typhae* (Pass.) Allesch. *in* Rabenh. Krypt.-Flora [ed. 2], Pilze **6** [Lief. 61]: 166. 1898 [vol. dated '1901'] [also erroneously as '(Pass.) Oudemans'].
= *Phoma typhicola* Oudem. *in* Versl. Gewone Vergad. Wisen natuurk. Afd. K. Akad. West. Amst. **9**: 298. 1900; Ned. Kruidk. Archf III, **2**(1): 246. 1900.
= *Phyllosticta coralliobola* Bubák & Kabát *in* Hedwigia 44: 350. 1905.

Diagn. data: Webster & Lucas (1959) (fig. 5A–G: drawings of ascocarps, asci, ascospores *in vivo*, pycnidia and conidia *in vivo* and *in vitro*; descriptions of both morphs *in vivo* and pycnidia in cultures from single ascospores), van der Aa & Vanev (2002) (documentation of holotype and other collections of anamorph; *in vivo* and *in vitro* descriptions; notes on misapplications and synonymy), Boerema (2003) (fig. 1B: drawings of conidiogenous cells and conidia *in vivo* and *in vitro*, reproduced in present Fig. 51B; cultural descriptions, partly adopted below).

Description in vitro

Pycnidia globose to subglobose, (80–)120–200(–240) μm diam., densely surrounded by short dark hyphae, closed or with one inconspicuous pore (usually only visible as a light spot in crushed pycnidia), honey/sienna, later olivaceous to olivaceous black; often containing only a compact mass of cells (pycnosclerotia). Walls very thin to rather thick in old cultures, composed of 1–8 outer layers of dark brown cells and 1–10 inner layers of smaller, often radially arranged cells. With Lugol's iodine the cell contents may become red. Conidial matrix white to salmon.

Conidia subglobose to ellipsoidal/allantoid, eguttulate or with some small guttules, always aseptate (3.5–)4.5–7(–9) × 1.5–2.5(–4) μm.

Colonies on OA rather fast growing, 7–8 cm after 7 days, regular, grey olivaceous/olivaceous to dull green, with floccose to woolly, white to pale olivaceous grey aerial mycelium; reverse olivaceous/olivaceous grey to greenish grey. Pycnidia scattered, both on and in the agar, solitary or confluent. Dark swollen cells may occur in the hyphae. Representative cultures CBS 132.69 and CBS 602.72.

In vivo (*esp.* Typha latifolia)

Pycnidia (subepidermal, then half-free on leaf spots and dead leaves, leaf-sheaths and stems) globose or depressed globose, variable in size, but commonly 150–200 μm diam., glabrous with a flat pore. Conidia as *in vitro*, but more variable in shape and size, arising from a layer of small hyaline cells lining the cavity.

Pseudothecia (also subepidermal and later almost superficial, on dead leaf-sheaths and stems; often occurring together with pycnidia) globose or irregularly globose, 280–480 μm diam. Asci cylindrical or broadly clavate, 200–240

× 44–48 μm, 8-spored, irregularly biseriate, separated by numerous paraphyses. Ascospores oblong, rounded at the ends, 3-septate with 1 longitudinal septum, running down the length of the spore in one plane, 46–56 × 18–24 μm,
strongly constricted at the septa; each cell of the spore is rounded and may separate slightly from adjacent cells. Mature spores are surrounded by a gelatinous
sheath (for detailed description and illustration see Webster & Lucas, 1959).

Ecology and distribution

In Europe frequently recorded in association with leaf spots and on dead leaves,
leaf-sheaths and stems of *Typha latifolia* and *Typha angustifolia* (Typhaceae).
Probably occurring everywhere the hosts are growing. The leaf spots are rather
indefinite with pale brown/white centre and broad brown/reddish border.

On account of the occurrence of thick-walled pycnidia and pycnosclerotia the
anamorph is also incorporated in the key to species of sect. *Sclerophomella* (**F**).

17 Miscellaneous

Some *Phoma*-like taxa known from cultural studies, but not yet classified

This is a group of anamorphs referred to as *Phoma*, but beyond the scope of the present classification of sections. Detailed comparative study of these anamorphs is still needed before their position can be decided. Most of these species could not be studied by ourselves in the usual way. Their cultural descriptions have been adopted from published data in other cultural studies, referred to under 'Diagn. data'.

The first species noted represents the *Phoma*-like anamorphs of various members of genera belonging to the Sporormiaceae, especially *Sporormia*, *Sporormiella* and *Preussia* spp. Those anamorphs all look very similar and their conidia may function as spermatia.

A further eleven *Phoma*-like anamorphs of *Leptosphaeria* spp., which do not fit into the section *Plenodomus* (**G**), have been placed in this miscellaneous group. Some of them show conidial dimorphism characteristic of sect. *Heterospora* (**B**). Five of these being validly described as *Phoma* species are treated in alphabetical order of their anamorphic names. The others, however, have not yet been provided with a name, and are therefore listed according to the teleomorphic names.

Finally an unnamed *Phoma*-like anamorph of a *Pleospora* sp., not agreeing with sect. *Pilosa* (**I**), is included.

As done elsewhere in this book, notes on observations *in vivo*, including short descriptions of the teleomorphs concerned, are added to the cultural characteristics of the validly described *Phoma* species of this Miscellaneous group. This is not done with the unnamed *Phoma*-like anamorphs. Entries via teleomorphs *in vivo* can be found in Sivanesan (1984) and other literature referred to under 'Diagn. data'.

Key to Listed Taxa Based on Characters *In Vitro*

1. a. Conidia small, not exceeding 4 μm in length, aseptate...................................**2**
   b. Conidia larger..**3**

2. a. Conidia cylindrical-ellipsoidal, 2–3 × 1–2 μm...............................*Ph. deserticola*
      ....(representing the anamorphs of Sporormiaceae)
      [Isolated from excrement of various herbivorous mammals in the deserts of
      Central Africa.]
   b. Conidia nearly spherical...**13**

3. a. Conidia always aseptate-phomoid..**4**
   b. Conidia sometimes also septate: phomoid and stagonosporoid........................**8**

4. a. Conidia 9–15 × 3–4 μm..............................*Ph. annulata*, teleom. *Lept. sacchari*
      [Common pathogen of sugarcane, *Saccharum officinarum* in Africa, Asia and
      South America: ring spot of leaves and leaf-sheaths. Pycnidia and pseudothecia
      often found together.]
   b. Conidia smaller...**5**

5. a. Conidia 4–5 × 1–2 μm...**6**
   b. Conidial length between 4 and 6 μm, width between 1.5 and 3.5 μm..............**7**

6. a. Colonies exudating a yellow brown pigment, discolouring the agar......................
      ....*Phoma*-anamorph of *Lept. biglobosa*
      [Common weakly virulent (or non-aggressive) pathogen of *Brassica* spp. in
      Australia, Europe, North America and Africa: blackleg.]
   b. Colonies exudating a purple-red pigment, discolouring the agar...........................
      ....*Ph. sanguinolenta*, teleom. *Lept. purpurea*
      (Conidia similar to those of *Ph. cruenta* in *vivo*, see note under *Ph. haematites*.)
      [Frequently found in reddened patches on dead stems of Compositae in Europe
      and North America (*Ph. cruenta* occurs in such patches on *Thalictrum flavum*,
      Ranunculaceae).]

7. a. Conidial length 5–6 μm..**9**
   b. Conidial length 4–6 μm..**12**

8. a. Phomoid conidial length not exceeding 6 μm..**10**
   b. Phomoid conidia also longer..**11**

9. a. Conidia 5–6 × 2.5–3 μm.....................*Ph. haematites*, teleom. *Lept. haematites*
      [Common in red coloured areas on stems of *Clematis vitalba* (Ranunculaceae) in
      Europe.]
   b. Conidia 5–6 × 1.5–2.5 μm..........................*Ph. rostruppii*, teleom. *Lept. libanotis*
      [Teleomorph frequently recorded on stems and roots of various Umbelliferae in
      Europe and North America. Anamorph reported in the past as serious pathogen
      of carrots, *Daucus carota*: phoma root rot, carrot phomosis, dieback of
      seedlings.]

10. a. Phomoid conidia 4–6 × 2–3 μm....................*Ph. meliloti*, teleom. *Lept. weimeri*
       Stagonosporoid conidia 8–20 × 2.5–3 μm, 1–3-septate....................................
       ....synanam. *Stagonospora meliloti*

[Recorded on various Leguminosae in Australia, Europe and North America.
Causal organism of leaf spot, stem canker and root rot of forage legumes.]
  b. Phomoid conidia 3.5–6 × 1 μm
      Stagonosporoid conidia 18–32 × 2–4 μm...........Anamorphs[28] of *Lept. pontiformis*
      [Teleomorph on *Agropyron* spp. and other grasses in Europe.]

11. a. Phomoid conidia 6–9 × 2.5–3 μm
        Stagonosporoid conidia 14–32 × 2–3 μm...........Anamorphs[28] of *Lept. dumetorum*
        [Teleomorph records refer to *Urtica dioica* and various other plants.]
    b. Phomoid conidia 4.5–13 × 2–5 μm (av. 6.7 × 3 μm)
        Stagonosporoid conidia 28–40.5 × 11.5–16 μm...Anamorphs[28] of *Lept. taiwanensis*
        [Teleomorph only recorded so far from Taiwan: leaf blight of sugarcane,
        Saccharum officinarum.]

12. a. Conidia 4–6 × 2–2.5 μm ....Anamorph[28] of *Lept. solani*
        [Teleomorph on *Solanum dulcamara* in Europe.]
    b. Conidia 4–6 × 3–3.5 μm ....Anamorph[28] of *Lept. spartinae*
        [Pycnidia have also been found *in vivo*, in association with the teleomorph, on
        dead culms of *Spartina townsendii* in England. Other European records refer to
        the teleomorph only.]

13. a. Conidia 3–4 × 2.5–3 μm ....Anamorph[28] of *Lept. typhicola*
        [Teleomorph recorded from various monocotyledonous marsh plants (esp.
        Cyperaceae, Gramineae and Typhaceae) in Europe. The conidia resemble those
        of *Phoma filarszkyana* from *Luzula spadicea*, Juncaceae.]
    b. Conidia about 2 μm diam. ....Anamorph[28] of *Pleospora rubicunda*
        [Pycnidia also occurred *in vivo* in association with pseudothecia on plant debris
        near the sea.]

Distribution and Facultative Host Relations

Plurivorous	Coprophilous in desert areas of Central Africa:
	Ph. deserticola
	[Representing the anamorphs of *Sordariaceae*; recorded on excrement of various herbivorous mammals, e.g. camels, goats and antelopes.]
	On monocotyledons in marshy areas of Europe:
	Phoma-anamorph of *Lept. typhicola*
	[? *Ph. filarszkyana*]
	[Teleomorph reported from Cyperaceae, Gramineae, Juncaceae and Typhaceae.]

With special host relations:

| Compositae | *Ph. sanguinolenta* |
| | (teleom. *Lept. purpurea*) |

---

[28] Without anamorphic names, listed under teleomorph name (anamorphs often not yet discovered in nature).

[In Europe and North America, in reddened patches on
dead stems.]

Cruciferae   *Phoma*-anamorph of
Brassica spp.   *Lept. biglobosa*
                                    [In Australia, Europe, North America and Africa known as
                                    the weakly virulent (or non-aggressive) pathogen causing
                                    blackleg of brassicas.]

Gramineae
Saccharum officinarum   *Ph. annulata*
                                       (teleom. *Lept. sacchari*)
                                    [In Africa, Asia and South America, frequently as causal
                                    organism of ring spot of leaves and leaf-sheaths.]

Spartina spp.   *Phoma*-anamorph of
esp. *S. townsendii*   *Lept. spartinae*
                                    [In the UK the *Phoma*-anamorph has been found in associ-
                                    ation with teleomorph on dead culms of *S. townsendii*. In
                                    other European countries only the teleomorph is recorded.]

Papillionaceae,   *Ph. meliloti*
e.g. species of *Medicago*,   (teleom. *Lept. weimeri*)
Melilotus, Trifolium and *Ulex*   (synanam. *Stagonospora meliloti*)
                                    [In Australia, Europe and North America, as causal organism
                                    of leaf spot, stem canker and root rot of forage legumes.]

Ranunculaceae
Clematis vitalba   *Ph. haematites*
                                       (teleom. *Lept. haematites*)
                                    [In Europe, frequently in association with red discoloured
                                    stem parts.]
[*Thalictrum flavum*   *Ph. cruenta*]

Umbelliferae,   *Ph. rostruppii*
e.g. *Daucus carota*   (teleom. *Lept. libanotis*)
                                    [In Europe and North America, the teleomorph on stems
                                    and roots of various umbellifers. In the past the anamorph
                                    was known as a serious pathogen of carrots: phoma root
                                    rot, carrot phomosis and dieback of seedlings.]

Phoma-like Anamorphs of *Sporormiaceae*

Representative species:

Phoma deserticola Faurel & Schotter

Phoma deserticola Faurel & Schotter *in* Revue Mycol. **30**: 153. 1965.

Diagn. data: Faurel & Schotter (1965a, b) (fig. 5a, b: shape of pycnidium with conidia; original description from excrement of goats: pycnidia brown to black, spherical, 70–80 μm diam., with 1 conspicuous, wide ostiole, 20–25 μm diam., glabrous, solitary or aggregated, thin-walled with undifferentiated conidiogenous cells ('not seen'), conidia 3 × 1.5 μm; recorded in the deserts of Central Africa, common on excrement of herbivorous mammals, including goats, rock goats, camels and antelopes. NB: Ascomata of seven different species of *Sporormia* were also found on such excrement).

General characteristics of similar anamorphs in vitro

Pycnidia resembling *Ph. deserticola* are frequently observed in isolates of different members of the Sporormiaceae, especially species of *Sporormia*, *Sporormiella* and *Preussia*; they are small, globose, mostly 70–75 μm diam., with relatively wide ostioles, 20–30 μm diam., glabrous, brown to black and thin-walled. Conidial matrix whitish.

Conidia small, cylindrical to ellipsoidal or spherical, often with 1 small guttule, mostly 2–3 × 1–2 μm, usually arising from less differentiated cells, but sometimes also from elongated skittle-shaped conidiogenous cells.

Colonies different, but usually flat, on OA faster growing than on MA, regular with less aerial mycelium; reverse on OA often shows a pink-yellow-red discoloration, which on addition of a drop of NaOH becomes pale green. Pycnidia often developing in abundance on and in the agar. As they age, ascomata of the teleomorph may be produced (sometimes only after some months). Example of culture: CBS 287.67 sub *Preussia* (*Sporormia*) *aemulans* (Rehm) Arx, see von Arx & Storm (1967).

Differentiation to species level using only anamorphic and cultural characteristics, is so far regarded as impossible.

Phoma-like Anamorphs of *Leptosphaeria* sensu lato, Not Fitting sect. *Plenodomus* (G) (Compare Boerema *et al.*, 1996)

Representative species:

Phoma annulata N. Pons

Teleomorph: ***Leptosphaeria sacchari*** **Breda de Haan** (*Phaeosphaeria* cf. Shoemaker & Babcock, 1989).

Phoma annulata N. Pons in Fitopatol. Venez. **3**(2): 38–39. 1990.
≡ *Phyllosticta saccharicola* Henn. in Ann. Mus. Congo Bot. **5**(2): 100. 1907/1908; not *Phoma saccharicola* S. Ahmad in Biologia, Lahore **6**(2): 131. 1960 [= *Phoma sorghina* (Sacc.) Boerema *et al.*, sect. *Peyronellaea*, **D**].

Diagn. data: Hudson (1960) (anamorph-teleomorph connection; anamorph description *in vivo* and *in vitro* and teleomorph description *in vivo*; fig. 2A, B asci and ascospores, fig. 2C, D, E conidia *in vitro* and *in vivo*, fig. 2J, K shape of pycnidia *in vitro* and *in vivo*), Pons (1990) (fig. 3: type of anamorph; figs 4, 5: pycnidia, conidia and conidiogenous cells), Sivanesan (1984) (teleomorph description, adopted below; fig. 271: ascus and ascospores, conidia and coni-diogenous cells).

Characteristics in vitro

Pycnidia occur as single or multilocular structures 180–350 μm diam., i.e. larger than on the natural host. Conidial matrix pale brown.

Conidia ellipsoidal-subfusoid, with 2 or many small guttules, aseptate, 9–14(–15) × (2–)3–4(–5) μm.

Colonies on OA slow growing, whitish but soon turning pink to umber with sparse aerial mycelium. Pycnidia may be produced all over the surface [pseudothecia did not develop in culture].

In vivo *(Saccharum officinarum)*

Pycnidia (subepidermal within ring spots on leaves, often together with pseudothecia, usually in rows) 65–135 μm diam., subglobose, papillate. Conidia as *in vitro*. Representative specimens VIA 5645, VIA 4866.

Pseudothecia (also subepidermal in the ring spots) mostly 110–140 μm diam., globose-papillate. Wall pseudoparenchymatous. Asci mostly 60–75 × 9–13 μm. Ascospores 18–25 × 3.5–6 μm, ellipsoidal, 3-septate, second cell from the apex enlarged, each cell with a single guttule (for detailed description see Hudson, 1960).

Ecology and distribution

In Africa, Asia and South America, a well-known pathogen of sugarcane, *Saccharum officinarum* (Gramineae): ring spot of leaves and leaf-sheaths. Usually of little economic importance.

Phoma haematites (Petr.) Boerema *et al.*

Teleomorph: *Leptosphaeria haematites* (Roberge ex Desm.) Niessl

> *Phoma haematites* (Petr.) Boerema, Loer. & Hamers in Persoonia **16**(2): 164. 1996.
>
> > ≡ *Plenodomus haematites* Petr. in Sydowia **1**: 135–136. 1947.

Diagn. data: Lucas & Webster (1967) (relation anamorph-teleomorph; charac-teristics in culture, adopted below; fig. 7C, D: pycnidium and conidia *in vitro*), Sivanesan (1984) (description teleomorph, adopted below; fig. 267: ascus and ascospores, conidia; description and illustration teleomorph), Shoemaker (1984) (teleomorph), Boerema *et al.* (1996) (type anamorph).

Characteristics in vitro

Pycnidia (sub)globose, up to 150 µm diam., slightly papillate. Conidiogenous cells globose to bottle shaped.

Conidia oblong to ellipsoidal, with an inconspicuous guttule at each end, mostly 5–6 × 2.5–3 µm.

Colonies on OA relatively slow growing, producing pinkish aerial mycelium, reverse reddish. Pycnidia were found in 2-month-old cultures [pseudothecia did not occur]. Representative dried cultures SHD (Sheffield) 2353, SHD 2367.

In vivo *(Clematis vitalba)*

Pycnidia (in red coloured areas on the stems, sometimes together with pseudothecia) similar to those *in vitro* but usually larger, 150–250 µm diam., often with some red pigmentation around the ostioles. Conidial dimensions (5–)5.5–7(–7.5) × 2–2.5(–3) µm.

Pseudothecia (densely gregarious in large red areas on the stems; occasionally together with pycnidia) up to 420 µm diam. with a cylindrical red-tipped neck. Asci 70–125 × 8–10 µm. Ascospores ellipsoidal, 3-septate, constricted at the middle, yellowish, second cell from apex swollen, 18–27 × 4–7 µm (for detailed descriptions see, e.g. Lucas & Webster, 1967, and Shoemaker, 1984).

Ecology and distribution

In Europe common in red coloured areas on stems of *Clematis vitalba* (Ranunculaceae).

Note

Ph. haematites is obviously close to *Phoma cruenta* (Syd. & P. Syd.) Boerema *et al.* (Boerema *et al.*, 1996: 160–161), found in reddened patches on stems of *Thalictrum flavum* (Ranunculaceae) in Europe. *Ph. cruenta* produces smaller conidia, (3–)3.5–4.5(–5) × 1–2 µm, arising from globose conidiogenous cells with conspicuous collarettes. It has been recorded in association with immature pseudothecia of *Leptosphaeria cruenta* Sacc., but its single identity has not been checked in pure culture.

Phoma meliloti **Allesch.**

Teleomorph: ***Leptosphaeria weimeri* Shoemaker *et al.*** (formerly confused with *L. pratensis* Sacc. & Briard and *L. viridella* (Peck) Sacc.)
Synanamorph: ***Stagonospora meliloti* (Lasch) Petr.**

Phoma meliloti Allesch. *in* Ber. bot. Ver. Landshut **12**: 19. 1892 [type in herb. Allescher showed that conidia were considerably broader than stated in description].

Diagn. data: Jones & Weimer (1938) (anamorph-teleomorph connection), Lucas & Webster (1967) (anamorph-teleomorph connection; cultural characteristics, adopted below; fig. 14C: conidia *in vitro*), Boerema *et al.* (1994) (*in vivo* and *in vitro* description; fig. 10B: conidial dimorphism), Shoemaker *et al.* (1991) (detailed teleomorph description).

Characteristics in vitro

Pycnidia usually coalescing, subglobose to irregular, 150–200 μm diam., thin-walled. They may contain conidia of *Phoma*-type or conidia of 'Stagonospora-type' and incidentally conidia of both types may be produced in the same pycnidium, though this has never been observed *in vivo*.

Conidia of the *Phoma*-type, oblong-ellipsoidal without guttules, 4–6 × 2–3 μm; *Stagonospora*-like macroconidia 8–20 × 2.5–3 μm, truncate at the base, tapering to the rounded apex, 1–3-septate, eguttulate.

Colonies on OA fast growing, producing a dark green mycelium with patches of white and orange aerial hyphae; reverse dark green with orange margin. Pycnidia were found within 3 months [pseudothecia did not develop]. Representative dried culture SHD 2050 (pycnidia on OA).

In vivo *(especially* Melilotus alba*)*

Pycnidia (immersed in lesions on stems and leaves, usually solitary) mostly 150–200 μm diam., subglobose with a distinct neck of variable length, usually narrower at the base than at the apex, thin-walled. They may contain conidia of the *Phoma*-type or conidia of the *Stagonospora*-type, similar to those *in vitro*.

Pseudothecia (subepidermal on last year's dead stems) mostly 250–350 μm diam., depressed globose with flattened base and conical neck composed of scleroplectenchyma cells. Asci mostly 70–80(–100) × 10–11 μm, 8-spored, overlapping biseriate. Ascospores 20–24 × 5–6 μm, fusoid, 3-septate, second cell from the apex enlarged, with or without small guttules (for detailed description see Shoemaker *et al.*, 1991).

Ecology and distribution

In Australia, Europe and North America, on various Leguminosae, e.g. *Medicago*, *Melilotus*, *Trifolium* and *Ulex* spp. Causes leaf spot, stem canker and root rot of forage legumes. Low temperatures (± 8°C) induce the development of the *Phoma*-anamorph; higher temperatures the *Stagonospora*-anamorph.

Phoma rostrupii Sacc.

Teleomorph: **Leptosphaeria libanotis (Fuckel) Niessl** [also quoted as 'libanotidis' and as '(Fuckel) Sacc.'].

Phoma rostrupii Sacc. in Sylloge Fung. **11**: 490. 1895.
≡ *Phoma sanguinolenta* Rostr. in Tidsskr. Landøkon. **5**(7): 384. 1888; not *Phoma sanguinolenta* Grove in J. Bot., Lond. **23**: 164. 1885 [see below].

Diagn. data: Lucas & Webster (1967) (anamorph-teleomorph connection; cultural characteristics adopted below; fig. 9C, D: pycnidium and conidia *in vitro*), Sivanesan (1984) (fig. 268: ascus and ascospores, conidia; teleomorph description), Boerema *et al.* (1994) (*in vivo* and *in vitro* description, adopted below; fig. 10C: conidial shape).

Characteristics in vitro

Pycnidia subglobose or depressed globose, 240–320 μm diam., papillate, thin-walled. Conidial matrix bright red.

Conidia oblong ellipsoidal or dumb-bell-shaped, biguttulate, 5–6.5 × 1.5–2.5 μm.

Colonies on OA fast growing; forming an olive-green mycelial mat with copious yellowish cream surface hyphae; reverse dark yellow to brown. In 4-week-old cultures numerous pycnidia were found in all isolates [pseudothecia did not develop in culture]. Representative dried cultures SHD 2009, SHD 2163.

In vivo *(especially on* Daucus carota*)*

Pycnidia (on stems and inflorescences, lesions on roots; usually densely grouped) subglobose apparently always thin-walled, exudate bright red. Conidial dimensions (4–)5–6(–6.5) × 1.5–2.5 μm.

Pseudothecia (on dead stems) mostly 360–415 μm diam., globose or conical globose with flattened base. Wall scleroplectenchymatous. Asci 100–125 × 8–10(–11) μm, 8-spored, uniseriate. Ascospores mostly 18–20 × 5.5–7 μm, ellipsoidal, 3-septate, second cell from the apex enlarged, eguttulate (for detailed description see Holm, 1957).

Ecology and distribution

In Europe the teleomorph has been recorded on stems and roots of various Umbelliferae, especially *Angelica sylvestris*. In northern Europe *Ph. rostrupii* was known in the past as a serious pathogen of carrots, *Daucus carota*: *Phoma* root rot, carrot phomosis, dieback of seedlings.

Note

The teleomorph, *L. libanotis*, is a characteristic representative of the scleroplectenchymatous 'Group *doliolum*' of the genus *Leptosphaeria*. However, so far only thin-walled pseudoparenchymatous pycnidia of *Ph. rostrupii* are known. It should also be noted that recent East European records of *Ph. rostrupii* on carrots refer to two entirely different species of *Phoma*, namely *Ph. exigua* Desm. var. *exigua* [sect. *Phyllostictoides*, **E**] and *Ph. complanata* (Tode : Fr.) Desm. [sect. *Sclerophomella*, **F**].

Phoma sanguinolenta Grove

Teleomorph: ***Leptosphaeria purpurea* Rehm**

> *Phoma sanguinolenta* Grove *in* J. Bot., Lond. **23**: 162. 1885; not *Phoma sanguinolenta* Rostr. *in* Tidsskr. Landøkon. **5**(7): 384. 1888 [≡ *Phoma rostrupii* Sacc., *see above*].
>> = *Phoma rubella* Grove *in* J. Bot., Lond. **23**: 162. 1885 [June]; not *Phoma rubella* Cooke *in* Grevillea **14**: 3. 1885 [Sept.; ≡ *Phoma porphyrogena* Cooke].
>> ≡ *Phoma grovei* Berl. & Voglino *in* Sylloge Fung. **10**: 168. 1892 [superfluous new name].

Diagn. data: Lucas & Webster (1967) (anamorph-teleomorph connection; characteristics in culture, partly adopted below; fig. 15D–G: conidia in culture and on host), Sivanesan (1984) (teleomorph description; fig. 270: ascus and ascospores, conidiogenous cells and conidia), Shoemaker (1984) (teleomorph description adopted below), Boerema *et al.* (1994) (*in vitro* and *in vivo* description adopted below; fig. 10D: conidial shape).

Characteristics in vitro

Pycnidia depressed globose, 280–400 μm diam., thin-walled.
Conidia oblong-ellipsoidal, biguttulate, (3.5–)4–5.5 × 1.5–2 μm.
Colonies on OA fast growing; producing a pale olive-green mycelial mat with patches of whitish aerial hyphae, reverse purplish. Pycnidia were discovered within 6 weeks. Representative dried cultures SHD 1928, SHD 1929.

In vivo *(especially* Cirsium *spp.)*

Pycnidia (at the base of rotting stems, loosely gregarious) resembling those *in vitro*, often occurring in association with pseudothecia, surrounded by purplish hyphae. Conidial dimensions 4–5.5 × 1.5–2.5 μm.
Pseudothecia (subepidermal on dead stems) mostly 250–350 μm diam., globose to depressed-globose with flattened base and truncate-conical neck, composed of red polygonal cells. Wall of ascocarp scleroplectenchymatous, Asci 55–95 × 8–11 μm, 8-spored, overlapping biseriate to tetraseriate. Ascospores (22–)27–31(–35) × 4.5–5.5 μm, narrowly fusiform, 3-septate, central cells slightly shorter than end cells, distinctly guttulate (for recent detailed description see Shoemaker, 1984).

Ecology and distribution

In Europe and North America (Canada) frequently found in reddened or purple coloured patches on dead stems of various Compositae, esp. *Cirsium* spp.; occasionally also recorded on plants of other families.

Unnamed *Phoma*-anamorphs of *Leptosphaeria* sensu lato and *Pleospora*

Anamorph of *Leptosphaeria biglobosa* Shoemaker & H. Brun, Colour Plate I-L

Diagn. data: McGee & Petrie (1978) (fig. 1A: culture on V-8 juice agar as 'avirulent isolate of *Lept. maculans*'; 'in Czapek' broth medium a very distinct brownish-yellow pigment developed'), Boerema (1981) (fig. 7B: pycnidia on *Brassica* seed, broadly papillate or with a neck, conspicuous inflated at the upper part, as 'avirulent strain of *Ph. lingam*'), Shoemaker & Brun (2001) (figs 1–7: photographs of pseudothecia and asci with ascospores, swollen germinated conidium; detailed description of pseudothecium and pycnidium *in vivo*; note on bizarre pycnidium formation at the end of cylindrical beak of pseudothecium).

Characteristics in vitro

Pycnidia thin-walled, pseudoparenchymatous, often in sectors with different dimensions, on and in the agar: relatively large, globose-papillate up to 330–400 μm diam., or smaller, 150–250 μm diam. and then often with a cylindrical neck 150–200 μm long, usually conspicuous inflated at the upper part, black or greyish brown. Conidial matrix reddish brown.

Conidia subcylindrical, straight, biguttulate, mostly 4–5 × 1.5–2 μm, hyaline.

Colonies on OA fast growing, c. 5(–7) cm diam. after 7 days, characterized by a yellow-brown diffusible pigment with intensity of colour varying from pale straw to cinnamon; horizontal growth regular with a little white or greyish aerial mycelium. Representative cultures DAOM 229269, DAOM 229270 (Ottawa, Canada).

Records refer to *Brassica* spp. This fungus was formerly interpreted as a weakly aggressive (avirulent) strain of *Lept. maculans* (Desm.) Ces. & De Not., anam. *Phoma lingam* (Tode : Fr.) Desm., sect. *Plenodomus* (**G**). So far only pseudoparenchymatous pycnidia of *Lept. biglobosa* are known.

Anamorph of *Leptosphaeria dumetorum* Niessl

Diagn. data: Lucas & Webster (1967) (pycnidial development in cultures derived from single ascospores; fig. 6C: conidial dimorphism: conidia of *Phoma*-type and *Stagonospora*-type from a single ascospore culture (occasionally in the same pycnidium; description and illustration of teleomorph *in vivo*)), Sivanesan (1984) (fig. 279: ascus and ascospores, conidia of both phenotypes; teleomorph description and illustration), Shoemaker (1984) (teleomorph).

Characteristics in vitro

Pycnidia globose, 240–400 μm diam., thick-walled; the smaller ones may contain conidia of *Phoma*-type (Sivanesan, 1984), but often only conidia of *Stagonospora*-type are found (Lucas & Webster, 1967). Conidial matrix pinkish.

Conidia of the *Phoma*-type oblong-ellipsoidal, biguttulate 6–9 × 2.5–3(–4) μm; *Stagonospora*-like macroconidia 1–3-septate, 14–32 × 2–3 μm, obclavate, pale brown (*Hendersonia* sp.).

Colonies on OA consisted of white submerged mycelium with yellowish aerial mycelium. The medium was yellow to emerald green with pale orange tinges at the border. Pycnidia were found in 10-week-old cultures [pseudothecia did not develop in culture]. Representative dried culture SHD 2043 (pycnidia on OA).

Records of the teleomorph refer to *Urtica dioica* and various other plants, i.e. a plurivorous fungus.

Anamorph of *Leptosphaeria pontiformis* (Fuckel) Sacc. (*Phaeosphaeria* cf. Leuchtmann, 1984)

Diagn. data: Webster & Hudson (1957), Sivanesan (1984).

Description in vitro

Pycnidia up to 800 μm broad often multilocular, containing *Phoma*-type conidia; smaller pycnidia, up to 300 μm contain conidia of '*Stagonospora*-type'.

Conidia 3.5–6 × 1 μm from 'tapering phialides'; *Stagonospora*-like macroconidia 18–32 × 2–4 μm, obclavate, straight or slightly curved (*Hendersonia* sp.).

Colonies on OA produced both *Phoma*- and *Stagonospora*-type anamorphs.

Ecology and distribution

Recorded on *Agropyrum* spp. and other grasses (Gramineae) in Europe.

Anamorph of *Leptosphaeria solani* Romell ex Berl.

Diagn. data: Lucas & Webster (1967) (pycnidial development in cultures derived from single ascospores; fig. 17C, D: pycnidium and conidia *in vitro*), Sivanesan (1984) (fig. 272: ascus and ascospores, conidia and conidiogenous cells; teleomorph description and illustration).

Characteristics in vitro

Pycnidia subglobose or irregular in shape, relatively large, up to 500 μm diam., wall relatively thick, 25–40 μm diam.

Conidia oblong, ellipsoidal or ovoid, 4–6 × 2–2.5 μm.

Colonies on OA rapidly growing; producing greyish submerged mycelium with pinkish aerial hyphae; reverse salmon coloured at margin of colony. Numerous pycnidia were found within a month [pseudothecia have also been discovered in 10-week-old cultures]. Representative dried culture SHD 2376 (pycnidia on OA).

Ecology and distribution

Records of the teleomorph indicate that this fungus is common on *Solanum dulcamara* (Solanaceae) in various European countries.

Anamorph of *Leptosphaeria spartinae* Ellis & Everh. (*Phaeosphaeria* cf. Shoemaker & Babcock, 1989)

Diagn. data: Lucas & Webster (1967) (pycnidial development in cultures derived from single ascospores; fig. 18 C, D, E: pycnidial shape, conidia *in vivo* and *in vitro*; description and illustration of teleomorph *in vivo*), Sivanesan (1984) (fig. 273: ascus and ascospores, conidia and conidiogenous cells; description of teleomorph).

Characteristics in vitro

Pycnidia formed in groups, globose-papillate, 120–360 μm diam., with walls up to 30 μm diam., surrounded by pale brown hyphae.

Conidia ellipsoidal to globose or ovoid, 4–6 × 3–3.5(–4) μm; produced on prismatic conidiogenous cells.

Colonies on OA grew fairly rapidly, producing an olive-green mycelium with brownish purple aerial hyphae; tinting the medium yellowish-green (black at the surface). Pycnidia were found within 7-week-old cultures [pseudothecia did not develop in culture]. Representative dried cultures SHD 2015, SHD 2016.

Ecology and distribution

Anamorph and teleomorph were found on dead culms of *Spartina townsendii* (Gramineae) in the UK. The teleomorph is also recorded in other European countries and in North America. Pycnidia similar both *in vivo* and *in vitro*, globose, up to 240 μm diam.

Anamorph of *Leptosphaeria taiwanensis* sensu Hsieh (see Note)

Diagn. data: Hsieh (1979) (pycnidial development in cultures from ascospores; conidial dimorphism in the same pycnidium: *Phoma*-type and *Stagonospora*-type; figs 1, 15: cultures on PDA; figs 5–13: pycnidium with both conidial dimorphs; illustration of teleomorph on sugarcane-leaf-decoction agar).

Description in vitro

Pycnidia globose to subglobose, 120–190 μm diam., containing *Phoma*-type as well as *Stagonospora*-type conidia.

Conidia ovoid to rod-shaped, straight or slightly curved, often with one or two guttules, 4.5–13 × 2–5 μm, average 6.7 × 3 μm; *Stagonospora*-like conidia 28–40.5 × 11.5–16 μm, 1–3-septate, straight or slightly curved, constricted at the septa, rounded or truncated at the apical end and slender towards the basal end.

Colonies growing well on different agar media, producing a loose mycelial mat. Production of pycnidia within 5 days at 26°C [pseudothecia with mature ascospores were produced together with pycnidia on potato dextrose agar and sugarcane-leaf-decoction agar]. Representative culture ATCC (American Type Culture Collection, Rockville, USA) 38204.

Ecology and distribution

The anamorphs are so far only known *in vitro*. The fungus proved to be the causal organism of an important leaf blight of sugarcane, *Saccharum officinarum* (Gramineae) in Taiwan.

Note

The name *Leptosphaeria taiwanensis* W.Y. Yen & C.C. Chi used by Hsieh (1979) for the teleomorph must be questioned, as the type of the latter concerns a species of *Didymella*, see Shoemaker & Babcock (1989).

Anamorph of *Leptosphaeria typhicola* P. Karst. (see Note)

Diagn. data: Lucas & Webster (1967) (pycnidial development in cultures derived from single ascospores of the teleomorph on culms of *Phragmites communis*; fig. 20C: conidia *in vitro*; description and illustration of teleomorph *in vivo*), Sivanesan (1984) (fig. 274: ascus and ascospores, conidia and conidiogenous cells; description of teleomorph).

[Pycnidial anamorph may be identical with *Phoma filarszkyana* (Moesz) Boerema *et al. in* Persoonia **16**(2): 162. 1996.]

Description in vitro

Pycnidia globose, 320–400 μm diam., with thick walls, 32–48 μm diam., surrounded by dark brown hyphae.

Conidia spherical ovoid-ellipsoidal, 3–4 × 2.5–3 μm.

Colonies slowly growing, forming a smoky grey mycelium and a purple colouration at the margins. Pycnidia were discovered in 12-week-old cultures [pseudothecia did not develop]. Representative dried cultures SHD 2102, SHD 2251.

[On dried stems of *Luzula spadicea* (Juncaceae) *Phoma filarszkyana* produced subglobose-ellipsoidal pycnidia with flattened base and a gradually developing ostiolate neck, variable in size, 100–900 μm diam., wall 40–50 μm, neck mostly 130–150 μm long. Conidia nearly globular, 2.5–3 μm diam.]

Ecology and distribution

The teleomorph is recorded from various monocotyledons, esp. marsh plants belonging to the Cyperaceae, Gramineae, Juncaceae and Typhaceae. The records are from Europe, and South and North America.

Note

This Ascomycete is plurivorous and very variable; synonyms *Massariosphaeria typhicola* (P. Karst.) Leuchtm. and *Chaetomastia typhicola* (P. Karst.) M.E. Barr. However, some of the records probably refer to another species, because isolates made by Leuchtmann (1984) developed a different anamorph with much larger conidia than listed here, cf. studies by Lucas & Webster (1967) and Sivanesan (1984).

Anamorph of *Pleospora rubicunda* Niessl

Diagn. data: Webster (1957) (pycnidial development in cultures derived from single ascospores; fig. 2B, D, E, F: pycnidia *in vivo* and *in vitro*, conidia and conidiogenous cells; description and illustration of teleomorph *in vivo*).

Characteristics in vitro

Pycnidia subglobose, up to 500 μm diam., thin-walled: 10–15 μm diam. Conidial matrix flesh-coloured.

Conidia spherical to broadly ellipsoidal, about 2 μm diam.; produced on skittle-shaped conidiogenous cells.

Colonies on OA (and PDA) slow growing (5 cm diam. in 2 months) with a characteristic purple colour in the medium and low pale aerial mycelium. Pycnidia were found in 2-month-old cultures [pseudothecia have been found only once in a multi-ascospore culture, they were in poor condition and contained collapsed and distorted ascospores]. Dried culture SHD 336.

Ecology and distribution

Teleomorph in Europe on various kinds of plant debris near the sea. Pycnidia may occur in association with pseudothecia; they are smaller than those *in vitro*, globose-papillate up to 200 μm diam.

References

Aa, H.A. van der (1973) Studies in *Phyllosticta* I. *Studies in Mycology* 5.

Aa, H.A. van der and Kesteren, H.A. van (1971) The identity of *Phyllosticta destructiva* Desm. and similar *Phoma*-like fungi described from Malvaceae and *Lycium halimifolium*. *Acta botanica neerlandica* 20, 552–563.

Aa, H.A. van der and Kesteren, H.A. van (1979) Some pycnidial fungi occurring on *Atriplex* and *Chenopodium*. *Persoonia* 10(2), 267–276.

Aa, H.A. van der and Kesteren, H.A. van (1980) *Phoma heteromorphospora* nom. nov. *Persoonia* 10(4), 542.

Aa, H.A. van der and Vanev, S. (2002) *A Revision of the Species Described in Phyllosticta*. Centraalbureau voor Schimmelcultures, Utrecht.

Aa, H.A. van der, Noordeloos, M.E. and Gruyter, J. de (1990) Species concepts in some larger genera of the Coelomycetes. *Studies in Mycology* 32, 3–19.

Aa, H.A. van der, Boerema, G.H. and Gruyter, J. de (2000) Contributions towards a monograph of *Phoma* (Coelomycetes) — VI-1. Section *Phyllostictoides*: Characteristics and nomenclature of its type species *Phoma exigua*. *Persoonia* 17(3), 435–456. [Errata (2002) in *Persoonia* 18(1), 53.]

Abeln, E.C.A., Stax, A.M., Gruyter, J. de and Aa, H.A. van der (2002) Genetic differentiation of *Phoma exigua* varieties by means of AFLP fingerprints. *Mycological Research* 106(4), 419–427.

Aderkas, P. von and Brewer, D. (1983) Gangrene of the ostrich fern caused by *Phoma exigua* var. *foveata*. *Canadian Journal of Plant Pathology* 5, 164–167.

Aderkas, P. von, Gruyter, J. de, Noordeloos, M.E. and Strongman, D.B. (1992) *Phoma matteuccicola* sp. nov., the causal agent of gangrene disease of ostrich fern. *Canadian Journal of Plant Pathology* 14, 227–228.

Ahmad, S. (1948) Fungi of Pakistan — I. *Sydowia* 2, 72–79.

Allescher, A. (1898–1901) Fungi imperfecti: Hyalinsporige Sphaeroideen. *Rabenh. Krypt.-Flora* [ed. 2], *Pilze* 6, 1–1016 (for dates of publication see *Pilze* 8 (1907), 85).

Anonymous (1960) Index of plant diseases in the USA. *Agriculture Handbook U.S. Department of Agriculture* 165.

Anonymous (1969) Title abbreviations for some common mycological taxonomic publications. *Bibliography of Systemic Mycology* 4, supplement.

Arenal, F., Platas, G., Monte, E. and Peláez, F. (2000) ITS sequencing support for *Epicoccum nigrum* and *Phoma epicoccina* being the same biological species. *Mycological Research* 104(3), 301–303.

Arny, D.C. and Nelson, R.R. (1971) *Phyllosticta maydis* species nova, the incitant of yellow leaf blight of maize. *Phytopathology* 61, 1170–1172.

Arx, J.A. von (1957) Revision der zu *Gloeosporium* gestellten Pilze. *Verh. K. ned. Akad. Wet. [Afd. Natuurk.] reeks* 2, 51(3), 1–153.

Arx, J.A. von (1970) A revision of the fungi classified as *Gloeosporium. Bibliotheca Mycologica* 24, J. Cramer, Lehre.

Arx, J.A. von and Storm, P.K. (1967) Über einige aus dem Erdboden isolierte, zu *Sporormia, Preussia* und *Westerdykella* gehörende Ascomyceten. *Persoonia* 4(4), 407–415.

Baker, K.F., Clark, L.H. and Kimball, M.H. (1949a) Association of *Deuterophoma* spp. with a chrysanthemum stunt disease in California. *Plant Disease Reporter* 33, 2–8.

Baker, K.F., Dimock, A.W. and Davis, L.H. (1949b) Life history and control of the *Ascochyta* ray blight of chrysanthemum. *Phytopathology* 29, 789–805.

Baker, K.F., Davis[-Clark], L.H., Wilhelm, S. and Snyder, W.C. (1985) An aggressive vascular-inhabiting *Phoma* (*Phoma tracheiphila* f.sp. *chrysanthemi* nov. f. sp.) weakly pathogenic to chrysanthemum. *Canadian Journal of Botany* 63, 1730–1735.

Balasubramanian, P. and Narayanasamy, P. (1980) A note on a new blight disease of groundnut caused by *Phoma microspora* sp. nov. in Tamil Nadu. *Indian Phytopathology* (phytopathological notes) 33, 133–136.

Bessi, G. and Carolis, D. de (1974) Funghi correlati al marciume basale del riso. *Riso* 23, 299–307.

Bick, I.R.C. and Rhee, C. (1966) Anthraquinone pigments from *Phoma foveata* Foister. *Biochemical Journal* 98, 112–116.

Bitancourt, A.A. (1938) '*Pyrenochaeta sacchari* n.sp.' e uma mancha da folha da cana de açucar. *Arquivos do Instituto biológico São Paulo* 9, 299–302.

Boerema, G.H. (1962) Zaadonderzoek. *Versl. Meded. plziektenk. Dienst Wageningen* 136 (Jaarb. 1961), 111.

Boerema, G.H. (1964) *Phoma herbarum* Westend., the type-species of the form-genus *Phoma* Sacc. *Persoonia* 3(1), 9–16.

Boerema, G.H. (1965) Spore development in the form-genus *Phoma. Persoonia* 3(4), 413–417.

Boerema, G.H. (1967a) Über die Identität der '*Phoma radicis*-Arten'. *Nova Hedwigia* 14, 57–60.

Boerema, G.H. (1967b) The *Phoma* organisms causing gangrene of potatoes. *Netherlands Journal of Plant Pathology* 73, 190–192.

Boerema, G.H. (1969) The use of the term forma specialis for *Phoma*-like fungi. *Transactions of the British Mycological Society* 52, 509–513.

Boerema, G.H. (1970) Additional notes on *Phoma herbarum. Persoonia* 6(1), 15–48.

Boerema, G.H. (1972) *Ascochyta phaseolorum* synonymous with *Phoma exigua. Netherlands Journal of Plant Pathology* 78, 113–115.

Boerema, G.H. (1973) Voortgezet mycologisch taxonomisch onderzoek. *Verslagen en Mededelingen Plantenziektenkundige Dienst Wageningen* 147 (Jaarb. 1972), 17–20.

Boerema, G.H. (1975) Pathogeniteits- en waardplantaspecten bij mycologisch-diagnostisch onderzoek. *Verslagen en Mededelingen Plantenziektenkundige Dienst Wageningen* 149 (Jaarb. 1974), 23–29.

Boerema, G.H. (1976) The *Phoma* species studied in culture by Dr R.W.G. Dennis. *Transactions of the British Mycological Society* 67, 289–319.

Boerema, G.H. (1977) De veroorzakers van gangreen bij aardappel. *Gewasbescherming* 8, 91–94.

Boerema, G.H. (1978) Mycologisch-taxonomisch onderzoek. *Verslagen en Mededelingen Plantenziektenkundige Dienst Wageningen* 152 (Jaarb. 1977), 16–19.

Boerema, G.H. (1980) Damping off of lilac seedlings caused by *Phoma exigua* var. *lilacis* (Sacc.) comb. nov. *Phytopathologia Mediterranea* 18(1979), 105–106.

Boerema, G.H. (1981) Mycologisch-taxonomisch onderzoek. *Phoma*- en *Leptosphaeria*-soorten bij cruciferen. *Verslagen en Mededelingen Plantenziektenkundige Dienst Wageningen* 157 (Jaarb. 1980), 21–24.

Boerema, G.H. (1982a) Mycologisch-taxonomisch onderzoek. Echte *Phoma*-soorten (= sectie *Phoma*). *Verslagen en Mededelingen Plantenziektenkundige Dienst Wageningen* 158 (Jaarb. 1981), 25–28.

Boerema, G.H. (1982b) Mycologisch-taxonomisch onderzoek. *Phoma*-soorten van de sectie *Plenodomus*. *Verslagen en Mededelingen Plantenziektenkundige Dienst Wageningen* 158 (Jaarb. 1981), 28–30.

Boerema, G.H. (1983a) Mycologisch-taxonomisch onderzoek. Echte *Phoma*-soorten (= sectie *Phoma*). *Verslagen en Mededelingen Plantenziektenkundige Dienst Wageningen* 159 (Jaarb. 1982), 21–25.

Boerema, G.H. (1983b) Mycologisch-taxonomisch onderzoek. *Phoma*-soorten van de sectie *Peyronellaea*. *Verslagen en Mededelingen Plantenziektenkundige Dienst Wageningen* 159 (Jaarb. 1982), 25–27.

Boerema, G.H. (1984a) Mycologisch-taxonomisch onderzoek. Echte *Ascochyta*-soorten (met hyaliene conidiën) die pathogeen zijn voor Papilionaceae. *Verslagen en Mededelingen Plantenziektenkundige Dienst Wageningen* 162 (Jaarb. 1983), 23–31.

Boerema, G.H. (1984b) Mycologisch-taxonomisch onderzoek. *Ascochyta*'s met lichtbruine conidiën die pathogeen zijn voor Chenopodiaceae. *Verslagen en Mededelingen Plantenziektenkundige Dienst Wageningen* 162 (Jaarb. 1983), 31–34.

Boerema, G.H. (1985) Mycologisch-taxonomisch onderzoek. Bodem-*Phoma*'s. *Verslagen en Mededelingen Plantenziektenkundige Dienst Wageningen* 163 (Jaarb. 1984), 34–40.

Boerema, G.H. (1986a) Een 'stress-schimmel' van loofbomen en struiken. *Verslagen en Mededelingen Plantenziektenkundige Dienst Wageningen* 164 (Jaarb. 1985), 26–28.

Boerema, G.H. (1986b) Mycologisch-taxonomisch onderzoek. Een subtropische bodemschimmel die zich ook thuis voelt in onze kassen. *Verslagen en Mededelingen Plantenziektenkundige Dienst Wageningen* 164 (Jaarb. 1985), 28–32.

Boerema, G.H. (1993) Contributions towards a monograph of *Phoma* (Coelomycetes) — II. Section *Peyronellaea*. *Persoonia* 15(2), 197–221.

Boerema, G.H. (1997) Contributions towards a monograph of *Phoma* (Coelomycetes) — V. Subdivision of the genus in sections. *Mycotaxon* 64, 321–333.

Boerema, G.H. (2003) Contributions towards a monograph of *Phoma* (Coelomycetes) — X. Section *Pilosa* (taxa with a *Pleospora* teleomorph) and nomenclatural notes on some other taxa. *Persoonia* 18(2), 153–161.

Boerema, G.H. and Bollen, G.J. (1975) Conidiogenesis and conidial septation as differentiating criteria between *Phoma* and *Ascochyta*. *Persoonia* 8(2), 111–144.

Boerema, G.H. and Dorenbosch, M.M.J. (1965) *Phoma*-achtige schimmels in associatie met appelbladvlekken. *Verslagen en Mededelingen Plantenziektenkundige Dienst Wageningen* 142 (Jaarb. 1964), 138–151.

Boerema, G.H. and Dorenbosch, M.M.J. (1968) Some *Phoma* species recently described from marine soils in India. *Transactions of the British Mycological Society* 51, 145–146.

Boerema, G.H. and Dorenbosch, M.M.J. (1970) On *Phoma macrostomum* Mont., a ubiquitous species on woody plants. *Persoonia* 6(1), 49–58.

Boerema, G.H. and Dorenbosch, M.M.J. (1973) The *Phoma* and *Ascochyta* species described by Wollenweber and Hochapfel in their study on fruit-rotting. *Studies in Mycology* 3.

Boerema, G.H. and Dorenbosch, M.M.J. (1977) Mycologisch-taxonomisch onderzoek. *Verslagen en Mededelingen Plantenziektenkundige Dienst Wageningen* 151 (Jaarb. 1976), 25–27.

Boerema, G.H. and Dorenbosch, M.M.J. (1979) Mycologisch-taxonomisch onderzoek. *Verslagen en Mededelingen Plantenziektenkundige Dienst Wageningen* 153 (Jaarb. 1978), 17–21.

Boerema, G.H. and Dorenbosch, M.M.J. (1980) Mycologisch-taxonomisch onderzoek. Identificaties t.b.v. fytopathologisch onderzoek in ontwikkelingslanden. *Verslagen en Mededelingen Plantenziektenkundige Dienst Wageningen* 156 (Jaarb. 1979), 21–27.

Boerema, G.H. and Dorenbosch, M.M.J. (1981) *Stagonosporopsis*: een geslacht met extreem variabele conidiën. *Verslagen en Mededelingen Plantenziektenkundige Dienst Wageningen* 157 (Jaarb. 1980), 19–20.

Boerema, G.H. and Gams, W. (1995) What is *Sphaeria acuta* Hoffm. : Fr.? *Mycotaxon* 53, 355–360.

Boerema, G.H. and Griffin, M.J. (1974) *Phoma* species from *Viburnum*. *Transactions of the British Mycological Society* 63, 109–114.

Boerema, G.H. and Gruyter, J. de (1998) Contributions towards a monograph of *Phoma* (Coelomycetes) — VII. Section *Sclerophomella*: Taxa with thick-walled pseudoparenchymatous pycnidia. *Persoonia* 17(1), 81–95.

Boerema, G.H. and Gruyter, J. de (1999) Contributions towards a monograph of *Phoma* (Coelomycetes) — III–Supplement: Additional species of section *Plenodomus*. *Persoonia* 17(2), 273–280.

Boerema, G.H. and Hamers, M.E.C. (1989) Check-list for scientific names of common parasitic fungi. Series 3b: Fungi on bulbs: Amaryllidaceae and Iridaceae. *Netherlands Journal of Plant Pathology* 95, Suppl. 3.

Boerema, G.H. and Hamers, M.E.C. (1990) Check-list for scientific names of common parasitic fungi. Series 3c: Fungi on bulbs: 'additional crops' belonging to the Araceae, Begoniaceae, Compositae, Oxalidaceae and Ranunculaceae. *Netherlands Journal of Plant Pathology* 96, Suppl. 1.

Boerema, G.H. and Höweler, L.H. (1967) *Phoma exigua* Desm. and its varieties. *Persoonia* 5(1), 15–28.

Boerema, G.H. and Jong, C.B. de (1968) Über die samenbürtigen *Phoma*-Arten von *Valerianella*. *Phytopathologische Zeitschrift* 61, 362–371.

Boerema, G.H. and Kesteren, H.A. van (1962) *Phoma*-achtige schimmels bij aardappel. *Verslagen en Mededelingen Plantenziektenkundige Dienst Wageningen* 136 (Jaarb. 1961), 201–209.

Boerema, G.H. and Kesteren, H.A. van (1964) The nomenclature of two fungi parasitizing *Brassica*. *Persoonia* 3(1), 17–28.

Boerema, G.H. and Kesteren, H.A. van (1972) Enkele bijzondere schimmelaantastingen IV. (Mycologisch waarnemingen no. 16.) *Gewasbescherming* 3, 65–69.

Boerema, G.H. and Kesteren, H.A. van (1974) Enkele bijzondere schimmelaantastingen V. (Mycologische waarnemingen no. 17.) *Gewasbescherming* 5, 119–125.

Boerema, G.H. and Kesteren, H.A. van (1980) Vermeldenswaardig schimmelaantastingen in de periode 1975–1979. A. Aantastingen door schimmels die in Nederland
niet bekend waren. *Gewasbescherming* 11(1), 115–127.

Boerema, G.H. and Kesteren, H.A. van (1981a) Mycological diagnostic problems.
EPPO Bulletin 11, 113–118.

Boerema, G.H. and Kesteren, H.A. van (1981b) Nomenclatural notes on some species
of *Phoma* sect. *Plenodomus. Persoonia* 11(3), 317–331.

Boerema, G.H. and Loerakker, W.M. (1981) *Phoma piskorzii* (Petrak) comb. nov., the
anamorph of *Leptosphaeria acuta* (Fuckel) P. Karst. *Persoonia* 11(3), 311–315.

Boerema, G.H. and Loerakker, W.M. (1985) Notes on *Phoma* 2. *Transactions of the
British Mycological Society* 84, 289–302.

Boerema, G.H. and Valckx, A.G.M. (1970) Enkele bijzondere schimmelaantastingen III.
(Mycologische waarnemingen no. 15.) *Gewasbescherming* 1, 65–68.

Boerema, G.H. and Verhoeven, A.A. (1972) Check-list for scientific names of common
parasitic fungi. Series 1a: Fungi on trees and shrubs. *Netherlands Journal of Plant
Pathology* 78, Suppl. 1.

Boerema, G.H. and Verhoeven, A.A. (1979) Check-list for scientific names of common
parasitic fungi. Series 2c: Fungi on field crops: pulse (legumes) and forage crops
(herbage legumes). *Netherlands Journal of Plant Pathology* 85, 151–185.

Boerema, G.H., Dorenbosch, M.M.J. and Kesteren, H.A. van (1965a) Remarks on
species of *Phoma* referred to *Peyronellaea. Persoonia* 4(1), 47–68.

Boerema, G.H., Dorenbosch, M.M.J. and Leffring, L. (1965b) A comparative study of
the black stem fungi on lucerne and red clover and the footrot fungus on pea.
Netherlands Journal of Plant Pathology 71, 79–89.

Boerema, G.H., Dorenbosch, M.M.J. and Kesteren, H.A. van (1968) Remarks on
species of *Phoma* referred to *Peyronellaea* II. *Persoonia* 5(2), 201–205.

Boerema, G.H., Dorenbosch, M.M.J. and Kesteren, H.A. van (1971) Remarks on
species of *Phoma* referred to *Peyronellaea* III. *Persoonia* 6(2), 171–177.

Boerema, G.H., Dorenbosch, M.M.J. and Kesteren, H.A. van (1973) Remarks on
species of *Phoma* referred to *Peyronellaea* IV. *Persoonia* 7(2), 131–139.

Boerema, G.H., Dorenbosch, M.M.J. and Kesteren, H.A. van (1977) Remarks on
species of *Phoma* referred to *Peyronellaea* V. *Kew Bulletin* 31(3, '1976'), 533–544.

Boerema, G.H., Loerakker, W.M. and Laundon, G.F. (1980) *Phoma rumicicola* sp. nov.,
a cause of leaf spots on *Rumex obtusifolius. New Zealand Journal of Botany* 18,
470–476.

Boerema, G.H., Crüger, G., Gerlagh, M. and Nirenberg, H. (1981a) *Phoma exigua* var.
diversispora and related fungi on *Phaseolus* beans. *Zeitschrift für
Pflanzenkrankheiten und Pflanzenschutz* 88, 597–607.

Boerema, G.H., Kesteren, H.A. van and Loerakker, W.M. (1981b) Notes on *Phoma.
Transactions of the British Mycological Society* 77, 61–74.

Boerema, G.H., Kesteren, H.A. van and Loerakker, W.M. (1984) Vermeldenswaardige
schimmelaantastingen in de periode 1980–1984. A. Aantastingen door schimmels
die in Nederland niet bekend waren. (Mycologische waarnemingen no. 21.)
Gewasbescherming 15, 163–177.

Boerema, G.H., Loerakker, W.M. and Wittern, I. (1986) Zum Auftreten von *Phoma
nigrificans* (P. Karst.) comb. nov. (Teleomorph *Didymella macropodii* Petrak) an
Winterraps (*Brassica napus* L. var. *oleifera* Metzger). *Journal of Phytopathology*
115, 267–273.

Boerema, G.H., Loerakker, W.M. and Hamers, M.E.C. (1987) Check-list for scientific
names of common parasitic fungi. Supplement Series 2a (additions and corrections): Fungi on field crops: beet and potato; caraway, flax and oilseed poppy.
Netherlands Journal of Plant Pathology 93, Suppl. 1, 1–20.

Boerema, G.H. and Coworkers (1993a) Check-list for scientific names of common parasitic fungi. *Libri Botanici* 10. IHW-Verlag Eching, D.

Boerema, G.H., Pieters, R. and Hamers, M.E.C. (1993b) Check-list for scientific names of common parasitic fungi. Supplement Series 2c,d (additions and corrections): Fungi on field crops: pulse (legumes), forage crops (herbage legumes), vegetables and cruciferous crops. *Netherlands Journal of Plant Pathology* 99, Suppl. 1.

Boerema, G.H., Gruyter, J. de and Kesteren, H.A. van (1994) Contributions towards a monograph of *Phoma* (Coelomycetes) — III-1. Section *Plenodomus*: Taxa often with a *Leptosphaeria* teleomorph. *Persoonia* 15(4), 431–487. [Errata (1996) *Persoonia* 16(2), 190.]

Boerema, G.H., Gruyter, J. de and Noordeloos, M.E. (1995) New names in *Phoma*. *Persoonia* 16(1), 131.

Boerema, G.H., Loerakker, W.M. and Hamers, M.E.C. (1996) Contributions towards a monograph of *Phoma* (Coelomycetes) — III-2. Misapplications of the type species-name and the generic synonyms of section *Plenodomus* (Excluded species). *Persoonia* 16(2), 141–190.

Boerema, G.H., Gruyter, J. de and Noordeloos, M.E. (1997) Contributions towards a monograph of *Phoma* (Coelomycetes) — IV. Section *Heterospora*: Taxa with large sized conidial dimorphs, *in vivo* sometimes as *Stagonosporopsis* synanamorphs. *Persoonia* 16(3), 335–371.

Boerema, G.H., Gruyter, J. de and Graaf, P. van de (1999) Contributions towards a monograph of *Phoma* (Coelomycetes) — IV–Supplement: An addition to section *Heterospora*: *Phoma schneiderae* spec. nov., synanamorph *Stagonosporopsis lupini* (Boerema and R. Schneider) comb. nov. *Persoonia* 17(2), 281–285.

Bontea, V. (1964) Beiträge zum Studium der Bildung und Entwicklung der *Phoma lingam* (Tode) Desm. Pycnidien. *Revue roum. Biol., Sér. Bot.* 9, 441–443.

Booth, C. (1967) *Pleospora björlingii*. *CMI Descr. pathog. Fungi Bact.* 149.

Bowen, J.K., Lewis, B.G. and Matthews, P. (1997) Discovery of the teleomorph of *Phoma medicaginis* var. *pinodella* in culture. *Mycological Research* 101(1), 80–84.

Brewer, J.G. and Boerema, G.H. (1965) Electron microscope observations on the development of pycnidiospores in *Phoma* and *Ascochyta* spp. *Proceedings K. Nederlandse Akademie van Wetenschappen C* 68, 86–97.

Brooks, F.T. (1932) A disease of the Arum lily caused by *Phyllosticta Richardiae*, n.sp. *Annals of Applied Biology* 19, 16–20.

Brown, P. and Stratton, G.B. [eds] (1963–1965) *World List of Scientific Periodicals*, 4th edn. Butterworths, London.

Brummitt, R.K. and Powell, C.E. [eds] (1992) *Authors of Plant Names*. Royal Botanic Gardens, Kew.

Buchanan, P.K. (1987) A reappraisal of *Ascochytula* and *Ascochytella* (Coelomycetes). *Mycological Papers* 156.

Butler, E.E. and Mann, M.P. (1959) Use of cellophane tape for mounting and photographing phytopathogenic fungi. *Phytopathology* 49, 231–232.

Camyon, S. and Gerhardson, B. (1997) Formation of pseudosclerotia and bacteria-induced chlamydospores in *Phoma foveata*. *European Journal of Plant Pathology* 103, 467–470.

Carter, J.C. (1941) Preliminary investigations on oak diseases in Illinois. *Bulletin of the Illinois State Natural History Survey* 21(6), 195–230.

Cejp, K. (1965) The occurrence of some *Phyllostictas* on ornamental plants. I. *Preslia (Praha)* 37, 345–352.

Cerkauskas, R.F. (1985) Canker of parsnip caused by *Phoma complanata*. *Canadian Journal of Plant Pathology* 7, 135–138.

Cerkauskas, R.F. (1987) Pathogenicity and survival of *Phoma complanata*. *Canadian Journal of Plant Pathology* 9, 63–67.

Chandra, S. and Tandon, R.N. (1966) Three new leaf-infecting fungi from Allahabad. *Mycopathologia et mycologia applicata* 29, 273–276.

Chen, S.Y., Dickson, D.W. and Kimbrough, J.W. (1996) *Phoma heteroderae* sp. nov. isolated from eggs of *Heterodera glycines*. *Mycologia* 88(6), 885–891.

Chevassut, G. (1965) Récoltes phytopathologiques du massif de l'Aigoual. *Bulletin de la Société Mycologique de France* 81, 34–41.

Ciccarone, A. and Russo, M. (1969) First contribution to the systemics and morphology of the causal agent of the 'mal secco' disease of citrus (*Deuterophoma tracheiphila* Petri). In: Chapman, H.P. (ed.) *Proceedings of the First International Citrus Symposium* 1, Vol 3, pp. 1239–1249.

Clements, F.E. and Shear, C.L. (1931) *The Genera of Fungi*. Wilson Comp., New York.

Colotelo, N. and Netolitzky, H. (1964) Pycnidial development and spore discharge of *Plenodomus meliloti*. *Canadian Journal of Botany* 42, 1467–1469.

Constantinescu, O. and Aa, H.A. van der (1982) *Phoma flavigena* sp. nov., from fresh water in Romania. *Transactions of the British Mycological Society* 79, 343–345.

Corbaz, R. (1957) Récherches sur le genre *Didymella* Sacc. *Phytopathologische Zeitschrift* 28, 375–414.

Corlett, M. (1974) *Didymella applanata*. *Fungi Canadenses* 49.

Corlett, M. (1981) A taxonomic survey of some species of *Didymella* and *Didymella*-like species. *Canadian Journal of Botany* 59, 2016–2092.

Corlett, M., Jarvis, W.R. and Maclatchy, I.A. (1986) *Didymella bryoniae*. *Fungi Canadenses* 303.

Creager, D.B. (1933) Leaf scorch of narcissus. *Phytopathology* 23, 770–786.

Curzi, M. (1933) L' '*Ascochyta heteromorpha*' N.C. nella necrosi dell' oleandro e nell' inoculazione sperimentale. *Bollettino della R. Stazione di Patologia Vegetale di Roma* II, 13, 380–422.

Dennis, R.W.G. (1946) Notes on some British fungi ascribed to *Phoma* and related genera. *Transactions of the British Mycological Society* 29, 11–42.

Diedicke, H. (1912) Die Abteilung Hyalodidymae der Sphaerioideen. *Annales Mycologici* 10, 135–152.

Diedicke, H. (1912–1915) *Kryptogamenflora der Mark Brandenburg IX, Pilze VII (Sphaeropsideae, Melanconieae)*. Leipzig, Verlag Gebr. Borntraeger (for dates of publication see bond copy of Vol. IX).

Dippenaar, B.J. (1931) Descriptions of some new species of South African fungi and of species not previously recorded from South Africa — II. *South African Journal of Science* 28, 284–289.

Domsch, K.H., Gams, W. and Anderson, T.H. (1980) *Compendium of Soil Fungi*. Vol. 1. Academic Press, London.

Donald, P.A., Bugbee, W.M. and Venette, J.R. (1986) First report of *Leptosphaeria lindquistii* (sexual stage of *Phoma macdonaldii*) on sunflower in North Dakota and Minnesota. *Plant Disease Reporter* 70(4), 352.

Dorenbosch, M.M.J. (1970) Key to nine ubiquitous soil-borne *Phoma*-like fungi. *Persoonia* 6(1), 1–14.

Dorenbosch, M.M.J. and Boerema, G.H. (1973) About *Phoma liliana* Chandra and Tandon II. *Mycopathologia et mycologia applicata* 50, 255–256.

Dorenbosch, M.M.J. and Höweler, L.H. (1968) About *Phoma liliana* Chandra and Tandon. *Mycopathologia et mycologia applicata* 35, 265–267.

Dorenbosch, M.M.J. and Kesteren, H.A. van (1978) Bladvlekkenziekte bij tomaat. *Verslagen en Mededelingen Plantenziektenkundige Dienst Wageningen* 152 (Jaarb. 1977), 12–13.

Dring, D.M. (1961) Studies on *Mycosphaerella brassicicola* (Duby) Oudem. *Transactions of the British Mycological Society* 44, 253–264.

Ebben, M.H. and Last, F.T. (1966) 2. Clematis wilt. *Report of the Glasshouse Crops Research Institute* 1965, 128–131.

Echandi, E. (1957) La quema de los cafetos causada por *Phoma costarricensis* n. sp. *Revista de Biologia Tropicale* 5(1), 81–102.

EPPO (1980a) Data sheets on quarantine organisms List A2. *Didymella chrysanthemi* (Tassi) Garibaldi and Guillini (also part of Set 5 of data sheets distributed by EPPO in 1982, see below). Paris, rue Le Nôtre.

EPPO (1980b) Data sheets on quarantine organisms List A2. *Phoma exigua* Desm. var. *foveata* (Foister) Boerema (also part of Set 5 of data sheets distributed by EPPO in 1982, see below). Paris, rue Le Nôtre.

EPPO (1982) Data sheets on quarantine organisms Set 5. List A2 (Quarantine organisms present in some EPPO countries). *EPPO Bulletin* 12(1) (no consecutive page-numbering).

EPPO (1984) Data sheets on quarantine organisms List A1 (pathogens not yet recorded in the EPPO countries). No. 141. *Phoma andina* Turkensteen. *EPPO Bulletin* 14(1), 45–47.

Fabricatore, J.A. (1951) *Ascochyta trachelospermi* n. sp. e considerazione relative ad una possibile revisione del genere *Peyronellaea*. *Annali della Sperimentazione Agraria* II, 5, 1433–1446.

Faurel, L. and Schotter, G. (1965a) Notes mycologiques IV. Champignons coprophiles du Sahara central et notamment de la Tefedest. *Revue de Mycologie* 30, 141–165.

Faurel, L. and Schotter, G. (1965b) Notes mycologiques V. Champignons coprophiles du Tibesti. *Revue de Mycologie* 30, 330–351.

Feekes, F.H. (1931) Onderzoekingen over schimmelziekten in bolgewassen. Thesis Utrecht/Baarn.

Frey, F. and Yabar, E. (1983) Enfermedades y plagas de lupinos en el Peru. *SchrReihe dt. Ges. Tech. Zusammenarb. (GTZ) Eschborn* 142.

Frezzi, M.J. (1964) Especie del género *Phoma* parasita de *Helianthus annuus* L. en Manfredi (Córdoba), República Argentina. *Idia* 1964(19), 37–40.

Frezzi, M.J. (1968) *Leptosphaeria lindquistii* n. sp. forma sexual de *Phoma oleracea* var. *helianthi-tuberosi* Sacc., hongo causal de la 'mancha negra del tallo' del girasol (*Helianthus annuus* L.), en Argentina. *Revista de Investigaciones agropecurias* V, 5, 73–80.

Fries, E.M. (1828) Elenchus Fungorum 2. Gryphiswaldiae. Reprinted in 1952 by Johnsons Reprint Corporation, New York.

Gams, W., Hoekstra, E.S. and Aptroot, A. [eds] (1998) *CBS Course of Mycology*. Centraalbureau voor Schimmelcultures, Baarn.

Gaudet, M.D. and Schulz, J.T. (1984) Association between a sunflower fungal pathogen, *Phoma macdonaldii*, and a stem weevil, *Apion occidentale* (Coleoptera: Curculionidae). *Canadian Entomologist* 116, 1267–1273.

Gilman, J.C. and Abbott, E.V. (1927) A summary of the soil fungi. *Iowa State College Journal of Science* 1(3), 225–344.

Gindrat, D., Semecnik, A. and Bolay, A. (1967) Une nouvelle maladie cryptogamique de la mâche découverte en Suisse romande. *Revue horticole Suisse* 40, 347–351.

Gloyer, W.O. (1915) *Ascochyta clematidina*, the cause of stem-rot and leaf-spot of *Clematis*. *Journal of Agricultural Research* 4, 331–342.

Goidànich, G. and Ruggieri, G. (1947) Recenti osservazioni sulla biologia della 'Deuterophoma tracheiphila' Petri e considerazioni sull'eziologia del 'mal secco' degli agrumi. *Atti dell' Accademia nazionale dei Lincei. Rendiconti. (Classe di scienze fisiche, matematiche e naturali)* VIII, 3, 395–402.

Goossens, J.A.A.M.H. (1928) Onderzoek over de door *Phoma Apiicola* Klebahn veroorzaakte schurftziekte van de knolselderij en over synergetische vormen en locale rassen van deze zwam. *Tijdschrift over Plantenziekten* 34, 271–348.

Gorenz, A.M., Walker, J.C. and Larson, R.H. (1948) Morphology and taxonomy of the onion pink-root fungus. *Phytopathology* 38, 831–839.

Graniti, A. (1955) Morfologia di *Deuterophoma tracheiphila* Petri e considerazioni sul genere *Deuterophoma* Petri. *Bollettino delle Sedute dell' Accademia Gioenia di Scienze Naturali in Catania IV*, 3(3), 1–18.

Greene, H.C. (1960) Notes on Wisconsin parasitic fungi XXVI. *Wisconsin Academy of Sciences, Arts and Letters* 49, 85–111.

Grigoriu, A.C. (1975) Contribution à l'étude de la biologie du champignon *Stagonospora bolthauseri* (Sacc.) comb. nov. provoquant une maladie du haricot, nouvelle pour la Grèce. *Annales de l'Institut phytopathologique Benaki II* [Nouvelle Série], 11, 109–126.

Grimm, R. and Vögeli, M. (2000) Ferns in urban horticulture-production, uses and phytopathological problems. *Mitteilungen der Biologischen Bundesanstalt für Land-u. Forstwirtschaft* 370, 109–110.

Grove, W.B. (1935) *British stem- and leaf-fungi (Coelomycetes) Vol. 1 Sphaeropsidales.* Cambridge University Press, Cambridge.

Gruyter, J. de (2002) Contributions towards a monograph of *Phoma* (Coelomycetes) — IX. Section *Macrospora*. *Persoonia* 18(1), 85–102.

Gruyter, J. de and Boerema, G.H. (1998) Taxonomy of pathogenic *Phoma* species on *Delphinium* spp. In: *Abstracts 6th Intern. Mycological Congress, 23–28 August 1998, Jerusalem, Israel*, p.59.

Gruyter, J. de and Boerema, G.H. (2002) Contributions towards a monograph of *Phoma* (Coelomycetes) — VIII. Section *Paraphoma*: Taxa with setose pycnidia. *Persoonia* 17(4) ['2001'], 541–561.

Gruyter, J. de and Noordeloos, M.E. (1992) Contributions towards a monograph of *Phoma* (Coelomycetes) — I-1. Section *Phoma*: Taxa with very small conidia *in vitro*. *Persoonia* 15(1), 71–92. [Errata (1993) *Persoonia* 15(2), 221.]

Gruyter, J. de and Scheer, P. (1998) Taxonomy and pathogenicity of *Phoma exigua* var. *populi* var. nov. causing necrotic bark lesions on poplars. *Journal of Phytopathology* 146, 411–415.

Gruyter, J. de, Noordeloos, M.E. and Boerema, G.H. (1993) Contributions towards a monograph of *Phoma* (Coelomycetes) — I-2. Section *Phoma*: Additional taxa with very small conidia and taxa with conidia up to 7 μm long. *Persoonia* 15(3), 369–400. [Errata (1998) *Persoonia* 16(4), 490.]

Gruyter, J. de, Noordeloos, M.E. and Boerema, G.H. (1998) Contributions towards a monograph of *Phoma* (Coelomycetes) — I-3. Section *Phoma*: Taxa with conidia longer than 7 μm. *Persoonia* 16(4), 471–490.

Gruyter, J. de, Boerema, G.H. and Aa, H.A. van der (2002) Contributions towards a monograph of *Phoma* (Coelomycetes) — VI-2. Section *Phyllostictoides*: Outline of its taxa. *Persoonia* 18(1), 1–53.

Guba, E.F. and Anderson, P.J. (1919) *Phyllosticta* leaf spot and damping off of snapdragons. *Phytopathology* 9, 315–325.

Gutner, A.S. (1933) Fungal parasites of glasshouse plants in the towns of Leningrad and Dyetskoye Selo. *Trudy̆ bot. Inst. Akad. Nauk SSSR Leningrad Ser. II (Plantae Cryptogamae)*, 1, 285–323 [Russian with German summary].

Hall, R. (1992) Epidemiology of blackleg of oilseed rape. *Canadian Journal of Plant Pathology* 14, 46–55.

Hansen, H.N. (1929) Etiology of the pink-root disease of onions. *Phytopathology* 19, 691–704.

Hansen, H.N. (1938) The dual phenomenon in imperfect fungi. *Mycologia* 30, 442–455.

Harter, L.L. (1913) Foot rot, a new disease of the sweet potato. *Phytopathology* 3, 243–245.

Hatai, K., Fujimaki, Y. and Egusa, S. (1986) A visceral mycosis in ayu fry, *Plecoglossus altivelis* Temmink and Schlegel, caused by a species of *Phoma*. *Journal of Fish Diseases* 9, 111–116.

Hauptmann, G. and Schickedanz, F. (1986) *In vitro*-Untersuchungen und mikroskopische Studien zur Entwicklung von *Phoma chrysanthemicola* Hollós f. sp. *chrysanthemicola*, dem Erreger einer Wurzel- und Stengelgrundfäule der Chrysantheme. *Journal of Phytopathology* 116, 289–298.

Hawksworth, D.L. (1974) *Mycologist's Handbook. An Introduction to the Principles of Taxonomy and Nomenclature in the Fungi and Lichens*. CMI, Kew.

Höhnel, F. von (1918) Fungi imperfecti. Beiträge zur Kenntnis derselben. *Hedwigia* 59, 236–284.

Holliday, P. and Punithalingam, E. (1970a) *Ascochyta gossypii*. *C.M.I. Descriptions of pathogenic Fungi and Bacteria* 271.

Holliday, P. and Punithalingam, E. (1970b) *Didymella lycopersici*. *C.M.I. Descriptions of pathogenic Fungi and Bacteria* 272.

Holliday, P. and Punithalingam, E. (1970c) *Macrophomina phaseolina*. *C.M.I. Descriptions of pathogenic Fungi and Bacteria* 275.

Holm, L. (1957) Études taxonomiques sur les Pléosporacées. *Symbolae botanicae Upsalienses* 14(3), 5–188.

Hooker, W.J. [ed.] (1980) *Compendium of Potato Diseases*. American Phytopathological Society, St Paul, Minnesota.

Horner, C.E. (1971) Rhizome and stem rot of peppermint caused by *Phoma strasseri*. *Plant Disease Reporter* 55, 814–816.

Hosford, R.M. (1975) *Phoma glomerata*, a new pathogen of wheat and triticales, cultivar resistance related to wet period. *Phytopathology* 65, 1236–1239.

Houten, J.G. Ten (1939) Kiemplantenziekten van coniferen. Thesis Utrecht.

Hsieh, W.H. (1979) The causal organism of sugarcane leaf blight. *Mycologia* 71, 892–898.

Hsu, C.H. and Sun, S.K. (1969) Stem blight of asparagus in Taiwan. I. Distribution of the disease and cultural characteristics and spore germination of the causal organism *Phoma asparagi*. *Plant Protection Bulletin, Taiwan* 11, 47–60.

Hudson, H.J. (1960) Pyrenomycetes of sugar cane and other grasses in Jamaica. I. Conidia of *Apiospora camptospora* and *Leptosphaeria sacchari*. *Transactions of the British Mycological Society* 43(4), 607–616.

Hughes, S.J. (1953) Conidiophores, conidia and classification. *Canadian Journal of Botany* 31, 577–659.

Hutchison, L.J., Chakravarty, P., Kawchuk, L.M. and Hiratsuka, Y. (1994) *Phoma etheridgei* spec. nov. from galls and cankers of trembling aspen (*Populus tremuloides*) and its potential role as a bioprotectant against the aspen decay pathogen *Phellinus tremulae*. *Canadian Journal of Botany* 72, 1424–1431.

Jaczewski, A. (1898) Monographie du genre *Sphaeronema* Fr. *Nouveaux mémoires de la Société (impériale) des Naturalistes de Moscou* 15, 275–386.

Janse, J.D. (1981) The bacterial disease of ash (*Fraxinus excelsior*), caused by *Pseudomonas syringae* subsp. *savastanoi* pv. *fraxini*. III. Pathogenesis. *European Journal of Forest Pathology* 12, 218–231.

Jansen, M.J.C.M. (1965) Bladvlekken en scheutsterfte bij *Vinca* species veroorzaakt door *Phoma exigua*. *Versl. Meded. plziektenk. Dienst Wageningen* 142 (Jaarb. 1964), 153–154.

Jauch, C.I. (1947) Una nuera enfermedad de las calas en la Argentina. *An. Soc. cient. argend.* 144, 447–456.

Jedrycza, M., Lewarttowska, E. and Frencel, I. (1995) *Phoma nigrificans*, the alternative pathogen of oilseed rape. In: *Proc. 9th Intern. Rapeseed Congress, 4–7 July 1995, Cambridge, UK.*

Johnston, P.R. (1981) *Phoma* in New Zealand grasses and pasture legumes. *New Zealand Journal of Botany* 19, 173–186.

Johnston, P.R. and Boerema, G.H. (1981) *Phoma nigricans* sp. nov. and *Ph. pratorum* sp. nov., two common saprophytes from New Zealand. *New Zealand Journal of Botany* 19, 393–396.

Jones, F.R. and Weimer, J.L. (1938) *Stagonospora* leaf spot and root rot of forage legumes. *Journal of Agricultural Research* 57, 791–812.

Jones, J.P. (1976) Ultrastructure of conidium ontogeny in *Phoma pomorum*, *Microsphaeropsis olivaceum* and *Coniothyrium fuckelii*. *Canadian Journal of Botany* 54, 831–851.

Jones, L.K. (1927) Studies of the nature and control of blight, leaf and pod spot, and footrot of peas caused by species of *Ascochyta*. *Bulletin of the New York State Agricultural Experiment Station* 547.

Jooste, W.J. and Papendorf, M.C. (1981) *Phoma cyanea* sp. nov. from wheat debris. *Mycotaxon* 12(2), 444–448.

Kaiser, W.J. (1997) Inter- and intranational spread of *Ascochyta* pathogens of chickpea, faba beans and lentil. *Canadian Journal of Plant Pathology* 19, 215–224.

Keim, R. (1979) Pruning wound dieback of oleander. *Plant Disease Reporter* 63, 499–501.

Keinath, A.P., Farnham, M.W. and Zitter, T.A. (1995) Morphological, pathological and genetic differentiation of *Didymella bryoniae* and *Phoma* spp. isolated from cucurbits. *Phytopathology* 85(3), 364–369.

Kendrick, W.B. (1971) Taxonomy of fungi imperfecti. *Proceedings of the First International Specialists' Workshop-Conference, Kananaskis, Alberta, Canada.* University of Toronto Press, Toronto.

Kesteren, H.A. van (1972) The causal organism of purple blotch disease on *Sedum*. *Netherlands Journal of Plant Pathology* 78, 116–118.

Khune, N.N. and Kapoor, J.N. (1980) *Ascochyta rabiei* synonymous with *Phoma rabiei*. *Indian Phytopathology* 33, 119–120.

Khune, N.N. and Kapoor, J.N. (1981) A new disease of pigeon pea. *Indian Phytopathology* 34, 258–260.

Kidd, M.N. and Beaumont, A. (1924) Apple rot fungi in storage. *Transactions of the British Mycological Society* 10, 98–118.

Klebahn, H. (1905) Untersuchungen über einige Fungi imperfecti und die zugehörigen Ascomyceten-formen. I. u. II. *Jahrbuch für Wissenschaftliche Botanik* 41(4), 485–560.

Koch, L.W. (1931) Spur blight of raspberries in Ontario caused by *Didymella applanata*. *Phytopathology* 21, 247–287.

Kovatschevski, I.C. (1936) *The Blight of Chickpea (*Cicer arietinum*)*, Mycosphaerella rabiei *n.sp.* Ministry of Agriculture and Natural Domains, Plant Protection Institute, Sofia, Bulgaria.

Kövics, G.J., Gruyter, J. de and Aa, H.A. van der (1999) *Phoma sojicola* comb. nov. and other hyaline-spored Coelomycetes pathogenic on soybean. *Mycological Research* 103, 1065–1070.

Kranz, J. (1963) Vergleichende Untersuchungen an *Phoma*-Isolierungen von der Kartoffel (*Solanum tuberosum*). *Sydowia* 16, 1–16.

Lacoste, L. (1965) Biologie naturelle et culturale du genre *Leptosphaeria* Cesati et de Notaris. Déterminisme de la reproduction sexuelle. Thèse Toulouse.

Laskaris, T. (1950) The *Diplodina* disease of *Delphinium*. *Phytopathology* 40, 615–626.

Lee, D.-H., Mathur, S.B. and Neergaard, P. (1984) Detection and location of seed-borne inoculum of *Didymella bryoniae* and its transmission in seedlings of cucumber and pumpkin. *Phytopathologische Zeitschrift* 109, 301–308.

Leuchtmann, A. (1984) Über *Phaeosphaeria* Miyake und andere bitunicate Ascomyceten mit mehrfach querseptierten Ascosporen. *Sydowia* 37, 75–194.

Liang, L. (1991) A new variety of *Phoma exigua* [Chinese with English summary]. *Acta Microbiologica Sinica* 31(2), 160–162.

Loerakker, W.M. (1986) Bastvlekkenziekte bij linden. *Verslagen en Mededelingen Plantenziektenkundige Dienst* Wageningen 164 (Jaarb. 1985), 17–21.

Loerakker, W.M. and Boerema, G.H. (1987) Mycologisch-taxonomisch onderzoek. Blakering: een nieuwe Phoma-ziekte bij tomaat en aardappel in Zuid-Amerika. *Verslagen en Mededelingen Plantenziektenkundige Dienst Wageningen* 165 (Jaarb. 1986), 53–56.

Loerakker, W.M., Navarro, R., Lobo, M. and Turkensteen, L.J. (1986) *Phoma andina* var. *crystalliniformis* var. nov., un patógeno nuevo del tomate y de la papa en los Andes. *Fitopatología* 21(2), 99–102.

Logan, C. and O'neill, R. (1970) Production of an antibiotic by *Phoma exigua*. *Transactions of the British Mycological Society* 55, 67–75.

Lowen, R. and Sivanesan, A. (1989) *Leptosphaeria pimpinellae* and its *Phoma* anamorph. *Mycotaxon* 35(2), 205–210.

Lucas, M.T. (1963) Culture studies on Portugese species of *Leptosphaeria* I. *Transactions of the British Mycological Society* 46, 361–376.

Lucas, M.T. and Webster, J. (1967) Conidial states of British species of *Leptosphaeria*. *Transactions of the British Mycological Society* 50, 85–121.

Maas, P.W.T. (1965) The identity of the footrot fungus of flax. *Netherlands Journal of Plant Pathology* 71, 113–121.

Magnani, G. (1966) Alterazioni de *Phoma* sp. su *Populus nigra* L. coltivato in vivaio. *Agricultura Italiana* 66, 197–199.

Magnani, G. (1969) Danni da *Phoma urens* Ell. et Ev. su pioppi coltivati in viviao. *Pubblicazioni del Centro di Sperimentazione agricola e foristale* 10(1), 1–10.

Maiello, J.M. (1978) The influence of ultraviolet light and carbohydrate nutrition on pycnidium formation in *Phyllosticta antirrhini*. *Mycologia* 69 ['1977'], 349–354.

Malathrakis, N.E. (1979) *A Study of an Olive Tree Disease Caused by the Fungus* Phoma incompta Sacc. et Mart. [in Greek]. Agricultural College, Athens.

Malone, J.P. (1982) Flax, Linseed. *Linum usitatissimum*. Foot rot. *Phoma exigua* var. *linicola* (Naumov and Vass.) Maas, P.W.T. In: *ISTA Handbook on Seed Health Testing*, Working Sheet 47.

Man in't Veld, W.A. and Gruyter, J. de (1995) The enzyme stain 6-phosphogluconate dehydrogenase distinguishes *Phoma urens* from *Phoma exigua* var. *exigua*. *Verslagen en Mededelingen Plantenziektenkundige Dienst Wageningen* 179 (A. Rep. Diagn. Cent. 1995), 85–90.

Marasas, W.F.O., Pauer, G.D. and Boerema, G.H. (1974) A serious leaf blotch disease of groundnuts (*Arachis hypogaea* L.) in southern Africa caused by *Phoma arachidicola* sp. nov. *Phytophylactica* 6, 195–202.

Marchionatto, J.B. (1948) El 'Tizon' o 'Podredumbre del tallo' del conejito. *Revista de la Facultad de agronomia y veterinaria, Universidad de Buenos Aires* 12, 3–7.

Marcinkowska, J. and Gruyter, J. de (1996) *Phoma nigrificans* (P. Karst.) Boerema *et al.*, a new species for Poland. *Journal of Phytopathology* 144, 53–54.

Marič, A. and Schneider, R. (1979) Die Schwarzfleckenkrankheit der Sonnenblume in Jugoslawien und ihr Erreger *Phoma macdonaldii* Boerema. *Phytopathologische Zeitschrift* 94, 226–233.

Marič, A., Maširević, S. and Fayzalla, S. (1981) Pojava *Leptosphaeria lindquistii* Frezzi, savršenog stadija gljive *Phoma macdonaldi* Boerema prouzrokovača crne pegavosti suncokreta u Jugoslaviji. *Zaštita Bilja* 32(4), 329–334.

Mathur, P.N. and Thirumalachar, M.J. (1959) Studies on some Indian soil fungi 1. Some new or noteworthy Sphaeropsidales. *Sydowia* 13, 143–147.

Mathur, R.S. (1979) *The Coelomycetes of India*. Grover Press, Meerut, India.

Mathur, S.B. (1981) Chickpea anthracnose, ascochytosis. *Cicer arietinum. Ascochyta rabiei* (Pass.) Labr. Perfect state: *Mycosphaerella rabiei* (Pass.) Kovach. In: *ISTA Handbook on Seed Health Testing*. Working sheet 38.

Maublanc, A. (1905) Champignon coprophiles. *Bulletin de la Société Mycologique de France* 21, 87–94.

McCoy, R.E. and Blakeman, J.P. (1976) Distribution of *Mycosphaerella ligulicola* and selection for environmental races. *Phytopathology* 66, 1310–1312.

McDonald, W.C. (1964) Phoma black stem of sunflowers. *Phytopathology* 54, 492–493.

McGee, D.C. and Petrie, G.A. (1978) Variability of *Leptosphaeria maculans* in relation to blackleg of oilseed rape. *Phytopathology* 68, 625–630.

McPartland, J.M. (1994) *Cannabis* pathogens X: *Phoma, Ascochyta* and *Didymella* species. *Mycologia* 86(6), 870–878.

Mel'nik, V.A. (1977) *Opredelitel' gribov roda* Ascochyta *Lib.* Akademiya Nauk SSSR Leningrad [see below].

Mel'nik, V.A. (2000) Key to the fungi of the genus *Ascochyta* Lib. (Coelomycetes) [translation of Mel'nik 1977 (see above) by Mel'nik, V.A., Braun, U. and Hagedorn, G.]. In: *Mitteilungen der Biologischen Bundesanstalt für Land-u Forstwirtschaft* 379.

Melouk, H.A. and Horner, C.E. (1972) Growth in culture and pathogenicity of *Phoma strassseri* to peppermint. *Phytopathology* 62, 576–578.

Mercier, S. and Metay, C. (1977) Le dépérissement du Laurier-rose. *Horticulture Française* 86, 9–14.

Minter, D.W., Kirk, P.M. and Sutton, B.C. (1982) Holoblastic phialides. *Transactions of the British Mycological Society* 79, 75–93.

Monte, E. and Garcia-Acha, I. (1988a) Vegetative and reproductive structures of *Phoma betae in vitro. Transactions of the British Mycological Society* 90, 233–245.

Monte, E. and Garcia-Acha, I. (1988b) Conidiogenesis in *Phoma betae. Transactions of the British Mycological Society* 90, 659–662.

Monte, E., Martin, P.M. and Garcia-Acha, I. (1989) Pycnidial development in *Phoma betae. Mycological Research* 92, 369–372.

Monte, E., Bridge, P.D. and Sutton, B.C. (1990) Physiological and biochemical studies in Coelomycetes *Phoma. Studies in Mycology* 32, 21–28.

Monte, E., Bridge, P.D. and Sutton, B.C. (1991) An integrated approach to *Phoma* systematics. *Mycopathologia* 115, 89–103.

Morgan-Jones, G. (1967) *Phoma prunicola. C.M.I. Descriptions of pathogenic Fungi and Bacteria* 134.

Morgan-Jones, G. (1988a) Studies in the genus *Phoma*. XIV. Concerning *Phoma herbarum*, the type species, a widespread saprophyte. *Mycotaxon* 33, 81–90.

Morgan-Jones, G. (1988b) Studies in the genus *Phoma*. XV. Concerning *Phoma multirostrata*, a leaf spot-inducing and soil-borne species in warm climates. *Mycotaxon* 33, 339–351.

Morgan-Jones, G. and Burch, K.B. (1987a) Studies in the genus *Phoma*. VIII. Concerning *Phoma medicaginis* var. *medicaginis. Mycotaxon* 29, 477–487.

Morgan-Jones, G. and Burch, K.B. (1987b) Studies in the genus *Phoma*. IX. Concerning *Phoma jolyana. Mycotaxon* 30, 239–246.

Morgan-Jones, G. and Burch, K.B. (1988a) Studies in the genus *Phoma*. X. Concerning *Phoma eupyrena*, an ubiquitous soil-borne species. *Mycotaxon* 31(2), 427–434.

Morgan-Jones, G. and Burch, K.B. (1988b) Studies in the genus *Phoma*. XI. Concerning *Phoma lycopersici*, the anamorph of *Didymella lycopersici*, causal organism of stem canker and fruit rot of tomato. *Mycotaxon* 32(1), 133–142.

Morgan-Jones, G. and Burch, K.B. (1988c) Studies in the genus *Phoma*. XII. Concerning *Phoma destructiva*, a second species implicated as a pathogen of tomato. *Mycotaxon* 32, 253–265.

Morgan-Jones, G. and Burch, K.B. (1988d) Studies in the genus *Phoma*. XIII. Concerning *Phoma exigua* var. *exigua*, a cosmopolitan, ubiquitous fungus on diseased and dead plant material. *Mycotaxon* 32, 477–490.

Morgan-Jones, G. and White, J.F. (1983a) Studies in the genus *Phoma*. I. *Phoma americana* sp.nov. *Mycotaxon* 16(2), 403–413.

Morgan-Jones, G. and White, J.F. (1983b) Studies in the genus *Phoma*. III. *Paraphoma*, a new genus to accommodate *Phoma radicina*. *Mycotaxon* 18(1), 57–65.

Morgan-Jones, G., Owsley, M.R. and Krupinsky, M.R. (1991) Notes on Coelomycetes. IV. *Phyllosticta andropogonivora*, causal organism of leafspot disease of *Schizachyrium scoparium* (little bluestem) and *Andropogon gerardii* (big and sand bluestem), reclassified in *Ascochyta*. *Mycotaxon* 42, 53–61.

Mukunya, D.M. and Boothroyd, C.W. (1973) *Mycosphaerella zeae-maydis* sp. n., the sexual stage of *Phyllosticta maydis*. *Phytopathology* 63, 529–532.

Müller, E. and Arx, J.A. von (1962) Die Gattungen der didymosporen Pyrenocmyceten. *Beiträge zur Kryptogamenflora der Schweiz* 11(2).

Munk, A. (1957) Danish Pyrenomycetes: a preliminary flora. *Dansk botanisk Arkiv* 17(1).

Nachmias, A., Barash, I., Buchner, V., Solel, Z. and Strobel, G.A. (1997) A phytotoxic glycopeptide from lemon leaves infected with *Phoma tracheiphila*. *Physiological Plant Pathology* 14, 135–140.

Nag Raj, T.R. (1993) *Coelomycetous Anamorphs with Appendage-Bearing Conidia*. Mycologue Publications, Waterloo, Canada.

Nagai, M., Shishido, M. and Tsuyama, H. (1970) Leaf blight of the aconite caused by *Phoma aconiticola* sp. nov. *Journal of the Faculty of Agriculture Iwate University* 10(1), 19–25.

Ndimande, B. (1976) Studies on *Phoma lingam* (Tode ex Fr.) Desm. and the dry rot on oil seed rape, *Brassica napus* (L.) var. *oleifera* Metzger. Thesis, Uppsala, Agricultural College Sweden.

Neergaard, P. (1938) Angreb af *Phoma nemophilae* Neergaard paa *Nemophila insignis* Bentham og *Nemophila atomaria* Fisher and Meyer. *Årsberetning fra J.E. Ohlsens enkes plantepatologiske Laboratorium* 3 (1 April 1937–31 Marts 1938), 7–10.

Neergaard, P. (1950) Mycological notes. III. 7. *Colletotrichum godetiae* Neerg. 8. *Phoma bellidis* Neerg. 9. *Zygosporium parasiticum* (Grove) Bunting and Mason. 10. *Peronospora dianthicola* Barthelet. *Friesia* 4, 72–80.

Neergaard, P. (1956) Nye Angreb. *Årsberetning vedrørende Frøpatologisk Kontrol Statens Plantetilsyn* 7, 15–16.

Neergaard, P. (1977) *Seed Pathology*. Volume I. The Macmillan Press, London.

Netolitzky, H. and Colotelo, N. (1965) Conidiophores of *Plenodomus meliloti*. *Canadian Journal of Botany* 43, 615–616.

Noordeloos, M.E. and Boerema, G.H. (1989a) Mycologisch-taxonomisch onderzoek. Veroorzakers van bladvlekken, bladnecroses en scheutinsterving bij oleander (*Nerium oleander*). *Verslagen en Mededelingen Plantenziektenkundige Dienst Wageningen* 166 (Jaarb. 1987), 108–109.

Noordeloos, M.E. and Boerema, G.H. (1989b) Mycologisch-taxonomisch onderzoek. Over lenticel-rot bij appel (*Malus pumila*) en kweepeer (*Cydonia*). *Verslagen en Mededelingen Plantenziektenkundige Dienst Wageningen* 166 (Jaarb. 1987), 110–111.

Noordeloos, M.E., Gruyter, J. de, Eijk, G.W. van and Roeijmans, H.J. (1993) Production of dendritic crystals in pure cultures of *Phoma* and *Ascochyta* and its value as a taxonomic character relative to morphology, pathology and cultural characteristics. *Mycological Research* 97, 1343–1350.

Obando Rojas, L. (1989) Contribution à l'étude de l'ascochytose du haricot *Phaseolus vulgaris* L.: identification de souches fongiques en cause et recherche de sources de résistance au sein du genre *Phaseolus*. Thesis Fac. Sci. agron. Gembloux.

Olembo, T.W. (1972) *Phoma herbarum* Westend.: a pathogen of *Acacia mearnsii* de Wild. in Kenya. *East African Agricultural and Forestry Journal* 38: 201–206.

Ondřej, M. (1968) Prispevek k poznáni fytopatogennick hub roda *Ascochyta* (Lib.) Sacc. na leguminózách. *Biológia, Bratislava* 23, 803–818.

Ondřej, M. (1970) Srovnávací studium nékolika druhu imperfektních hub rodu *Ascochyta* Lib. na zivnýchpudach. *Biológia, Bratislava* 25, 679–690.

Otazú, V., Boerema, G.H., Mooi, J.C. and Salas, B. (1979) Possible geographical origin of *Phoma exigua* var. *foveata*, the principal causal organism of potato gangrene. *Potato Research* 22, 333–338.

Padwick, G.W. (1950) *Manual of Rice Disease*. CMI, Kew.

Pag, H. (1965) Zur Ätiologie des 'Roten Brenners' an *Hippeastrum*. *Mitt. biol. BundAnst. Ld- u. Forstw.* 115, 225–229.

Patil, A.S. (1987) Leaf spot of groundnut caused by *Phyllosticta arachidis-hypogaea* Rao in Maharashtra State. *Maharasthra Agricultural University* 12, 205–209.

Patil, A.S. and Rao, V.G. (1974) Studies into *Phyllosticta* leaf-spot disease of garden gerbera. *The Research Journal of Mahatma Phule Agricultural University* 5, 47–56.

Paulitz, T.C. and Cote, E. (1991) First report of *Phoma* spp. on white lupine in North America. *Plant Disease Reporter* 75, 862.

Pawar, V.H., Mathur, P.N. and Thirumalachar, M.J. (1967) Species of *Phoma* isolated from marine soils in India. *Transactions of the British Mycological Society* 50, 259–265.

Pennycook, S.R. (1989) *Plant Diseases Recorded in New Zealand*, Vol. 2. Pl. Dis. Div. DSIR, Auckland.

Petrak, F. (1929) Mykologische Beiträge zur Flora von Siberien. *Hedwigia* 68, 203–241.

Petrak, F. (1943) Über die systematische Stellung und Nomenklatur von *Ascochyta Bolthauseri* Sacc. und *Stagonospora Curtisii* (Berk.) Curt. *Annales Mycologici* 41, 190–195.

Petrak, F. (1947) Ein kleiner Beitrag zur Pilzflora von Südfrankreich. *Sydowia* 1, 206–231.

Petrie, G.A. and Vanterpool, T.C. (1965) Diseases of rape and cruciferous weeds in Saskatchewan in 1965. *Canadian Plant Disease Survey* 45, 111–112.

Petrie, G.A. and Vanterpool, T.C. (1968) The occurrence of *Leptosphaeria maculans* on *Thlaspi arvense*. *Canadian Journal of Botany* 46, 869–871.

Phipps, P.M. (1985) Web blotch of peanut in Virginia. *Plant Disease Reporter* 69, 1097–1099.

Pons, N. (1990) Estudio taxonómico de especies de *Phoma* y *Phyllosticta* sobre caña de azúcar (*Saccharum* sp.). *Fitopatología Venezolana* 3, 34–43.

Pound, G.S. (1947) Variability in *Phoma lingam*. *Journal of Agricultural Research* 75, 113–133.

Punithalingam, E. (1976) *Phoma oculi-hominis* sp. nov. from corneal ulcer. *Transactions of the British Mycological Society* 67, 142–143.

Punithalingam, E. (1979a) *Ascochyta* I. Graminicolous *Ascochyta* species. *Mycological Papers* 142.

Punithalingam, E. (1979b) *Phoma caricae*. *C.M.I. Descriptions of pathogenic Fungi and Bacteria* 634.

Punithalingam, E. (1979c) Sphaeropsidales in culture from humans. *Nova Hedwigia* 31, 119–158.

Punithalingam, E. (1980a) A combination in *Phoma* for *Ascochyta caricae-papayae*. *Transactions of the British Mycological Society* 75, 340–341.

Punithalingam, E. (1980b) *Didymella chrysanthemi. C.M.I. Descriptions of pathogenic Fungi and Bacteria* 662.

Punithalingam, E. (1982a) *Didymella applanata. C.M.I. Descriptions of pathogenic Fungi and Bacteria* 735.

Punithalingam, E. (1982b) *Didymosphaeria arachidicola. C.M.I. Descriptions of pathogenic Fungi and Bacteria* 736.

Punithalingam, E. (1982c) *Phoma epicoccina. C.M.I. Descriptions of pathogenic Fungi and Bacteria* 738.

Punithalingam, E. (1982d) Conidiation and appendage formation in *Macrophomina phaseolina* (Tassi) Goid. *Nova Hedwigia* 36, [249–]266–269[–290].

Punithalingam, E. (1985) *Phoma sorghina. C.M.I. Descriptions of pathogenic Fungi and Bacteria* 825.

Punithalingam, E. (1990) *Mycosphaerella zeae-maydis. C.M.I. Descriptions of pathogenic Fungi and Bacteria* 1015; in: *Mycopathologia* 112, 49–50.

Punithalingam, E. and Gibson, I.A.S. (1976) *Phoma medicaginis* var. *pinodella. C.M.I. Descriptions of pathogenic Fungi and Bacteria* 518.

Punithalingam, E. and Harling, R. (1993) *Phoma gentianae-sino-ornatae* sp. nov. from *Gentiana sino-ornata* with root rot. *Mycological Research* 97(11), 1299–1304.

Punithalingam, E. and Holliday, P. (1972a) *Leptosphaeria maculans. C.M.I. Descriptions of pathogenic Fungi and Bacteria* 331.

Punithalingam, E. and Holliday, P. (1972b) *Didymella bryoniae. C.M.I. Descriptions of pathogenic Fungi and Bacteria* 332.

Punithalingam, E. and Holliday, P. (1972c) *Phoma insidiosa. C.M.I. Descriptions of pathogenic Fungi and Bacteria* 333.

Punithalingam, E. and Holliday, P. (1972d) *Ascochyta rabiei. C.M.I. Descriptions of pathogenic Fungi and Bacteria* 337.

Punithalingam, E. and Holliday, P. (1973a) *Pyrenochaeta terrestris. C.M.I. Descriptions of pathogenic Fungi and Bacteria* 397.

Punithalingam, E. and Holliday, P. (1973b) *Deuterophoma tracheiphila. C.M.I. Descriptions of pathogenic Fungi and Bacteria* 399.

Punithalingam, E. and Spooner, B.M. (2002) New taxa and new records of Coelomycetes for the U.K. *Kew Bulletin* 57, 533–563.

Punithalingam, E., Tulloch, M. and Leach, G.M. (1972) *Phoma epicoccina* sp. nov. on *Dactylis glomerata. Transactions of the British Mycological Society* 59, 341–345.

Rai, M.K. (1985) Taxonomic studies on species of *Phoma* isolated from air. *J. Econ. Taxon. Bot.* 7, 645–647.

Rai, M.K. (1989) *Phoma sorghina* infection in human being. *Mycopathologia* 105, 167–170.

Rai, M.K. (1998) *The Genus* Phoma *(Identity and Taxonomy)*. International Book Distributors, Dehra Dun, India.

Rai, M.K. and Rajak, R.C. (1993) Distinguishing characteristics of selected *Phoma* species. *Mycotaxon* 48, 389–414.

Rajak, R.C. and Rai, M.K. (1983a) Cholesterol content in twenty species of *Phoma*. *National Academy of Sciences and Letters* 6, 113–114.

Rajak, R.C. and Rai, M.K. (1983b) Effect of different factors on the morphology and cultural characters of 18 species and 5 varieties of *Phoma* I. Effect of different media. *Bibliotheca Mycologica* 91, 301–317.

Rajak, R.C. and Rai, M.K. (1984) Effect of different factors on the morphology and cultural characters of *Phoma*. II. Effect of different hydrogen-ion-concentrations. *Nova Hedwigia* 40, 299–311.

Rayner, M.C. (1915) Obligate symbiosis in *Calluna vulgaris*. *Annals of Botany, London* 29, 97–132.

Rayner, R.W. (1970) *A Mycological Colour Chart*. CMI, Kew and British Mycological Society.

Reifschneider, F.J.B. and Lopes, C.A. (1982) *Phoma asparagi* on asparagus. *FAO Plant Protection Bulletin* 30, 57.

Reisinger, O. (1972) Contribution à l'étude ultrastructurale de l'appareil sporifère chez quelques Hyphomycètes à paroi mélanisée. Genèse, modifications et décomposition. Thèse, University Nancy.

Richter, H. (1933) Fragekasten, Frage 557. Triebsterben an Pappeln. *Mitteilungen der Deutschen dendrologischen Gesellschaft* 45 (Jahrbuch), 401.

Riedl, H. (1959) Kulturversuche zum Pleomorphismus einiger Pyrenomyzeten. *Öst. bot. Z.* 106, 477–545.

Robertson, N.F. (1967) A slow-wilt of pyrethrums (*Chrysanthemum coccineum* Willd.) caused by a species of *Cephalosporium*. *Plant Pathology* 16, 31–36.

Röder, K. (1937) *Phyllosticta cannabis* (Kirchner?) Speg. eine Nebenfruchtform von *Mycosphaerella cannabis* (Winter) n.c. *Zeitschrift für Pflanzenkrankheiten (Pflanzen pathologie) und Pflanzenschutz* 47, 526–531.

Rössner, H. (1968) Untersuchungen über den Erreger der Phomakrankheit der Luzerne (*Phoma medicaginis* Malbr. and Roum.). *Phytopathologische Zeitschrift* 63, 101–123.

Saccardo, P.A. (1884) Sylloge Sphaeropsidearum et Melanconiearum omnium hucusque cognitorum. *Sylloge Fungorum* 3.

Salas, B. (1987) *Phoma lupini* on lupine in the Peruvian Southern highlands. *Phytopathology* 77, 1774.

Salonen, A. (1962) *Plenodomus meliloti* Dearness and Sanford in Finnish Lapland. *Maataloustieleelinen Aikakauskirja* 34, 169–172.

Sandford, G.B. (1933) A root rot of sweet clover and related crops caused by *Plenodomus meliloti* Dearness and Sanford. *Canadian Journal of Research* 8, 337–348.

Sauthoff, W. (1962) *Ascochyta bohemica* Kab. et Bub. als Erreger einer Blattfleckenkrankheit an *Campanula isophylla* Moretti. *Phytopathologische Zeitschrift* 45, 160–168.

Schneider, R. (1976) Taxonomie der Pyknidienpilzegattung *Pyrenochaeta*. *Bericht der Deutschen botanischen Gesellschaft* 89, 507–514.

Schneider, R. (1979) Die Gattung *Pyrenochaeta* De Notaris. *Mitteilungen der Biologischen Bundesanstalt für Land-u. Forstwirtschaft* 189.

Schneider, R. and Boerema, G.H. (1975a) Nachweis einer spezialisierten Form von *Phoma chrysanthemicola* (*Phoma chrysanthemicola* Hollós f. sp. *chrysanthemicola*). *Phytopathologische Zeitschrift* 83, 239–243.

Schneider, R. and Boerema, G.H. (1975b) *Phoma tropica* n. sp., ein an Gewächshauspflanzen häufig vorkommender, nicht pathogener Pilz. *Phytopathologische Zeitschrift* 83, 360–366.

Schwarz, M.B. (1922) Das Zweigsterben der Ulmen, Trauerweiden und Pfirsichbäume. *Mededelingen uit het Phytopathologisch Laboratorium 'Willie Commelin Scholten'* 5, 1–73.

Schweinitz, L.D. von (1822) Synopsis fungorum Carolinae superioris. *Schriften der Naturforschenden Gesellschaft in Leipzig* 1(1), 1–105.

Seaver, F.J. (1922) Phyllostictales. Phyllostictaceae (pars). *North American Flora* 6, 1 (reprinted 1961).

Sharma, K.D., Bhattacharya, S. and Bhadauria, S. (1990) Two new records of fungi from India. *Phytopathology* 43, 122–123.

Shoemaker, R.A. (1984) Canadian and some extralimital *Leptosphaeria* species. *Canadian Journal of Botany* 62, 2688–2729.

Shoemaker, R.A. and Babcock, C.E. (1989) *Phaeosphaeria*. *Canadian Journal of Botany* 67, 1500–1599.

Shoemaker, R.A. and Bissett, J. (1998) *Pleospora betae*. Fungi Canadenses 339. *Canadian Journal of Plant Pathology* 20, 206–209.

Shoemaker, R.A. and Brun, H. (2001) The teleomorph of the weakly aggressive segregate of *Leptosphaeria maculans*. *Canadian Journal of Botany* 79, 412–419.

Shoemaker, R.A. and Redhead, S.A. (1999) Proposals to conserve the names of four species of Fungi (*Phoma betae, Helminthosporium avenae, Pyrenophora avenae* and *Pleospora tritici-repentis*) against competing earlier synonyms. *Taxon* 48, 381–384.

Shoemaker, R.A., Babcock, C.E. and Irwin, J.A.G. (1991) *Massarina walkeri* n.sp., the teleomorph of *Acrocalymma medicaginis* from *Medicago sativa* contrasted with *Leptosphaeria pratensis, L. weimeri* n.sp. and *L. viridella*. *Canadian Journal of Botany* 69, 569–573.

Shreemali, J.L. (1972) Two new pathogenic fungi causing diseases on Indian medical plants. *Indian Journal of Mycology and Plant Pathology* 2, 84–85.

Shukla, N.P., Rajak, R.K., Agarwal, G.P. and Gupta, D.K. (1984) *Phoma minutispora* as a human pathogen. *Mykosen* 27(5), 255–258.

Siemaszko, W. (1923) Fungi caucasici novi vel minus cogniti. II. Diagnoses specierum novarum ex Abchazia Adzariaque provenientium. *Acta Societatis botanicorum Poloniae* 1, 19–28.

Singh, P.J., Pal, M. and Prakash, N. (1997) Ultrastructural studies of conidiogenesis of *Ascochyta rabiei*, the causal organism of chickpea blight. *Phytoparasitica* 25(4), 291–304.

Sivanesan, A. (1984) *The Bitunicate Ascomycetes and their Anamorphs*. J. Cramer, Fl-9490 Vaduz.

Smiley, E.M. (1920) The *Phyllosticta* blight of snapdragon. *Phytopathology* 10, 232–248.

Smith, C.O. (1935) Inoculations of *Stagonospora curtisii* on the Amaryllidaceae in California. *Phytopathology* 25, 262–268.

Smith, H.C. and Sutton, B.C. (1964) *Leptosphaeria maculans* the ascogenous state of *Phoma lingam*. *Transactions of the British Mycological Society* 47, 159–165.

Smith, J.D. (1987) Winter-hardiness and overwintering diseases of amenity turfgrasses with special reference to the Canadian prairies. *Technical Bulletin Research Branch Agriculture Canada* 1987, 12.

Sprague, R. (1935) *Ascochyta bolthauseri* on beans in Oregon. *Phytopathology* 25, 416–420.

Stadelmann, F.X. and Schwinn, F.J. (1982) Beitrag zur Biologie von *Venturia inaequalis* und *Venturia pirina*. *Zeitschrift für Pflanzenkrankheiten und Phlanzenschutz* 89, 96–109.

Stewart, R.B. (1957) An undescribed species of *Pyrenochaeta* on soybean. *Mycologia* 49, 115–117.

Sutton, B.C. (1977) Coelomycetes VI. Nomenclature of generic names proposed for Coelomycetes. *Mycological Papers* 141, 1–253.

Sutton, B.C. (1980) *The Coelomycetes. Fungi Imperfecti with Pycnidia, Acervuli and Stromata*. CMI, Kew.

Sutton, B.C. and Pirozynski, K.A. (1963) Notes on British microfungi. I. *Transactions of the British Mycological Society* 46, 505–525.

Taber, R.A., Pettit, R.E. and Philley, G.L. (1984) Peanut web blotch: I. Cultural characteristics and identity of causal fungus. *Peanut Science* 11, 109–114.

Taylor, R.H. (1962) *Deuterophoma* – a fungal pathogen of chrysanthemum previously unrecorded in Australia. *Australian Journal of Experimental Agriculture and Animal Husbandry* 2, 90–91.

Ternetz, C. (1907) Über die Assimilation des atmosphärischen Stickstoffes durch Pilze. *Jahrbücher für Wissenschaftliche Botanik* 44, 353–395.

Tichelaar, G.M. (1974) The use of thiophanate-methyl for distinguishing between the two *Phoma* varieties causing gangrene of potatoes. *Netherlands Journal of Plant Pathology* 80, 169–170.

Tosi, L. and Zazzerini, A. (1994) *Phoma incompta*, a new olive parasite in Italy. *Petria (Giornale di patologia delle piante)* 4, 161–170.

Trotter, A. (1972) Supplementum universale Sylloge Fungorum. Pars XI *Syll. Fung.* 26. [Johnsons Reprint Corporation, New York; revised by E.K. Cash.]

Turkensteen, L.J. (1978) Tizón foliar de la papa en el Perú: I Especies de *Phoma* asociadas. *Fitopatología* 13(1), 67–69.

Turkensteen, L.J. (1980) *Phoma* leaf spot. In: Hooker (ed.) *Compendium of Potato Diseases*. Am. phytopath. Soc., St Paul, Minnesota, pp.47–48.

Tuset, J.J. and Portilla, M.T. (1985) Contribucion al conocimiento de los *Phoma* presentes en los frutales de hueso mediterraneos. *Anales del Instituto Nacional de Investigaciónes Agronomicas serie Agricola* 28, N. Extr. 181–210.

Vanev, S.G. and Aa, H.A. van der (1998) An annotated list of the published names in *Asteromella*. *Persoonia* 17(1), 47–67.

Vegh, I., Bourgeois, M., Bousquet, J.F. and Velastegui, J. (1974) Contribution à l'étude du *Phoma exigua* Desm., champignon pathogène associé au dépérissement de la pervenche mineure (*Vinca minor* L.) médicinale. *Bulletin trimestriel de la Société mycologíque de France* 90, 121–133.

Vestergren, T. (1897) Bidrag til en monografi éfver Sveriges Sphaeropsideer. *Öfversigt af K. Vetenskapsakademiens förhandlingar* 1897, 35–46.

Walker, J. and Baker, K.F. (1983) The correct binomial for the chrysanthemum ray blight pathogen in relation to its geographical distribution. *Transactions of the British Mycological Society* 80, 31–38.

Webster, J. (1957) *Pleospora straminis*, *P. rubelloides* and *P. rubicunda*: three fungi causing purple-staining of decaying tissues. *Transactions of the British Mycological Society* 40, 177–186.

Webster, J. and Hudson, H.J. (1957) Graminicolous Pyrenomycetes VI. Conidia of *Ophiobolus herpotrichus*, *Leptosphaeria lactuosa*, *L. fuckelii*, *L. pontiformis* and *L. eustomoides*. *Transactions of the British Mycological Society* 40, 509–522.

Webster, J. and Lucas, M.T. (1959) Observations on British species of *Pleospora*. I. *Transactions of the British Mycological Society* 42, 332–342.

Wehmeyer, L.E. (1946) Studies on some fungi from Northwestern Wyoming. II. Fungi imperfecti. *Mycologia* 38, 306–330.

Westhuizen, G.C.A. van der (1980) *Phyllosticta maydis* on maize in South Africa. *Phytophylactica* 12, 27–29.

White, J.F. and Morgan-Jones, G. (1983) Studies in the genus *Phoma*. II. Concerning *Phoma sorghina*. *Mycotaxon* 18(1), 5–13.

White, J.F. and Morgan-Jones, G. (1984) Studies in the genus *Phoma*. IV. Concerning *Phoma macrostoma*. *Mycotaxon* 20(1), 197–204.

White, J.F. and Morgan-Jones, G. (1986) Studies in the genus *Phoma*. V. Concerning *Phoma pomorum*. *Mycotaxon* 25(2), 461–466.

White, J.F. and Morgan-Jones, G. (1987a) Studies in the genus *Phoma*. VI. Concerning *Phoma medicaginis* var. *pinodella*. *Mycotaxon* 28(1), 241–248.

White, J.F. and Morgan-Jones, G. (1987b) Studies in the genus *Phoma*. VII. Concerning *Phoma glomerata*. *Mycotaxon* 28(2), 437–445.

Williams, R.H. and Fitt, B.D.L. (1999) Differentiating A and B groups of *Leptosphaeria maculans*, causal agent of stem canker (blackleg) of oilseed rape. *Plant Pathology* 48, 161–175.

Wollenweber, H.W. and Hochapfel, H. (1936) Beiträge zur Kenntnis parasitärer und saprophytischer Pilze. I. *Phomopsis, Dendrophoma, Phoma* und *Ascochyta* und ihre Beziehung zur Fruchtfäule. *Zeitschrift für Parasitenkunde* 8, 561–605.

Wollenweber, H.W. and Hochapfel, H. (1937) Beiträge zur Kenntniss parasitärer und saprophytischer Pilze. IV. *Coniothyrium* und seine Beziehung zur Fruchtfäule. *Zeitschrift für Parasitenkunde* 9, 600–637.

Yáñez-Morales, M. de J., Korf, R.P. and Babcock, J.F. (1998) Fungi on *Epifagus* (Orobanchaceae) — I. On *Sclerotium orobanches* and its *Phoma* synanamorph. *Mycotaxon* 67, 275–286.

Zachos, D.G., Constantinou, P.T. and Panagopoulos, C.G. (1960) Une trachémycose du chrysanthème causée par une nouvelle espèce de *Cephalosporium*. *Annales de l'Institut phytopathologique Benaki* II [N.S.], 3, 50–59.

Zafar, S.I. and Colotelo, N. (1978) Influence of various inorganic oxidants and organic compounds on mycelial growth and pycnidial production of *Plenodomus meliloti* in light and darkness. *Canadian Journal of Botany* 56, 1588–1593.

Žerbele, I. (1962) O psecializacii gribov roda *Ascochyta*. *Bot. Issle dov. (ser. bot.)* 2, 108–120.

Žerbele, I.Ya. (1971) Ob izmenchivosti gribov roda *Ascochyta*. *Trudy̆ Vsesoyuznogo nauchno-issledovatel'-skogo instituta zashchitȳ rasteniĭ* 32('29'), 12–21.

Index of Fungus Names

The accepted names of the *Phoma* species and varieties treated, their teleomorphs and synanamorphs, are printed in **bold**.

Phoma friesii Brunaud 276
Phoma fumaginoides Peyronel 195, 197
Phoma funkiae-albomarginatae Punith. 155
Phoma fusispora Wehm. 151
Phoma gardeniae (S. Chandra & Tandon) Boerema **168**
Phoma garflorida S. Chandra & Tandon 97
Phoma genistaecola Hollós 342
Phoma gentianae J.G. Kühn 275
Phoma gentianae-sino-ornatae Punith. & R. Harling **311**
Phoma glaucii Brunaud [V] **142**
Phoma glaucii Therry [Θ] 142
Phoma glaucispora (Delacr.) Noordel. & Boerema **79**
Phoma glycines Sawada [‡] 169
Phoma glycinicola Gruyter & Boerema **169**
Phoma glomerata (Corda) Wollenw. & Hochapfel 179, **194**, 211, 213, 363, 385
Phoma glumarum Ellis & Tracy 26
Phoma glumicola Speg. 207
Phoma gossypii Sacc. 396
Phoma gossypiicola Gruyter **396**
Phoma grovei Berl. & Voglino 420
Phoma gutneri N. Pons 176
Phoma haematites (Petr.) Boerema *et al.* 385, **416**
Phoma haematocycla (Berk.) Aa & Boerema **80**
Phoma hedericola (Durieu & Mont.) Boerema **81**
Phoma heliopsidis (H.C. Greene) Aa & Boerema **268**
Phoma henningsii Sacc. **81**
Phoma herbarum Westend. 3, 8, 32, 34, 60, 65, 82, **82**, 113, 238, 243, 363, 366
Phoma herbarum f. *ansoniae-salicifoliae* Berl. & Roum. [Θ] 151
Phoma herbarum f. *antherici* Hollós 250
Phoma herbarum f. *antirrhini* Sacc. 243
Phoma herbarum f. *aristolochiae-siphonis* Sacc. 243
Phoma herbarum f. *brassicae* Sacc. 244
Phoma herbarum f. *calystegiae* Sacc. 243
Phoma herbarum f. *cannabis* Allesch. 248
Phoma herbarum f. *capparidis* Sacc. 203
Phoma herbarum f. *catalpae-capsularum* Sacc. 275
Phoma herbarum f. *chenopodii-albi* Roum. [Θ] 83

Phoma herbarum f. *chrysanthemi-corymbosi* Allesch. 195
Phoma herbarum f. *dahliae* Sacc. 243
Phoma herbarum f. *datiscae-cannabinae* Berl. & Roum. [Θ] 245
Phoma herbarum f. *dipsaci* Sacc. 243
Phoma herbarum f. *eupatorii-sessilifolii* Berl. & Roum. [Θ] 245
Phoma herbarum f. *euphorbiae-gayonianae* Pat. 195
Phoma herbarum f. *euphrasiae* Bres. [H] 247
Phoma herbarum f. *euphrasiae* Sacc. 244, 247
Phoma herbarum f. *foeniculi* Sacc. 243
Phoma herbarum f. *helichrysi* Sacc. 244
Phoma herbarum f. *humuli* Gonz. Frag. [H] 83, 243
Phoma herbarum f. *humuli* Sacc. 83, 243
Phoma herbarum f. *hyoscyami* Sacc. 244
Phoma herbarum f. *lactucae* Sacc. ex Sacc. 244
Phoma herbarum f. *lilacis* Sacc. 262
Phoma herbarum f. *marrubii* Sacc. 243
Phoma herbarum f. *medicaginea* Sacc. 243
Phoma herbarum f. *medicaginis* Sacc. 243
Phoma herbarum f. *medicaginum* Westend. ex Fuckel [often listed as '*medicaginis* Fuckel'] 243, 282
Phoma herbarum f. *melampyri* Westend. [Θ] 316
Phoma herbarum f. *mercurialis* Sacc. 243
Phoma herbarum f. *minor* Unamuno 83
Phoma herbarum f. *nicotianae* Roum. 247
Phoma herbarum f. *parietariae* Brunaud 245
Phoma herbarum f. *phytolaccae* Sacc. 243
Phoma herbarum f. *rubi* Sacc. 275
Phoma herbarum f. *salicariae* Sacc. 151
Phoma herbarum f. *sambuci-nigrae* Sacc. 291
Phoma herbarum f. *schoberiae* Sacc. 243
Phoma herbarum f. *sempervivi-tectorum* Berl. & Roum. [Θ] 245
Phoma herbarum f. *solani-nigricantis* Berl. & Roum. [Θ] 245
Phoma herbarum f. *solidaginis* Sacc. 243
Phoma herbarum f. *stramonii* Thüm. [Θ, V] 241
Phoma herbarum f. *urticae* Sacc. 243
Phoma herbarum f. *valerianae* Sacc. 295
Phoma herbarum f. *verbasci* Gonz. Frag. 250
Phoma herbarum f. *verbenae-paniculatae* Berl. & Roum. [Θ] 245

Index of Substrata

Plant family and genus names; see also under Inorganic, Fish, Insects, Mammal (human), Nematoda, Nutritional, Plurivorous, Soil and Water.

Acacia 82
Achillea 346
Aconitum 132–133, 138, 364
Actaea 130
Adiantum 54
Agavaceae 73
Agropyrum 422
Allium 177
Amaryllidaceae 58, 70, 148, 172
Ambrosia 268
Andropogon 392
Angelica 339, 348, 419
Anigozanthus 58
Antirrhinum 108, 109
Apium 60
Apocynaceae 80, 261
Aquilegia 132–133
Araceae 213
Arachis 61, 99, 168, 231
Araliaceae 81
Armoracia 314
Artemisia 234
Asparagus 29
Astragalus 62, 342
Atriplex 395
Aubrietia 62

Ballota 358
Bellis 63, 64

Berberidaceae 57
Berberis 57
Berteroa (*Farsetia*) 344–345
Beta 178, 395, 408
Bignoniaceae 352
Blechnum 281
Boerhavia 306
Brassica 25, 314, 362, 380

Cactaceae 104
Cajanus 65
Callistephus 72
Campanula 157
Campanulaceae 157
Cannabaceae 57, 235
Cannabis 235
Caprifoliaceae 57, 118, 266, 292, 352
Capsicum 71, 239
Cardamine 319
Carica 236
Caricaceae 236, 238
Carya 64
Caryophyllaceae 71, 319
Catalpa 352
Celastraceae 57
Chamaespartium 70
Chelidonium 143–144

Chenopodiaceae 68, 88, 142, 145, 178, 267, 395, 408
Chenopodium 68, 88, 142, 145, 266–267, 395
Chrysanthemum (Dendranthema) 117, 191, 272
Chrysanthemum (Tanacetum) 273
Cicer 400
Cichorium 256
Cimicifuga 130
Cirsium 420
Citrullus 238
Citrus 384
Clematis 136, 417
Coffea 69, 294
Commelina 396
Commelinaceae 396
Compositae 48–49, 140, 191, 225–226, 329–331, 420
Consolida 138
Cordyline 73
Corydalis 144
Cotoneaster 57
Crassulaceae 295
Crinum 69–70
Cruciferae 62, 314, 319, 344, 362, 380
Cucumis 201, 238
Cucurbita 238
Cucurbitaceae 201, 238
Cyperaceae 425

Dactylis 105, 109
Dahlia 256
Daucus 102, 256, 308
Delphinium 56, 132–133, 138, 401
Dendranthema 117, 191, 272
Dicentra 144
Dictamnus 309
Digitalis 240
Dracaena 73
Dryopteris 281

Elettaria 172
Epifagus 357
Ericaceae 206
Erigeron 346
Eucalyptus 74
Eugenia 74
Euonymus 57

Eupatorium 75, 341
Euphorbiaceae 367

Fagaceae 168
Fagus 357
Ficus 94
Fish 85
Foeniculum 348
Forsythia 260, 262
Fragaria 206
Fraxinus 26

Gardenia 168
Gentiana 311, 351, 369
Gentianaceae 311, 351, 369
Glaucium 144
Glycine 169, 175, 292
Gossypium 397
Gramineae 88, 94, 105, 110, 113, 164, 187, 390, 422–425

Hedera 81
Helianthus 366
Heliopsis 268
Hippeastrum 148
Humulus 57
Hydrophyllaceae 284

Inorganic substrata 85, 87, 198, 206, 210
Insects 210
Ipomoea 29
Iridaceae 173
Iris 173

Juglandaceae 65
Juncaceae 425

Labiatae 72, 285, 293, 318, 358
Lactuca 256
Leguminosae 49–50, 126, 163, 226–227, 332, 394, 400, 418
Leonurus 358
Liliaceae 80, 177
Linaceae 91, 263
Linum 91, 263

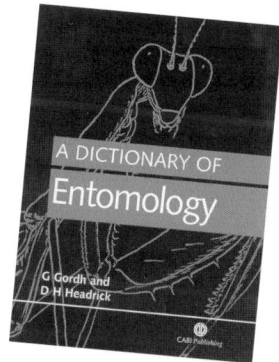

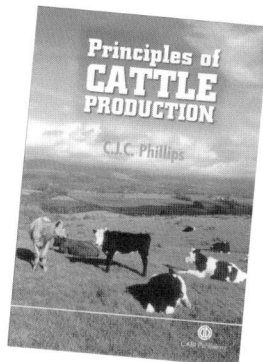

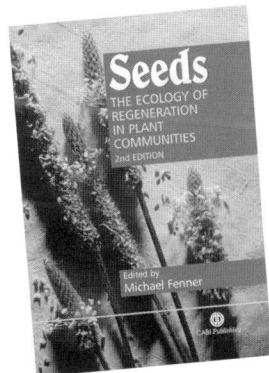